# Aquaculture Engineering

# Aquaculture Engineering

Editor: Ellie Ward

Callisto Reference
www.callistoreference.com

**Callisto Reference,**
118-35 Queens Blvd., Suite 400,
Forest Hills, NY 11375, USA

Visit us on the World Wide Web at:
www.callistoreference.com



ISBN: 978-1-63239-855-0 (Hardback)

The publisher's policy is to use permanent paper from mills that operate a sustainable forestry policy. Furthermore, the publisher ensures that the text paper and cover boards used have met acceptable environmental accreditation standards.



Printed in China.

**Cataloging-in-publication Data**

Aquaculture engineering / edited by Ellie Ward.
p. cm.
Includes bibliographical references and index.
ISBN 978-1-63239-855-0
1. Aquacultural engineering. 2. Aquaculture. 3. Fish culture--Technological innovations. 4. Fisheries. I. Ward, Ellie.
SH137 .A68 2017
639.8--dc23

# Table of Contents

# Preface

Aquaculture engineering is an emerging field of study. It aims to device new and sustainable techniques for aquaculture and its management. Some of the most widely practiced forms of aquaculture include fish farming, mariculture, shrimp farming, etc. This book brings forth some of the most innovative concepts and elucidates the unexplored aspects of aquaculture engineering. Some of the diverse topics covered in this book address the varied branches that fall under this category. It is compiled in such a manner, that it will provide in-depth knowledge about the theory and practice of aquaculture engineering. Students, researchers, professionals, experts and anyone else associated with this field will benefit alike from this book. It aims to serve as a resource guide to a broad spectrum of readers and contribute to the growth of aquaculture engineering.

This book aims to highlight the current researches and provides a platform to further the scope of innovations in this area. This book is a product of the combined efforts of many researchers and scientists from different parts of the world. The objective of this book is to provide the readers with the latest information in the field.

I would like to express my sincere thanks to the authors for their dedicated efforts in the completion of this book. I acknowledge the efforts of the publisher for providing constant support. Lastly, I would like to thank my family for their support in all academic endeavors.

**Editor**

# 1

# Modulation of Habitat-Based Conservation Plans by Fishery Opportunity Costs: A New Caledonia Case Study Using Fine-Scale Catch Data

**Marilyn Deas[1], Serge Andréfouët[1]*, Marc Léopold[2,3], Nicolas Guillemot[1,4]**

**1** UR-CoRéUs, Institut de Recherche pour le Développement, Laboratoire d'Excellence, CORAIL, Nouméa, New Caledonia, **2** UR-CoRéUs, Institut de Recherche pour le Développement, Laboratoire d'Excellence, CORAIL, Port-Vila, Vanuatu, **3** Fisheries Department of Vanuatu, Port-Vila, Vanuatu, **4** Nicolas Guillemot Consultant, Noumea, New Caledonia

## Abstract

Numerous threats impact coral reefs and conservation actions are urgently needed. Fast production of marine habitat maps promotes the use of habitat-only conservation plans, where a given percentage of the area of each habitat is set as conservation objectives. However, marine reserves can impact access to fishing grounds and generate opportunity costs for fishers that need to be minimized. In New Caledonia (Southwest Pacific), we used fine-scale fishery catch maps to define nineteen opportunity costs layers (expressed as biomass catch loss) considering i) total catches, ii) target fish families, iii) local marine tenure, and iv) gear type. The expected lower impacts on fishery catch when using the different cost constraints were ranked according to effectiveness in decreasing the costs generated by the habitat-only scenarios. The exercise was done for two habitat maps with different thematic richness. In most cases, habitat conservation objectives remained achievable, but effectiveness varied widely between scenarios and between habitat maps. The results provide practical guidelines for coral reef conservation and management. Habitat-only scenarios can be used to initiate conservation projects with stakeholders but the costs induced by such scenarios can be lowered by up to 50–60% when detailed exhaustive fishery data are used. When using partial data, the gain would be only in the 15–25% range. The best compromises are achieved when using local data.

**Editor:** Judi Hewitt, University of Waikato (National Institute of Water and Atmospheric Research), New Zealand

**Funding:** Data acquisition funded by: New Caledonia ZONECO Program, the LITEAU French national Program (COGERON Project) and the Koniambo Nickel SAS mining company. The Millennium map is a product funded by NASA and the IRD. The funders had no role in study design, data collection and analysis, decision to publish, or preparation of the manuscript.

**Competing Interests:** The authors declare that no competing interest or financial disclosure exists between the authors and Nicolas Guillemot Consultant, and that the affiliation of one author to this company does not alter in any way the adherence of all authors to PLOS ONE policies on sharing data and materials.

* E-mail: serge.andrefouet@ird.fr

## Introduction

In designing reserve networks to meet international, national and local targets, planners typically set specific goals for habitat conservation, with targets for a given percentage of the total area of each habitat found inside the focal domain. When such given percentage of the total area of each habitat completely defines the conservation objectives, without any other objectives, this is what we call a habitat-based conservation plan. This type of plan is attractive because it should theoretically include all the biological species and functional groups found in these habitats, whether these species and groups are known or not [1]. These plans are also straightforward to compute once habitat maps are available, which is routine task with remote sensing technology [2]. As a consequence, these habitat-only designs are not data demanding, reserve networks can be easily produced, and as such they can be extremely useful to initiate discussions with the various stakeholders with a first exploratory plan in hand. The proposed plans can be enhanced iteratively when new data become available and when new objectives are defined [3].

However, such plans are idealistic in the sense that they ignore local socio-economic constraints [4,5,6,7,8]. There is a wide consensus that when avoiding socio-economic realities, conservation actions are likely doomed, with poor compliance if conservation brings significant pressure on local community livelihoods [9]. In the marine realm, a habitat-only design brings penalties, and costs, to fishers. It is therefore needed to evaluate the socio-economic impact when using a simple (in terms of required data) habitat-based conservation design and compare with marine reserve scenarios that minimize costs. Better understanding of how socio-economic constraints modulate habitat-based conservation plans would help managers knowing their limits and building confidence in using such plans. This is especially true in coral reef environments where spatially-explicit socio-economic data are often lacking and conservation actions are urgent. Indeed, better conservation measures are widely needed for many reefs worldwide, at risk from coastal development, unsustainable resource use and climate change [10]. Therefore, we here aim to test in a coral reef environment if habitat-based plans are robust to socio-economic constraints and how valid they would be when socio-economic data become available. Specifically, we test how fishery opportunity costs modulate habitat-based plans.

Opportunity costs are defined as the costs of foregone opportunities; in other words, they are a measure of what could have been gained via the next-best use of a resource had it not been put to the current use [11]. They are increasingly used to account for the socio-economic costs of conservation, especially for fisheries [12]. Opportunity costs for fisheries have been previously considered in coral reef environment: Weeks et al. [13] have used socio-economic proxies to infer opportunity costs in Philippines and Adams et al. [14] have estimated opportunity costs in Fiji both from fishing catch per unit effort surveys and from what could be the production of a reef using underwater fish visual census. Availability of fine-scale socio-economic data are expected to better optimize the costs of conservation [4], particularly where local tenure or resource use rights over marine spaces covers small areas, a common feature of many countries in the Coral Triangle and Southwest Pacific. Otherwise, for coral reefs, fishery data have also been used to assess trade-offs between conservation and fishers activities, but no opportunity costs were estimated or mapped [7]. This gap is explained by the difficulty, or lack of resources, devoted to achieving a precise mapping of the fishing effort and catch in coral reef environment. In fact, this information is sorely missing for virtually all reefs of the planet that are under stress due to fishing activity [14,15]. Data are more frequently available for temperate commercial fisheries that are well monitored using observers and logbooks [12]. Most previous studies that have collected spatially-explicit information on coral reef fisheries have only included locations of fishing grounds, sometimes detailed by targeted species and gear types [7,16,17,18,19,20].

Use of fine-scale maps of fishery catch to directly infer opportunity costs would be a substantial improvement over use of models or proxies of socio-economic costs. Recently, Guillemot and Léopold [21] released a thorough compilation of coral reef fishery activities for one region of New Caledonia, in the Voh-Koné-Pouembout area. The result is a fishery atlas with fine scale maps of total production (or catch), and detailed catch by key fish family, gear type and tenure. To the author's knowledge, no similar data set has been used in any given coral reef locations to analyze conflicts between conservation and fishery activities. The associated database is a gold mine of spatially explicit information that we used together with different types of habitat maps to answer the following questions:

- How do fishery opportunity costs modulate habitat-based marine reserve plans in a coral reef environment?
- How does the thematic richness of the habitat maps influence the results?
- How does the type of opportunity costs (fish family-related, location-related, gear-related) influence the results, and do these different types provide redundant information? Conversely, can data from one location be used to efficiently make decisions for another location?
- What are the practical lessons for coral reef managers in charge of implementing conservation actions in heavily fished areas?

To answer these questions, we compare habitat-based only conservation plans with habitat-based conservation plans constrained by opportunity costs. This quantitative comparative analysis was made possible due to the unique availability of the Voh-Koné-Pouembout fishery atlas, and two types of habitat maps.

## Materials and Methods

### Ethics statement

This work did not involve manipulations of animals, thus it did not involve endangered or protected species. Modeling methods did not require specific permissions from any relevant body as they are harmless and meet all applicable standards for the ethics of experimentation and research integrity.

### Study site

Voh Koné Pouembout (VKP) is a rural area of the north-west coast of New Caledonia. The barrier reef system, at a distance of 6 km on average from the coast, includes four passes that delimitate the 211 $km^2$ study area (Fig. 1). Nickel mining is a main driver of the New Caledonia economy [22]. A nickel mining project in the VKP region recently boosted the local economy leading to urban development and a population increase of 18% between 2004 and 2009. Around 6400 newcomers are expected to settle by 2015 [23,24] and to increase the demand for fishery resources. This was the reason why fishery data were collected in 2007, and compiled in a comprehensive atlas of fishery maps, to serve as a future baseline [15].

### Habitat maps

The lagoon and reefs include a broad range of habitats. To evaluate the possible influence of habitat map thematic resolution and complexity on the results, our study design included two distinct habitat maps. The first map included 23 different habitats, defined by their geomorphological attributes as listed by the Millennium Coral Reef Mapping Project [25]. The interest of using this map is that similar products are available for many countries and regions worldwide [26,27], and lessons learned with this product can provide guidance elsewhere. The second map, much more complex, included 106 habitats defined by both geomorphological and benthic cover attributes (substrate, percentage cover, and architecture of numerous coral, algae and seagrass communities). This map was created following a user approach [26], with a Quickbird satellite image at 2.4 meter spatial resolution. Figure 1 illustrates the differences between these two maps. Geographical Information Systems (GIS) shapefiles of both products are available on request to the corresponding author.

### VKP reef fishery data

Methods for the production of the fishery atlas are detailed in Léopold et al. [15]. These authors describe four different mangrove and coral reef fisheries targeting invertebrates or finfish around New Caledonia. The fisheries at these sites were mapped following a five-step framework: 1) stratified random sampling of regular fishers; 2) collection of fishers' knowledge on fishing areas, fishing effort, and catch characteristics through map-based interviews; 3) data integration into a spatial geodatabase; 4) statistical extrapolation of fisher data to the fishery scale; and 5) mapping of catch, effort, and catch per unit of effort (CPUE) for each fishery. An example of product is provided Figure 2. Fishery maps are representations of catch data and are therefore also shaped by indirect factors affecting fishing locations, such as clan disputes, customary reserves and taboos, sea conditions, or fishers' personal preferences. Furthermore, catch maps indirectly reflect to some extent the preference habitat distribution and home range of each target species over a one-year period.

Most regular fishers in the area were present in two small multicultural and multi-ethnic small towns (Koné and Pouembout) and one Melanesian village (Oundjo) that totaled about 310 fishing boats. Fishing practices differed between these localities and communities. In Pouembout and Koné, fishing was mostly a recreational week-end activity, involving few fishers using large boats and selective gears like spear gun and handline. Annual fishing effort catches and mean yield were consequently relatively

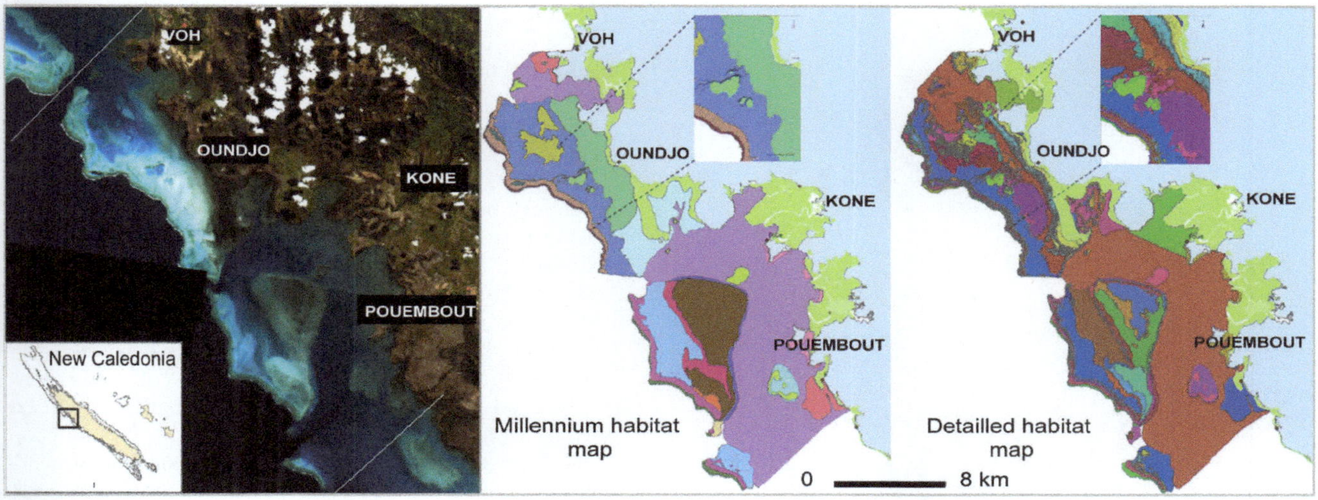


**Figure 1. View and maps of the study area.** Left panel: view of the Voh-Koné-Pouembout coral reefs and lagoons. The white lines show the extent of the study area. Right and Central panels: habitat maps used for this study, Millennium habitat map and Detailed habitat map respectively. Legends are not shown here given their length (23 and 106 classes) but the differences in levels of details are visible.

low. Conversely, Oundjo fishers used highly efficient but non-selective gears like gillnets, with small boats during the week, leading to high fishing yield and annual fishing effort. About 30% of catches were unofficially sold directly to local markets. The activity was thus driven by both income and food needs, resulting in more intense exploitation than subsistence fishing only [28].

Socioeconomic and cultural differences between fishers explained the spatial partitioning of the lagoon fishing area. Noteworthy is the Oundjo's customary exclusive fishing area, a 61 km$^2$ restricted access area closed to outsiders. Oundjo fishers are also active outside this area but since they used small boats, they favored shore-fishing on inner and fringing reefs near their village. Fishers from Koné and Pouembout consequently use only the southern part of the lagoon area, and travel further from the coast, including to the barrier reef [23].

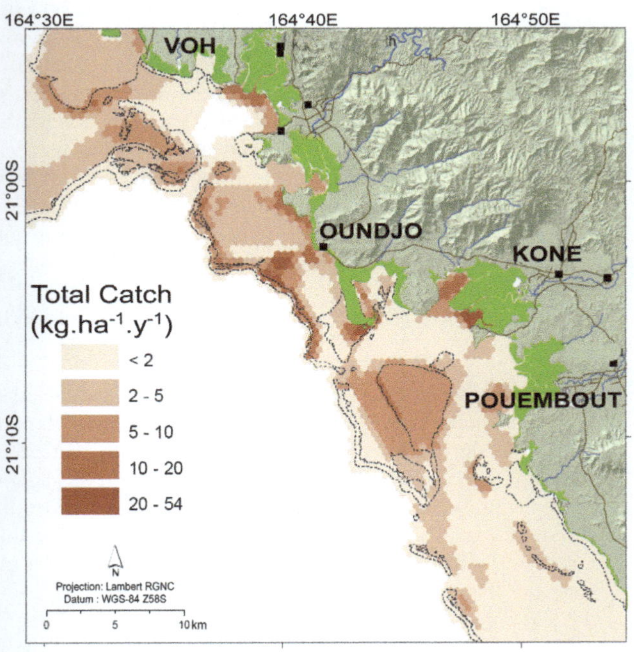


**Figure 2. Map of *Total Catch* for the Voh-Koné-Pouembout area.** Example data included in the fishery atlas used to compute opportunity costs (From Guillemot and Léopold [21]).

Overall fishing pressure was quite low in 2007, with an estimated average fish catch of 0.24 t.km$^{-2}$.year$^{-1}$. However, in restricted sites of the barrier reef, mangroves and fringing reef areas inside Oundjo's exclusive area, fishing pressure reached 5 t.km$^{-2}$.year$^{-1}$, which was considered as a threshold for sustainable reef fisheries [29].

## Reef fishery opportunity costs

Estimated annual mean yield, total effort, total catch, catch by gear type, catch by locality and catch by fish family were mapped by Guillemot and Léopold throughout the study area using a hexagonal grid [21]. Each of these data layers offers the possibility to run a different scenario, and directly compute an opportunity cost for each grid cell. For this; we assumed that if a grid cell is turned into a no take area, the associated catch loss (total, by locality, by gear type, by fish family) for fishers represents the opportunity cost.

From the fishery atlas, we considered 19 catch data layers to define opportunity costs and conservation scenarios. These are (Table 1):

1. total catches,
2. catches by locality, to compare the effectiveness of conservation plans designed according to local fishery data with plans designed according to fishery data from another area (or for the entire area),
3. catches by gear, to compare the effectiveness of conservation plans designed according to a sub-set of fishermen with plans designed according to the entire fishing population,
4. catches by fish family, to compare the effectiveness of conservation plans designed using limited set of data with plans designed using an exhaustive list of target fish families and total catch.

**Table 1.** Opportunity costs and their related data layers, used to define the different scenarios, identified by indexes.

| General | Data and Cost index | Catches by locality | Data and Cost index | Catches by gear | Data and Cost index | Catches by fish family | Data and Cost index |
|---|---|---|---|---|---|---|---|
| Total catches | 1 | Koné | 2 | Gillnet | 5 | Lethrinidae | 8 |
| | | Oundjo | 3 | Spear gun | 6 | Acanthuridae | 9 |
| | | Pouembout | 4 | Hand line | 7 | Mugilidae | 10 |
| | | | | | | Scaridae | 11 |
| | | | | | | Serranidae | 12 |
| | | | | | | Siganidae | 13 |
| | | | | | | Lutjanidae | 14 |
| | | | | | | Haemulidae | 15 |
| | | | | | | Carangidae | 16 |
| | | | | | | Sparidae | 17 |
| | | | | | | Gerreidae | 18 |
| | | | | | | Kyphosidae | 19 |

## Conservation designs

Planning units were defined by a grid of hexagons (side length = 500 m, surface of 21.6ha) mimicking the grid used for the fishery maps (Fig. 2). The number of planning unit was 1525 and 1531 respectively for the Millennium and detailed map.

Our conservation objective was to represent at least 20% of all mapped habitats in the area. This threshold was used as a compromise between the Convention on Biological Diversity 11$^{th}$ target of 10% established in 2010 and the 20 to 30% target more likely needed in the longer term for effective conservation in the Pacific [7]. First, a "habitat-only plan" was computed, to define a habitat-based reserve network without using any cost constraint. Then, we included the constraint to minimize opportunity costs for each of the 19 cost layers, corresponding to 19 cost scenarios. Each cost was used independently, and no combination of costs was used.

The Marxan software was used to compute the networks,. Marxan is a freely available conservation planning software based on simulated annealing. It solves a minimum-set problem where the objective is to minimize the cost of a reserve network while ensuring conservation feature targets are met [30]. We searched for an optimal boundary length modifier (BLM) following Stewart and Possingham [31]. This parameter controls the trade-offs between minimizing costs and allowing for a more compact and realistic reserve. The optimal BLM was in the same range, 0.0025, for the habitat and the *Total Catch* cost scenarios, and we applied the same BLM value for all scenarios. Each scenario was run for both habitat maps. Eventually, this resulted in forty different reserve networks (twenty designs for each map, including one habitat-only design, and 19 scenarios with cost constraints).

## Scenario evaluation and comparison

Efficient cost scenarios minimize opportunity costs while meeting all conservation targets. In other words, an efficient scenario minimizes fishery catch loss while meeting the 20% habitat representation target for each habitat. First, we studied the effect of each scenario on the matching cost, which means how the inclusion of a cost as a constraint in the design influenced its own production. For instance we looked at how using a cost constraint on "Lethrinidae" influences the Lethrinidae catch that is locked in the reserves. Second, we studied the effects of each scenario on the total catch loss. For instance we looked at how using the cost constraint on "Lethrinidae" influences the total catch that is locked in the reserves. Indeed, total catch is a global representation of the potential production on the area, without distinction between fish families, gears or groups of fishermen, and this level of catch was used here as a reference.

To achieve these objectives, we described the reserve networks according to:

1. conservation targets met (at least 20% of each habitat included in the network). Specifically, the variable $Mv$ from Marxan represents the achieved percentage of habitat conservation target. $Mv$ can range from 0 (habitat not included in the design) to 1 (conservation target fully satisfied). The $Mv>0.95$ threshold marks the limit where the conservation target was considered satisfactory. We then counted how many targets ($N$) were not met to rank the scenarios.
2. $P_i$, as the initial catch for data layer $i$ ($i = 1$ to 19, Table 1).
3. $h$, as the remaining catch when using the habitat-only scenario, and $H = P\text{-}h$ as the resulting opportunity cost of the habitat-only scenario. $H$ ($H$ for *H*abitat) can be expressed for Total Catches ($H_1$) but also for any of the other 18 other data layers ($H_2$ to $H_{19}$) that can be used to define opportunity costs.

4. *c*, as the remaining catch for the same data layer *i* that was used to define opportunity costs, and $C = P_i - c_i$, as the resulting opportunity cost. Catches *C* (*C* for *Cost* layer) can be computed for each of the 19 selected data layers. For instance, the variable $C_4$ will quantify how the scenario based on *Oundjo* cost (scenario 4 in Table 1) reduces *Oundjo* catch loss compared to the habitat-only scenario $H_4$.
5. *t*, as the total remaining catch in a scenario with opportunity cost and $T = P_1 - t$ as the resulting opportunity cost. *T* can also be computed for each of the 19 selected data layers. For instance, the variable $T_8$ will quantify how the scenario based on *Lethrinidae* cost (scenario 8 in Table 1) reduces total catches loss compared to the habitat-only scenario $H_1$. Note that $T_1 = C_1$.

All metrics above (*P, H, T, C*) are expressed in kg.ha-$^1$.yr$^{-1}$ and are computed for the entire area. Table 2 summarizes these variables.

More metrics are theoretically available. In particular, we could cross-analyze each scenario and cost, for instance by looking at the influence of the *Lethrinidae* cost scenario on *Scaridae* opportunity cost. For simplicity sake, we did not present here these results, as the main trends and results can be described from the above metrics.

To evaluate and compare the effectiveness of the different scenarios, we used different indices summarized in Table 3:

- First, *%H* expresses the percentage of potential annual catch unavailable to fishers based on the habitat only scenario, or, in other words, the initial production "locked-in" by the habitat-only scenario. A $\%H_i$ can be computed for any of the 19 possible scenarios *i*. The higher $\%H_i$, the more costly is the habitat-based design for fishers that yielded the production $P_i$.
- Second, *%C* expresses the percentage of potential annual catch unavailable to fishers based on the corresponding cost scenario. A $\%C_i$ can be computed for any of the 19 possible scenarios *i*. The higher $\%C_i$, the more costly is the scenario for fishers that yielded the production $P_i$.
- Third, *%T* expresses the percentage of *Total Catch* locked in, for any considered scenario. Among the 19 different possibilities in selecting a reference, we favored the comparison of the different cost scenarios relatively to the *Total Catch* information ($i = 1$). A $\%T_i$ can be computed for any of the 19 possible scenarios *i*. The higher $\%T_i$, the more costly is the scenario for fishers that yielded the production $P_1$.

Finally, we define the effectiveness *E* of a scenario as the reduction of opportunity cost between a cost-based scenario and the habitat-only, scenario, with $E = 1\text{-}C/H$ (Table 3). Then, the effectiveness $E_1$ is similar to *E* but considering *Total Caches* as the reference, or $E_1 = 1\text{-}T/H_1$ (Table 3). These metrics provide different references (catch for scenario *i* or *Total Catch*) on the percentage of opportunity costs induced by the habitat-only scenario that have been effectively reduced when using fishery cost constraints. *E* and $E_1 \in [0,1]$ and are unitless (or expressed as percentage). The higher the *E* or $E_1$ value, the more efficient the cost scenario is compared to the habitat-based scenario. Note that *E* and $E_1$ can be negative if the cost scenario does not reduce the opportunity costs below the levels reached by the habitat-based scenario (i.e. $C>H$ or $T>H_1$). In this case, the scenario is not efficient.

Figure 3 shows a flow diagram with input data, output scenarios and relationships between the different variables.

## Results

### Networks of protected areas

For each design, we selected the best solution computed by Marxan as our final networks. Best solutions were consistent with planning unit selection frequencies (Fig. 4). Examples of output networks are presented in Figure 4. Figure 4 also illustrates 1) the compactness of the different reserves, due to the selection of BLM = 0.0025, 2) the differences achieved between a Millennium habitat analysis and a detailed habitat map analysis, 3) the influence of an opportunity cost constraint on a habitat-only design. In each case, this influence is high and substantially changes the distribution of the reserves in the network, as shown in Figure 4.

All metrics are provided in Tables 4 and 5. Catch values and effectiveness are provided for each scenario.

### Comparison of scenarios based on conservation objectives

Table 4 presents the number of missing habitats *N* in each scenario for both habitat maps.

For the habitat-only scenario, only one habitat could not be represented for each habitat map ($N = 1$). Including costs generally led to trade-offs to satisfy conservation objectives compared to the habitat-only scenarios (i.e., $N>1$). However, the best scenarios overall ($N = 0$) were for *Pouembout*, *Haemulidae* and *Gerreidae* scenarios for the Millennium and detailed map respectively. The worst scenarios were the *Mugilidae* and *Handline* scenarios with respectively 4 and 8 missing habitats, from Millennium map and detailed map respectively.

When comparing the different values for *N* (Table 4), the detailed map plan was generally more efficient in satisfying conservation objectives than the Millennium one, considering the highest complexity and fragmentation of habitats in the detailed

**Table 2.** Variables used to describe the different scenarios.

| Descriptor | Meaning of descriptor |
|---|---|
| *N* | Number of habitats with *Mv* <0.95 (*i.e.,* conservation objective not met) |
| i ∈[1;19] | Index of cost layer (cf Table 1) |
| $P_i$ | Initial production available for data layer *i* |
| $H_i$ | Value of opportunity cost for data layer *i* in the habitat-only design |
| $C_i$ | Value of opportunity cost for data layer *i* with the corresponding cost *i* scenarios |
| $T_i$ | Value of opportunity cost for Total Catch, for the cost *i* scenario ($T_1 = C_1$) |

**Table 3.** Selected metrics of scenario effectiveness, privileging *Total Catch* ($i = 1$) for reference.

| Variable | Description |
|---|---|
| $\%H_i = H_i/P_i$ | Ratio opportunity cost *vs* initial catch of data layer *i* ($i = 1$ to 19, cf. Table 1), for the habitat only design |
| $\%C_i = C_i/P_i$ | Ratio opportunity cost *vs* initial catch of data layer *i* ($i = 1$ to 19, cf. Table 1), for the opportunity cost *i* scenario |
| $\%T_i = T_i/P_1$ | Ratio opportunity cost of Total Catch *vs* initial Total Catch, for the opportunity cost *i* scenario |
| $E_i = 1-(C_i/H_i)$ | Effectiveness of cost *i* scenario considering opportunity cost *i (*100 if given in %)*. |
| $E_{1,i} = 1-(T_i/H_1)$ | Effectiveness of cost *i* scenario considering total catches opportunity cost *(*100 if given in %)* |

map. These characteristics intuitively should lead to more frequent trade-offs when trying to include 20% of the surface of all habitats.

## Comparison of scenarios based on costs

Table 4 show that for both habitat maps, almost all cost-based scenarios yielded a decreased opportunity cost (i.e., $H_i > C_i$), or in other words an increase in available biomass for fishers, when shifting from a habitat-only to a cost-based scenario. This result was expected, but the differences achieved when using the various opportunity costs appear clearly.

Noteworthy are the scenarios that surprisingly increased the opportunity costs. These were the *Haemulidae, Sparidae, Gerreidae* and *Kyphosidae* scenarios for Millennium maps, and *Carangidae* with the detailed map. These scenarios used very low catch values, and the catches appeared scattered across the maps in few habitat patches. Small changes in how these patches are included or not in the network led to high variations of effectiveness, even if the absolute costs remained low. This suggests that the effectiveness of a network constrained by low catches scattered in the seascape would not be robust to small spatial changes in the design. These scenarios are hereafter flagged as "non-efficient".

Table 5 presents *%C, %H* and *%T* achieved for each scenario, including the non-efficient ones (in italics). For efficient scenarios *%C* is systematically lower than *%H*, as expected. The scenario that generated the lowest *%H* was *Lutjanidae* using the Millennium map (6.2%). In other words, the remaining *Lutjanidae* catch available to fishers was the least affected by the habitat-based scenario. The lowest *%C* is generated by the *Gillnet* scenario with only 1.8% of initial catch immobilized, also using the Millennium map.

Lowest *%T* is achieved for *Total Catch* scenario, for both maps (5.7% and 10.6% for the Millennium and the detailed map respectively). This implies that, to decrease the habitat-only opportunity cost, using *Total Catch* data, as the largest picture on the catch distribution, leads to better outcomes for fishers than using only a fraction of the catch information (from a sublist of species, selected catch gear or specific locality). This was also expected, but the ratio *%T* clarifies the amount of benefits for the fishers between the different scenarios. Using the *Total Catch* scenario proves to be 2 and 3 times more interesting for fishers than any other scenario.

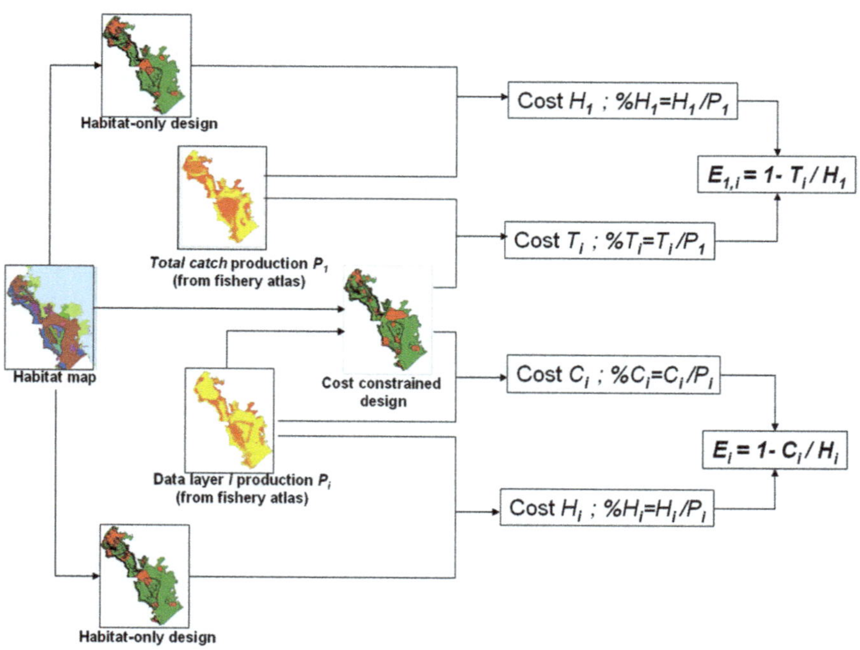


**Figure 3. Flow chart showing how the different metrics were obtained.** Habitat map and fishery Catch (or production) data are in input; then different conservation designs can be computed, with or without cost optimization. The various metrics (see text for details) can be computed and provide different outputs relative to the effectiveness of the different scenarios. The "Habitat only design" icon is repeated for clarity, to avoid crossing lines in the chart.

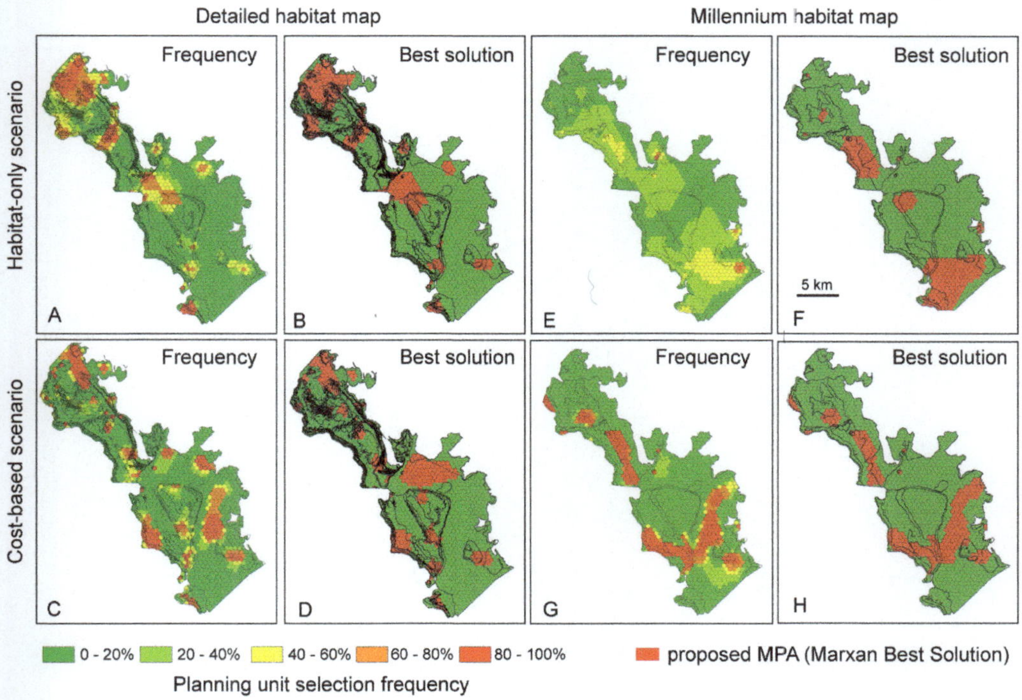


**Figure 4. Examples of conservation designs, with best solutions given by Marxan and selection frequencies of planning units.** For the two different habitat maps, the changes in proposed reserve locations are shown between a habitat-only scenario and a cost-based scenario, based here on *Total Catch* data (see Figure 2). High frequencies and best solutions are in strong agreement in some cases (panels A and B; panels G and H). When they differ in specific areas, it is because of habitats particularly abundant, such as the deep lagoon, that offer flexibility in the selection of management units (panels C and D), or because of lack of prioritary units (as shown also in low contrasted frequencies) (panels E and F).

Among the scenarios that are effective with their own corresponding costs ($E$), some are not efficient anymore compared to their effect on *Total Catch* costs ($E_I$). This includes the *Lutjanidae, Carangidae, Koné* and *Pouembout* scenarios for the Millennium map. For instance, *Pouembout* scenario shows the second best $E$ (76.9%) but also one of the worst $E_I$ (−49.8%). In other words, to decrease opportunity costs of the Millennium habitat-only design for the entire area, using only information on *Pouembout* catch is not effective at all. Similarly, the highest $E$ is reached for *Gillnet* (86.6%) and *Oundjo* (71.4%) for the Millennium and the detailed map respectively. However, for $E_I$, *Gillnet* and *Oundjo* drop to 22.2% and 34.3% respectively. Overall, on average, $E_I$ is about twice lower then $E$, with an average $E_I$ of 22.3% and 29.8% for the Millennium and detailed maps respectively, and an average $E$ of 52.3 and 45.8% for the Millennium and detailed maps respectively. This means that when using a particular catch layer $i$ to constrain the design, the effectiveness in decreasing the *Total Catch* loss is always lower than decreasing the corresponding catch $i$ loss.

## Influence of the type of habitat map

According to $H$, $C$, $\%H$ and $\%C$ (Tables 4 and 5), using the Millennium map generates lower opportunity costs than the detailed map, for all but one (*Pouembout* scenario) of the possible type of catches. This is observed with (when using $\%C$) or without (when using $\%H$) the constraint of minimizing opportunity cost in the design. Therefore, all habitat maps are not equal in generating costs when using them for a habitat-only design, or for a cost-constrained design. Here, we found that using the Millennium map, less thematically rich, is less costly for fishers in all situations.

The pattern is slightly different with $E$, i.e. when comparing the relative benefits of a cost-based approach to decrease the opportunity costs generated by the habitat only approach (Table 5, Figure 5). $E$ shows a less uniform hierarchy between map types. Specifically, the benefits of using cost constraints are higher with *Lethrinidae, Serranidae* and *Lutjanidae* when using the detailed habitat map compared to the Millennium map. The Figure 5 ranks the performances of each scenario according to $E$ and $E_I$ and for each map. If a manager has the choice of the scenario and can mix at will the types of productions and habitat maps, Figure 5 suggests that Millennium map-based scenarios will offer better combination than the detailed map-based scenarios, with higher $E$ values overall. But the pattern is opposite for $E_I$. In other words, a manager would be less efficient in decreasing the loss of *Total Catch* production (due to a habitat-based scenario) with a cost-based scenario if he has to use the Millennium maps.

# Discussion

## Sensitivity of habitat-based conservation plans to fishery opportunity costs

Effective and accepted conservation will be achieved if socio-economic realities are taken into account, but conservation actions may be channeled more quickly when using easily accessible data, such as habitat maps created anew with remote sensing or from already existing archives (e.g., Millennium maps). As such, we assessed here the ability of different opportunity costs to reduce the impact of habitat-based conservation plans, using fine-scale catch data. Indirectly, we measure the cost of conservation for fishers when usng only habitat-based scenarios. In contrast with Adams et al. [14] who modeled opportunity costs using the potential

**Table 4.** Level of conservation objectives met (*N*) and lost catch (opportunity costs *H, C, T*).

| | | | **Costs** *(kg.km$^{-2}$.yr$^{-1}$)* | | | | | | | |
|---|---|---|---|---|---|---|---|---|---|---|
| | *N* | | *P* | | *H* | | *C* | | *T* | |
| **Data layer** | **M** | **D** | **M** | **D** | **M** | **D** | **M** | **D** | **M** | **D** |
| Habitat only | 1 | 1 | | | | | | | | |
| Total catches | 3 | 7 | 5052 | 5058 | 736 | 1211 | 286 | 538 | 286 | 538 |
| Oundjo | 2 | 7 | 2552 | 2553 | 389 | 839 | 97 | 240 | 522 | 796 |
| Gillnet | 2 | 5+ | 2140 | 2141 | 281 | 550 | 38 | 162 | 572 | 883 |
| Handline | 2 | **8** | 1668 | 1672 | 209 | 361 | 84 | 156 | 505 | 816 |
| Koné | 2+ | 5 | 1411 | 1411 | 206 | 191 | 81 | 92 | 990 | 1613 |
| Speargun | 3 | 5+ | 1181 | 1181 | 231 | 284 | 120 | 153 | 644 | 888 |
| Lethrinidae | 3 | 6 | 1055 | 1058 | 158 | 264 | 79 | 125 | 511 | 870 |
| Acanthuridae | 1 | 3+ | 824 | 824 | 147 | 238 | 86 | 136 | 623 | 852 |
| Mugilidae | **4** | 4 | 759 | 759 | 74 | 156 | 22 | 77 | 712 | 1302 |
| Scaridae | 2 | 4 | 520 | 520 | 104 | 163 | 48 | 89 | 644 | 917 |
| Serranidae | 1 | 4 | 517 | 518 | 73 | 109 | 66 | 85 | 611 | 1095 |
| Siganidae | 1 | 2 | 498 | 498 | 87 | 120 | 58 | 99 | 661 | 1244 |
| Pouembout | 0 | 3 | 329 | 329 | 103 | 43 | 24 | 33 | 1102 | 1744 |
| Lutjanidae | **4** | 3 | 270 | 270 | 17 | 56 | 16 | 36 | 737 | 1252 |
| Haemulidae | 2 | 0 | 192 | 192 | 11 | 16 | 13 | 19 | 1079 | 1718 |
| Carangidae | 1 | 2 | 169 | 169 | 23 | 41 | 17 | 43 | 738 | 1676 |
| Sparidae | 1 | 1 | 100 | 100 | 3 | 6 | 4 | 7 | 1202 | 1753 |
| Gerreidae | 1 | 0 | 34 | 34 | 0 | 0 | 6 | 0 | 917 | 1915 |
| Kyphosidae | 1 | 4+ | 21 | 21 | 3 | 7 | 3 | 10 | 1259 | 1844 |

Results are provided for both maps (M for Millennium, D for the detailed map) and for all fishery data layers and scenarios, sorted and presented by decreasing *P*. In the "*N*" columns, "+" next to a number indicates that at least one habitat could not be represented at all in the design. The slight differences in *P* between the Millennium and Detailed map are due to a resolution-edge effect at the boundary of the respective map domains.

**Table 5.** Results for all metrics (*%H, %C, %T, E,* $E_1$).

| Data layer | *%H* | | *%C* | | *E (%)* | | *%T* | | $E_1$ *(%)* | |
|---|---|---|---|---|---|---|---|---|---|---|
| | **M** | **D** | **M** | **D** | **M** | **D** | **M** | **D** | **M** | **D** |
| Habitat-only | | | | | | | | | - | - |
| Total catches | 14.6 | 24 | 5.7 | 10.6 | 61.1 | 55.6 | **5.7** | **10.6** | **61.1** | **55.6** |
| Oundjo | 15.3 | **32.8** | 3.8 | 9.4 | 75.2 | **71.4** | 10.3 | 15.7 | 29.0 | 34.3 |
| Gillnet | 13.1 | 25.7 | **1.8** | 7.5 | **86.6** | 70.6 | 11.3 | 17.5 | 22.2 | 27.1 |
| Handline | 12.5 | 21.6 | 5.1 | 9.3 | 59.6 | 56.7 | 10.0 | 16.1 | 31.3 | 32.7 |
| Koné | 14.6 | 13.5 | 5.7 | **6.5** | 60.9 | 51.7 | 19.6 | *31.9* | *−34.5* | *−33.2* |
| Speargun | 19.5 | 24.0 | 10.1 | 12.9 | 48.2 | 46.2 | 12.7 | 17.6 | 12.4 | 26.7 |
| Lethrinidae | 15.0 | 24.9 | 7.4 | 11.8 | 50.2 | 52.7 | 10.1 | 17.2 | 30.5 | 28.2 |
| Acanthuridae | 17.9 | 28.9 | 10.4 | 16.5 | 41.8 | 42.9 | 12.3 | 16.9 | 15.2 | 29.6 |
| Mugilidae | 9.8 | 20.6 | 2.9 | 10.2 | 70.6 | 50.4 | 14.1 | *25.7* | **3.3** | *−7.5* |
| Scaridae | **20.1** | 31.4 | 9.2 | 17.1 | 53.9 | 45.5 | 12.7 | 18.1 | 12.5 | 24.3 |
| Serranidae | 14.1 | 21.1 | 12.8 | 16.4 | **9.1** | 22.3 | 12.1 | 21.6 | 16.9 | **9.6** |
| Siganidae | 17.5 | 24.1 | 11.7 | 19.9 | 33.2 | **17.5** | 13.1 | *24.6* | 10.1 | *−2.7* |
| Pouembout | 31.4 | **13.0** | 7.3 | 10.1 | 76.9 | 22.2 | *21.8* | *34.5* | *−49.8* | *−44.0* |
| Lutjanidae | **6.2** | 20.8 | 5.9 | 13.5 | 4.8 | 35.3 | *14.6* | *24.8* | *−0.2* | *−3.4* |
| *Haemulidae* | 5.7 | 8.1 | 6.8 | 10.0 | *−20.9* | *−23.9* | *21.4* | *34.0* | *−46.7* | *−41.8* |
| *Carangidae* | 13.4 | 24.2 | 10.2 | 25.3 | 23.7 | *−4.8* | *14.6* | *33.1* | *−0.3* | *−38.3* |
| *Sparidae* | 3.2 | 5.7 | 4.4 | 6.5 | *−39.8* | *−15.4* | *23.8* | *34.7* | *−63.5* | *−44.7* |
| *Gerreidae* | 0.2 | 0.6 | 17.0 | 0.6 | *−10476* | 7.0 | *18.1* | *37.9* | *−24.6* | *−58.1* |
| *Kyphosidae* | 14.4 | 34.7 | 14.8 | 47.1 | *−3.4* | *−35.8* | *24.9* | *36.5* | *−71.1* | *−52.2* |

Results are provided for both maps (M for Millennium, D for the detailed map) and for all fishery data layers and scenarios, sorted in the same order as in Table 4 (by decreasing *P*). Non-efficient scenarios are indicated by the data layer in italics. Best results for efficient scenarios (maximum values for *E* and $E_1$, minimum values for other indicators) are in bold underlined, while worst results are in bold only.

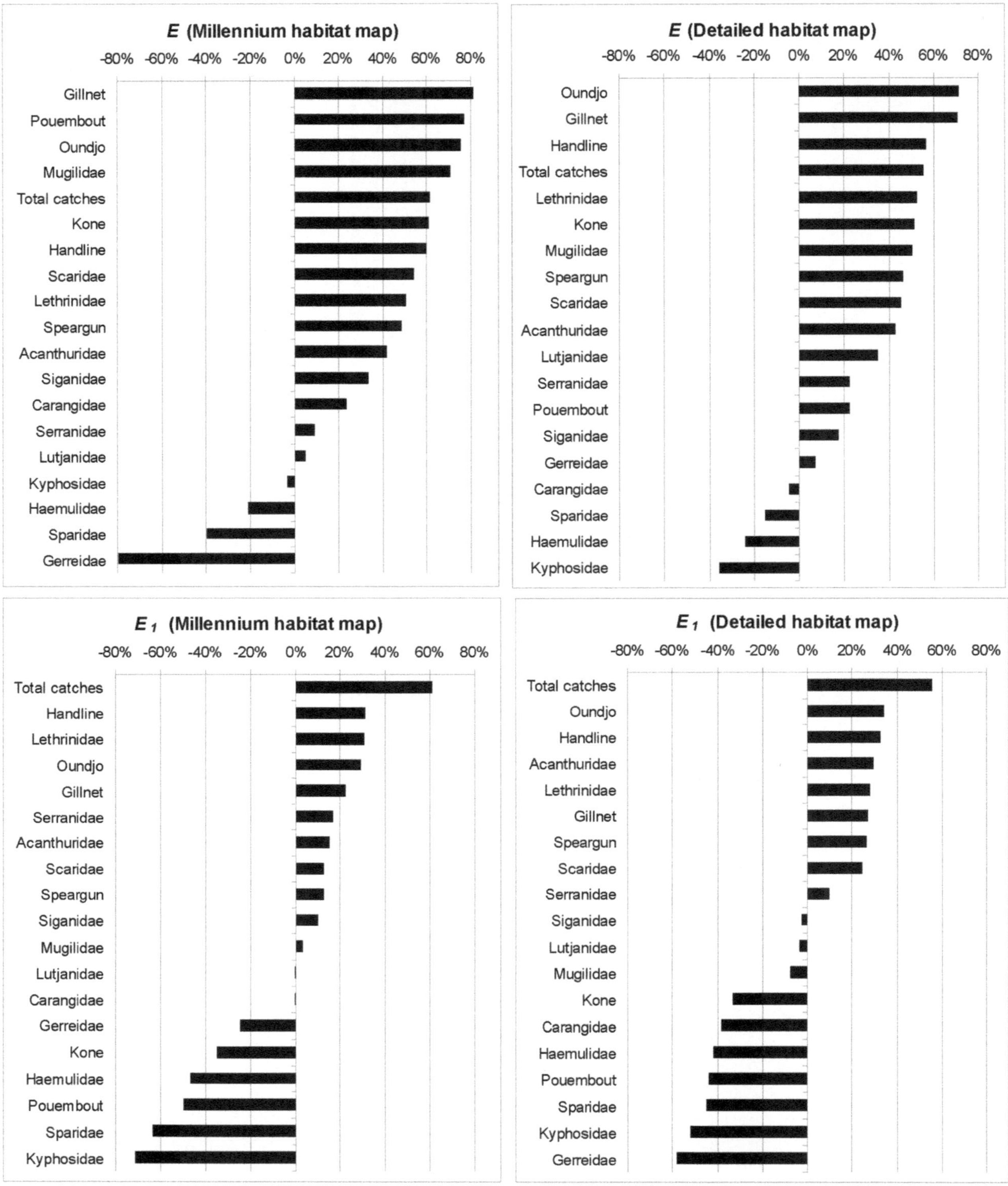


**Figure 5. Comparison of *E* and $E_1$ achieved between the two habitat maps and the various cost-based scenarios. Negative values <−100% (see Table 5, *Gerreidae* family) are forced to −80%.**

production of a site, including unfished areas, we used here directly the estimated spatial catch to directly derive opportunity costs. Here, the entire focal area was fished [15].

According to our New-Caledonia case study, minimizing fishery opportunity costs modulate the habitat-only conservation plan in various ways. First, the spatial distribution of the potential protected areas changes substantially (Fig. 4), and the cost constraints generally did not inhibit the ability of reserve designs to meet the conservation objectives (20% of the surface area of each habitat) even for the most complex habitat map. Second, the potential catch loss following habitat-driven conservation can be minimized, but there are many options. The effectiveness in achieving this goal will largely depend on the type of fishery catch used and the type of habitat map as well. Our results (Tables 4, 5; Fig. 5) indicate the sensitivity of the different {fishery data, habitat map} combinations in minimizing the opportunity costs of habitat-driven scenario. The results also show that it is possible to generate costs higher than that of the habitat-only scenarios when using a poor choice of fishery data (the non-efficient scenarios). These poor data were identified by their negative $E$ and $E_I$. We learnt that using low production sectors of fishery activity is inefficient in minimizing the cost of a habitat-based design although they are widely distributed throughout the area (Figure 5).

From Tables 4 and 5, we infer that results achieved with one habitat map are not immediately transposable to another habitat map. Using one fishery cost does not have the same effect depending on the map, and even the ranking in terms of effectiveness can be different. We also infer that each cost layer is more effective when it is used to reduce the corresponding catch loss. For instance, to minimize the *Koné* production locked in the conservation network, it is more efficient to optimize the network using the *Koné* production/cost data, and not any other cost layer. Thus, when looking at a given marine area, local and specific information is important to avoid unnecessary catch loss for conservation reasons. It is obviously possible to minimize the opportunity cost of a design by using any of the available production layers, but this is not the most effective way, and it can even be counter-productive. This conclusion also suggests that crude catch proxies should be used with caution, as they are likely to be a poor representation of the actual catch level and distribution if they are crudely mapped. Weeks et al. [13] have stressed, with proxies, the importance of using the most accurate and fine-scale possible data, but our results also suggest that most proxies, or generalization from one site to another, could lead to costly conservation mistakes [32].

We conclude here that local data are recommended. Next, a manager could ask if individual reserves need to be designed more locally with local production data and then scaled up together; or whether considerations of overall network habitat representation is more important? In other words, where is the limit, or trade-off, in terms of spatial domains? This is a study in itself that warrant further investigations. The results and recommendations will likely be dependent on reef and habitat structures, differences of livelihood between coastal communities and type of fishing practices. Here, we found that local data are needed, but this also likely reflects the Oundjo, Koné and Pouembout socio-economic differences and habitat configurations in the survey area.

Although our results confirm that habitat-driven plans would be useful for discussing the possible location of fishery restrictions and marine reserves in particular with stakeholders, conservation plans would certainly need to be revised to minimize the impacts on fishers. *Total Catches* opportunity costs of a habitat-only plan can be decreased by up to 55–61% depending on the type of habitat maps if criteria on *Total Catches* are accounted for. Although effectiveness will certainly vary in a different location than the VKP site, *Total Catches* is the most robust criteria to minimize costs as it is the highest and most widely distributed of all types of fish catches. Using a data set more easily compiled for a specific production or type of fishermen activities, would also decrease the cost of habitat-based only conservation, but far less effectively than using *Total Catches*. Here, the benefits dropped to a range of 15–25%, instead of 55–61% when using anything else than *Total Catch* information. On the other hand, mapping a *Total Catch* distribution can be a time-consuming costly task [15].

Here, only opportunity costs estimated from recent fishery production maps were considered. However other types of opportunity costs can affect fishers when setting a conservation plan [33]. This can include greater travelling distances, update of fishing gears and boats, time to rebuild knowledge on different areas, etc. Consequently, a manager must keep in mind that only one aspect of the equation was presently clarified, and that other costs need to be considered as well. In particular, Van de Geer et al. [33] emphasize taking into account the temporal variability of fishing effort. This is clearly an important aspect in a developing area like VKP. In VKP, the entire area can be considered fished and this will not change. However, catch per unit effort may change due to overfishing or better equipment coming with a booming economic activity and higher household incomes.

Finally, the 20% threshold is obviously one factor that could change the level of opportunity costs achieved by the habitat-only scenario, and the effectiveness of the cost-based scenarios. Decreasing the 20% threshold will decrease the opportunity costs, but it will also decrease the effectiveness of the network in preserving biodiversity and services. The sensitivity of the results presented here to this threshold warrants further investigation.

## Lessons for other sites

Generalizing the VKP findings to other Pacific Islands fisheries needs caution. The town of Koné is now a fairly large town by New Caledonia standards and compared to most rural locations. The level of fishing equipments used is also up-scaled compared to many Pacific rural areas [28,34,35,36]. The configuration of Oundjo village is more representative of the level of activities and equipment found in most rural New Caledonia and elsewhere in Melanesia. Oundjo fishers also use a reef that offer a fairly typical Pacific island geomorphological zonation with a fringe of seagrass beds onshore, patch reefs in the lagoon, a wide sedimentary lagoon, and a barrier reef that is 4–9 km away from the village. When limiting the VKP analysis to the Oundjo fishing ground, $E$ reaches up to 80% and effectiveness $E_I$ is overall higher than for the entire VKP area. For several fish families, effectiveness is much higher for Oundjo than VKP. This includes the *Haemulidae*, *Scaridae*, *Acanthuridae*, *Serranidae* and *Siganidae* families.

The effectiveness of different types of production (by fish families, by gear, by locations) in minimizing the *Total Catches* opportunity costs ($T$, $\%T$, and $E_I$) are thus likely site-dependent, in addition to being habitat map dependent as shown here (Figure 5). Additional analysis as performed for the entire VKP area, but only considering the effects on Oundjo or Koné catches, shows indeed different rankings among sites and between habitat maps (Table 6). It is therefore difficult to prioritize or recommend the systematic acquisition of any particular production data set. If one aims to produce a simplified fishery atlas (e.g., for at least one fish family, or one gear or one location) for the sake of minimizing data collection, the expected effectiveness $E_I$ would be around 25% at most, when using the detailed map and its typology of habitats for the 20% conservation objectives.

One may suggest that to avoid extensive data collection, it may be possible to apply to each type of reefs and habitats a "standard" catch production values. This would entail a review on published production per habitats, either measured or modeled, which is likely to be unsuccessful beyond a few number of broadly defined habitats [14]. Assigning published mean production rates is a simple way to map potential productions. For instance, Newton et al. [29] suggested the 3–5 $t.km^{-2}.y^{-1}$ threshold as the limit of sustainable fishing rate for most reefs. Here, in New-Caledonia, we observed this level of catch rate on the most exploited areas, but much less elsewhere. Sustainability thresholds and actual produc-

**Table 6.** Ranking of scenarios according to their effectiveness in minimizing Oundjo and Koné catch loss generated by the habitat-based scenario.

| | **Detailled map** | | **Millennium map** | |
|---|---|---|---|---|
| | **Fishery ground** | | **Fishery ground** | |
| Rank ($E_i$) | Oundjo | Koné | Oundjo | Koné |
| 1 | Total catches | Speargun | Total catches | Total catches |
| 2 | Gillnet | Total catches | Gillnet | Speargun |
| 3 | Lethrinidae | Scaridae | Lethrinidae | Handline |
| 4 | Handline | Siganidae | Handline | Scaridae |
| 5 | Acanthuridae | Pouembout | Mugilidae | Lethrinidae |
| 6 | Scaridae | Acanthuridae | Serranidae | Siganidae |
| 7 | Speargun | Lutjanidae | Acanthuridae | Acanthuridae |
| 8 | Serranidae | Handline | Scaridae | Serranidae |
| 9 | Mugilidae | Lethrinidae | Lutjanidae | Pouembout |
| 10 | Lutjanidae | Serranidae | Siganidae | Oundjo |
| 11 | Siganidae | Mugilidae | Speargun | Lutjanidae |
| 12 | Haemulidae | Gillnet | Haemulidae | Gillnet |
| 13 | Koné | Oundjo | Koné | Mugilidae |
| 14 | Pouembout | Haemulidae | Pouembout | Haemulidae |

For each type of habitat maps, cost-based scenarios are ranked by decreasing $E_i$ (i.e., by decreasing effectiveness in minimizing the catch loss for Oundjo and Koné generated by a habitat-based scenario). In particular, we note that minimizing catch loss on one area (e.g., Koné) using the catch for the other area (e.g., Oundjo catch) is among the least effective approach.

tion do not necessarily match and it is not recommended to apply one for another, as the opportunity costs would be over-estimated. Further, habitats as they can be recognized by remote sensing and ground-truthing do not necessarily convey a sense of quality and conditions, and therefore catch potential. How habitat conditions are taken into account may substantially change a conservation plan [8]. This aspect is difficult to tackle without local data. It is also another reason for avoiding generalization from one site to another.

The systematic lower costs when using Millennium maps can tentatively be explained by a greater flexibility in achieving less stringent habitat targets (Fig. 4). However, we lack more evidences from other case studies that would confirm that thematically richer maps always have worse effects for fishers. In fact, intuitively, habitat maps reflecting precisely targeted species' distribution would be optimal to minimize costs, and designing a habitat map to represent a number of species habitats will also likely lead to a more complex habitat map than the Millennium map. One way to investigate precisely this hypothesis would be to evaluate the exact link between mapped habitats and catch distribution, for different species. The distribution of catch for several key species may be related to some types of coarse habitats found in the Millennium typology (e.g. barrier reef forereef), while more detailed habitat maps may provide an artificially fragmented view of the habitat of the same species, and then less effective results. This warrants further investigation: how does mapping the exact habitat of fished species minimize costs for fishers in a habitat-based conservation plan? The fishery atlas that was used for this study would be again useful to answer this question.

## Conclusions

Using a comprehensive fine-scale coral reef fishery data set and different habitat maps, we could quantify how conservation costs can be minimized compared to a habitat-only conservation approach. The analysis was realistic, with a full range of scenarios that reflect the diversity of fishing practices found in a rural, but developing, area.

The main observation in resource-limited management situations is that habitat-based conservation schemes can be used to initiate discussions around a conservation project at a cost that is not necessarily prohibitive compared to scenario with cost optimization constraints. In terms of management recommendation, a habitat-only scenario would help move forward rapidly, but at the disadvantages of fishers in most cases. A thematically simple geomorphological (e.g., Millennium) map can provide rapidly a first low cost solution. Then, once fishery data and more detailed habitat maps become available, it would be possible to optimize the spatial design of the conservation plan to minimize opportunity costs while still meeting conservation objectives. However, depending on which fishery data set is used the benefits vary widely, and can even be negative when using marginal low-catch fisheries. Knowledge of the spatial distribution of all types of fishing activities and catch (i.e., *Total Catch*) is the most effective way, but this requires intensive data collection. The effort appears so intensive that very few coral reef fisheries are quantitatively characterized. Using instead for convenience a fraction of the *Total Catch* information may lead to non negligible improvement and lower costs for fishers (<25%), but the benefit may be seen as relatively poor, compared to habitat-only scenarios that can be used fairly quickly. It is difficult to generalize these recommendations to every configuration, depending on the urgency of the situation, the distribution of the human population, and the extent and complexity of the reefs. This is the first case study to tackle this issue and more configurations need to be investigated. Unfortunately, this study may remain isolated if no other fishery atlas are developed.

Finally, it is also recommended to use local fishery data, specific to the place, when optimizing a local conservation plan. This may sound obvious, but lack of local data may promote the use of data acquired elsewhere to fill gaps, which is likely to lead to non-optimal choices (Table 6), failure to maintain adequate incomes for fishers, disinterest of local communities (and managers) for conservation on the long run, and lack of compliance. Our results provide some guidance on the sensitivity to different types of opportunity costs (Tables 4, 5, 6), which can be interpreted as the range of risk that a manager takes when using one data layer or another to define costs. The risk can be immediately translated in economic statements, since the metrics used here (in particular $T$, $E_I$) are ratio of biomass catch.

## Acknowledgments

We acknowledge the two reviewers for their constructive comments and suggestions.

## Author Contributions

Conceived and designed the experiments: MD SA ML. Performed the experiments: MD SA. Analyzed the data: MD SA ML. Contributed reagents/materials/analysis tools: MD SA ML NG. Wrote the paper: MD SA ML NG.

## References

1. Dalleau M, Andréfouët S, Wabnitz CCC, Payri C, Wantiez L, et al F (2010) Use of Habitats as Surrogates of Biodiversity for Efficient Coral Reef Conservation Planning in Pacific Ocean Islands. Conservation Biology 24: 541–552.
2. Andréfouët S, Chagnaud N, Chauvin C, Kranenburg C (2008) Atlas des récifs coralliens de France Outre-Mer. Centre IRD de Nouméa, Nouméa, Nouvelle-Calédonie. 153 p.
3. Weeks R, Jupiter SD (2013) Adaptive comanagement of a marine protected area network in Fiji. Conservation Biology 27:1234–1244
4. Richardson E, Kaiser M, Edwards-Jones G, Possingham H (2006) Sensitivity of marine-reserve design to the spatial resolution of socioeconomic data. Conservation Biology 20: 1191–1202.
5. Ban NC, Hansen GJA, Jones M, Vincent ACJ (2009) Systematic marine conservation planning in data-poor regions: Socioeconomic data is essential. Marine Policy 33: 794–800.
6. Ban NC, Adams VM, Almany GR, Ban S, Cinner JE, et al. (2011) Designing, implementing and managing marine protected areas: Emerging trends and opportunities for coral reef nations. Journal of Experimental Marine Biology and Ecology 408: 21–31.
7. Hamel M, Andréfouët S, Pressey RL (2013) Compromises between international habitat conservation guidelines and small-scale fisheries in Pacific island countries. Conservation Letters 6: 46–57.
8. Klein CJ, Tulloch VJ, Halpern BS, Selkoe KA, Watts ME, et al. (2013) Tradeoffs in marine reserve design: habitat condition, representation, and socioeconomic costs. Conservation Letters. doi: 10.1111/conl.12005.
9. Walmsley SF, White AT (2003) Influence of social, management and enforcement factors on the long-term ecological effects of marine sanctuaries. Environmental Conservation 30: 388–407.
10. Bell JD, Johnson JE, Hobday AJ (2011) Vulnerability of tropical Pacific fisheries and aquaculture to climate change. Secretariat of the Pacific Community, Noumea, New Caledonia.
11. Naidoo R, Balmford A, Ferraro PJ, Polaski S, Ricketts TH, et al. (2006) Integrating economic costs into conservation planning. Trends in Ecology and Evolution 21, 7.
12. Klein CJ, Chan A, Kircher L, Cundiff AJ, Gardner N, et al. (2008) Striking a balance between biodiversity conservation and socioeconomic viability in the design of marine protected areas. Conservation Biology 22: 691–700.
13. Weeks R, Russ GR, Bucol AA, Alcala AC (2010) Shortcuts for marine conservation planning: The effectiveness of socioeconomic data surrogates. Biological Conservation 143: 1236–1244.
14. Adams VM, Mills M, Jupiter SD, Pressey RL (2011) Improving social acceptability of marine protected area networks: A method for estimating opportunity costs to multiple gear types in both fished and currently unfished areas. Biological Conservation 144: 350–361.
15. Léopold M, Guillemot N, Rocklin D, Chen C (2014). A framework for mapping small-scale coastal fisheries using fishers' knowledge. ICES Journal of Marine Science. doi:10.1093/icesjms/fst204
16. Anuchiracheeva S, Demaine H, Shivakoti GP, Ruddle K (2003) Systematizing local knowledge using GIS: fisheries management in Bang Saphan Bay, Thailand. Ocean and Coastal Management 46: 1049–1068.
17. Aswani S, Lauer M (2006) Incorporating fishermen's local knowledge and behaviour into geographical information systems for designing marine protected areas in Oceania. Human Organization 65: 81–102.
18. Hall GB, Close CH (2007) Local knowledge assessment for a small-scale fishery using geographic information systems. Fisheries Research 83: 11–22.
19. De Freitas DM, Tagliani PRA (2009) The use of GIS for the integration of traditional and scientific knowledge in supporting artisanal fisheries management in southern Brazil. Journal of Environmental Management 90: 2071–2080.
20. Hamel M, Andréfouët S (2010) Using very high resolution remote sensing for the management of coral reef fisheries: Review and perspectives. Marine Pollution Bulletin 60: 1397–1405.
21. Guillemot N, Léopold M (2009) Atlas de la pêche non professionnelle de poissons du lagon dans la zone de Vook-Koohnê-Pwëëbuu (Voh-Koné-Poumebout), Nouvelle-Calédonie. Données 2007. IRD, Nouméa (http://cogeren.ird.nc/index.php/download_file/view/136/133/ access August 2013)
22. David G, Léopold M, Dumas PS, Ferraris J, Herrenschmidt JB, et al. (2010) Integrated coastal zone management perspectives to ensure the sustainability of coral reefs in New Caledonia. Mar Pollut Bull 61: 323–334.
23. Léopold M, Sourisseau JM, Cornuet N, David C, Bonmarchand A, et al. (2013) La gestion d'un lagon en mutation: acteurs, enjeux et recherche-action en Nouvelle-Calédonie (Pacifique Sud). Vertigo 13, doi: 10.4000/vertigo.13297
24. Guillemot N (2009) Reef Fish assemblages and their exploitation in the Voh-Koné-Pouembout area (New Caledonia): characterization, Indicators and Monitoring Issues. PhD Thesis, IRD/Agrocampus ouest, Rennes.
25. Andréfouët S, Muller-Karger FE, Robinson JA, Kranenburg CJ, Torres-Pulliza D, et al. (2006) Global assessment of modern coral reef extent and diversity for regional science and management applications: a view from space, In 10th Int. Coral Reef Symposium. eds Y. Suzuki, T. Nakamori, M. Hidaka, H. Kayanne et al. Japanese Coral Reef Society, Okinawa, Japan. pp. 1732–1745.
26. Andréfouët S (2008) Coral reef habitat mapping using remote sensing: a user *vs* producer perspective. Implications for research, management and capacity building. Journal of Spatial Science 53: 113–129.
27. Andréfouët S, Chagnaud N, Kranenburg C (2009) Atlas des récifs coralliens de l'Océan Indien Ouest. Atlas of Western Indian Ocean Coral Reefs. Centre IRD de Nouméa, Nouméa, Nouvelle-Calédonie. 102 p.
28. Guillemot N, Léopold M, Chabanet P, Cuif M (2009) Characterization and management of informal fisheries confronted with socio-economic changes in New Caledonia (South Pacific). Fisheries research 98: 51–61.
29. Newton K, Côté I, Pilling G, Jennings S, Dulvy N (2007) Current and future sustainability of island coral reef fisheries. Current Biology 17: 655–658.
30. Ball IR, Possingham HP, Watts M (2009). Marxan and relatives: Software for spatial conservation prioritisation. Chapter 14: Pages 185–195 in Spatial conservation prioritisation: Quantitative methods and computational tools. Eds Moilanen A, Wilson KA, and Possingham HP. Oxford University Press, Oxford, UK.
31. Stewart RR, Possingham HP (2005) Efficiency, costs and trade-offs in marine reserve system design. Environmental Modeling and Assessment 10: 203–213.
32. Carwardine J, Wilson KA, Watts ME, Etter A, Klein CJ, et al. (2008) Avoiding costly conservation mistakes: the importance of defining actions and costs in spatial priority setting. PLoS One. doi: 10.1371/journal.pone.0002586
33. Van de Geer C, Mills M, Adams VM, Pressey RL, McPhee D (2013) Impacts of the Moreton Bay Marine Park rezoning on commercial fishermen. Marine Policy 39: 248-256.
34. Bell J, Kronen M, Vunisea A, Nash WGK, Demnke A, et al. (2009) Planning the use of fish for food security in the Pacific. Marine Policy 33: 64–76.
35. Pinca S, Kronen M, Magron F, McArdle B, Vigliola L, et al. (2012) Relative importance of habitat and fishing in influencing reef fish communities across seventeen Pacific Island Countries and Territories. Fish and Fisheries 13: 361–379.
36. Kronen M, Pinca S, Magron F, McArdle B, Vunisea A, et al. (2012) Socio-economic and fishery indicators to identify and monitor artisanal finfishing pressure in Pacific Island countries and territories. Ocean Coastal Management 55: 63–73.

# 2

# Depth and Medium-Scale Spatial Processes Influence Fish Assemblage Structure of Unconsolidated Habitats in a Subtropical Marine Park

**Arthur L. Schultz[1,3]*, Hamish A. Malcolm[2], Daniel J. Bucher[3], Michelle Linklater[4], Stephen D. A. Smith[1,3]**

**1** National Marine Science Centre, Southern Cross University, Charlesworth Bay, Coffs Harbour, New South Wales, Australia, **2** Marine Ecosystem Research, New South Wales Department of Primary Industries, Coffs Harbour, New South Wales, Australia, **3** Marine Ecology Research Centre, School of Environment, Science and Engineering, Southern Cross University, Lismore, New South Wales, Australia, **4** Coastal and Marine Unit, Science Division, Office of Environment and Heritage, Woolongong, New South Wales, Australia

## Abstract

Where biological datasets are spatially limited, abiotic surrogates have been advocated to inform objective planning for Marine Protected Areas. However, this approach assumes close correlation between abiotic and biotic patterns. The Solitary Islands Marine Park, northern NSW, Australia, currently uses a habitat classification system (HCS) to assist with planning, but this is based only on data for reefs. We used Baited Remote Underwater Videos (BRUVs) to survey fish assemblages of unconsolidated substrata at different depths, distances from shore, and across an along-shore spatial scale of 10 s of km (2 transects) to examine how well the HCS works for this dominant habitat. We used multivariate regression modelling to examine the importance of these, and other environmental factors (backscatter intensity, fine-scale bathymetric variation and rugosity), in structuring fish assemblages. There were significant differences in fish assemblages across depths, distance from shore, and over the medium spatial scale of the study: together, these factors generated the optimum model in multivariate regression. However, marginal tests suggested that backscatter intensity, which itself is a surrogate for sediment type and hardness, might also influence fish assemblages and needs further investigation. Species richness was significantly different across all factors: however, total MaxN only differed significantly between locations. This study demonstrates that the pre-existing abiotic HCS only partially represents the range of fish assemblages of unconsolidated habitats in the region.

**Editor:** Dennis M. Higgs, University of Windsor, Canada

**Funding:** Funding was supplied by a postgraduate (PhD) research budget from Southern Cross University. The funders had no role in study design, data collection and analysis, decision to publish, or preparation of the manuscript.

**Competing Interests:** The authors have declared that no competing interests exist.

* E-mail: a.schultz.22@student.edu.au

## Introduction

To adequately represent the entire range of biota present, conservation planning for sub-tidal marine ecosystems should be based on comprehensive understanding of species, habitats, ecosystems and associated ecological processes [1–2]. However, in most cases, these data are not available, and hence much of the decision-making for planning and management of Marine Protected Areas (MPAs) relies on incomplete species inventories and distributions [1,3–5]. New approaches that better integrate abiotic data into overall planning have consequently been advocated [1,4,6–7]. Physical and biophysical surrogates for biotic communities are now commonly used in planning of MPAs as they are more easily measured, categorised and geo-referenced [8–9]. Recent improvements in sonar equipment and techniques now provide high-resolution imagery of the physical properties and topography of the seabed [10–12]. However, the use of abiotic characteristics in planning carries the assumption that biotic patterns, which are the end-points of conservation management, consistently correlate with abiotic variables. While abiotic variables have sometimes been shown to accurately predict biotic patterns [13–16], this is not always the case [8]. Testing needs to be on a scale relevant to MPA planning [8,16–17], and better information on biotic distributions is required for integration into existing planning and management of MPAs [4,16].

Sub-tidal sediments are the most extensive benthic habitat worldwide [18]. However, there is a paucity of data for fish assemblages when compared with adjacent reef habitats [19]. The fact that unconsolidated habitats are critical for many commercial fisheries worldwide further emphasises the need for greater understanding of these habitats and their associated biota [20–21]. Often, the use of abiotic surrogates is an unavoidable strategy, due to the lack of biotic data [1,3–4,22], and the lack of fish assemblage data in unconsolidated habitats restricts objective selection of ecologically important areas for protection.

Most data on demersal fish assemblages of unconsolidated habitats have been collected by trawling [20,23–24], an approach that is generally incompatible with MPA objectives [25]. In this study, we used Baited Remote Underwater Video (BRUV), which is a low-impact, remote method with increasing application to a range of sub-tidal habitats [25–33], to assess the effectiveness of a Habitat Classification Scheme (HCS) for representing fish assemblages of unconsolidated habitats. The current HCS for

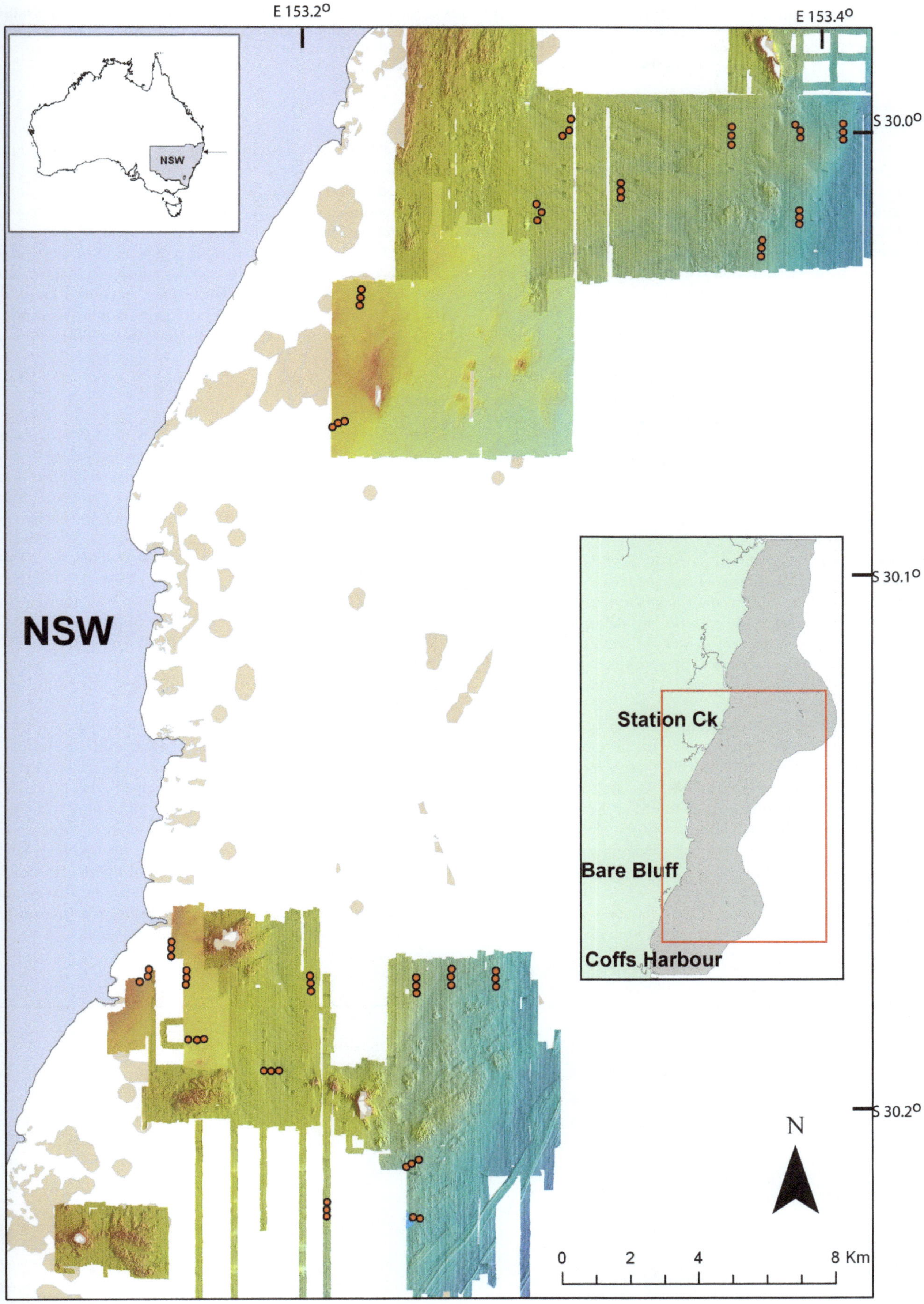
E 153.2°
E 153.4°
S 30.0°
S 30.1°
S 30.2°
NSW
NSW
Station Ck
Bare Bluff
Coffs Harbour
N
0
2
4
8 Km

**Figure 1. The location of the sixty-five individual BRUVs deployments in the Solitary Islands Marine Park (SIMP), New South Wales, Australia.** The northern sampling site is approximately adjacent to Station Creek, and the southern sampling site is approximately adjacent to Bare Bluff. The insert box (right side of image) displays the full extent of the SIMP in grey, with the complete sampling area indicated within the red box. The second insert box (top of image) shows the study position in relation to Australia.

marine parks in New South Wales (NSW), Australia, is an important tool for the design and planning of multi-purpose MPAs [34] and uses a hierarchical classification, primarily based on abiotic factors: habitat type (unconsolidated sediments and hard substrata) and depth (shallow: <25 m, intermediate: 25–50 m, and deep: >50 m) [16,34].

The Solitary Islands Marine Park (SIMP) in the Tweed-Moreton Bioregion of northern New South Wales, eastern Australia, comprises both State waters (to 3 nautical miles from shore), and the adjacent Solitary Islands Marine Reserve (SIMR) in Commonwealth waters. The SIMP is unique among NSW marine parks in that it includes several islands up to 12 km from the mainland coastline. The dominant system of currents, which includes strong influence of the East Australian Current (EAC) [35–36], cooler coastal counter-currents, and periodic upwelling [37], results in an overlap between temperate, tropical and subtropical endemic biota in the region [38–39]. The SIMP has extensive areas of both rocky reef and unconsolidated sediments from intertidal areas to depths >70 m [40]. Comprehensive surveys of fish associated with hard substrata have detected strong depth-related patterns in assemblage structure [16]: depth categories for the HCS for the SIMP (shallow: <25 m, intermediate: 25–50 m, deep: >50 m) were consequently refined to reflect these patterns. Distance from the mainland coast, which is independent of depth due to the presence of shallow fringing reef around the islands, also influences patterns of reef fish assemblages, and HCS therefore incorporates categories to represent these patterns (Inshore <1.5 km, Mid-shelf = 1.5–6 km, Offshore > 6 km).

Unconsolidated sediments dominate sub-tidal habitats in the SIMP [40], but there are currently no data to test whether the HCS adequately represents the diversity of fish communities of these habitats. However, preliminary BRUVs data, both from an examination of the effects of proximity to reef on fish assemblages [30], and from haphazard deployment across the SIMP (Malcolm, unpublished data), suggest a substantially different suite of species to those found on rocky reefs.

The focus of this research, therefore, was to examine fish assemblage structure of unconsolidated habitats to determine patterns over factors known to affect their reef counterparts (depth, distance from shore), and to determine how consistent they are over medium spatial scales (km). As data on the broad composition of unconsolidated habitats has recently been generated using swath acoustic multi-beam sidescan sonar [40], we also explored the relationship between fish assemblages and physical habitat characteristics. The specific aims were to: 1) quantify patterns of fish assemblages across different depths, distances from shore and over a medium spatial scale; and 2) determine which of these, and other environmental factors (backscatter intensity, fine-scale bathymetric range, and rugosity), are important in structuring fish assemblages of unconsolidated habitats in the SIMP using a multivariate regression model.

## Methods

The SIMP covers an area of approximately 71,000 hectares, within a broad ecotone between tropical and temperate assemblages [16,40]. The marine park is zoned for multiple use, with sections either fully protected (Sanctuary Zone), protected from commercial trawling (Habitat Protection Zone), or having no protection (General Use). Sixty-five Baited Remote Underwater Video (BRUV) deployments were completed within the SIMP during the Austral winter of 2011 (Fig. 1, Table 1). The New South Wales Marine Park Authority provided the permit for this research. As all deployments were within the SIMP boundaries, no specific permit was required for this research regarding private land access. Similarly, as all sampling was non-extractive and no collection of endangered or threatened species or other vertebrate fauna was performed, no specific permission was required in this context. BRUVs deployments were spread between two broad cross-shelf transects ('locations') located in the north and south of the SIMP, where the SIMP is longitudinally at its widest, and on seafloor that had previously been swath mapped using a 125 kHz GeoSwath interferometric sidescan sonar [40]. BRUV deployments were also spread across depth and distance from shore gradients that provided replication across the existing HCS categories. The design was not fully balanced due to constraints around areas that had been swath mapped: however, this did not preclude statistical modelling. Prior to the field work, coordinates for each deployment were uploaded to a handheld GPS to

**Table 1.** Factor groups, levels and number of BRUVs replicates sampled.

| Factor | Level | Number of Replicates |
|---|---|---|
| Depth | 10–30 m | 18 |
| | 30–50 m | 24 |
| | 50+m | 23 |
| Location | North | 29 |
| | South | 35 |
| Backscatter Intensity | 0–50 | 50 |
| | 50–100 | 6 |
| | 100–150 | 8 |
| | 150+ | 1 |

**Table 2.** Marginal and sequential test results for the distance based linear modelling (DistLM) procedure using forward selection and 4999 permutations, examining four factors of interest within the Adjusted $R^2$ model, and two within the AIC model.

**MARGINAL TESTS**

| Selection criterion | Variable | SS(trace) | Pseudo-F | P | Prop. |
|---|---|---|---|---|---|
| **Adjusted $R^2$** | Depth | 8086.4 | 8.8324 | **0.0002** | 0.12296 |
| | Location | 4927.5 | 5.1026 | **0.0006** | 7.49E-02 |
| | Backscatter Range 10 m | 3364.4 | 3.3966 | **0.0092** | 5.12E-02 |
| | Rugosity | 1205.7 | 1.1766 | 0.3168 | 1.83E-02 |
| **AIC** | Depth | 8086.4 | 8.8324 | **0.001** | 0.12296 |
| | Location | 4927.5 | 5.1026 | **0.001** | 7.49E-02 |
| | Backscatter Range 10 m | 3364.4 | 3.3966 | 0.011 | 5.12E-02 |
| | Rugosity | 1205.7 | 1.1766 | 0.314 | 1.83E-02 |

**SEQUENTIAL TESTS**

| Adjusted $R^2$ | Variable | SS(trace) | Pseudo-F | P | Prop. (%) | Cumul. (%) |
|---|---|---|---|---|---|---|
| 0.10904 | Depth | 8086.4 | 8.8324 | **0.0002** | 0.12296 | 0.12296 |
| 0.17173 | Location | 4909.8 | 5.7687 | **0.0002** | 7.47E-02 | 0.19761 |
| 0.17258 | Rugosity | 904.3 | 1.0636 | 0.3836 | 1.38E-02 | 0.21137 |
| 0.17937 | Backscatter Range 10 m | 1269.2 | 1.5051 | 0.1998 | 1.93E-02 | 0.23066 |
| **AIC** | **Variable** | **SS(trace)** | **Pseudo-F** | **P** | **Prop.** | **Cumul.** |
| 445.24 | Depth | 8086.4 | 8.8324 | **0.001** | 1.23E-01 | 0.12296 |
| 441.45 | Location | 4909.8 | 5.7687 | **0.001** | 7.47E-02 | 0.19761 |

Significant values are shown in bold.

accurately position BRUVs in the field. There was a minimum distance of 200 m between deployments [16,30,41] and all deployments were made at least 400 m from the nearest reef, a distance that has been shown to be beyond the foraging range of most reef-associated fish [30].

## Field methods

Each BRUV drop consisted of a mini-DV video camera with a wide angle lens, within an underwater housing with flat acrylic end ports, an attachment frame, a bait pole and mesh bag with bait, and a rope and float system linking the unit to the surface [30,41]. Approximately 1 kg of pilchard (*Sardinops neopilchardus*) was mashed into the bait bag, approximately 1.5 m from the lens of the camera, to attract fish to the viewing area in front of the camera. Each housing and bait pole was bolted to the attachment frame so that fish could be viewed in a horizontal orientation to the substratum. The field-of-view was standardised to approximately 2 m behind the bait, to minimise the effects of water clarity on measures of relative abundance [30,41]. All files were analysed by the same person (AS) to minimise variability in estimating the field-of-view. Tapes were converted to digital format (avi) using Adobe Premiere Elements (Version 10, Adobe Systems Pty Ltd), at a resolution suitable for fish identification (720×576) [30].

## Video interrogation

Files were analysed using Eventmeasure (SeaGIS Pty Ltd, Version 3.31), and the identity and MaxN of each fish species was recorded. MaxN is the maximum number of fish of each species within the field-of-view at any one time during the 30-min recording, which removes the possibility of recounting the same fish. As counts reflect relative abundance and not density, we expect our data to be robust to variability in observing the field-of-view [41].

## Habitat factors

The placement for all BRUV deployments was determined using GIS high-24 resolution imaging software (ArcGIS v9.3, ESRI software, USA) which incorporated metrics for the factors of primary interest: depth; distance from shore; and along-shore location (as distance from latitude -29 S – hereafter termed location). Additional factors, including fine-scale bathymetric range, rugosity and backscatter were also considered. Bathymetry and backscatter data (5 m cell size) were provided by the NSW Office of Environment and Heritage (2012). Rugosity was derived from the bathymetry layer using the Benthic Terrain Modeller extension 1.2 [42]. These datasets were interrogated at a 10-m radius surrounding the BRUV point location using the Buffer tool in ArcGIS (v9.3) and Zonal Statistics tool (ArcGIS Spatial Analyst). The Zonal Statistics tool used the 10-m buffer radius to calculate cell statistics (mean, minimum, maximum, range, standard deviation and sum) for each layer within this zone. Results were exported and incorporated into the variable data for subsequent statistical analyses.

## Statistical methods

Both multivariate and univariate analyses were performed using procedures in the PRIMER 6.0 software [43], with the PERMANOVA+ add-on [44–45]. Distance-based linear modelling (DistLM) was conducted to model the percentage of overall variation in fish assemblage structure accounted for by each environmental factor of interest [44–47]. This procedure utilised a permutational approach, whereby the similarity between samples was calculated using the Bray-Curtis similarity measure, and a forward-stepping selection procedure was applied to regress the suite of environmental factors on the resultant matrix. Actual values for depth, distance from shore, location, backscatter and bathymetric range (within 10-m radius), and rugosity (averaged

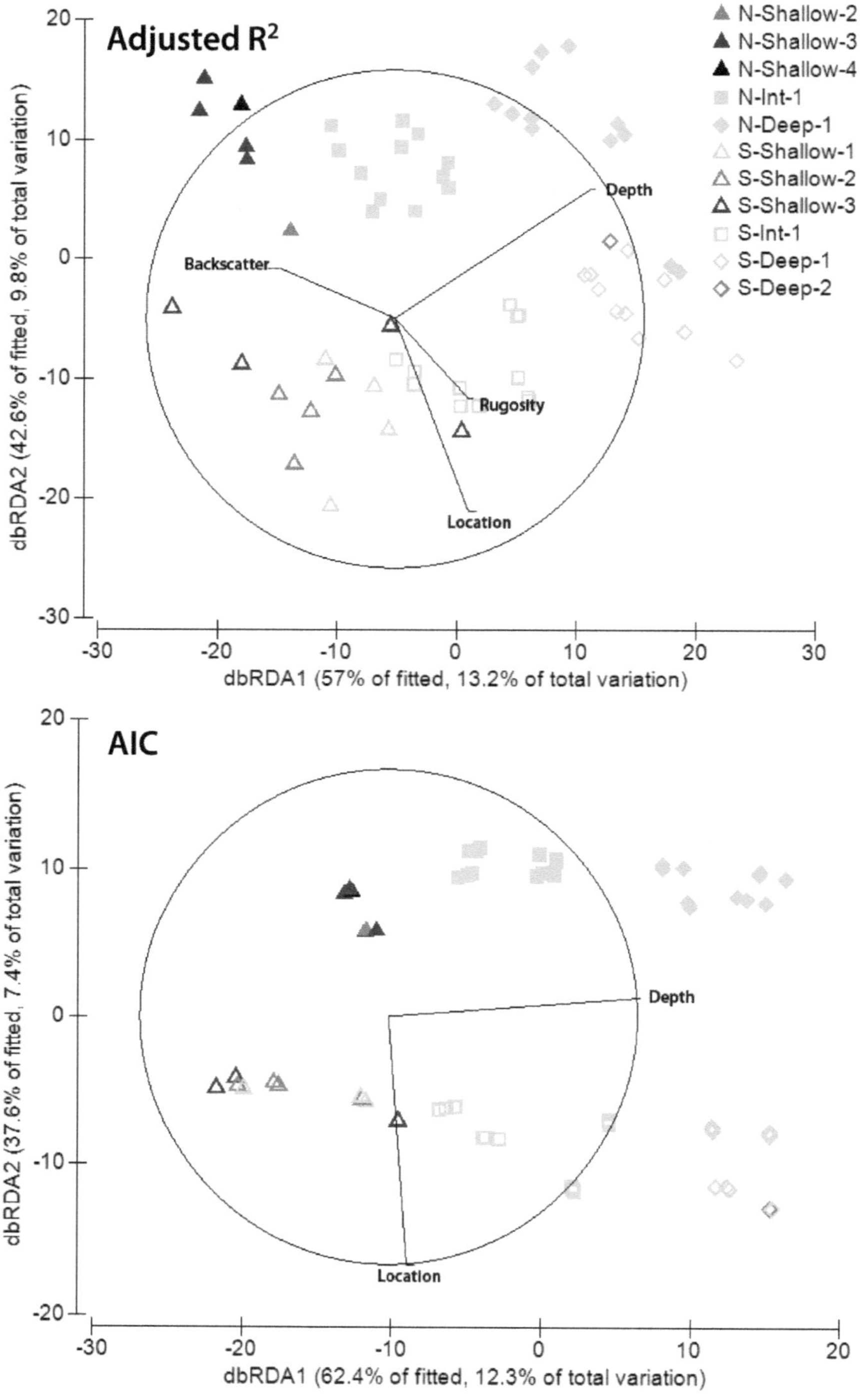


**Figure 2. Distance-based redundancy analysis biplots representing raw Pearson correlations for habitat variables.** Vectors are overlaid to represent the different environmental variables most important in each modelling approach. Length and direction of vectors indicate the strength and direction of the relationship. Numbers 1–4 within the key indicate backscatter intensity.

within 10-m radius) around each BRUV position, provided continuous abiotic data for the regression. The factor with the highest percentage influence on fish assemblage structure is selected first in this procedure, followed by the next highest ranked, in sequential order, until no further improvement to the model is made. Both adjusted $R^2$ and Akaike Information Criterion (AIC) [48] selection criterion were tested in the analysis to provide a comprehensive evaluation of appropriate predictor factors to include in the model. As regression-based models are sensitive to correlation between factors [49], environmental factors with a correlation coefficient of $\geq 0.7$ were noted, and one of the correlated pair was excluded from the analysis [49–50]. Such correlations were found between depth and distance from shore, and between bathymetric range and rugosity, so distance from shore and bathymetric range were discarded from the analysis, and depth and rugosity, along with location and backscatter, were

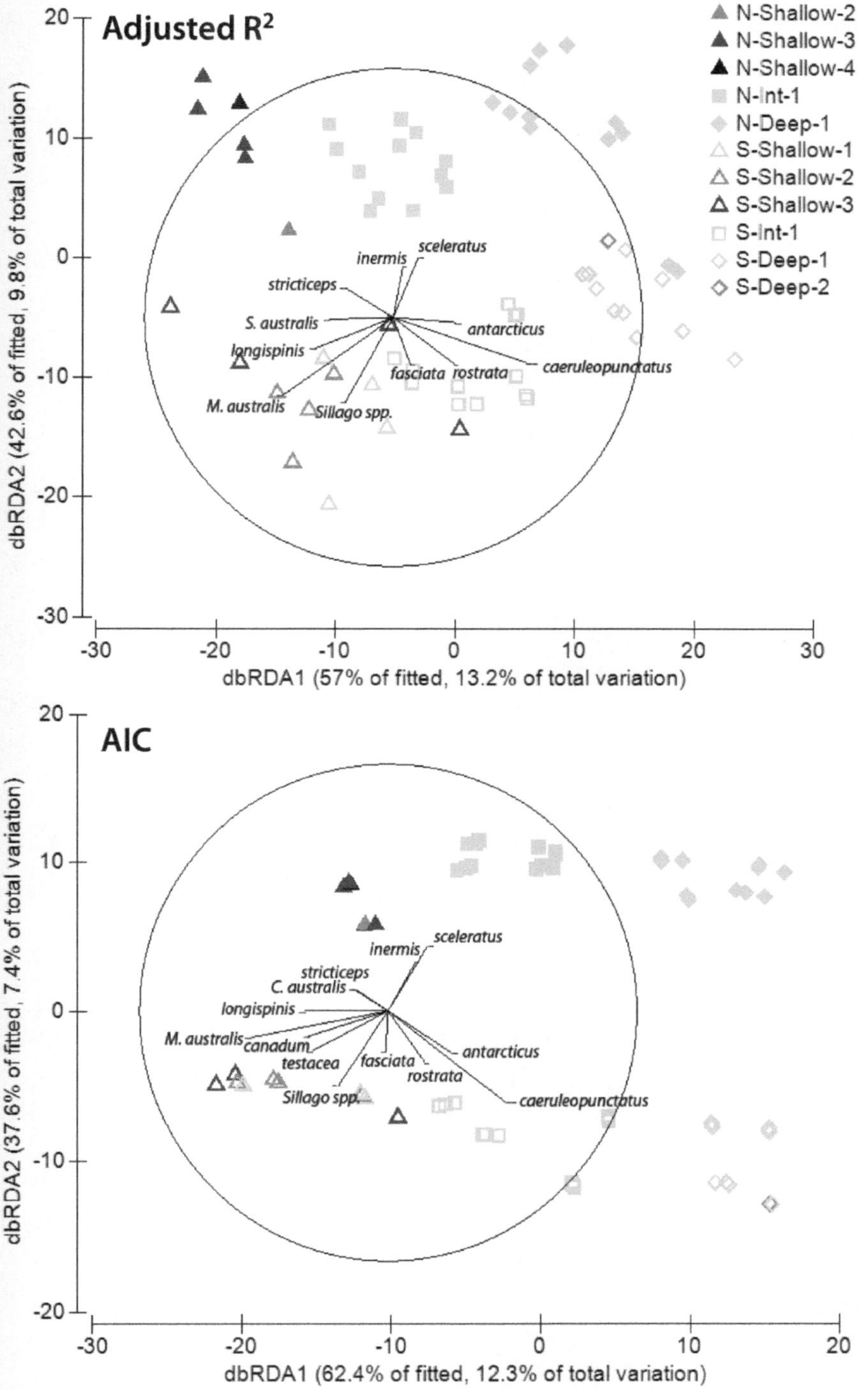


**Figure 3. Distance-based redundancy analysis biplots representing raw Pearson correlations for fish species.** This is as seen in Fig. 2, however here vectors for the 13 most influential fish species in the analysis are overlaid. Length and direction of vectors indicate the strength and direction of the relationship. Numbers 1–4 within the key indicate backscatter intensity.

retained. These variables were retained as they were considered to be the most intuitive and interpretable of the measures available.

Distance based redundancy analysis (dbRDA) biplots were generated to visually display the direction and magnitude of the relationship between habitat factors and individual fish species [47]. If DistLM indicated a significant depth effect, BRUV positions were categorised for that factor using the existing SIMP habitat classification system (shallow = 10–30 m, intermediate = 30–50, deep = 50+). If location was significant, BRUV positions were categorised as north or south. If backscatter was significant it was analysed by categorising intensity range (maximum value minus minimum value) into four levels (0–50, 50–100, 100–150, and 150+) as this could be considered most analogous to sediment complexity from the values available. Likewise, if rugosity was significant it was analysed by categorising intensity range into four categories (0.00001–0.0001, 0.00011–0.001, 0.00101–0.01 and

**Table 3.** Family, genus, species, common name, and Pearson correlations of environmental variables with each of the four dbRDA axes, for both the adjusted R2 and AIC models.

| Family | Genus | Species | Common Name | Correlation | |
|---|---|---|---|---|---|
| **RDA1** | | | | **Adjusted $R^2$** | **AIC** |
| Myliobatididae | *Myliobatis* | *australis* | Eagle Ray | 0.44 | 0.55 |
| Platycephalidae | *Platycephalus* | *longispinis* | Long-Spine Flathead | 0.31 | 0.33 |
| Scombridae | *Sarda* | *australis* | Australian Bonito | 0.26 | 0.22 |
| Urolophidae | *Trygonoptera* | *testacea* | Common Stingaree | 0.19 | 0.30 |
| Rachycentridae | *Rachycentron* | *canadum* | Black Kingfish | 0.19 | 0.32 |
| Platycephalidae | *Platycephalus* | *caeruleopunctatus* | Blue-Spotted Flathead | −0.54 | −0.47 |
| Triakidae | *Mustelus* | *antarcticus* | Gummy Shark | −0.25 | −0.25 |
| Rhinobatidae | *Aptychotrema* | *rostrata* | Eastern Shovelnose Ray | −0.25 | −0.15 |
| Aracanidae | *Anoplacapros* | *inermis* | Eastern Smooth Boxfish | −0.04 | −0.11 |
| Tetraodontidae | *Lagocephalus* | *sceleratus* | Silver Pufferfish | −0.10 | −0.16 |
| **RDA2** | | | | | |
| Tetraodontidae | *Lagocephalus* | *sceleratus* | Silver Pufferfish | 0.24 | 0.26 |
| Aracanidae | *Anoplacapros* | *inermis* | Eastern Smooth Boxfish | 0.21 | 0.19 |
| Pinguipedidae | *Parapercis* | *stricticeps* | White-Streaked Grubfish | 0.12 | 0.08 |
| Scorpaenidae | *Centropogon* | *australis* | Fortesque | 0.03 | 0.08 |
| Urolophidae | *Trygonoptera* | *testacea* | Common Stingaree | −0.34 | −0.16 |
| Rachycentridae | *Rachycentron* | *canadum* | Black Kingfish | −0.34 | −0.10 |
| Sillaginidae | *Sillago* | *spp.* | Whiting spp. | −0.34 | −0.29 |
| Myliobatididae | *Myliobatis* | *australis* | Eagle Ray | −0.31 | −0.10 |
| Rhinobatidae | *Aptychotrema* | *rostrata* | Eastern Shovelnose Ray | −0.19 | −0.20 |
| Rhinobatidae | *Trygonorrhina* | *fasciata* | Eastern Fiddler Ray | −0.19 | −0.16 |
| **RDA3** | | | | | |
| Rachycentridae | *Rachycentron* | *canadum* | Black Kingfish | 0.27 | |
| Urolophidae | *Trygonoptera* | *testacea* | Common Stingaree | 0.17 | |
| Tetraodontidae | *Lagocephalus* | *sceleratus* | Silver Pufferfish | 0.16 | |
| Aracanidae | *Anoplacapros* | *inermis* | Eastern Smooth Boxfish | 0.09 | |
| Paralichthyidae | *Pseudorhombus* | *arsius* | Large-Tooth Flounder | −0.20 | |
| Triakidae | *Mustelus* | *antarcticus* | Gummy Shark | −0.16 | |
| Platycephalidae | *Platycephalus* | *caeruleopunctatus* | Blue-Spotted Flathead | −0.13 | |
| Pinguipedidae | *Parapercis* | *stricticeps* | White-Streaked Grubfish | −0.12 | |
| **RDA4** | | | | | |
| Paralichthyidae | *Pseudorhombus* | *arsius* | Large-Tooth Flounder | 0.39 | |
| Myliobatididae | *Myliobatis* | *australis* | Eagle Ray | 0.17 | |
| Rhinobatidae | *Aptychotrema* | *rostrata* | Eastern Shovelnose Ray | 0.10 | |
| Pinguipedidae | *Parapercis* | *stricticeps* | White-Streaked Grubfish | 0.03 | |
| Platycephalidae | *Platycephalus* | *caeruleopunctatus* | Blue-Spotted Flathead | −0.21 | |
| Urolophidae | *Trygonoptera* | *testacea* | Common Stingaree | −0.10 | |
| Aracanidae | *Anoplacapros* | *inermis* | Eastern Smooth Boxfish | −0.08 | |
| Triakidae | *Mustelus* | *antarcticus* | Gummy Shark | −0.08 | |

0.01001+) as it could be considered most analogous to benthic topographic complexity. These biplots were used to visually assess the relationship between existing categories and significant factors.

Univariate permutational multivariate analysis of variance (PERMANOVA) (using Euclidean distance as the similarity measure) was performed to test for significant differences in Species Richness and Total MaxN between factors that were significant within the DistLM regression. Species Richness is calculated as the total number of different species viewed on each replicate video, and Total MaxN is calculated as the sum of MaxN for all species per replicate. Differences across bathymetry and rugosity categories were not analysed as the range of values was very low.

One-way PERMANOVA [44–45] was used to test for statistical differences in assemblage structure for those factors that were significant in the DistLM regression. Each test was run using 4999

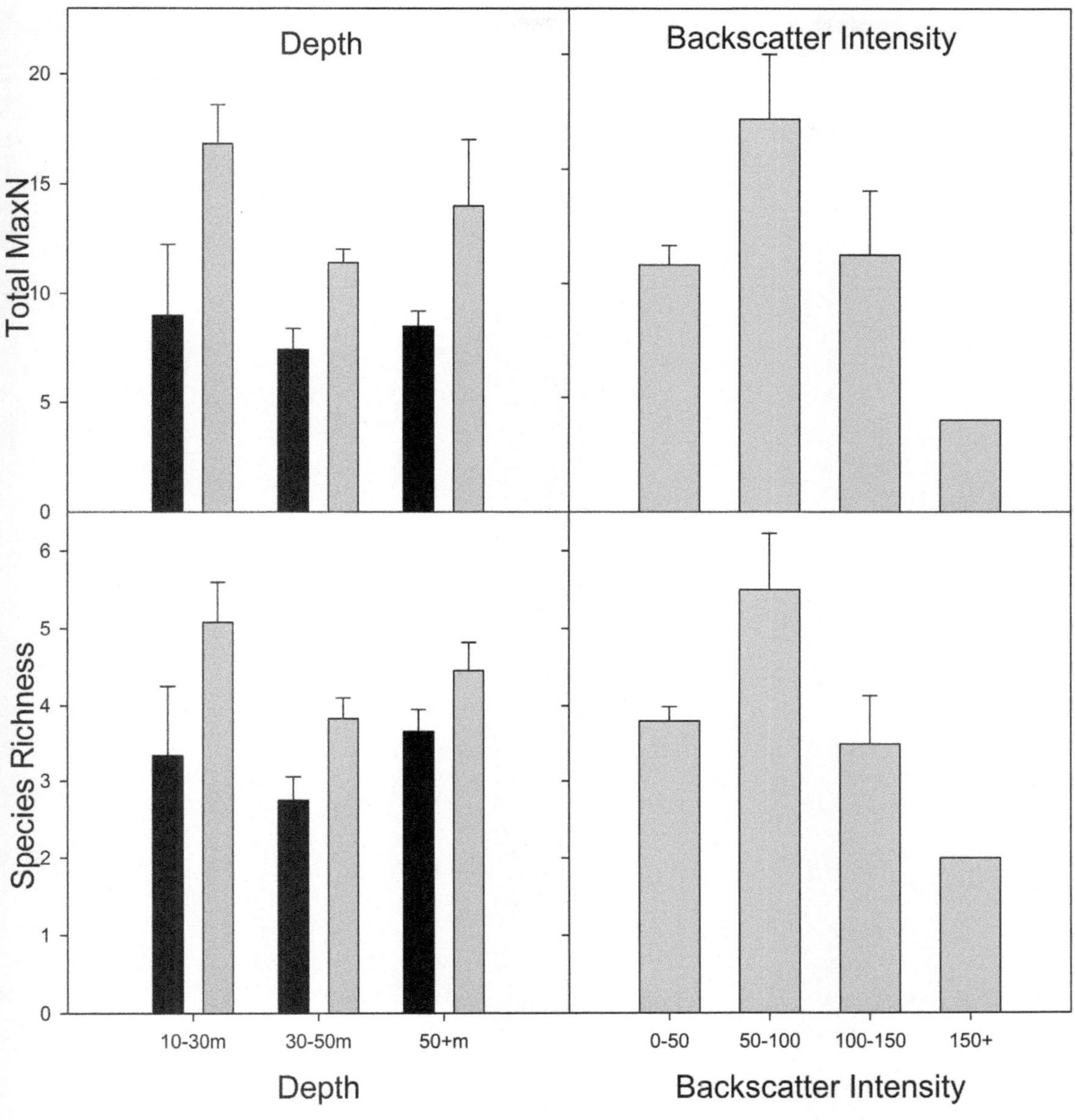


**Figure 4. Mean species richness and total MaxN (±SE) at three depths and four different levels of backscatter intensity.** The graph for depth also show differences between northern (black bars) and southern (grey bars) locations within these factors. Data from northern and southern locations has been combined for backscatter intensity due to unbalanced numbers of replicates within the four levels of intensity.

permutations. Where significant results were generated, *post-hoc* pair-wise contrasts were performed to establish where differences in assemblage structure were occurring.

Multivariate assemblage structure was visually examined using non-metric multi-dimensional scaling (nMDS) ordination. Data were square-root transformed, and a dummy variable was added, prior to the generation of Bray-Curtis similarities. The resultant nMDS was used to examine differences in assemblage structure across factors that were significant in the DistLM analysis. Depth-related patterns were visually examined to see how well they fitted the current depth-based habitat classification categories [16].

## Results

### General

A total of 16 teleost and elasmobranch species, from 12 families, was recorded from BRUVs deployments. Platycephalidae was the most speciose family (3 species). Three species were numerically abundant and ubiquitous across all BRUV positions in the study – *Platycephalus caeruleopunctatus, P. longispinis* and *Aptychotrema rostrata.* Schooling *Sillago* spp. were numerically abundant in some positions, and recorded across all depths, but were highly variable between positions and between depths.

### Correlations between habitat variables and assemblage structure

Marginal tests in distance-based linear modelling (DistLM) showed that, of the four environmental factors retained for analysis, three were significant in isolation (Table 2) for both the adjusted $R^2$ and AIC selection criteria. However, the optimum model for adjusted $R^2$ included all four factors: for AIC, only depth and location were included in the optimum model. The adjusted $R^2$ model explained 23.1% of the total variation, while

**Table 4.** Summary of one-way PERMANOVA results for the analysis of differences in species richness and total MaxN, and for differences in assemblage structure across the different factors.

| | Source | df | SS | MS | Pseudo-F | P(perm) |
|---|---|---|---|---|---|---|
| **Univariate Measures** | | | | | | |
| Species Richness | Depth | 2 | 0.9015 | 0.45075 | 3.1429 | **0.043** |
| | Res | 62 | 8.8921 | 0.14342 | | |
| | Total | 64 | 9.7936 | | | |
| | Location | 1 | 1.7317 | 1.7317 | 13.533 | **0.001** |
| | Res | 63 | 8.0618 | 0.12797 | | |
| | Total | 64 | 9.7936 | | | |
| | Backscatter | 3 | 1.2872 | 0.42908 | 3.077 | **0.034** |
| | Res | 61 | 8.5063 | 0.13945 | | |
| | Total | 64 | 9.7936 | | | |
| Total MaxN | Depth | 2 | 3.8573 | 1.9287 | 2.4472 | 0.092 |
| | Res | 62 | 48.862 | 0.7881 | | |
| | Total | 64 | 52.72 | | | |
| | Location | 1 | 12.859 | 12.859 | 20.323 | **0.001** |
| | Res | 63 | 39.861 | 0.63271 | | |
| | Total | 64 | 52.72 | | | |
| | Backscatter | 3 | 5.8483 | 1.9494 | 2.5371 | 0.065 |
| | Res | 61 | 46.871 | 0.76838 | | |
| | Total | 64 | 52.72 | | | |
| **Assemblage Structure** | Depth | 2 | 8532.2 | 4266.1 | 6.8736 | **0.0002** |
| | Res | 62 | 38480 | 620.65 | | |
| | Total | 64 | 47013 | | | |
| | Location | 1 | 5203.1 | 5203.1 | 7.8402 | **0.0002** |
| | Res | 63 | 41810 | 663.64 | | |
| | Total | 64 | 47013 | | | |
| | Backscatter | 3 | 5905 | 1968.3 | 2.9208 | **0.0016** |
| | Res | 61 | 41108 | 673.9 | | |
| | Total | 64 | 47013 | | | |

Significant results are shown in bold.

the AIC model explained 19.8% (Fig. 2, Table 2). The first two dbRDA axes explained 99.6% of the fitted variation for the adjusted $R^2$ model, and 100% for the AIC model. In the context of hypothesis generation, the more parsimonious AIC model does not allow for further exploration of factors outside depth and location, thus it was worthwhile to further explore the adjusted $R^2$ model despite it being less parsimonious. Backscatter was found to be significant in the marginal tests despite it being highly unbalanced across the study design, and thus also warranted further exploration. Raw Pearson correlations of each factor with each dbRDA axis for the adjusted $R^2$ model showed depth ($\rho = -0.90$) correlated closely with the first dbRDA axis, with backscatter intensity also partially correlated ($\rho = 0.62$). The second axis was a combination of location ($\rho = -0.85$) and depth ($\rho = -0.42$). For the AIC model depth ($\rho = -0.99$) was almost perfectly correlated with the first axis, while location ($\rho = -0.99$) was similarly correlated with the second axis.

By examining the raw Pearson correlations for different species, we could identify those most responsible for driving the assemblage response to each environmental factor: these are displayed as vectors in Fig. 3. The data for the adjusted $R^2$ model must be interpreted somewhat cautiously, as the first dbRDA axis represents a combination of depth and backscatter, while the second axis is a combination of location and depth. Therefore, determination of the factor primarily responsible for the distribution of each species is partially confounded. Some of the species and their respective correlation values for both the adjusted $R^2$ and AIC models are displayed in Table 3. Species with the four most positive and four most negative correlations with each axis are displayed, as these are the species most responsible for driving patterns in the assemblage.

## Univariate measures

Species richness was generally higher in the south of the park (Fig. 4). The highest species richness by depth categories was for shallow deployments in the south of the park and the second highest was for deep deployments in the south. As the number of replicates was unbalanced across backscatter intensity levels (Table 1), all deployments were pooled for graphical representation. The highest species richness occurred in the 50–100 category with the lowest being in the 150+ category.

PERMANOVA revealed a significant difference between species richness for all four factors of interest (Table 4). Pairwise

**Table 5.** Results of *post-hoc* pair-wise contrasts of species richness and total MaxN, and assemblage structure (PERMANOVA) for each pair of levels across the factors.

| | | Groups | t | P(perm) | perms |
|---|---|---|---|---|---|
| **Species Richness** | Depth | Shallow, Int. | 2.0833 | **0.039** | 850 |
| | | Shallow, Deep | 0.53341 | 0.598 | 833 |
| | | Int., Deep | 2.3135 | **0.029** | 543 |
| | Dist. From Sh. | Inshore, Midshelf | 1.7718 | 0.112 | 477 |
| | | Inshore, Offshore | 3.0799 | **0.003** | 378 |
| | | Midshelf, Offshore | 8.43E-02 | 0.917 | 517 |
| | Backscatter | 0–50, 50–100 | 2.6171 | **0.016** | 280 |
| | | 0–50, 100–150 | 0.7656 | 0.415 | 374 |
| | | 0–50, 150+ | 1.4333 | 0.211 | 7 |
| | | 50–100, 100–150 | 2.0349 | 0.06 | 222 |
| | | 50–100, 150+ | 2.1264 | 0.156 | 5 |
| | | 100–150, 150+ | 0.73973 | 0.784 | 6 |
| **Assemblage Structure** | Depth | Shallow, Int. | 2.377 | **0.0006** | 4989 |
| | | Shallow, Deep | 2.8837 | **0.0002** | 4992 |
| | | Int., Deep | 2.5972 | **0.0002** | 4990 |
| | Dist. From Sh. | Inshore, Midshelf | 1.9663 | **0.0114** | 4622 |
| | | Inshore, Offshore | 3.2945 | **0.0002** | 4990 |
| | | Midshelf, Offshore | 2.3576 | **0.0016** | 4987 |
| | Backscatter | 0–50, 50–100 | 2.5076 | **0.0002** | 4983 |
| | | 0–50, 100–150 | 1.6913 | **0.0258** | 4983 |
| | | 0–50, 150+ | 1.1757 | 0.2954 | 49 |
| | | 50–100, 100–150 | 1.0779 | 0.346 | 2408 |
| | | 50–100, 150+ | 1.414 | 0.1356 | 7 |
| | | 100–150, 150+ | 0.55622 | 1 | 9 |

Significant results are shown in bold.

contrasts revealed significant differences between shallow and intermediate, and intermediate and deep categories (Table 5). The only significant contrasts for backscatter were between levels 0–50 and 50–100 (Table 5).

Total MaxN values were slightly higher in the south of the park, and showed trends similar to those seen for species richness (Fig. 4). The highest Total MaxN was recorded in the shallow southern category and the second highest in the deep southern category. As with species richness, highest Total MaxN across backscatter categories was for 50–100. Of the three factors of interest, PERMANOVA found a significant effect only for the latitude (Table 4).

### Assemblage structure

PERMANOVA revealed significant effects for all three selected factors (Table 4). Pairwise contrasts between different levels of depth were all significant. Pairwise contrasts across backscatter categories were complicated by unbalanced numbers of samples in each level, generating a large number of unique permutations between some pairs of samples, and relatively few between other pairs (Table 4), making it impossible to make statistical inferences at a level $\leq 0.05$ for several pairs.

Some trends were evident in the nMDS ordination (Fig. 5) with the majority of shallow inshore and mid-shelf sites grouping in the top half of the plot, and the majority of the deep offshore sites in the lower half. In addition, there was some along-shore separation between assemblages, with those in northern locations tending to the left of the ordination, and those from southern locations tending to the right (Fig. 5).

## Discussion

Patterns of fish assemblage structure in unconsolidated habitats within the SIMP are clearly influenced both by depth and factors operating over medium spatial scales (i.e. the distances between our locations). Additionally, backscatter intensity, while not a primary driver, was highlighted as potentially explaining some of the differences in fish assemblages across the scales of the study.

### Assemblage patterns and correlations with environmental variables

Worldwide, fish assemblage patterns of both reef and unconsolidated habitats have often been found to be correlated with depth [16,21,23,28,32,51–52] and distance from shore [26,34,53], and these factors have consequently been used as biophysical surrogates for conservation planning in MPAs [16,34]. Fishes are an important ecological and socio-economic component of marine diversity in their own right [54] but can also reflect the distribution of benthic communities [16,34]. Distinct patterns in the structure of fish assemblages of unconsolidated habitats have also been documented in previous research in coastal and shelf waters of eastern Australia [14,19–20]. For example, [19] found distinct

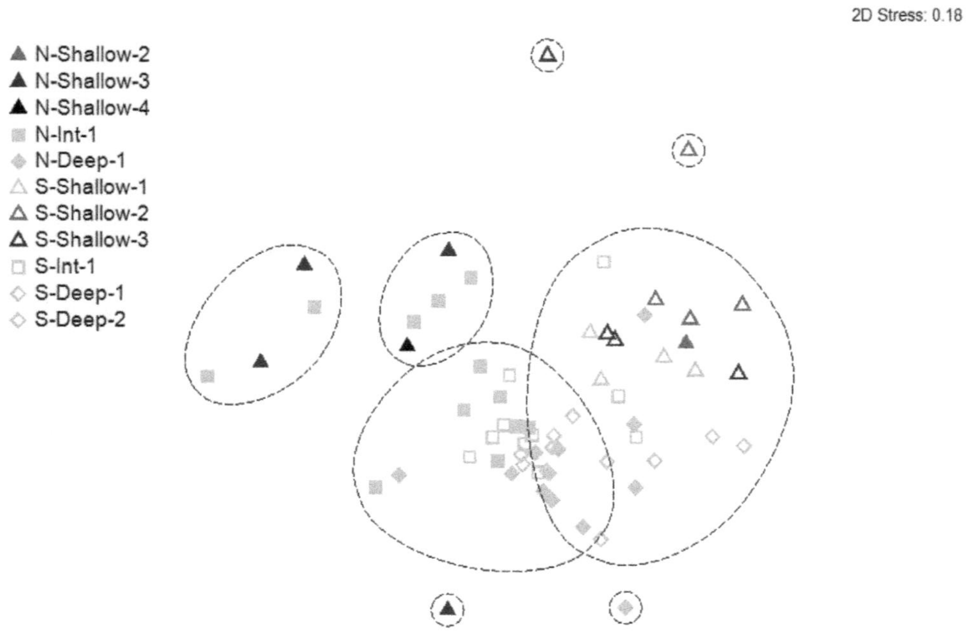


**Figure 5. Non-metric multi-dimensional scaling (nMDS) ordination showing the relationship among fish assemblages from three depths, three distances from shore, and two locations.** Data were square-root transformed prior to analysis. Lines represent 60% similarity. Numbers 1–4 within the key indicate backscatter intensity.

assemblages in three depth categories (20–30 m, 40–50 m and 60–70 m) on unconsolidated sediments offshore from Sydney. They also showed that: while broad patterns were evident, none of the abundant species consistently showed a preference for one depth category; species that were unique to a single depth category were found in low abundance (<3); and the numerically dominant species were found at all depths [19]. Similarly, [14] concluded that assemblage patterns were being driven mostly by differences in the abundance of common species that occurred across multiple habitats and depths. Our data show a similar pattern in that the species most responsible for differences in assemblage structure are also numerically abundant, and were recorded at all depths and within different locations in the SIMP.

The importance of depth in structuring assemblages reflects what is known from other habitats in the SIMP and elsewhere. Strong depth-related patterns have been detected in reef fishes [34], for motile invertebrates of natural [55] and artificial [56] substrata, and for habitat-forming benthos [57]. The shallow/intermediate boundary in the current HCS in the SIMP was defined at 25 m due to the disjunct patterns of reef fish and general benthic assemblages (from coral to sponge-dominated) below this depth [58].

With strong co-linearity between depth and distance from shore within our study, it is difficult to distinguish which factor primarily drives the observed patterns. Shelf position is included in the current HCS for the SIMP, as distinct cross-shelf patterns have been demonstrated for corals [59], molluscs [60] and reef fish [34]. Due to the broad morphology of the continental shelf region in the SIMP, and the mobile nature of unconsolidated habitats, this correlation between factors is not unexpected, and from a practical perspective, depth could be considered a proxy for distance from shore, or vice-versa, for management purposes.

The correlation with location in our modelling contradicts findings for reef fish communities in the SIMP [16,34] which showed no noticeable pattern between assemblage structure over similar spatial scales, and it is currently unclear as to the reasons for this variation. While broad-scale latitudinal effects have been observed in fish communities of unconsolidated sediments [24,26,61–62], these patterns are generally evident over a much larger spatial scale than in our study (e.g. [26]). It is likely that this variation is due to other factors not identified within this study, rather than actual latitudinal effects.

Our modelling also suggests the possibility that sediment grain size, for which backscatter intensity is an appropriate surrogate, may influence fish communities of unconsolidated habitats. Broad-scale bathymetric mapping of the seafloor within the SIMP has shown that unconsolidated habitats show distinct variability in sediment type, as revealed in acoustic backscatter [40]. Some areas show a darker, 'reef-like' backscatter, and are likely to contain varying amounts of pebbles and cobbles [40]. These patches of denser backscatter, while extensive in some sections of the SIMP, were not the primary focus of our study. During this study, 4 species (*Pseudorhombus arsius, Parapercis stricticeps, Centropogon australis, Ratabulus diversidens*) were found only at sites with a backscatter index ≥50, despite a low number of replicates with this index or greater (15) relative to the total (65, see Table 1). Additionally, while our sampling design did not consider backscatter intensity as an *a priori* factor, the marginal tests from the DistLM analysis generated a significant value for this factor. In the context of hypothesis generation, we recommended that habitat types should be examined at a finer scale to comprehensively assess the role of backscatter intensity (as a proxy for sediment type) in the distribution of different species. With obvious and extensive differences in sediment characteristics, it is likely that, in addition to the other factors identified in this study, sediment type and habitat complexity are important factors structuring fish assemblages. Recent research in the SIMP [30] has demonstrated that the presence of reef adjacent to sedimentary habitats, even at the scale of hundreds of metres, can have a substantial effect on fish assemblage structure. It is possible that gravels may act as a 'reef surrogate' for some species.

When compared to reefs, unconsolidated habitats, particularly in a high energy environment, have lower topographical complexity, support fewer sessile biota, and provide limited foraging opportunities and protection from predation [63–65]. The low species richness observed in this study reflects this, with the species complement comprising ~3% of the overall teleost and elasmobranch diversity recorded from the SIMP [34]. Despite this low species richness, statistical differences were evident over levels of each of the four factors we tested, and rare species were mostly restricted to single levels within each factor. In contrast, the only significant factor for total MaxN was location, with higher overall MaxN values at the southern location.

The current Habitat Classification System (HCS) for the SIMP has been developed and adjusted primarily based on fish assemblage patterns associated with hard substrata [16,34]. This HCS has been used to systematically examine representation of habitat categories as a surrogate for biotic pattern [66–68], under existing and different marine park zoning scenarios [69]. In the absence of specific data, the same categories have been used for conservation planning for sediment-associated biota. There is value, both ecologically and from a planning perspective, in recognising that broad assemblage patterns in fish assemblages of unconsolidated substrata do exist in the SIMP and require consideration for planning elsewhere. That these, at least in part, align with the current habitat categories used within this MPA is opportune. However, it also indicates that some further refinement of the current HCS in the SIMP is required to improve future systematic planning. Use of quantitative and systematic planning in an MPA, based on suitable and available datasets, is more-likely to achieve MPA objectives, as well as address some of the essential criteria necessary to effectively evaluate and select areas for protection [70–71]. It is also more likely to provide cost-effective solutions in terms of area required and/or social impacts [71–72]. Biotic data are often spatially constrained, whereas physical/environmental data are often less so [66] and, therefore, integrating these can achieve better planning outcomes [73]. Without testing at the appropriate scale, however, it cannot be assumed that the patterns between fish and physical factors observed in this study will occur in other MPAs. Our study highlights not only the presence of these relationships, but also the fact that there is considerable overlap in the factors driving differences between different habitat types (i.e. unconsolidated habitats and reefs). Importantly, this study has identified additional factors, and in particular sediment morphology, which require further testing with a view to potential incorporation into the planning process.

## Conclusions

The availability of high resolution benthic mapping has allowed us to examine the influence of a range of factors, which would otherwise be difficult and expensive to assess, on the poorly described fish assemblage of unconsolidated habitats. The patterns evident within the study endorse the relevance of depth and distance from shore as categories currently used as biophysical surrogates to represent discrete assemblages in marine parks in NSW. However, the study also indicates that the inclusion of habitat type, based on backscatter and possibly other remotely-sensed metrics, will lead to better representation of assemblages of unconsolidated habitats.

## Acknowledgments

This paper comprises part of a PhD project by the first author. Field assistance was provided by Andrew Cox and Mitch Young. Logistic and administrative assistance was provided by the staff at the Solitary Islands Marine Park office and the National Marine Science Centre. We thank Peter Davies, Tim Ingleton and Edwina Foulsham from the NSW Office of Environment and Heritage for provision of bathymetric and habitat data. Thanks also to Alan Jordan and Nicola Johnstone for very helpful comments on this manuscript. Technical assistance was provided by Ken Cowden and Gary Shipley. We gratefully acknowledge Jim Seager (SeaGIS) for providing a student license for Eventmeasure for our video analysis.

## Author Contributions

Conceived and designed the experiments: AS HM DB SS. Performed the experiments: AS. Analyzed the data: AS HM ML SS. Wrote the paper: AS HM ML SS.

## References


1. Ward TJ, Vanderklift MA, Nicholls AO, Kenchington RA (1999) Selecting marine reserves using habitats and species assemblages as surrogates for biological diversity. Ecol Appl 9: 691–698
2. Curley BG, Kingsford MJ, Gillanders BM (2002) Spatial and habitat-related patterns of temperate reef fish assemblages: implications for the design of Marine Protected Areas. Mar Freshw Res 53: 1197–1210
3. Gladstone W (2002) The potential value of indicator groups in the selection of marine reserves. Biol Conserv 104: 211–220
4. Smith SDA (2005) Rapid assessment of invertebrate biodiversity on rocky shores: where there's a whelk there's a way. Biodivers Conserv 14: 3565–3576
5. Gladstone W (2007) Requirements for marine protected areas to conserve the biodiversity of rocky reef fishes. Aquat Conserv 17: 71–80
6. Stevens T (2002) Rigour and representativeness in Marine Protected Area design. Coast Manage 30: 237–248
7. Roberts CM, Andelman S, Branch G, Bustamante RH, Castilla JC, et al. (2003) Application of ecological criteria in selecting marine reserves and developing reserve networks. Ecol Appl 13(1) Supp: S215–S228
8. Stevens T, Connolly RM (2004) Testing the utility of abiotic surrogates for marine habitat mapping at scales relevant to management. Biol Conserv 119: 351–362
9. Dalleau M, Andrefouet S, Wabnitz CCC, Payri C, Wantiez L, et al. (2010) Use of Habitats as Surrogates of Biodiversity for Efficient Coral Reef Conservation Planning in Pacific Ocean Islands. Conserv Biol 24(2): 541–552.
10. Rattray A, Lerodiaconou D, Laurenson L, Burq S, Reston M (2009) Hydro-acoustic remote sensing of benthic biological communities on the shallow South East Australian continental shelf. Est Coast Shelf Sci 84: 237–245
11. Moore CH, Harvey ES, Van Niel KP (2009) Spatial prediction of demersal fish distributions: enhancing our understanding of species-environment relationships. ICES J. Mar. Sci 66: 2068–2075
12. Brown CJ, Smith SJ, Lawton P, Anderson JT (2011) Benthic habitat mapping: A review of progress towards improved understanding of the spatial ecology of the seafloor using acoustic techniques. Est Coast Shelf Sci 92: 502–520
13. Long BG, Bode L, Mason L, Pitcher CR (1997) Seabed current stress predicts the distribution and abundance of epibenthos in Torres Strait. Report to Australian Fisheries Management Authority, CSIRO, Australia.
14. Williams A, Bax NJ (2001) Delineating fish-habitat associations for spatially based management: an example from the south-eastern Australian continental shelf. Mar Freshw Res 52: 513–536
15. Zacharias MA, Roff JC (2001) Explanations of patterns of intertidal diversity at regional scales. J Biogeogr 28: 471–483
16. Malcolm HA, Jordan A, Smith SDA (2011a) Testing a depth based Habitat Classification System against reef fish assemblage patterns in a subtropical marine park. Aquat Conserv 21: 173–185
17. Roff JC, Evans SMJ (2002) Frameworks for marine conservation – non-heirarchical approaches and distinctive habitats. Aquat Conserv 12: 635–648.
18. Snelgrove PVR, Blackburn TH, Hutchings PA, Alongi DM, Grassle DM, et al. (1997) The importance of marine sediment biodiversity in ecosystem processes. Ambio 26: 578–583
19. Connell SD, Lincoln-Smith MP (1999) Depth and the structure of assemblages of demersal fish: experimental trawling along a temperate coast. Est Coast Shelf Sci 48: 483–495.
20. Gray CA, Otway NM (1994) Spatial and temporal differences in assemblages of demersal fishes on the inner continental shelf off Sydney, South-eastern Australia. Aust J Mar Freshw Res 45: 665–676.
21. Garcia CB, Duarte LO, von Schiller D (1998) Demersal fish assemblages of the Gulf of Salamanca, Columbia (southern Caribbean Sea) Mar Ecol Prog Ser 174: 13–25

22. Butler AJ, Rees T, Beesley P, Bax NJ (2010) Marine biodiversity in the Australian region. PLoS One 5(8): e11831
23. Hyndes GA, Platell ME, Potter IC, Lenanton RJC (1999) Does the composition of demersal fish assemblages in temperate coastal waters change with depth and undergo consistent seasonal change? Mar Biol 134: 335–352
24. Travers MJ, Potter IC, Clarke KR, Newman SJ, Hutchins JB (2010) The inshore fish faunas over soft substrates and reefs on the tropical west coast of Australia differ and change with latitude and bioregion. J Biogeogr 37: 148–169
25. Cappo M, Speare P, De'ath G (2004) Comparison of baited remote underwater video stations and prawn (shrimp) trawls for assessments of fish biodiversity in inter-reefal areas of the Great Barrier Reef Marine Park. J Exp Mar Biol Ecol 302(2): 123–152
26. Cappo M, De'ath G, Speare P (2007) Inter-reef vertebrate communities of the Great Barrier Reef Marine Park determined by baited remote underwater video stations. Mar Ecol Prog Ser 350: 209–221
27. Harvey ES, Cappo M, Butler J, Hall N, Kendrick GA (2007) How does the presence of bait as an attractant affect the performance of remote underwater video stations in assessments of demersal fish community structure? Mar Ecol Prog Ser 350: 245–254
28. Moore CH, Harvey ES, Van Neil K (2010) The application of predicted habitat models to investigate the spatial ecology of demersal fish assemblages. Mar Biol 157: 2717–2729
29. Owen V, Gladstone W (2010) The effectiveness of an abiotic classification of subtidal unvegetated habitat as a surrogate for the diversity of fish assemblages in a marine protected area. Program and Abstracts for the 2010 meeting of the Australian Marine Sciences Association (4–8 July 2010, Wollongong, New South Wales, Australia).
30. Schultz AL, Malcolm HA, Bucher DJ, Smith SDA (2012) Effects of Reef Proximity on the Structure of Fish Assemblages of Unconsolidated Substrata. PloS ONE 7(11): e49437
31. Harvey ES, Butler JJ, McLean DL, Shand J (2012) Contrasting habitat use of diurnal and nocturnal fish assemblages in temperate Western Australia. J Exp Mar Biol Ecol 426–427: 78–86
32. Fitzpatrick BM, Harvey ES, Heyward AJ, Twiggs EJ, Colquhoun J (2012) Habitat Specialisation in Tropical Continental Shelf Demersal Fish Assemblages. PloS ONE 7(6): e39634
33. White J, Simpfendorfer CA, Tobin AJ, Heupel MR (2013) Application of baited remote underwater video surveys to quantify spatial distribution of elasmobranchs at an ecosystem scale. J Exp Mar Biol Ecol 448: 281–288
34. Malcolm HA, Smith SDA, Jordan A (2010a) Using patterns of reef fish assemblages to refine a Habitat Classification System for marine parks in NSW, Australia. Aquat Conserv 20: 83–92
35. Zann LP (2000) The eastern Australian region: a dynamic tropical/temperate biotone. Mar Pollut Bull 41:188–203
36. Malcolm HA, Davies P, Jordan A, Smith SDA (2011b) Variation in sea temperature and the East Australian Current in the Solitary Islands region between 2001 to 2008. Deep Sea Res Part 2 Top Stud Oceanogr 58:616–627
37. Roughan M, Middleton JH (2002) A comparison of observed upwelling mechanisms off the east coast of Australia. Cont Shelf Res 22:2551–2572
38. Malcolm HA, Jordan A, Smith SDA (2010b) Biogeographical and cross-shelf patterns of reef fish in a transition zone. Mar Biodiv 40:181–193
39. Smith SDA, Rule MJ, Harrison M, Dalton SJ (2008) Monitoring the sea change: preliminary assessment of the conservation value of nearshore reefs, and existing impacts, in a high-growth, coastal region of subtropical eastern Australia. Mar Poll Bull 56:525–534
40. NSW Marine Park Authority (2010) Seabed habitat mapping in the Solitary Islands Marine Park and Jervis Bay Marine Park, 58 pp
41. Malcolm HA, Gladstone W, Lindfield S, Wraith J, Lynch TP (2007) Spatial and temporal variation in reef fish assemblages of marine parks in New South Wales, Australia – baited video observations. Mar Ecol Prog Ser 350:277–290
42. Wright DJ, Lundblad ER, Larkin EM, Rinehart RW, Murphy J, et al. (2005) ArcGIS Benthic Terrain Modeller, Corvallis, Oregon, Oregon State University, Davey Jones Locker Seafloor Mapping/Marine GIS Laboratory and NOAA Coastal Services Center. Available: http://www.csc.noaa.gov/products/btm/
43. Clarke KR, Gorley RN (2006) PRIMER v6: user manual/tutorial. PRIMER-E, Plymouth UK.
44. Anderson MJ (2001) A new method for non-parametric multivariate analysis of variance. Austral Ecol 26: 32–46
45. Anderson MJ, Gorley RN, Clarke KR (2008)PERMANOVA+ for PRIMER: guide to software and statistical methods. PRIMER-E, Plymouth UK
46. McArdle B, Anderson MJ (2001) Fitting multivariate models to community data: a comment on distance-based redundancy analysis. Ecology 82(1): 290–297
47. Legendre P, Anderson MJ (1999) Distance-based Redundancy analysis: Testing multispecies responses in multifactorial ecological experiments. Ecol Monogr 69: 1–24
48. Akaike H (1973) Information theory as an extension of the maximum likelihood principle. Pp267–281In: Petrov BN, Caski F, editors. Proceedings, 2nd International Symposium on Information Theory. Akademiai Kiado, Budapest.
49. Leathwick JR, Elith J, Hastie T (2006) Comparative performance of generalized additive models and multivariate adaptive regression splines for statistical modelling of species distributions. Ecol Modell 199: 188–196
50. Moore CH, Van Niel K, Harvey ES (2011) The effect of landscape composition and configuration on the spatial distribution of temperate demersal fish. Ecography 34: 425–435
51. Friedlander AM, Parrish JD (1998) Habitat characteristics affecting fish assemblages on a Hawaiian coral reef. J Exp Mar Biol Ecol 224: 1–30
52. Anderson MJ, Millar RB (2004) Spatial variation and effects of habitat on temperate reef fish assemblages in northeastern New Zealand. J Exp Mar Biol Ecol 305: 191–221
53. Kendall MS, Christensen JD, Caldow C, Coyne M, Jeffery C, et al. (2004) The influence of bottom type and shelf position on biodiversity of tropical fish inside a recently enlarged marine reserve. Aquat Conserv 14: 113–132
54. Shears NT, Babcock RC (2002) Marine reserves demonstrate top-down control of community structure on temperate reefs. Oecologica 132: 131–142
55. Smith SDA (1996) The macrofaunal community of *Ecklonia radiata* holdfasts: variation associated with sediment regime, sponge cover and depth. Aust J Ecol 21: 144–153
56. Rule M, Smith SDA (2007) Depth-associated patterns in the development of benthic assemblages on artificial substrata deployed on shallow, subtropical reefs. J Exp Mar Biol Ecol 345: 38–51
57. Mau R, Byrnes T, Wilson J, Zann L (1998) The distribution of selected continental shelf habitats and biotic communities in the Solitary Islands Marine Park. Southern Cross University Report to the NSW Marine Parks Authority. Lismore NSW
58. Jordan A, Davies P, Ingleton T, Mesley E, Neilson J, et al. (2010) Seabed habitat mapping of continental shelf waters of NSW. P. 209. NSW Department of Environment, Climate Change and Water.
59. Harriott VJ, Smith SDA, Harrison PL (1994) Patterns of coral community structure of subtropical reefs in the Solitary Islands Marine Reserve, Eastern Australia. Mar Ecol Prog Ser 109: 67–76
60. Harrison MA, Smith SDA (2012) Cross-shelf variation in the structure of molluscan assemblages on shallow, rocky reefs in subtropical eastern Australia. Mar Biodivers 42(2): 203–216
61. Beentjes MP, Bull B, Hurst RJ, Bagley NW (2002) Demersal fish assemblages along the continental shelf and upper slope of the east coast of the south island, New Zealand. NZ J Mar Freshw Res 36: 197–223
62. Travers MJ, Potter IC, Clarke RK, Newman SJ (2012) Relationships between latitude and environmental conditions and the species richness, abundance and composition of tropical fish assemblages over soft substrata. Mar Ecol Prog Ser 446: 221–241
63. Robertson AI, Lenanton RCJ (1984) Fish community structure and food chain dynamics in the surf-zone of sandy beaches: the role of detached macrophyte detritus. J Exp Mar Biol Ecol 84:265–283.
64. Ayvazian SG, Hyndes GA (1995) Surf-zone fish assemblages in south-western Australia: do adjacent nearshore habitats and the warm Leeuwin Current influence the characteristics of the fish fauna? Mar Biol 122: 527–536
65. Hyndes GA, Lavery PS (2005) Does transported seagrass provide an important trophic link in unvegetated, nearshore areas? Est Coast Shelf Sci 63:633–643
66. Williams A, Bax NJ (2001) Delineating fish-habitat associations for spatially-based management: an example from the south-eastern Australian continental shelf. Mar Freshw Res 52: 513–536
67. Stevens T, Connolly RM (2005) Local-scale mapping of benthic habitats to assess representation in a marine protected area. Mar Freshw Res 56: 111–123
68. Blamey LK, Branch GM (2009) Habitat diversity relative to wave action on rocky shores: implications for the selection of marine protected areas. Aquat Conserv 19: 645–657
69. Malcolm HA, Foulsham E, Pressey RL, Jordan A, Davies PL, et al. (2012) Selecting zones in a marine park: Early systematic planning improves cost efficiency, combining habitat and biotic data improves effectiveness. Ocean Coast Manage 59: 1–12
70. Roberts CM, Branch G, Bustamante RH, Castilla JC Dugan J, et al. (2003) Application of ecological criteria in selecting marine reserves and developing reserve networks. Ecol Appl 13(1): S215–S228
71. Possingham H, Ball I Andelman S (2000) Mathematical methods for identifying representative reserve networks. In: Ferson S, Bergman M, editors.Quantitative methods for conservation biology. Springer-Verlag, New York, pp 291–305
72. Fernandes L, Day J, Lewis A, Slegers S, Kerrigan B, et al. (2005) Establishing Representative No-Take Areas in the Great Barrier Reef: Large-Scale Implementation of Theory on Marine Protected Areas. Conserv Biol 19: 1733–1744
73. Ferrier S (2002) Mapping Spatial Pattern in Biodiversity for Regional Conservation Planning: Where to from Here? Syst Biol 51(2):331–363

# 3

# Fishers' Behaviour in Response to the Implementation of a Marine Protected Area

**Bárbara Horta e Costa[1,2], Marisa I. Batista[3], Leonel Gonçalves[1], Karim Erzini[2], Jennifer E. Caselle[4], Henrique N. Cabral[3], Emanuel J. Gonçalves[1]***

**1** Eco-Ethology Research Unit, ISPA – Instituto Universitário, Lisboa, Portugal, **2** Centre of Marine Sciences - CCMAR, University of Algarve, Campus de Gambelas, Faro, Portugal, **3** Centro de Oceanografia, Faculdade de Ciências, Universidade de Lisboa, Lisboa, Portugal, **4** Marine Science Institute, University of California, Santa Barbara, California, United States of America

## Abstract

Marine Protected Areas (MPAs) have been widely proposed as a fisheries management tool in addition to their conservation purposes. Despite this, few studies have satisfactorily assessed the dynamics of fishers' adaptations to the loss of fishing grounds. Here we used data from before, during and after the implementation of the management plan of a temperate Atlantic multiple-use MPA to examine the factors affecting the spatial and temporal distribution of different gears used by the artisanal fishing fleet. The position of vessels and gear types were obtained by visual surveys and related to spatial features of the marine park. A hotspot analysis was conducted to identify heavily utilized patches for each fishing gear and time period. The contribution of individual vessels to each significant cluster was assessed to better understand fishers' choices. Different fisheries responded differently to the implementation of protection measures, with preferred habitats of target species driving much of the fishers' choices. Within each fishery, individual fishers showed distinct strategies with some operating in a broader area whereas others kept preferred territories. Our findings are based on reliable methods that can easily be applied in coastal multipurpose MPAs to monitor and assess fisheries and fishers responses to different management rules and protection levels. This paper is the first in-depth empirical study where fishers' choices from artisanal fisheries were analysed before, during and after the implementation of a MPA, thereby allowing a clearer understanding of the dynamics of local fisheries and providing significant lessons for marine conservation and management of coastal systems.

**Editor:** Sebastian C. A. Ferse, Leibniz Center for Tropical Marine Ecology, Germany

**Funding:** This study was financed by the European Commission and Secil corporation through the Biomares LIFE project (LIFE06 NAT/P/000192), by the Fundação para a Ciência e a Tecnologia (FCT: www.fct.pt) through the PDCT/MAR/57934/2004 project and the Pluriannual Program (R&D Unit 331/94, PEst-OE/MAR/UI0331/2011), and by the Sesimbra Municipality. BHC and MIB were supported by PhD grants from FCT (SFRH/BD/41262/2007 and SFRH/BD/64395/2009, respectively). The funders had no role in study design, data collection and analysis, decision to publish, or preparation of the manuscript.

**Competing Interests:** The authors have declared that no competing interests exist.

* E-mail: emanuel@ispa.pt

## Introduction

Besides conservation purposes, marine protected areas (MPAs) have also been suggested as important fisheries management tools [1–3]. The expected effects from the exclusion of extractive activities in marine reserves (no-take) are an increase in abundance, size and fecundity of fished individuals, especially for those most impacted by fisheries [4]. This so-called "reserve effect" is expected to translate to biomass export of post-settlers to adjacent areas (spillover) which may, in turn, depend on density-dependent mechanisms and carrying capacity of protected and adjacent areas, as well as connectivity of suitable habitats [4,5]. Some authors have also suggested that fisheries are more likely to benefit through larval export from reserves to surrounding areas due to an increase in size and fecundity of adults inside the reserve [4,6], but these benefits have been much more difficult to detect [7,8]. Further to these direct responses, indirect effects may also occur and affect nearby areas after some time due to the build-up of top-predators and subsequent trophic cascades inside no-take areas [9,10].

While some of these effects are well documented, their magnitude depends not only on factors such as habitat connectivity, oceanographic characteristics, species life histories, environmental requirements and mobility patterns [10,11], but also on the enforcement of rules and compliance by local users [12]. Several reviews have focussed on the evaluation of the reserve effect [2,9,13], but fewer studies have empirically considered the patchy distribution of species and fishing effort [8,14–16], which might have a large influence on the assessment of fisheries benefits of a MPA. In fact, the loss of fishing grounds and the redistribution of fishing effort in adjacent areas may affect the magnitude of the reserve effect [12]. Hence, it is important to include and understand fishers' behaviour in relation to enforced management rules, habitat preferences of commercial species and other fishers or competing activities.

The concentration of fishing effort near boundaries of no-take areas (i.e. fishing-the-line) is not uncommon and can be interpreted as spillover benefits to adjacent fisheries [16,17]. On the other hand, very intense fishing-the-line behaviour may produce a sharp decrease in density adjacent to a reserve boundary [8]. This is intrinsically related to gear selectivity since species catchability influences the extent of spillover and the effects inside the reserve [16]. Traditional fishing grounds and travel costs may also influence fisheries allocation [18]. Recently, some studies have shown that the distance to borders of no-take areas, water depth and distance to the landing port are the most important factors explaining fisheries aggregations around MPAs, which can be associated, respectively, with fishery benefits, target species distribution and costs [14–16,19,20]. The responses of coastal [21,22] and recreational [23] fisheries distribution before and some years after rezoning have been reported for tropical MPAs. Fisheries displacement was assessed mainly based on face-to-face interviews, and the direct observations conducted after rezoning in one of the studies showed that fishers were reluctant to self-report spatial infringements [21]. Therefore, in spite of work on the redistribution of fishing effort in large-scale trawl fisheries [14,24], there are no empirical studies using direct observations to compare spatial fishing allocations before and after implementation of protection measures in coastal MPAs where artisanal fisheries dominate.

Here we provide the first in-depth assessment of spatial redistribution of fishers in response to MPA implementation. The Arrábida Marine Park is a multiple-use MPA containing a core no-take zone surrounded by several zones with intermediate levels of protection where some human activities are allowed (e.g. small-scale fisheries, diving, tourism and recreational fisheries). In this coastal area artisanal fisheries prevail, where fishers use multiple gears, including trammel and gill nets, traps, longlines and jigs [25]. This study aims to analyse density patterns of the main fishing gear types by comparing the spatial distribution of vessels and buoys before, during and after implementation of the MPA management plan. Density clusters of individual fishers in preferred fishing grounds were investigated through time to understand fishers' choices and adaptability to the MPA rules.

## Methods

### Study Area

The Arrábida Marine Park (AMP) is a 38 km stretch of coastline (53 $km^2$) on the west coast of Portugal, adjacent to a terrestrial nature park created in 1976– the Arrábida Nature Park. The marine park includes the rocky shores and adjacent mixed sandy substrata between north of the Espichel Cape (38°27′N, 9°12′W) and Portinho da Arrábida (38°29′N, 8°57′W) (Figure 1). This area is utilized year-round for commercial and recreational activities as it faces south and is protected from the prevailing north and northwest winds and waves. Nearby are the cities of Lisboa and Setúbal, the latter being an important fishing and commercial port located to the east of the park in the Sado estuary. In the middle of the park there is a small fishing town, Sesimbra, which has a long fishing tradition and is nowadays an important touristic area.

Nearshore, the subtidal shallow rocky reefs are dominated by boulders created by the erosion of the calcareous cliffs and by bedrock with fissures and crevices generating a complex habitat. This habitat is confined to the first 100–150 m from shore except on the west tip of the park where reefs extend beyond this range. Sand is the primary habitat covering the majority of the park from shallow (adjacent to rocky reefs and rocky outcrops) to deeper areas where it is replaced by mud.

The management plan was approved in August 2005 and multiple areas with differing levels of protection have been designated (Figure 1): a fully-protected area (FPA) totalling 4 $km^2$; four partially-protected areas (PPA) totalling 21 $km^2$; and three buffer areas (BA) encompassing 28 $km^2$. Commercial diving for bivalves or other marine organisms, spearfishing, trawling, dredging and purse seining are forbidden in the whole park. These activities were considered to be the ones with the largest impact on coastal marine communities. Commercial fishing licenses for the park were exclusively allocated to fishers from Sesimbra who owned vessels smaller than 7 m in length. The FPA is a no-take, no-go area (except for research, monitoring and education purposes). In the PPAs, artisanal fishing with traps and jigs is allowed, but only beyond 200 m from coast and no extractive recreational activities (i.e. angling) are permitted. In the BAs, licensed fishing vessels and authorised recreational fishing are allowed.

The park's management plan established a transitional period for fisheries, aimed at facilitating the adaptation by fishers to the changes in uses, in which the rules of the different protection zones were gradually implemented during the first four years. In August 2006, management measures were enforced in the BAs, the east half of the FPA (FPA1) began as a PPA and the Portinho PPA (PPA1) was implemented. One year later, the remaining PPAs were implemented and the west half of the FPA (FPA2) started as a PPA. In 2008, the east half of the (current) FPA changed from PPA to FPA. The west half of the FPA was enforced in the summer of 2009, ending the transition period (Portuguese legislation, Council of Ministers Resolution 141/2005) (see Figure 1).

The zoning and rules of the marine park were submitted to a consultation process as required by the Portuguese Law. This process involved NGOs, local authorities, professional fishers associations and other stakeholders. However, there is generally a low level of representation of artisanal fishers using small vessels in the fishers associations. This created problems in understanding the park objectives and accepting the management rules and it is still a focus of mistrust not only between the fishers association and the park authority, but also among fishers themselves. The exclusion of larger vessels from the Park was also very contentious, since the associations represent mainly these fishers. Zones were decided based on the MPA objectives and natural values present, with fishers' perceptions not influencing the zoning scheme. However, the initial plan (before consultation) was greatly changed to address the artisanal fishers' concerns, namely by including nets in the BAs (no nets were to be allowed in the MPA in the initial proposal) and reducing the level of protection in the PPAs with traps and jigs being allowed beyond 200 m from coast (in the initial proposal no fishing activities were considered in the PPAs). Nowadays, fishers with license to operate within the marine park appear to generally support it [26], possibly due to the decrease in fishing effort from competing gears (e.g. dredges) and larger vessels, but also to the exclusion of other competing fishing activities, such as spearfishing. However, most seem to disagree with several measures and enforcement strategies (depending on which type of gear they use), although poaching inside the no-take area is not supported, which suggests recognition of the benefits this area may provide.

### Sampling Surveys

Fishing vessels and buoys within the marine park limits were surveyed along transects by boat. During each sampling day (one sample), the location of vessels and buoys (using a Global Positioning System - GPS), fishing gear type and vessels' names were recorded for all vessels and fishing buoys surveyed (the Portuguese legislation requires that fishing buoys at sea have to be identified with a code for fishing gear type and vessel identification). Two transects were performed each day covering the entire marine park (except the area north of the Espichel cape due to frequent rough sea conditions). The first transect, focussed on

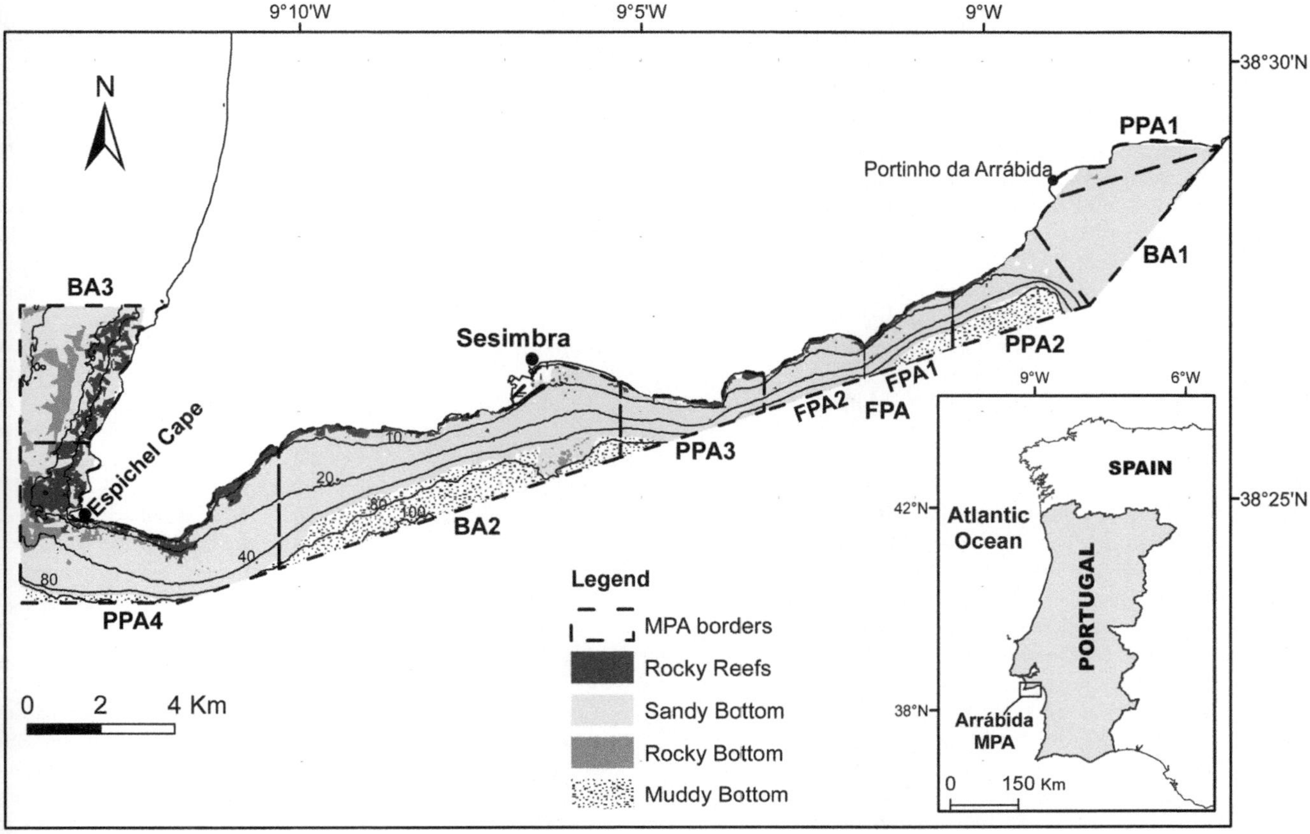


**Figure 1. Map of the Arrábida Marine Park with zoning implemented by the management plan.** Zoning: BA – Buffer areas; PPA – Partially-protected areas, FPA – Fully-protected area (divided in FPA1 and FPA2 due to the transitory phase of the management plan implementation – see Methods). Bathymetry and main habitat types are shown [26].

vessels, started early (6:45 to 7:45 am) in the east part of the Park, near the Portinho da Arrábida bay, and ended at Espichel cape. All buoys were then surveyed on the second transect in the opposite direction. Sampling was carried out inside the Arrábida Marine Park under a permit by the marine park authority (Parque Natural da Arrábida, Instituto da Conservação da Natureza e da Biodiversidade).

Sampling was carried out in five different periods corresponding to the 'before', 'implementation' and 'after' phases of the management plan (see above): 'Before' period – from April to November 2004 (7 samples for buoys and 28 samples for jig vessels); 'Implementation' period refers to Years 1, 2 and 3: Year 1– from March to August 2007 (15 complete samples: for both buoys and vessels), Year 2– September 2007 to February 2008 (14 complete samples), Year 3– November 2008 to August 2009 (16 complete samples); 'After' period – September 2009 to December 2009 (6 complete samples). This classification was used for all analyses. Buoy surveys in the Before period were not uniformly distributed over time, whereas in the Implementation and After periods an average of three and two samples/month were conducted, respectively. The small vessels using jigs were only identified in Year 3 and in the After period.

In the Before period, vessel surveys were shore based, with ten stations established on the high cliffs along the coast covering the entire marine park. Sampling was done early in the morning on a weekly basis, and vessels were georeferenced based on the topographic triangulation method [27], using an electronic theodolite (Topcon, model DT –30) and a GPS. This method has a high level of accuracy in terms of the spatial positioning of objects/features [28].

Three fishing gear types were analysed since they were identified as the most important in the study area: traps, trammel and gill nets, and jigs. Other fishing gears were recorded but were observed infrequently (longlines) or occurred only before the management plan was approved (purse seines). Data for vessel location was used for jigs, since this gear is operated manually directly below the fishing vessel, while buoy geographic coordinates were used for stationary gear (traps and nets).

Even though some vessels were seen few times or only in one of the periods (some fish infrequently, others did not maintained their license or were transferred to other ports), others were observed consistently over the course of the study, with some of them fishing with more than one gear. Three vessels were detected fishing with the three gear types, one with jigs and traps and twenty with nets and traps. Fifty four vessels were seen fishing exclusively with traps, fourteen with nets and a hundred and thirteen vessels were fishing only with jigs.

## Data Analysis

**Generalized additive models.** For all analyses data were aggregated by periods and the three fishing gear types. The spatial and temporal fishing dynamics in the AMP and possible explanatory variables were analysed combining geographic information system (GIS) techniques and generalized additive models (GAMs). The marine park limits and zoning (source: AMP authority) were superimposed onto a map of habitats and bathymetry (source: [26]) using a 500×500 m grid (0.25 $km^2$ cells), although some grids were smaller due to the coastal line and legal borders. Densities (counts per area) of the main fishing gears in the park were summarized by grid and for GAMs only fished grids (with recorded fishing activities) were included [15,16].

Fishing effort allocation was related to the following features using GIS to measure the shortest linear distance (m) from each feature to the mid-point of each grid: distance to Sesimbra port (DistPort), distance to coast (DistCoast), depth, distance to the partially-protected areas (DistPPA), distance to the fully protected area (DistFPA), distance to the 200 m line offshore of the coast (Dist200 m) and distance to the ¼ nautical mile line (Dist1/4 nm). The variable Dist200 m was only related to traps and jigs since it is a limit implemented inside the PPAs, where nets are excluded. Therefore, when these gears were used beyond the 200 m limit but inside the PPAs, the distance to the borders of these areas was negative, to distinguish from fishing gears operating outside these areas. On the other hand, Dist1/4 nm is a national legal limit only for bottom fixed nets (trammel nets and gillnets). The DistPPA1 (PPA1 refers to Portinho bay) was removed from the analyses since only forbidden small drift nets were found fishing there before the management plan started. Throughout the implementation period (Years 1, 2 and 3), DistPPA and DistFPA refer to the respective regime of each protected area in each period.

GAMs were used to explore the density response to the explanatory variables as non-linear relationships were expected and this non-parametric technique does not require linear trends [29]. Autocorrelations among spatial features were tested for each period and only variables with no or low levels of correlation were used to conduct these models. Choosing gamma as the exponential family and using a square root transformation of the response variable resulted in residuals showing a good approximation with normality. Several GAMs were run to test for the best fitted model for each gear type and period. Since some variables were highly correlated we selected those considered to better explain fishers' choices: DistPort, depth, DistPPA and DistFPA. Additionally, alternative models were run to test the influence of the current FPA (DistFPA) during the Before, Year 1 and Year 2 periods (i.e. before full protection was implemented) to evaluate if this was an area pre-selected for its specific characteristics. All explanatory variables included in the models were allowed to be non-linear (using smoothers). Approximate significance of the smooth terms and deviance explained were obtained from each GAM.

For all gears, depth was highly correlated to distance to the coast. For traps and jigs, depth was also highly correlated to Dist200 m and for nets to Dist1/4 nm. Therefore, significant results for depth should be interpreted with caution as they may also reflect significant effects of those other variables. Additionally, a bottom type was assigned to each grid cell using habitat maps. Bottom type by grid cell was classified into the following categories: sand, mud, rock (isolated rocky outcrops) and reefs (coastal shallow rocky reefs). Variables related to habitat were not included in GAMs due to co-linearity but since bottom type may influence both species and fishers' distribution, a Kruskal–Wallis test was conducted to assess the density of gear types (square root transformed) relative to bottom type in each period. Multiple comparison tests evaluated differences in density for each pair of habitat-types. These analyses were conducted with the R 2.14.1 software [30].

**Spatial hotspot analyses.** Fishing areas were analysed using area pattern statistics [31]. Specifically, hotspot analysis was performed in order to study the changes in uses in the main locations chosen by fishers across the five periods, for each fishing gear type. Spatial patterns were investigated using GIS modelling techniques with Arcgis 10.0 (ESRI) software. For this, a 250×250 m grid covering the marine park was superimposed to the fishing GPS points. Hotspot analysis was conducted separately for each of the main gear types with the geographic positions of each vessel occurring in each grid with the aim to study the patterns of use of fishing grounds by individual fishers. To determine statistically significant hotspots, Getis-Ord Gi* statistic (which gives a Z-score and a p-value) [32] was calculated for each grid cell. Statistical tests for significant spatial patterns in data (obtained by a Z score, which varied between -1.96 and +1.96), were compared with the null hypothesis of complete spatial randomness (CSR) with a 95% confidence level against the alternative hypothesis that events are spatially clustered or dispersed. The larger the Z score, the more intense is the clustering of high values (i.e. a hotspot) whereas for negative Z scores, the smaller the Z score, the more intense is the clustering of low values (cold spot) [33]. Significant clusters were defined as the aggregation of adjacent grid cells with a Z-score $\geq |1.96|$, consistent with spatial clustering. To understand the vessel composition in each aggregation, the number of vessels and their percent contribution to each significant cluster was calculated by period and fishing gear. However, since the identification of individual vessels using jigs was not always possible in the Before, Year 1 and Year 2periods, the contribution of these vessels was only evaluated for clusters from Year 3 and the After period.

To perform these analyses, the best distance band was chosen based on global Morans I statistics for spatial autocorrelation [33]. This tool provides a Z-score for the entire study area, measuring spatial autocorrelation based on feature locations and attribute values. To calculate Morans I, the 200 meters distance was used as the starting distance with a cut-off at 800 meters. The minimum distance was chosen based on grid size and the maximum observed dispersion of points. The conceptualization of spatial relationships used for the analysis was the zone of indifference. The final global Z-scores were plotted against the Euclidean distance values and when the increase of the distance caused a decrease in the Z-value (peak), that distance was selected as the best distance band to use in the hotspot analysis [33].

## Results

### Traps

The selected models for the density of traps by period explained between 16.5% and 53.2% of the total variability (Table S1). Overall, the distance to port significantly influenced fisher's behaviour in the Before period and Year 1 ($p<0.05$), whereas depth influenced effort density allocation in all periods (Year 1: $p<0.05$; Years 2, 3, After: $p<0.001$), although in the Before period it was marginally non-significant ($p = 0.055$). The distance to PPAs was not significant in all periods but, interestingly, after the two halves of the fully protection zone were implemented (in Year 3 and After periods), the distance to their borders significantly influenced the variance (Year 3: $p<0.01$; After: $p<0.001$).

In separate models (not shown), distance to the current FPA was tested for the periods before this protection level was effective (Before, Year 1 and Year 2). This variable did not influence the density of traps in the Before period but significant differences were found in both Year 1 ($p<0.05$) and Year 2 ($p<0.001$).

The additive fits of the significant predictor variables from all modelled time periods are shown in Figure 2. During the Before and Year 1 periods the density increased with distance to port showing two peaks, at around 5000 m and 13000 m (Figure 2a, b). Trap density decreased steeply with depth up to approximately 18–20 m, and then increased up to approximately 80–90 m,

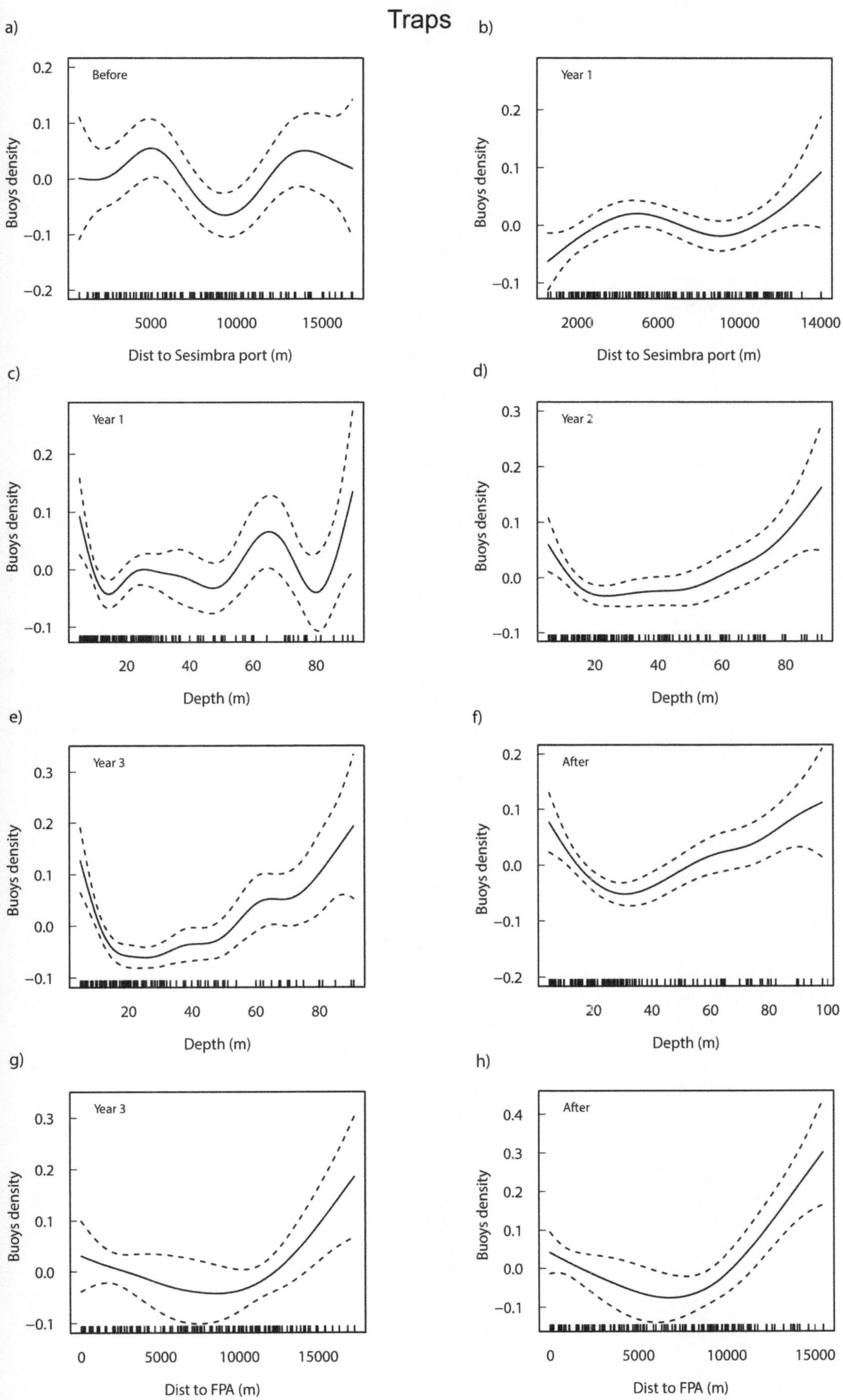
Traps
a)
Before
Buoys density
Dist to Sesimbra port (m)
b)
Year 1
Buoys density
Dist to Sesimbra port (m)
c)
Year 1
Buoys density
Depth (m)
d)
Year 2
Buoys density
Depth (m)
e)
Year 3
Buoys density
Depth (m)
f)
After
Buoys density
Depth (m)
g)
Year 3
Buoys density
Dist to FPA (m)
h)
After
Buoys density
Dist to FPA (m)

**Figure 2. Additive fits of the significant predictor variables to the density of trap buoys for each period.** Distance to Sesimbra port (a, b), depth (c–f), and distance to FPA (g, h) from the significant periods of the selected GAMs (see Table S1) are shown. Tick marks above the *x*-axis indicate the distribution of observations and the *y*-axis shows the contribution of the smoother to the fitted values. The solid line is the estimated smoothing curve and the dotted lines show the 95% confidence intervals.

although there are few observations at those deeper locations (Figure 2c–f) which were mainly situated in front of Sesimbra port where depth increases rapidly offshore (Figure 1). The density of traps also decreased with distance to the fully protected area (Figure 2g, h), but this trend shifted at around 8000 m from the FPA border, where density started to increase.

Density patterns were not influenced by habitat type in each of the five periods. However, aggregating data from all periods showed a significant effect of habitat on effort density allocation (Kruskal-Wallis chi-square = 16.6, $p<0.001$). Multiple comparisons revealed significantly ($p<0.05$) higher density of traps in sand compared to mud and rock, but not compared to reefs.

## Nets

The selected models for the density of nets by period explained between 37.6% and 64.30% of the total variability (Table S2). Overall, distance to port (except in Year 3; Before, Year 2: $p<0.001$; Year 1: $p<0.005$; After: $p<0.01$) and depth ($p<0.05$; except in Year 3 and the After period) had an important role in the spatial allocation of nets. Additionally, distance to PPA and to FPA started to have a significant influence in Year 2 and in the After period, respectively ($p<0.05$). Unlike traps, in the models testing the distance to the current FPA (not shown) for the periods before this protection level was effective, this descriptor was significant for the density of nets before the management plan was implemented ($p<0.01$) but lost significance after its implementation.

The density of nets increased with the distance to port (Figure 3a–d) although there are few observations beyond 10000 m (Before), 8000 m (Year 1) and 6000 m (Year 2). In the After period, there was a decrease in density between 3000 m and 6000 m. Nets generally decreased with depth up to approximately 20 meters, increasing afterwards (Figure 3e–g). In Year 1 a steep decrease in density was found at around 50 m. Density in relation to distance to PPA increased significantly in Year 2 (and was marginally non-significantly in Year 3) and in the After period, when distance to FPA also increased significantly (Figure 3h–j).

Habitat type was significantly related to the density of nets only in Year 3 (Kruskal-Wallis chi-square = 9.4, $p<0.05$), with multiple comparisons showing that density on rock was higher than on mud ($p<0.05$), with reefs and sand showing intermediate values. When aggregating data from all periods (Kruskal-Wallis chi-square = 12.9, $p<0.005$) the same pair of habitats differed significantly ($p<0.05$).

## Jigs

The selected models for the density of vessels fishing with jigs show a very high deviance varying between 56.6% and 84.4% which was much higher than that of the other gear types (Table S3). Depth was a highly significant factor associated to the density of jigs in all periods ($p<0.001$; except in Year 1 when it was marginally non-significant). In both Year 3 and the After period, it was the only significant factor in the model. Distance to port was also an important factor both in the Before ($p<0.05$) and Year 2 ($p<0.001$) periods. The only period where protection measures significantly influenced the density of jigs was in Year 1 ($p<0.05$) with a decreasing pattern with the distance to PPA. No significant influence was found on jigs allocation in relation to the distance to FPA. When the distance to the current FPA was tested for the first periods (Before, Year 1 and Year 2) in separated models (not shown) it was also highly significant before protection started ($p<0.001$) and in Year 2 ($p<0.005$).

Overall, jig density increased with the distance to port, especially in the first 5000–6000 m (Figure 4a, b). Additionally, in the Before period there was a decrease in density between 7000 m and 10000 m followed by a second increase further away from port. In the following periods, very few vessels were seen beyond 8000 m from port. Depth greatly influenced density (Figure 4c–f), decreasing to up to approximately 18 m with a subsequent increase to approximately 30 m (but where few vessels occurred). Fitted significant models showed a complex response of the density of jigs with the distance to the PPA during Year 1 (Figure 4g), when only the current FPA1 was enforced with a partial protection status.

Habitat significantly influenced density in the Before ($p<0.001$), Year 2 ($p<0.05$) and Year 3 ($p<0.05$) periods, but was marginally non-significant in Year 1 ($p = 0.06$) and no relation was detected in the After period. Similarly to the other fishing gears, when aggregating all periods, the density of jigs was highly influenced by habitat (Kruskal-Wallis chi-square = 26.9211, $p<0.001$), with significantly ($p<0.05$) higher values in rocky reefs comparing to sand and mud.

## Hotspot Analysis and Individual Vessels Trends

The hotspot analysis revealed the dynamics of significant fishing clusters throughout the different time periods and gear types analysed (Text S1). Traps followed closely the sequential enforcement of rules through the Implementation years, with some fishing effort displaced from no-fishing areas as shown in the rearrangement of clusters, some of which merged as a result of the MPA rezoning (Figure 5). On the other hand, the cluster closer to the no-take area was divided in two, with vessels surrounding its borders. The same rearrangement of clusters and changes in preferred areas (Text S1) as a consequence of the management plan implementation were also detected in nets (Figure 6) and jigs (Figure 7), although nets remained relatively stable through time in their main fishing grounds, which were already in fished areas. Jigs showed larger changes, with vessels generally moving towards home port but keeping close to the no-take zone. The contribution of individual vessels to each cluster in each time period was also analysed (Text S1) for traps (Figure S1), nets (Figure S2) and jigs (Figure S3).

# Discussion

Here we found that artisanal fisheries showed fisher- and fisheries-specific adaptations to multiple protection measures in a marine protected area (MPA). These findings suggest that artisanal fisheries from temperate systems have complex dynamics and that accounting for individual fishers' behaviour and preferences in exploiting fishing grounds is crucial to implement more successful and effective multiple-use MPAs (i.e. areas with different zones with different types of rules applied to uses).

Different fisheries responded differently to the implementation of protection measures, with preferred habitats of target species driving fishers' preferences in the selection of fishing grounds. Moreover, within each fishery individual fishers showed distinct

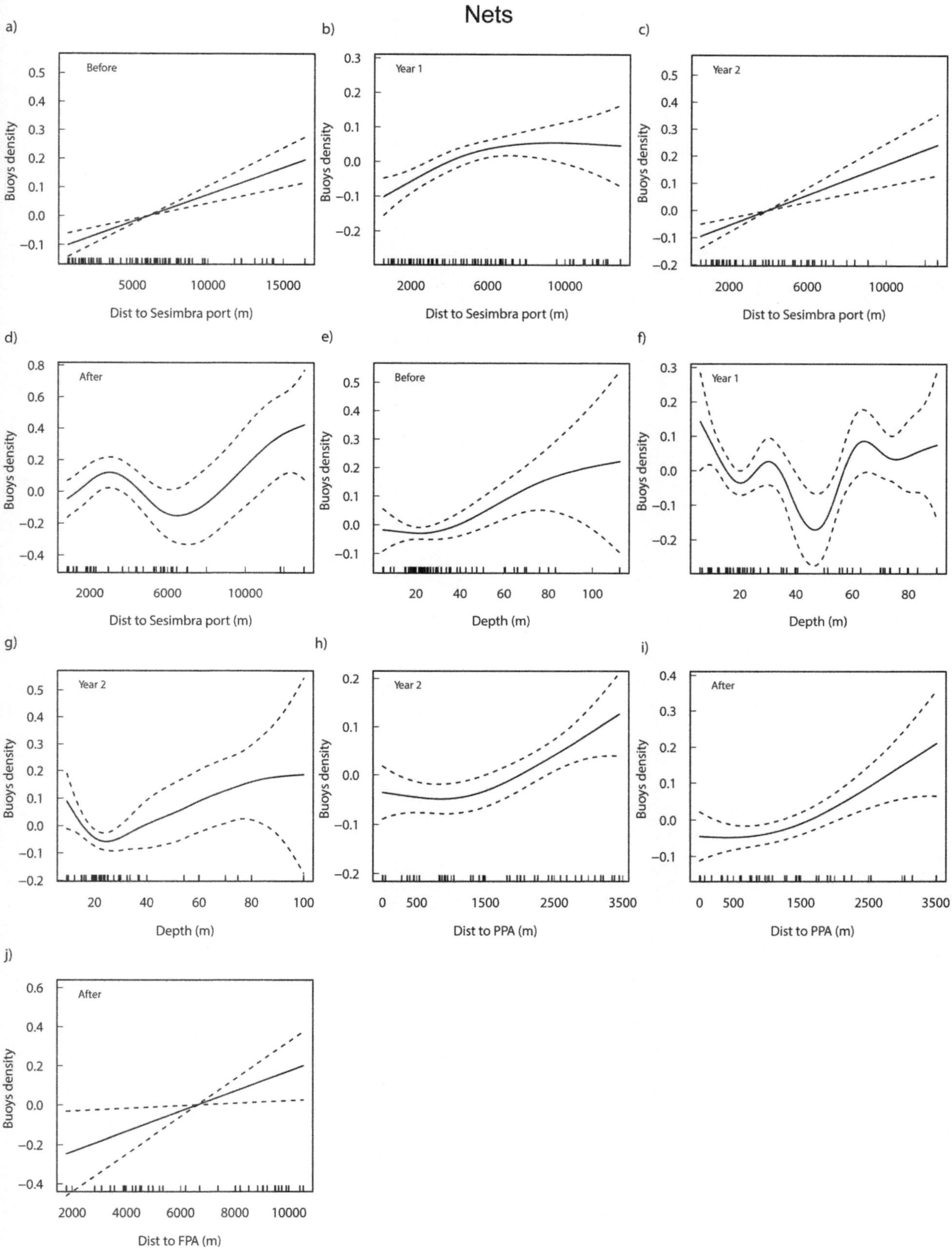


**Figure 3. Additive fits of the significant predictor variables to the density of nets buoys for each period.** Distance to Sesimbra port (a–d), depth (e–g) and distance to PPA (h, i) and to FPA (j) from the significant periods of the selected GAMs (see Table S2) are shown. Tick marks above the *x*-axis indicate the distribution of observations and the *y*-axis shows the contribution of the smoother to the fitted values. The solid line is the estimated smoothing curve and the dotted lines show the 95% confidence intervals.

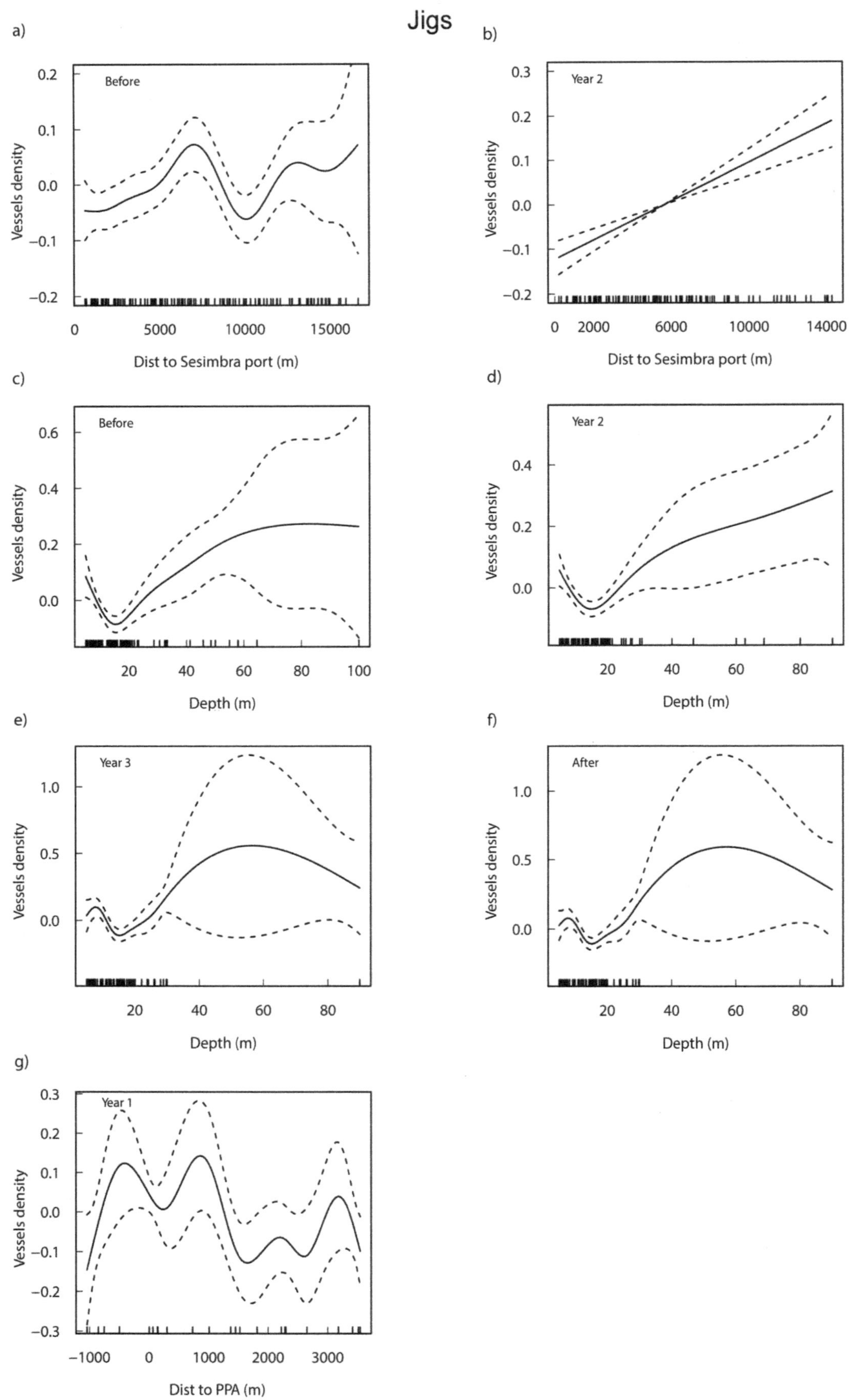
Jigs
a)
Before
0.2
0.1
0.0
−0.1
−0.2
Vessels density
0
5000
10000
15000
Dist to Sesimbra port (m)
b)
Year 2
0.3
0.2
0.1
0.0
−0.1
−0.2
Vessels density
0
2000
6000
10000
14000
Dist to Sesimbra port (m)
c)
Before
0.6
0.4
0.2
0.0
Vessels density
20
40
60
80
100
Depth (m)
d)
Year 2
0.4
0.2
0.0
Vessels density
20
40
60
80
Depth (m)
e)
Year 3
1.0
0.5
0.0
Vessels density
20
40
60
80
Depth (m)
f)
After
1.0
0.5
0.0
Vessels density
20
40
60
80
Depth (m)
g)
Year 1
0.3
0.2
0.1
0.0
−0.1
−0.2
−0.3
Vessels density
−1000
0
1000
2000
3000
Dist to PPA (m)

**Figure 4. Additive fits of the significant predictor variables to the density of jig vessels for each period.** Distance to Sesimbra port (a, b), depth (c–f), and distance to PPA (g) from the significant periods of the selected GAMs (see Table S3) are shown. Tick marks above the *x*-axis indicate the distribution of observations and the *y*-axis shows the contribution of the smoother to the fitted values. The solid line is the estimated smoothing curve and the dotted lines show the 95% confidence intervals. Negative distances refer to jig vessels fishing inside the PPA.

strategies, with some operating in a broader area whereas others kept preferred territories, some of them being adjacent to a no-take area. Spatial allocation of fishing grounds was well defined and apparently agreed upon among the most common fishers, supporting the occurrence of traditional routines. One of the possible consequences of effort reallocation inside multiple-use MPAs is an increase of spatial competition for setting fishing gears in buffer areas [22]. When fishing effort is very high, the catchability of each gear may be reduced, affecting the expected benefits from protection. Interestingly, when fishers have licences for multiple gears, adapting to management rules may be easier. In fact, in our study traps and jigs faced a smaller reduction of fishing grounds than nets, although jigs may have lost important areas close to shore. Several fishers can opt to operate with various gears with a preference for traps instead of nets, as revealed by the increasing trend in the number of vessels fishing with traps. This suggests fishing with traps was the least affected fishery and that fishers are adapting to other productive alternatives in response to the zoning and rules of the marine park.

Some recent studies addressed the allocation of fisheries before and in response to spatial closures (temperate trawl fisheries: [14,24]; tropical artisanal fisheries: [21,22]), although we could not find other empirical cases in the literature where artisanal fishers' distribution were analysed through direct observations before, during and after the implementation of a temperate MPA. Tracking the spatial position of vessels and fishing gears through time and analysing factors affecting the selection of a fishing ground may allow for a clearer understanding of the fishers' choices and adaptations to different situations as well as of the dynamics of small scale artisanal fisheries, which comprise a large percentage of the fishing communities throughout the world.

Jigging for cuttlefish (*Sepia officinalis*) and squid (*Loligo vulgaris*) from small wooden vessels is a traditional artisanal fishing activity in the region. Jigging takes place mainly close to shore and near rocky reefs at depths up to 20 m. Thus, the jigging effort distribution in shallow areas can be attributable to species occurrence, gear characteristics and safety for these small vessels. Jigs were mainly influenced by depth and habitat through time. They were significantly more associated to rocky reefs than to other habitat types. Some previously preferred fishing grounds located inside the reserve may have become off-limits to these fishers since association with nearshore habitats lost significance with time and there was an important effect of the FPA location on vessels' density before its implementation. Consequently, this fishery seems to have been impacted by the zoning as the fishers lost fishing grounds close to shore within the full and partially-protected areas. This may have negative consequences on the acceptance by fishers and on their attitudes towards the marine park [34].

A highly dynamic spatial distribution of jig vessels through time was detected with three main clusters identified. These were typically formed by a high number of vessels, sometimes with a large contribution of occasional fishers. In the After period, the three clusters that were previously scattered throughout the park merged into a single large cluster in front of the port where no restrictions apply to this fishery. The management plan implementation therefore caused some significant changes to the spatial distribution of this type of fishery, operated by small 3–4 m vessels, which take advantage of the very sheltered conditions of this coastline, mainly in front and to the east of their home port. They operate by drifting with the alongshore tides and target cephalopod species which occur in nearshore environments.

Benefits from protection may however have occurred since jig fishers remained in the area beyond 200 m in the PPAs, and near the western border of the FPA, which was the closest to their home port, even during the implementation phase of the fully protected area. This suggests that some fishers were able to profit by staying a little further away from shore, probably intercepting species over sandy bottoms adjacent to the shallow rocky reefs, rather than competing with other commercial and recreational fishers in the buffer zone.

Jig fishers' adaptations suggest they tried to keep as close as possible to their former fishing ground, possibly also benefiting from protection, whereas at the same time their displacement was towards their home port, revealing other additional concerns probably related to security and operating costs. Lédée et al. [22] found that fishers preferentially redistributed to areas already known before protection, suggesting that previous experience and tradition may play an important role in the site-fidelity behaviour, influencing the choice of a fishing location. However, similar to the present study, the authors also report that most of the fishers' displacement was towards their home port, mainly due to the lower costs, leading to an increase in the fishing pressure in areas that had already high density.

There were several factors explaining the spatial and temporal distribution of nets. Distance to port influenced effort density except in Year 3. The two main clusters occurred right in front of and to the west of the port and remained stable through time. A third cluster was detected in the initial periods in the east of the park encompassing part of the fully protected area but disappeared thereafter, with some fishing activities probably moving adjacently to the southern limit of the fully protected area outside the marine park limits. The proximity to both the partial and fully protected areas became important in the After period, with nets being located further away from these areas, which is consistent with the location of the main clusters.

The area in front and to the west of vessels' home port is an important fishing ground where the main clusters consisting of several vessels were detected. Those clusters did not relocate after protection started as they were already in an allowed area. This is an extensive shallow sandy area used by commercial fishers targeting soles, cuttlefish and fish species such as sparids by trammel and gill nets [35]. When all periods were combined, buoy density was significantly associated with rock suggesting that fishers prefer shallow habitats, especially those with the potential to attract fish such as rocky outcrops and adjacent sand or shallow reefs.

The trap fishery showed preferred sites with clusters close to the home port and on each side of the park. Depth was the strongest influence in trap allocation with higher densities in shallow waters (18–20 m) and at around 70 m (but with fewer vessels), suggesting the possibility that these fishers were targeting different habitats. There were more traps distributed on sand than on other habitats, except for shallow rocky reefs. This is consistent with the behaviour of octopus (*Octopus vulgaris*, Cuvier, 1797), the target species of this gear, which is found in mixed sandy habitats, from

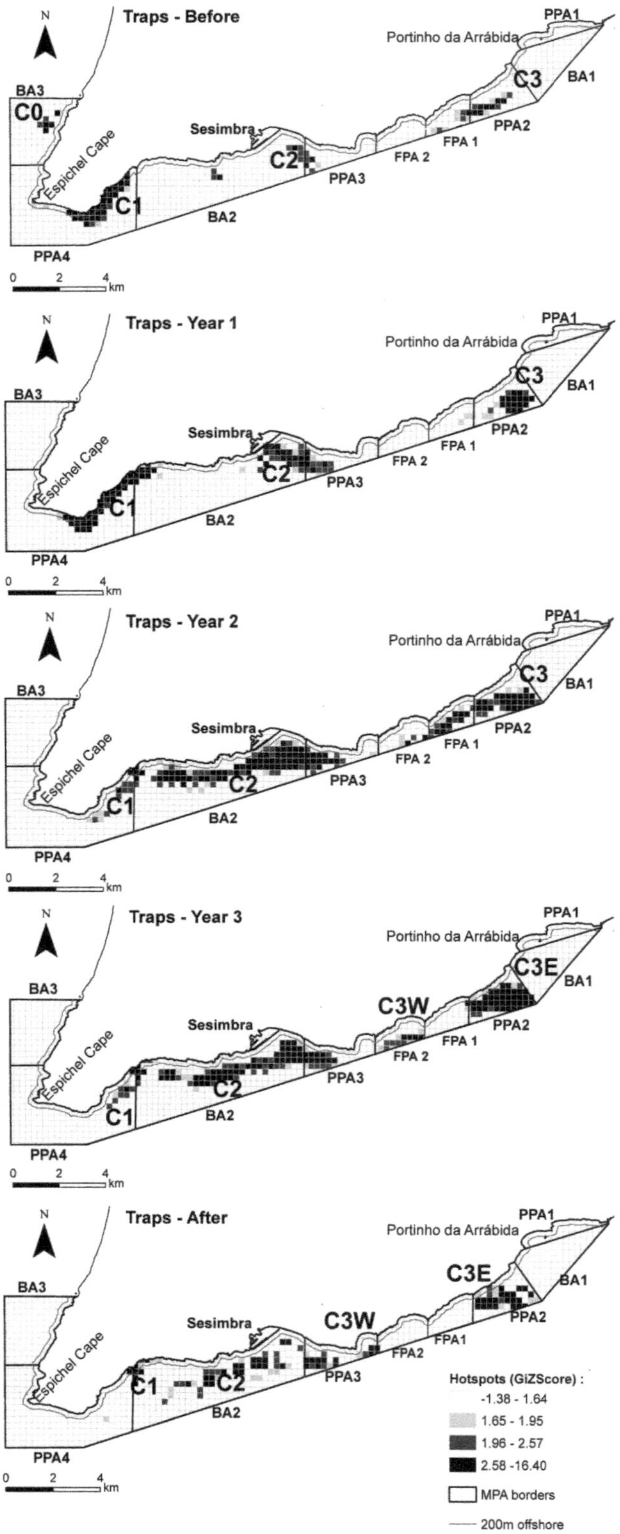


**Figure 5. Maps obtained from the hotspot analyses of trap buoys for each period.** The location of significant clusters (GIZScore >1.96) by period (a – Before, b – Year 1, c – Year 2, d – Year 3, e – After), and the different protection levels at the Arrábida Marine Park are shown: BA – buffer area; PPA – partial protection area; FPA – fully protected area (see Methods for a detailed description of the protection levels in the park and their implementation through time). The 200 m offshore line is also shown.

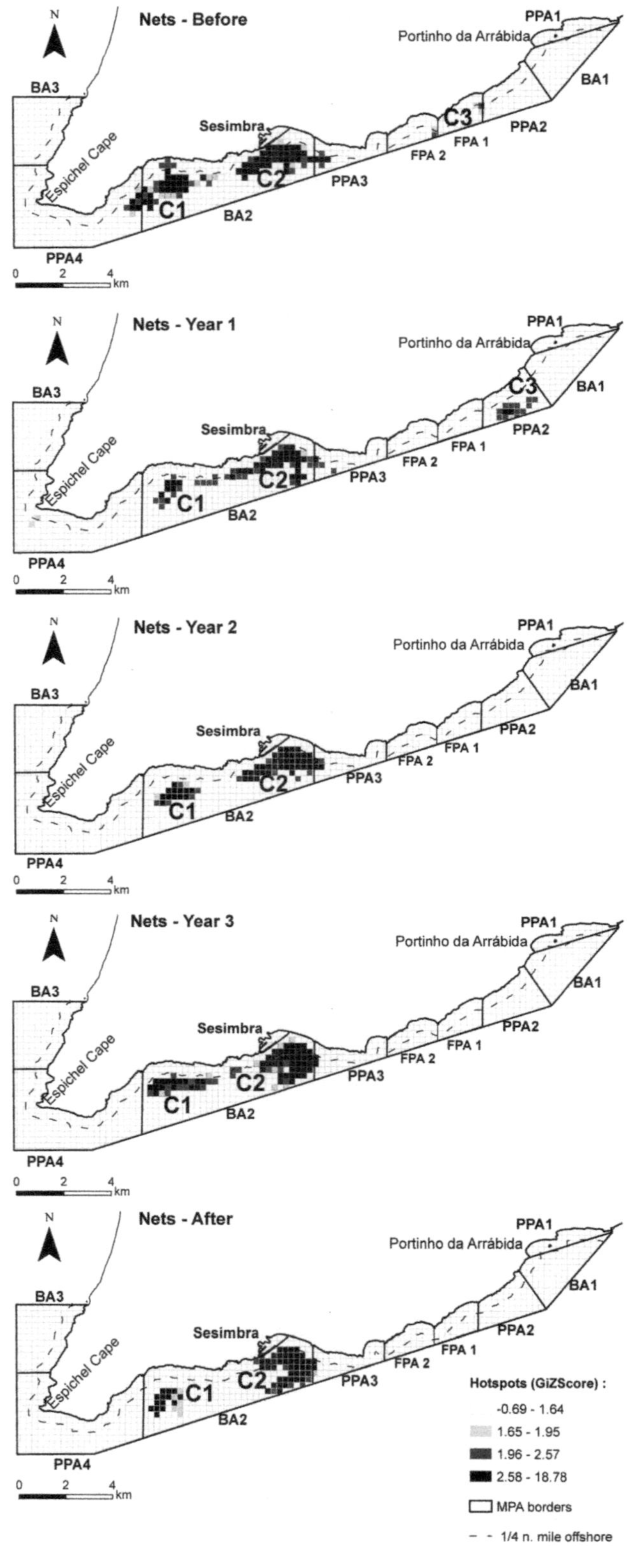


**Figure 6. Maps obtained from the hotspot analyses of nets buoys for each period.** The location of significant clusters (GIZScore >1.96) by period (a – Before, b – Year 1, c – Year 2, d – Year 3, e – After), and the different protection levels at the Arrábida Marine Park are shown: BA – buffer area; PPA – partial protection area; FPA – fully protected area (see Methods for a detailed description of the protection levels in the park and their implementation through time). The national legal limit for nets of the line of ¼ nautical miles offshore is also shown.

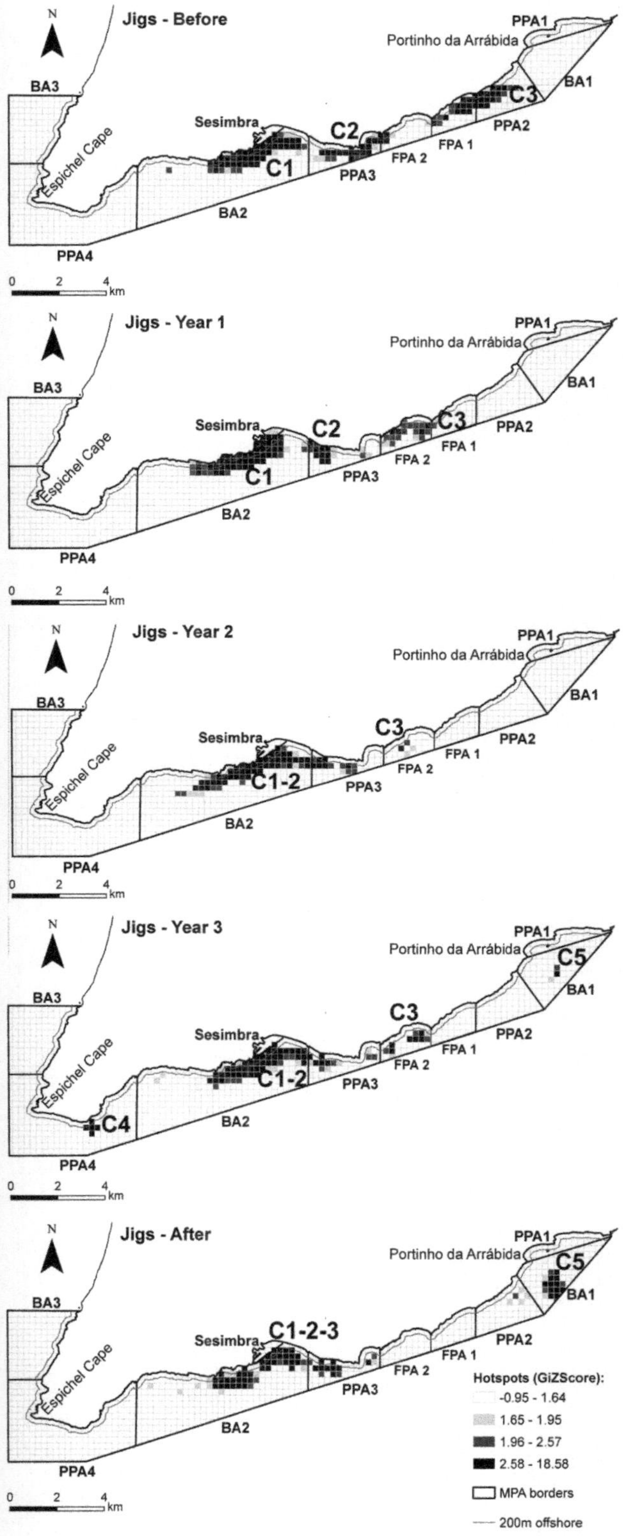


**Figure 7. Maps obtained from the hotspot analyses of jig vessels for each period.** The location of significant clusters (GIZScore >1.96) by period (a – Before, b – Year 1, c – Year 2, d – Year 3, e – After), and the different protection levels at the Arrábida Marine Park are shown: BA – buffer area; PPA – partial protection area; FPA – fully protected area (see Methods for a detailed description of the protection levels in the park and their implementation through time). The 200 m offshore line is also shown.

the coastline to depths of around 200 m, usually spending the winter in deeper waters and migrating inshore by early spring to breed [36,37].

The spatial dynamics of trap fishers showed a cluster close to their home port, which is advantageous for small vessels that cannot travel far for safety reasons, are limited by sea and weather conditions [20] and where operating costs are a significant burden. Another cluster was found near the most complex reefs of the park [38] which are also near the entrance of the Sado estuary, an important spawning and nursery area [39], with vessels extending their activity outside the park limits. Interestingly, the analysis of fishers' choices through time showed that in this cluster (which in the Before period partially occupied the future fully protection area), considerable changes occurred in both the spatial distribution of traps and composition of vessels dominating this area. Although in Year 1 no cluster was found in the fully protection area (FPA), in Year 2 the eastern cluster extended to this area with fishers likely trying to gain access to this fishing ground before it became off-limits. This interpretation is reinforced by the fact that distance to the current FPA was not significant in the Before period, but became an important explanatory variable in the model during Year 3 and After periods (when the FPA was fully implemented), indicating that fishers were attracted to this area possibly due to the expectation of future benefits.

A few (3–5) vessels dominated several of these clusters and their behaviour changed through time. The western cluster became a hotspot dominated by a single vessel which was able to secure this fishing ground, whereas the central cluster was characterized by a larger number of vessels with a more erratic behaviour (i.e. vessels joined other clusters through time). This may be related to high competition in this fishing ground. On the other hand, with the retrieval of one dominant vessel that did not receive a license from the park, the eastern cluster was taken over by two new dominant vessels showing specific territories and dominance in these fishing grounds. These two vessels fished mainly on the borders of the FPA adopting a strategy of "fishing the line" [17]. Several reasons can explain this increase in effort at the edge of a no-take area: the reallocation and aggregation of effort because of the reduction of fishing grounds or due to perceived or expected benefits from protection [8,40].

In spite of the loss of fishing ground as a consequence of MPA designation, the spatial competition between trap fisheries and, namely, nets decreased on important and traditional fishing grounds since nets became only allowed in the buffer areas. Moreover, before the management plan implementation, near-shore reefs were heavily exploited by spearfishing. The exclusion of this type of recreational fishery, which has a large impact on high trophic level species such as large sparids, seabass and octopus [41,42], likely contributed to the increase of such target species' biomass inside the marine park. In fact, the landings of octopus for vessels licensed to fish in the park have increased since protection started [43].

Other studies in Mediterranean MPAs found that the proximity to the reserve borders significantly affected the spatial distribution of fishing effort [15,16,20,44]. The loss of fishing grounds and the attraction to the reserves' boundaries when spillover effects are substantial, are important factors explaining the reallocation of fishing effort related to the implementation of MPAs. These effects are however influenced by the spatial distribution of habitats and target species inside and outside the reserve [45]. Thus, the proximity to no-take zones may not be involved in the choice of the fishing ground or may be due to the fishers' preference for being closer to their former fishing location [22,34].

Abesamis et al. [18] found that artisanal fishers tended to select traditional fishing grounds which were probably preferred due to their guarantee of higher stability in catches and a higher minimum average income. The experience and familiarity with fishing grounds, one component of traditional and ecological knowledge [46,34], may also help to minimize gear loss and enhance catches.

## Conclusion

To understand the complexity of impacts (both positive and negative) on fisheries related to marine protected areas, one needs to closely follow the dynamics of fisheries operating nearby. This is particularly challenging for coastal multiple-use MPAs where artisanal fisheries occur. Here we show an effective method for the study of fishing effort allocation and dynamics for artisanal fisheries using different gears by following individual fishers' choices before, during and after the implementation of protection. Our results have relevance to the vast majority of global MPA designs; that is, single, relatively small multiple-use areas utilized by local fishers using multiple gear types. Besides the importance of assessing fishing effort within and around MPAs, this study shows that gear type, habitat features and MPA design influence individual fishers' behaviour and this must be taken into account when planning MPA design and evaluating the effects of marine conservation measures. This type of information is lacking in most studies evaluating the effects of marine protected areas although it is central for an unbiased assessment of biological, social and cultural responses to marine protection.

## Supporting Information

**Figure S1 Proportion contribution of vessels using traps (T) to each of the significant clusters obtained in the hotspot analysis of trap buoys by period: Before, Year 1, Year 2, Year 3, After.** See the location of each cluster in Fig. 5. The number of vessels observed in each cluster is also shown.

**Figure S2 Proportion contribution of vessels using nets (N) to each of the significant clusters obtained in the hotspot analysis of nets buoys by period: Before, Year 1, Year 2, Year 3, After.** See the location of each cluster in Fig. 6. The number of vessels observed in each cluster is also shown.

**Figure S3 Proportion contribution in of jig vessels (J) to each of the significant clusters obtained in the hotspot analysis of jig vessels by period (jigs were only correctly identified in Year 3 and After periods).** See the location of each cluster in Fig. 7. The number of vessels observed in each cluster is also shown.

**Table S1 Results of the smoothing terms from the generalized additive models (GAM) testing the density of trap buoys relative to the distance to several spatial features in the Before, implementation (Years 1, 2 and 3) and After periods.**

**Table S2 Results of the smoothing terms from the generalized additive models (GAM) testing the density of nets buoys relative to the distance to several spatial features in the Before, implementation (Years 1, 2 and 3) and After periods.**

**Table S3 Results of the smoothing terms from the generalized additive models (GAM) testing the density of jig vessels relative to the distance to several spatial features in the Before, implementation (Years 1, 2 and 3) and After periods.**

**Text S1 Hotspot analysis and individual vessels trends.** Analysis of significant clusters for each fishing gear type in the different protection areas throughout the Before, Implementation and After periods. Contribution of individual vessels for the observed trends in each cluster.

## Acknowledgments

The authors thank the AMP/ICNB authority, park rangers C. Silva and A. Mourato and a number of volunteers for all the help in the field, as well as M. Henriques for all the marine park support. Part of the work of B.H.C. was carried out at the Marine Science Institute (University of California).

## Author Contributions

Conceived and designed the experiments: BHC MIB LG KE JEC HNC EJG. Performed the experiments: BHC MIB LG. Analyzed the data: BHC MIB LG KE JEC HNC EJG. Wrote the paper: BHC MIB LG KE JEC HNC EJG.

## References

1. Gell FR, Roberts CM (2003) Benefits beyond boundaries: the fishery effects of marine reserves. Trends in Ecology and Evolution 18: 448–456.
2. Claudet J, Guidetti P, Mouillot D, Shears NT, Micheli F (2011) Ecological effects of Marine Protected Areas: conservation, restoration, and functioning. In: Claudet J, editor. Marine Protected Areas - a multidisciplinary approach. Cambridge: Cambridge University Press. 37–71.
3. Goñi R, Badalamenti F, Tupper M (2011) Effects of Marine Protected Areas on local fisheries: evidence from empirical studies. In: Claudet J, editor. Marine Protected Areas - a multidisciplinary approach. Cambridge: Cambridge University Press. 72–98.
4. Russ G (2002) Yet another review of marine reserves as reef fishery management tools. In: Sale P, editor. Coral Reef Fishes. San Diego: Academic Press. 421–443.
5. Kramer D, Chapman M (1999) Implications of fish home range size and relocation for marine reserve function. Environmental Biology of Fishes 55: 65–79.
6. Tetreault I, Ambrose R (2007) Temperate marine reserves enhance targeted but not untargeted fishes in multiple no-take MPAS. Ecological Applications 17: 2251–2267.
7. Pelc RA, Warner RR, Gaines SD, Paris CB (2010) Detecting larval export from marine reserves. Proc Natl Acad Sci U S A 107: 18266–18271.
8. Goñi R, Hilborn R, Díaz D, Mallol S, Adlerstein S (2010) Net contribution of spillover from a marine reserve to fishery catches. Marine Ecology Progress Series 400: 233–243.
9. Micheli F, Halpern BS, Botsford LW, Warner RR (2004) Trajectories and correlates of community change in no-take marine reserves. Ecological Applications 14: 1709–1723.
10. Hamilton SL, Caselle JE, Malone DP, Carr MH (2010) Incorporating biogeography into evaluations of the Channel Islands marine reserve network. Proc Natl Acad Sci U S A 107: 18272–18277.
11. Claudet J, Osenberg CW, Domenici P, Badalamenti F, Milazzo M, et al. (2010) Marine reserves: fish life history and ecological traits matter. Ecological applications 20: 830–839.
12. Claudet J, Guidetti P (2010) Improving assessments of Marine Protected Areas. Aquatic Conservation: Marine and Freshwater Ecosystems 20: 239–242.
13. Lester SE, Halpern BS, Grorud-Colvert K, Lubchenco J, Ruttenberg BI, et al. (2009) Biological effects within no-take marine reserves: a global synthesis. Marine Ecology Progress Series 384: 33–46.
14. Murawski SA, Wigley SE, Fogarty MJ, Rago PJ, Mountain DG (2005) Effort distribution and catch patterns adjacent to temperate MPAs. ICES Journal of Marine Science 62: 1150–1167.

15. Stelzenmüller V, Maynou F, Bernard G, Cadiou G, Camilleri M, et al. (2008) Spatial assessment of fishing effort around European marine reserves: implications for successful fisheries management. Mar Pollut Bull 56: 2018–2026.
16. Goñi R, Adlerstein S, Alvarez-Berastegui D, Forcada A, Reñones O, et al. (2008) Spillover from six western Mediterranean Marine Protected Areas: evidence from artisanal fisheries. Marine Ecology Progress Series 366: 159–174.
17. Kellner JB, Tetreault I, Gaines SD, Nisbet RM (2007) Fishing the line near marine reserves in single and multispecies fisheries. Ecological Applications 17: 1039–1054.
18. Abesamis RA, Alcala AC, Russ GR (2006) How much does the fishery at Apo Island benefit from spillover of adult fish from the adjacent marine reserve? Fishery Bulletin 104: 360–375.
19. Wilcox C, Pomeroy C (2003) Do Commercial Fishers Aggregate around Marine Reserves? Evidence from Big Creek Marine Ecological Reserve, Central California. North American Journal of Fisheries Management 23: 241–250.
20. Forcada A, Valle C, Sanchéz-Lizaso JL, Bayle-Sempere JT, Corsi F (2010) Structure and spatial-temporal dynamics of artisanal fisheries around a Mediterranean Marine Protected Area. ICES Journal of Marine Science 67: 191–203.
21. Campbell SJ, Hoey AS, Maynard J, Kartawijaya T, Cinner J, et al. (2012) Weak compliance undermines the success of no-take zones in a large government-controlled marine protected area. PLoS One 7: e50074.
22. Lédée EJI, Sutton SG, Tobin RC, De Freitas DM (2012) Responses and adaptation strategies of commercial and charter fishers to zoning changes in the Great Barrier Reef Marine Park. Marine Policy 36: 226–234.
23. De Freitas DM, Sutton SG, Moloney JM, Lédée EJI, Tobin RC (2013) Spatial substitution strategies of recreational fishers in response to zoning changes in the Great Barrier Reef Marine Park. Marine Policy 40: 145–153.
24. Abbott JK, Hayne AC (2012) What are we protecting? Fisher behaviour and the unintended consequences of spatial closures as a fishery management tool. Ecological Applications 22: 762–777.
25. Batista MI, Baeta F, Costa MJ, Cabral HN (2011) MPA as management tools for small-scale fisheries: The case study of Arrábida Marine Protected Area (Portugal). Ocean & Coastal Management 54: 137–147.
26. Cunha AH, Erzini K, Serrão EA, Gonçalves E, Borges R, et al. (2011) Restoration and Management of Biodiversity in the Marine Park Site Arrábida-Espichel (PTCON0010) LIFE06 NAT/P/000192. Portugal: LIFE06 NAT/P/000192.
27. Davis RE, Foote FS, Anderson JM, Mikhail EM (1981) Surveying: Theory and Practice. New York: McGraw-Hill. 1024 p.
28. Singh R, Artman D, Taylor DW, Brinton D (2000) Basic surveying - theory and practice. Report presented during the Ninth Annual Seminar of Geometronics by the Oregon Department of Transportation - Geometronics Unit. Bend, Oregon. 162 p.
29. Zuur AF, Ieno EN, Walker NJ, Saveliev AA, Smith GM (2009) Mixed Effects Models and Extensions in Ecology with R. New York: Springer. 574 p.
30. R Core Team (2012) R: A language and environment for statistical computing. R Foundation for Statistical Computing, Vienna, Austria. www.R-project.org. Acessed 19 October 2012.
31. Fortin MJ, Dale M (2005) Spatial Analysis: A Guide for Ecologists. Cambridge: Cambridge University Press. 369 p.
32. Ord JK, Getis A (1995) Local Spatial Autocorrelation Statistics: Distributional Issues and an Application. Geographical Analysis 27: 286–306.
33. Ord JK, Getis A (2001) Testing for local spatial autocorrelation in the presence of global autocorrelation. Journal of Regional Science 41: 411–432.
34. Leleu K, Alban F, Pelletier D, Charbonnel E, Letourneur Y, et al. (2012) Fishers' perceptions as indicators of the performance of Marine Protected Areas (MPAs). Marine Policy 36: 414–422.
35. Batista MI, Teixeira CM, Cabral HN (2009) Catches of target species and bycatches of an artisanal fishery: The case study of a trammel net fishery in the Portuguese coast. Fisheries Research 100: 167–177.
36. Guerra A (1975) Determinatión de las diferentes fases del desarrollo sexual de *Octopus vulgaris* Lamarck, mediante un índice de madurez. Investigatigación Pesquera, Barcelona 39: 397–416.
37. Roper CFE, Sweeney MJ, Nauen CE (1984) Cephalopods of the world, an annotated and illustrated catalogue of species of interest to fisheries. Rome: FAO.
38. Gonçalves EJ, Henriques M, Almada VC (2003) Use of a temperate reef-fish community to identify priorities in the establishment of a marine protected area. In: Beumer JP, Grant, A & Smith, D C., editors. Aquatic Protected Areas: What works best and how do we know? Proceedings of the World Congress on Aquatic Protected Areas, Cairns, Australia – August 2002. 261–272.
39. Vasconcelos RP, Reis-Santos P, Maia A, Fonseca V, França S, et al. (2010) Nursery use patterns of commercially important marine fish species in estuarine systems along the Portuguese coast. Estuarine, Coastal and Shelf Science 86: 613–624.
40. McClanahan TR, Mangi S (2000) Spillover of exploitable fishes from a marine park and its effect on the adjacent fishery. Ecological Applications 10: 1792–1805.
41. Cooke SJ, Cowx I (2004) The role of recreational fishing in global fish crises. BioScience 54: 857–859.
42. Rocklin D, Tomasini JA, Culioli JM, Pelletier D, Mouillot D (2011) Spearfishing regulation benefits artisanal fisheries: the ReGS indicator and its application to a multiple-use Mediterranean marine protected area. PLoS One 6: e23820.
43. Horta e Costa B, Erzini K, Caselle JE, Folhas H, Gonçalves EJ (2013) The reserve effect within a temperate marine protected area in the north-eastern Atlantic (the Arrábida Marine Park, Portugal). Marine Ecology Progress Series In press.
44. Stelzenmüller V, Maynou F, Martín P (2007) Spatial assessment of benefits of a coastal Mediterranean Marine Protected Area. Biological Conservation 136: 571–583.
45. Forcada A, Bayle-Sempere JT, Valle C, Sanchez-Jerez P (2008) Habitat continuity effects on gradients of fish biomass across marine protected area boundaries. Mar Environ Res 66: 536–547.
46. Davis A, Hanson JM, Watts H, MacPherson H (2004) Local ecological knowledge and marine fisheries research: the case of white hake (Urophycis tenuis) predation on juvenile American lobster (Homarus americanus). Canadian Journal of Fisheries and Aquatic Sciences 61: 1191–1201.

# 4

# Size Selection of Antarctic Krill (*Euphausia superba*) in Trawls

**Ludvig A. Krag[1]*, Bent Herrmann[2], Svein A. Iversen[3], Arill Engås[3], Sigve Nordrum[4], Bjørn A. Krafft[3]**

**1** DTU Aqua, Technical University of Denmark, Hirtshals, Denmark, **2** SINTEF Fisheries and Aquaculture, Fishing Gear Technology, Hirtshals, Denmark, **3** Institute of Marine Research, Bergen, Norway, **4** Aker BioMarine Antarctic AS, Oslo, Norway

## Abstract

Trawlers involved in the Antarctic krill (*Euphausia superba*) fishery use different trawl designs, and very little is known about the size selectivity of the various gears. Size selectivity quantifies a given trawl's ability to catch different sizes of a harvested entity, and this information is crucial for the management of a sustainable fishery. We established a morphological description of krill and used it in a mathematical model (FISHSELECT) to predict the selective potential of diamond meshes measuring 5–40 mm with mesh opening angles (oa) ranging from 10 to 90°. We expected the majority of krill to encounter the trawl netting in random orientations due to high towing speeds and the assumed swimming capabilities of krill. However, our results indicated that size selectivity of krill is a well-defined process in which individuals encounter meshes at an optimal orientation for escapement. The simulation-based results were supported by data from experimental trawl hauls and underwater video images of the mesh geometry during fishing. Herein we present predictions for the size selectivity of a range of netting configurations relevant to the krill fishery. The methods developed and results described are important tools for selecting optimal trawl designs for krill fishing.

**Editor:** Konstantinos I. Stergiou, Aristotle University of Thessaloniki, Greece

**Funding:** This study is financed by the Research Council of Norway. The grant number (project number) is: 216574. The funders had no role in study design, data collection and analysis, decision to publish, or preparation of the manuscript.

**Competing Interests:** The academic editor of PLOS One has requested confirmation from Aker BioMarine Antarctic AS (AKBM) that the current manuscript and work represented in it do not involve any competing interests of Financial Disclosures. The authors hereby on behalf of AKBM declare that the current work do not involve any competing interests or any Financial Disclosure. Therefore this does not alter the authors' adherence to PLOS ONE policies on sharing data and materials.

* Email: lak@aqua.dtu.dk

## Introduction

The largest annual catch of Antarctic krill (*Euphausia superba*, hereafter called krill) in the Southern Ocean since the inception of a commercial fishery in 1972 was 528,000 tons. During the last decade, the annual catches in the Scotia Sea, where the commercial operations are located today, have ranged between 100,000 and 212,000 tons. Due to limited knowledge about this marine ecosystem and the potential negative effects of fishery activities, a precautionary catch limit for the Scotia Sea was set at 620,000 tons by the Commission of Antarctic Marine Living Resources (CCAMLR) in 1991 to avoid potential conflicts with land-based predators that depend on krill as prey such as penguins [1]. Based on a coordinated survey conducted by the CCAMLR in 2000, during which krill density was measured acoustically in the fishing areas [2], the biomass of krill was calculated to be 60.3 million tons [3]. A theoretical total allowable catch (TAC) limit based on this calculation was set at 5.61 million tons [3].

The pelagic trawlers involved in the krill industry use different trawl systems and designs, and very little is known about the size selectivity of the commercial trawls used to harvest krill. Pshenikov [4] reported that during the 1970s when the Soviets trawled for krill, only 10–20% of the krill that entered the trawl opening were retained in the codend. However, neither data nor analyses were provided in this paper. In the near future, the krill harvest is expected to reach the precautionary catch limit due to increased demand for krill and improved harvesting and processing technologies. The potential to increase the annual catch further (i.e., to the TAC limit in the Scotia Sea) is significant [3,5]. Krill have a circumpolar distribution [6], which will enable a spatial expansion of the fishery that potentially could contribute to increases in the total annual catch. The objectives of responsible harvesting of krill, rational management of the krill fishery, and economic profit for the industry demand development of fishing gear that reduces accidental (escape) mortality during the fishing process. Both the Commission and Scientific Committee of the CCAMLR strongly recommend member states that are fishing for krill to investigate the effects of different fishing gears on krill escapement to assess the total mortality of the krill stock caused by the fishery [7,8].

Studies of size selectivity for other commercially harvested species in towed fishing gear have traditionally been made by conducting a series of sea trails. Sea trials are economically costly and time consuming, and there is therefore a limit to the number of different gear designs that can be tested. In addition, some commercial krill fishing vessels use a continuous pumping technique to bring the catch from the trawl to the deck, which makes standard selectivity studies difficult. Due to these limitations, the FISHSELECT method was used in the present study to evaluate size selectivity of relevant gears for the krill fishery [9]. FISHSELECT is based on a combination of laboratory experiments and computer simulations which involves assessment of the

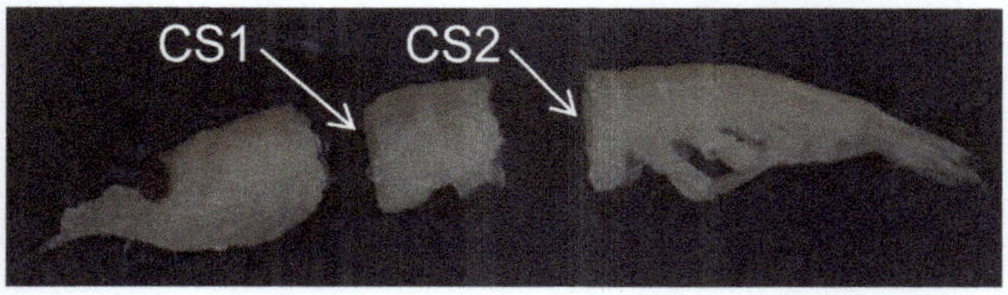


**Figure 1. Krill sliced at cross section 1 (CS1) and cross section 2 (CS2).**

morphology relevant for size selection. This method has been used successfully to describe, understand, and predict size selection of fish [9,10,11,12,13] and crustaceans, such as *Nephrops norvegicus* [14]. The essential first step of the method includes collecting morphometric measurements of the species investigated and estimating selectivity parameters based on comparing the morphology of the species and the geometry of relevant meshes. This procedure makes it possible to quantify the size selectivity of different trawl designs, including those commercially used today. Moreover, it constitutes an essential step towards evaluating escape mortality in the trawl-based krill fishery. The ability to quantify and predict size selectivity in the commercial fishery for krill will allow managers to explore the consequences of fishing in terms of catch efficiency and catch loss of different gear designs when harvesting populations with dissimilar size structures.

In general, fish targeted by trawls have good swimming ability relative to the towing speed used in these fisheries. Several species of fish also have been observed to orientate themselves in relation to the trawl netting during the capturing process e.g., [15]. In contrast to fish, smaller invertebrates such as prawns tend to display a more limited response to stimuli from the trawl [16,17]. Krill are similar in size to the smaller species of commercially fished prawns, but they are fished using towing speeds similar to those used for targeting fish (i.e., 2.5–3.0 knots). Therefore, the selectivity process for krill in trawls is expected to resemble a sieving process in which the individual krill may meet the trawl netting with a more random orientation (in contrast to what has been observed for fish). Trawls designed to catch small crustaceans like krill or shrimps are designed with small meshes in the entire trawl indicating similar expectations from the fishermen.

The objective of this study was to establish a morphology based model for krill to predict size selectivity of a range of netting configurations relevant to the krill fishery. Simulation-based results (FISHSELECT, [9]) were compared with results from a selectivity experiment that involved hauling two different trawl gear design, a macroplankton survey trawl and commercial trawl simultaneously through a krill swarm. To explore behavioral patterns of krill during fishing, underwater video observations were made.

## Materials and Methods

### Ethics statement

This study did not involve endangered or protected species. Experimental fishing was conducted onboard a Norwegian commercial trawler. No permit was required to conduct the study.

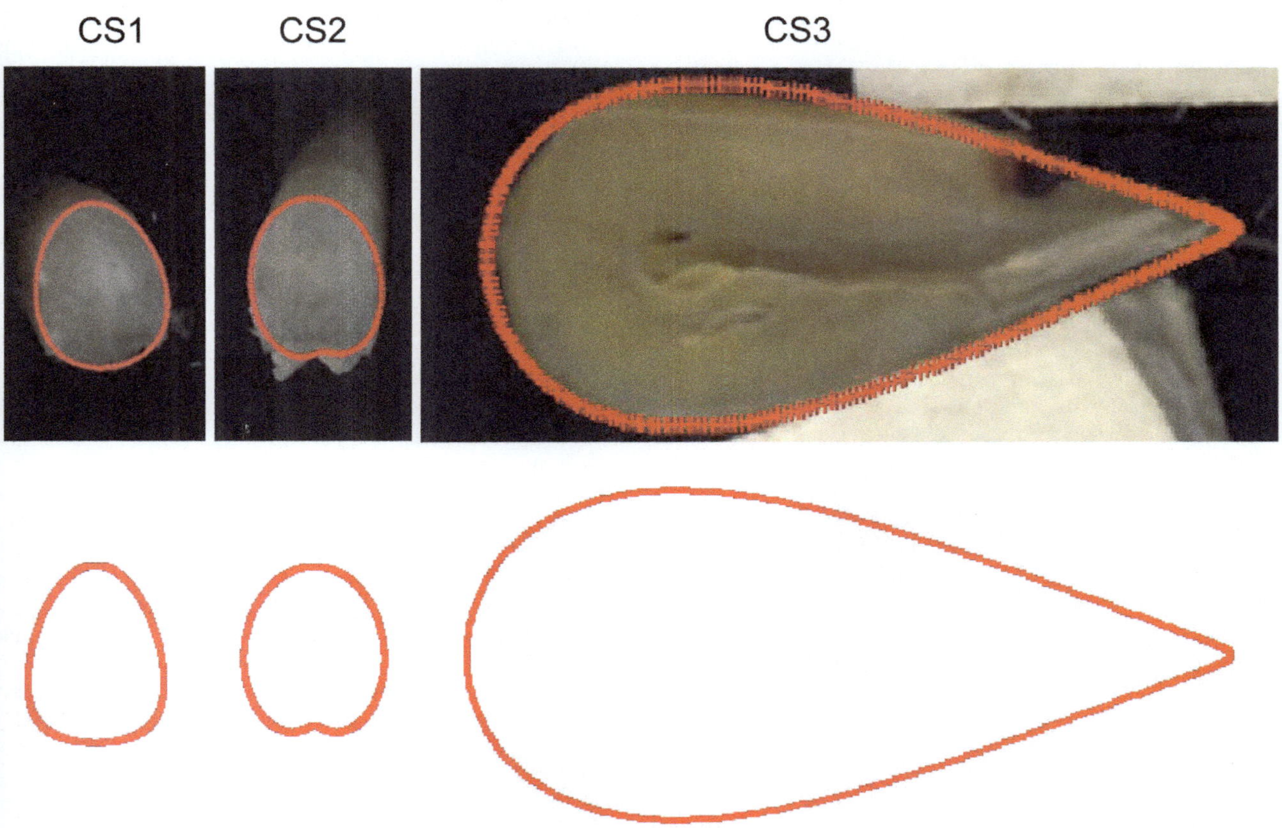


**Figure 2. Shapes of CS1, CS2 and CS3. CS1 is described by a flexellipse_1, CS2 is described by a flexellipse_3 and CS3 is described by a flexdrope_2 (see appendix S1).**

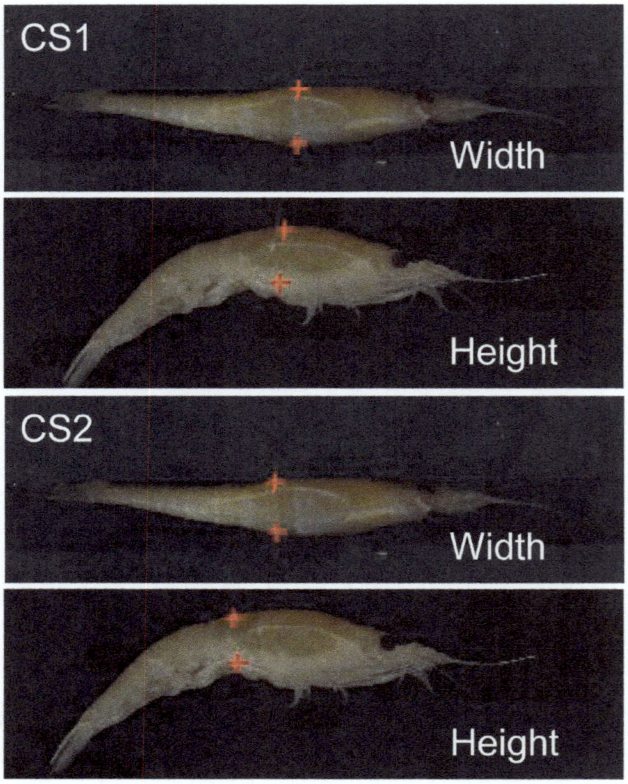


**Figure 3. Scanned pictures of krill with markings to illustrate the width and height at CS1 and CS2.**

## Morphological description of krill

We used the FISHSELECT method with some modifications to establish a morphological description of krill. The FISHSELECT method aims to identify and parameterize the morphological properties of a given species that determine the species size selection of towed fishing gear. For some fish species the vital morphology has been identified as one or two cross sections along the length axis that contains the maximum compressed height and/or the maximum width due to the cylindrical shape of fish such as gadoids [9,10]. Descriptions of these cross sections are then used to predict the size selectivity by examining whether or not the cross section is able to pass through the mesh geometry, which can be of a variety of sizes and shapes. Traditionally, the FISHSELECT method includes a series of "fall-through trials" [9] to estimate the compressibility of the identified cross sections for fish, as fish are flexible to a certain point and can compress their cross section shape during mesh penetration. However, crustaceans have a hard exoskeleton, so this procedure is not necessary for krill [14]. To date, the FISHSELECT methodology and the measuring tools have not been applied to crustaceans as small as krill, which can grow to a maximum total body length of 60 mm (e.g., [18]). Thus, application of the FISHSELECT method to krill requires further development of some of the procedures. The major steps involved in the method are described in detail in Herrmann et al. [9], and the customization for using the method on krill is elaborated below.

## Collecting krill and studying the effect of preservation

Fresh krill were not available for the experimental laboratory trials needed for this study, so preserved animals were used in these trials. Determining if and how the preservation treatment changed the morphology of krill were important for accurate interpretation of the results. When Germany first initiated its krill research program in the Southern Ocean in 1976, researchers compared fresh, frozen, ethanol-preserved, and formalin-preserved krill samples to evaluate potential effects of preservation on morphological properties. They observed shrinkage in samples preserved in ethanol, but no statistically significant differences were detected in formalin fixed or frozen animals (Volker Siegel, pers comm.). Because these factors are fundamental for the reliability of our results, and because these historical results never were published, this result required corroboration.

During a survey conducted aboard the Norwegian fishing vessel Juvel (Emerald Fisheries ASA) off the South Orkney Islands in late January to early February 2012, krill were collected fresh from the catch using a Macroplankton trawl (see [18] for additional descriptions of the trawl). The trawl was lowered from the sea surface to 200 m depth and hauled at 2.5–3 knots. Sex and maturity stages of krill were determined using the classification methods outlined by [19]. Total body length was measured (±1 mm) from the anterior margin of the eye to the tip of telson, excluding the setae, according to the "Discovery method" used in [20]. Carapace width was measured using a caliper at its widest cross section point. A total of 30 krill including juveniles, sub adults, adults with gravid females at stage FIIID were preserved individually in borax-buffered formalin (4%), and body length and carapace width were measured again after 2 and 10 months.

Comparisons of any temporal change in the morphology measurements was made using an analysis of variance test (Proc NPAR1WAY, SAS Institute Inc., Box 8000, Cary, N.C., U.S.A.) with the 0.05 level accepted as indicating statistical significance.

## Morphological measurements

Before initiation of the laboratory trials, the krill body was examined to identify the relevant morphology that potentially would determine the animal's ability to penetrate different meshes. We first determined the body orientation that would allow the largest individual to escape through a given mesh, as this orientation is indicative of the selective potential for krill for that mesh type. The optimal orientation for krill is when the body is stretched and meets the mesh opening head or tail first. We next identified morphological measures that could describe this optimal orientation. Two cross sections, CS1 and CS2, were identified along the length axis of the krill body as being decisive for size selection, as these cross sections contain the maximum height (h) and width (b) measures of animals in the optimal orientation (Fig. 1).

We also considered the different ways in which krill can meet the meshes in a trawl (i.e., head first, tail first, curled, stretched), as this determines which additional measurements should be made during data collection. In this study we included a second contact mode, the curled shape, CS3 (Fig. 2). We expect that krill are more passive in a trawl compared with fish and thus might encounter the meshes in more random orientations. CS3 represents a contact mode very different from the optimal mode, and it is expected to result in low selectivity due to the large cross section shape. At this stage we chose to include only one contact mode in addition to the optimal contact mode. If the observed size selectivity of krill is difficult to explain by simulation the process using the optimal contact mode is a more detailed approach, involving more contact modes needed.

A two-step process was used to take morphological measurements. The first step involved taking measurements of 30 individuals covering the length span from 27 to 55 mm. Each individual was placed on a flatbed scanner to measure width and

**Table 1.** Model parameters and fit-statistics for model enabling that descriptions of CS1 and CS2 can be created based on height and width values of the cross sections alone.

| | | CS1 | | | CS2 | | |
|---|---|---|---|---|---|---|---|
| | | $c_1$ | $c_2$ | $c_3$ | $c_1$ | $c_2$ | $c_3$ |
| **Model** | | $a_1 \times b$ | $a_1 \times b + a_2 \times h$ | $a_2 \times h + a_3 \times b \times h$ | $a_1 \times b + a_2 \times h$ | $a_1 \times b + a_2 \times h$ | $a_2 \times h + a_3 \times b \times h$ |
| DF | | 29 | 28 | 28 | 28 | 28 | 28 |
| $R^2$ | | 0.9892 | 0.9887 | 0.6831 | 0.9929 | 0.9902 | 0.8713 |
| a0 | Value | * | * | * | * | * | * |
| | SE | * | * | * | * | * | * |
| | p-value | * | * | * | * | * | * |
| a1 | Value | 0.5224 | 0.4403 | * | 0.2768 | 0.3456 | * |
| | SE | 0.0010 | 0.1027 | * | 0.0644 | 0.1000 | * |
| | p-value | <2e-16 | 1.9e-4 | * | 1.9e-4 | 1.8e-3 | * |
| a2 | Value | * | 0.2069 | 0.1584 | 0.1473 | 0.2126 | 0.1247 |
| | SE | * | 0.0932 | 0.0423 | 0.0566 | 0.0878 | 0.0267 |
| | p-value | * | 3.5e-2 | 8.2e-4 | 1.5e-2 | 2.2e-2 | 4.7e-5 |
| a3 | Value | * | * | -0.0190 | * | * | -0.0129 |
| | SE | * | * | 0.0082 | * | * | 0.0056 |
| | p-value | * | * | 2.8e-2 | * | * | 2.8e-2 |

*not in model

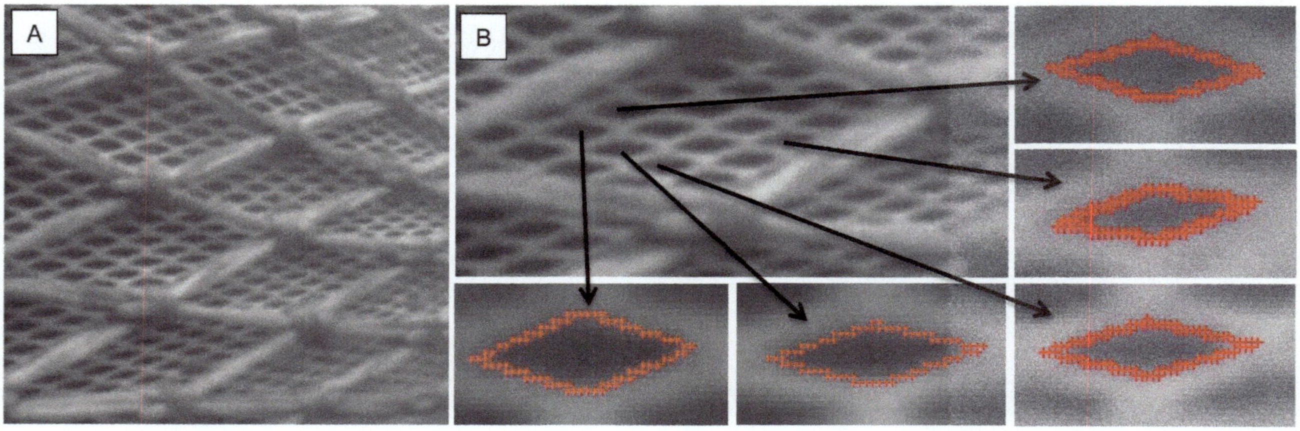


**Figure 4. Image of the netting wall (mesh size15.4 mm) of the krill trawl captured during fishing operations (A).** Digitizition of selected meshes to establish realistic values for mesh opening angles (oa) (B). The camera is located 10 m from the codline-end, pointing backwards. The intire 15.4 mm trawl was covered with 200 mm double 5 mm PE diamond netting for protection.

height at CS1 and CS2. Width and height were extracted using the image analysis in the FISHSELECT software (Fig. 3). After these measurements were taken, each individual was frozen on a metal plate at −80°C to ensure rapid freezing. The frozen individuals were cut with a scalpel at CS1 and CS2 perpendicular to the length of the body. The frozen condition prevented the specimens from deforming during the slicing, which would cause deviation from their natural morphology. These sliced cross sections were photographed on a flatbed scanner. Then based on the shape describtion by analyzing the cross section images, we found relations between the three parameters describing the cross section shapes and size and the height and width at the cross sections. This enabled us to assess the cross section shapes and sizes established for CS1 and CS2 for a larger number of individuals based on only measuring height and width at CS1 and CS2. In the second step, the length of 83 individuals with a length range from 19 to 54 mm was measured and the established length to cross-section relationship applied. Without cutting we assessed the cross sections CS1, CS2, and CS3 by placing a given specimen on the scanner for the different measures (height, width, and curled).

## Modeling of cross sections

To describe the shapes of the the cut cross sections of CS1 and CS2 (fig 2), the best parametric model was selected for each based on $R^2$ and Akaike's Information Criterion (AIC) [21]. Ten different geometric models were tested for both CS1 and CS2 (see Appendix S1). Models that fit the data well have a high $R^2$ value. Comparison between models with different numbers of parameters can be made using AIC values following the procedure described in Sistiaga et al. [11], and the model with the lowest AIC value is the one that best fits the data. A more detailed mathematical description of the chosen models is given in the Appendix S1. Each model is defined by three parameters (c1, c2, and c3). For each cross section (CS1 and CS2), we have a dataset consisting of measurements from the 30 individuals that includes the values for b, h, c1, c2, c3 and a cross section model. To be able to estimate a cross section of a krill specimen based on

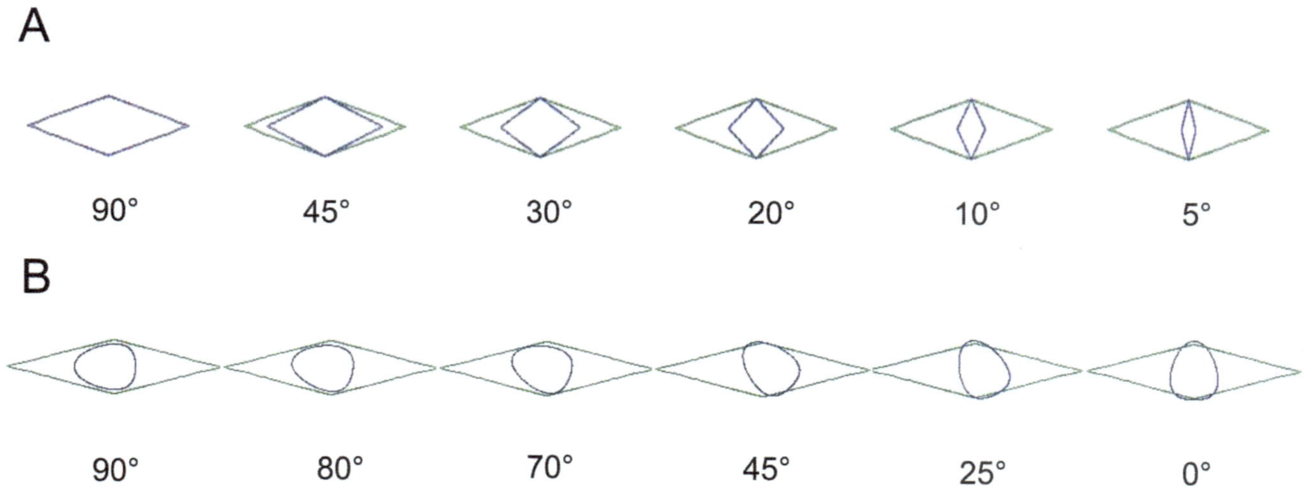


**Figure 5. Illustration of how the effective mesh openings in a trawl can vary depending of the attack angle of krill (5–90°) (A).** The light mesh is the real mesh in 90° view and the dark mesh is the effective mesh opening when projecting the mesh opening to a plan perpendicular to the towing direction. The lower mesh series (B) shows the potential effect of rotation of the cross section (CS1_CS2) in a 30° open mesh.

**Table 2.** Mesh size measurements based on image analysis (mesh id 1–5) and fit statistics for using a diamond mesh description ($R^2$) and measured mesh opening angles (oa values in degrees) based on underwater video recordings during commercial fishing (mesh id 6–10).

| Mesh id | Mesh size (mm) | Mesh id | R2 | oa (degrees) |
|---|---|---|---|---|
| 1 | 15.08 | 6 | 0.9511 | 44 |
| 2 | 15.4 | 7 | 0.9881 | 34 |
| 3 | 15.81 | 8 | 0.9167 | 28 |
| 4 | 15.94 | 9 | 0.8709 | 31 |
| 5 | 14.66 | 10 | 0.9762 | 33 |
| **Mean** | **15.38** | | **0.9406** | **34** |

measurements of $b$ and $h$ at that cross section requires knowing how c1, c2, and c3 depend on b and h for the selected models. Therefore, for each cross section we assumed that cx (x = {1,2,3} can be modeled by a simple bilinear function in b and h including interaction. Thus we use:

$$c_x = a_{0x} + a_{1x} \times b + a_{2x} \times h + a_{3x} \times b \times h \qquad (1)$$

Model (1) was applied using the lm function in statistical package R (version 2.15.2; www.r-project.org) to establish the models for c1, c2, and c3 for both CS1 and CS2. Each model term in (1) that was found to be non-significant was removed to establish the final models for how c1, c2, and c3 can be estimated based on b and h measured for each cross section (CS1 and CS2). The resulting models (Table 1) were subsequently applied to estimate the cross section shapes of other krill individuals based on measuring only b and h at CS1 and CS2.

## Assessment of mesh shape and size of trawl netting

In addition to a morphological description of krill, we needed a precise description of sizes and shapes of the meshes used in commercial trawls for which experimental size selectivity data were collected. Weaved diamond mesh polyamide (PA) netting with a nominal mesh size of 16 mm (stretched inside measure) is commonly used in the commercial krill fishery. A small sample of this netting was placed on a flatbed scanner with no tension in the netting together with a measuring unit to determine the precise mesh size. Individual meshes in the picture were analyzed in FISHSELECT using the built-in image analysis function, and mesh size was assessed following the procedures described in Sistiaga et al. [11]. Standard mesh measuring methods (e.g., the OMEGA measuring gauge [22]), which are applied for larger mesh sizes, could not be used in this study because the measuring jaws are too large for the small mesh sizes used in the krill fishery. We used underwater video recordings made during commercial fishing operations onboard the Norwegian vessel Antarctic Sea during the 2013 season to assess the shapes of the meshes during fishing. Following Sistiaga et al. [11], the digitized images were used to extract the mesh opening angle (oa) to identify the best shape description of the meshes (e.g., diamond, hexagonal, square) (Fig. 4).

## Predicting potential size selectivity of krill

We generated a combined model of CS1 and CS2 because together they contained the maximum cross section width (CS1) and height (CS2). CS3 represents a different body shape mode by which krill can make contact with the trawl netting during fishing, so it was not combined with the other cross sections and instead was treated separately. Using the established cross section descriptions (CS1, CS2, and CS3) we conducted simulations to predict the basic selective properties for a variety of different mesh sizes and shapes using the optimal orientation of CS1_CS2 and CS3. Optimal orientation includes optimal rotation of the cross section description in the given mesh using a 90° attack angle relative to the mesh opening. The optimal orientation therefore gives the absolute potential and the upper limit for the potential size selection of the mesh sizes and shapes investigated. The standard FISHSELECT predictions of size selectivity are based on optimal orientation of the identified cross section descriptions. Such predictions of size selectivity are based only on the species morphology and on the given mesh shape and size and do not take into account any behavioral effects that may affect size selectivity of the species in question.

Herein we made selective predictions over a relevant range of diamond meshes with varying sizes and opening angles. To test a sufficient number of individuals in the selective range of all relevant mesh sizes, we created a large virtual population. The empirically established relationship between body length and the cross section shape parameters (c1, c2, c3) was used to define the properties of the individuals in the virtual population. We estimated L50 by assuming a logistic selection curve and treated the simulated penetration data as covered codend data [23]. L50 is the length at which there is 50% retention likelihood for the individual [23], and we used this and the selection range (SR).These basic selective properties were assessed for the relevant mesh sizes and shapes for each mesh type, and they are presented as iso-L50 plots known as design guides (see [9] for further details about this procedure).

## Selective effect of cross section orientation and attack angle

In previous studies carried out using FISHSELECT to predict size selectivity for different species of fish, it was assumed that each individual is optimally orientated when attempting to pass through the meshes. However, we expect krill to have a lower probability of meeting the meshes in the optimal orientation with optimal attack angle due to the relatively high towing speed compared to the size of the animal. In addition, the tapering of commercial krill trawls differs from that of traditional fish trawls, and this can affect how the individuals in the trawl meet the mesh. The commercial trawl used to collect the experimental data had a mouth opening of about 20×20 m, and the trawl was about 200 m long. This small amount of tapering resulted in an attack angle of less than 3° between the direction of the flow and the netting wall.

**Table 3.** Dates, body length and carapax width measurements of 30 juvenile, sub adult and adult krill collected fresh from the trawl catch (25. January) off South Orkney Islands and after two and 10 months preserved in borax-buffered formalin (4%).

| Date in 2012 | Range body length (mm) | Mean (SD) body length (mm) | Range body width (mm) | Mean (SD) body width (mm) |
|---|---|---|---|---|
| 25. January | 27.0–55.0 | 44.7 (7.7) | 2.4–7.1 | 4.7 (1.1) |
| 21. March | 27.0–55.0 | 44.3 (7.6) | 2.2–7.0 | 4.7 (1.1) |
| 14. November | 27.0–54.0 | 43.8 (7.4) | 2.5–7.2 | 4.8 (1.1) |

We explored the potential effect of orientation of CS1_CS2 in the mesh opening and the effect of attack angles between krill and the netting for the specific gear for which we collected experimental size selection data. In principle, orientation of the cross section and the attack angle work in combination, but for simplicity we assessed their potential effects separately. We therefore examined the effect of orientation of CS1_CS2 using an optimal attack angle; in this scenario, the description of CS1_CS2 was presented perpendicular to the mesh (90° in Fig. 5 Panel A) and then was rotated stepwise from 0 to 90° (Fig. 5 Panel B). To evaluate the effect of the attack angle was the mesh shape stepwise projected to a plan perpendicular to the towing direction in steps of 10° (Fig. 5). The description of CS1_CS2 during this procedure was orientated optimally in the projected mesh shape. This analysis was conducted using the FISHSELECT software.

## Collecting experimental selectivity data

Experimental size selectivity data were collected in February 2011 onboard the Norwegian commercial krill trawler Saga Sea (Aker Biomarine ASA) off the South Orkney Islands. The Saga Sea was equipped with a twin trawl beam system. One beam was rigged with a 7 mm macroplankton trawl with a 38 m$^2$ mouth opening. The other beam was rigged with a commercial trawl with a 400 m$^2$ mouth opening and a mesh size of 15.4 mm from the trawl opening to the rear end. The trawls were lowered from the sea surface to 200 m depth simultaneously and then slowly hauled at a vessel speed of 2.5–3.0 knots, which corresponds to the towing speed used during commercial fishing. At the surface, one trawl was taken onboard before the other trawl. The order in which each trawl that was taken onboard first was alternated between a total of four hauls. When a trawl catch was landed on deck, the body length of about 100 individual krill from each trawl was measured (±1 mm) from the anterior margin of the eye to the tip of telson, excluding the setae. A total of 416 individuals were measured from the macroplankton trawl and 393 individuals from the commercial 15.4 mm trawl.

## Analyzing experimental selectivity data

The experimentally collected data were analyzed using SELNET [12] and paired gear analysis [23]. Data were modeled by the traditional logistic model (2) with parameters L50 and SR.

$$r(l,L50,SR)=\frac{\exp\left(\frac{\ln(9)}{SR}\times(l-L50)\right)}{1.0+\exp\left(\frac{\ln(9)}{SR}\times(l-L50)\right)} \quad (2)$$

The following function (3) was minimized with respect to the parameters L50, SR, and SP. SP denotes the so-called split between the fishing power of the test and control trawls. Although it is of no real interest here, it is necessary to assess the value of SP to obtain the value of the selection parameters L50 and SR (see Wileman et al. [23] for further information on this subject). Function (3) is written as:

$$\sum_l \left\{ nt_l \times \ln\left(\frac{sp\times r(l,L50,SR)}{1-sp+sp\times r(l,L50,SR)}\right) + nc_l \times \ln\left(\frac{1-sp}{1-sp+sp\times r(l,L50,SR)}\right)\right\} \quad (3)$$

where the summation is over length classes l. $nt_l$ denotes the number of individuals found in the test gear (15.4 mm trawl) of length l, whereas $nc_l$ denotes the number found in the control gear (7 mm trawl). Data were pooled for the four pairs of hauls prior to conducting the analyses to obtain the average size selection estimation for the 15.4 mm trawl.

Based on the estimated selection parameters L50 and SR, the length Li by which the retention likelihood is in % can be calculated by:

$$Li=L50+\frac{SR}{\ln(9)}\times\ln\left(\frac{0.01\times i}{1.0-0.01\times i}\right) \quad (4)$$

Estimation of L05 to L95 based on (4) is required for the comparison between experimental- and simulation-based results by means of the method described in Herrmann et al. [13].

**Table 4.** Fit statistic for the used models to describe CS1, CS2 and CS3.

| Cross section (CS) | Model | R2 | AIC |
|---|---|---|---|
| 1 | flexellipse_1 | 0.8334 | −53.63 |
| 2 | flexellipse_3 | 0.7427 | −37.32 |
| 3 | flexdrope_2 | 0.8670 | 77.97 |

The model description for each cross section is based the highest $R^2$ value and the lowest AIC value among the tested models (see appendix S1).

**Table 5.** Estimated regression coefficients for CS1 and CS2.

| | CS1 | | CS2 | |
|---|---|---|---|---|
| | Length vs. width | Length vs. height | Length vs. width | Length vs. height |
| a | 0.1031 | 0.1444 | 0.0995 | 0.1282 |
| b | 0.0033 | 0.7278 | 0.0833 | 0.2461 |
| R2 | 0.8478 | 0.8220 | 0.8822 | 0.8419 |

Estimation of 95% confidence intervals for the selection parameters and the entire selection curve as well as for L05 to L95 is carried out using a bootstrapping method implemented in the SELNET software.

## Simulation of size selection of krill

Based on the experimentally obtained size selection results for the commercial trawl and using CS1 and CS2 optimally rotated with optimal attack angle, we can explain the experimental results

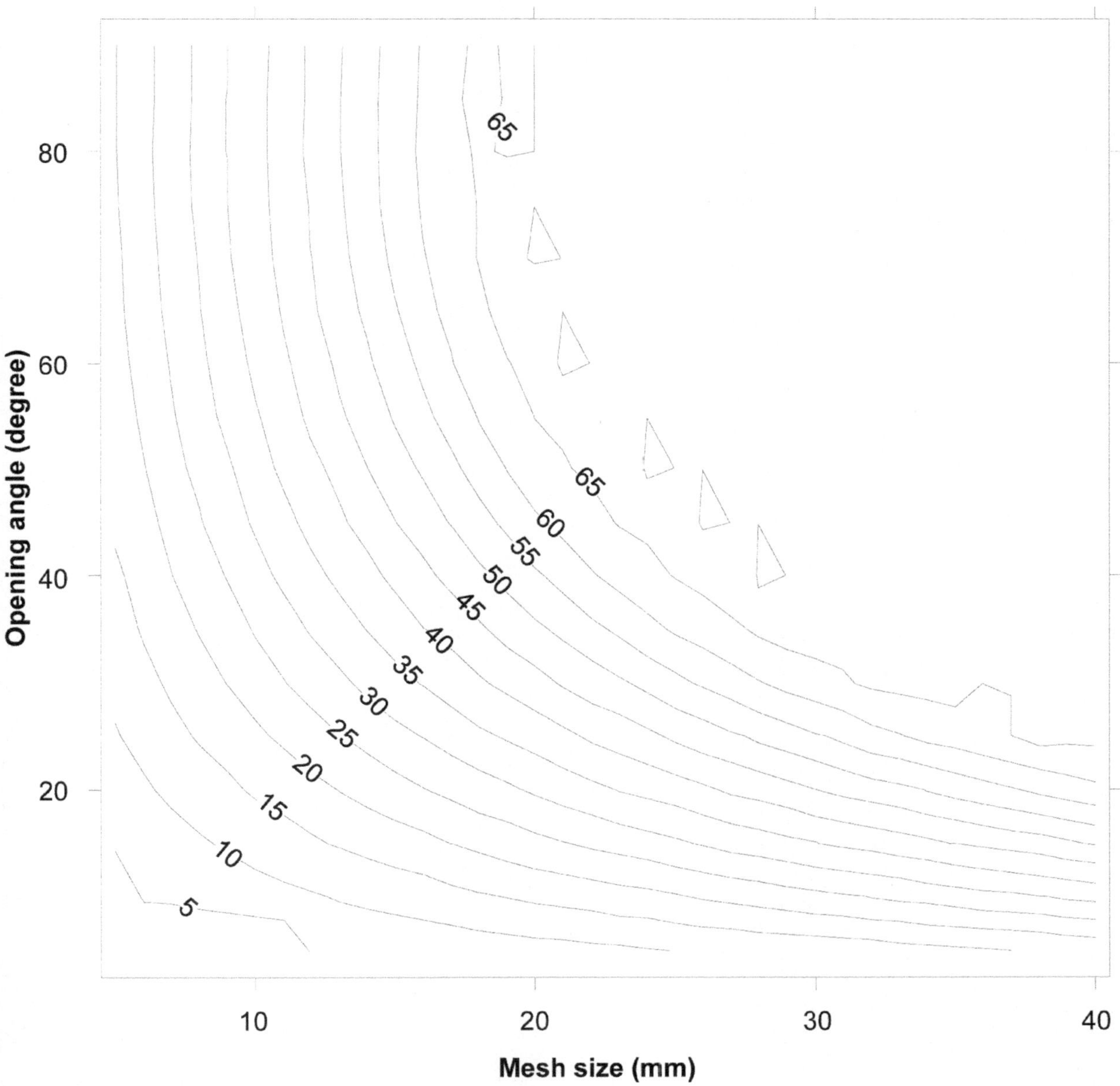


**Figure 6. Diamond mesh design guide for Krill, based in combination of CS1 and CS2.** The plot gives iso-L50 curves as a function of mesh size (mm) and mesh opening angle (oa).

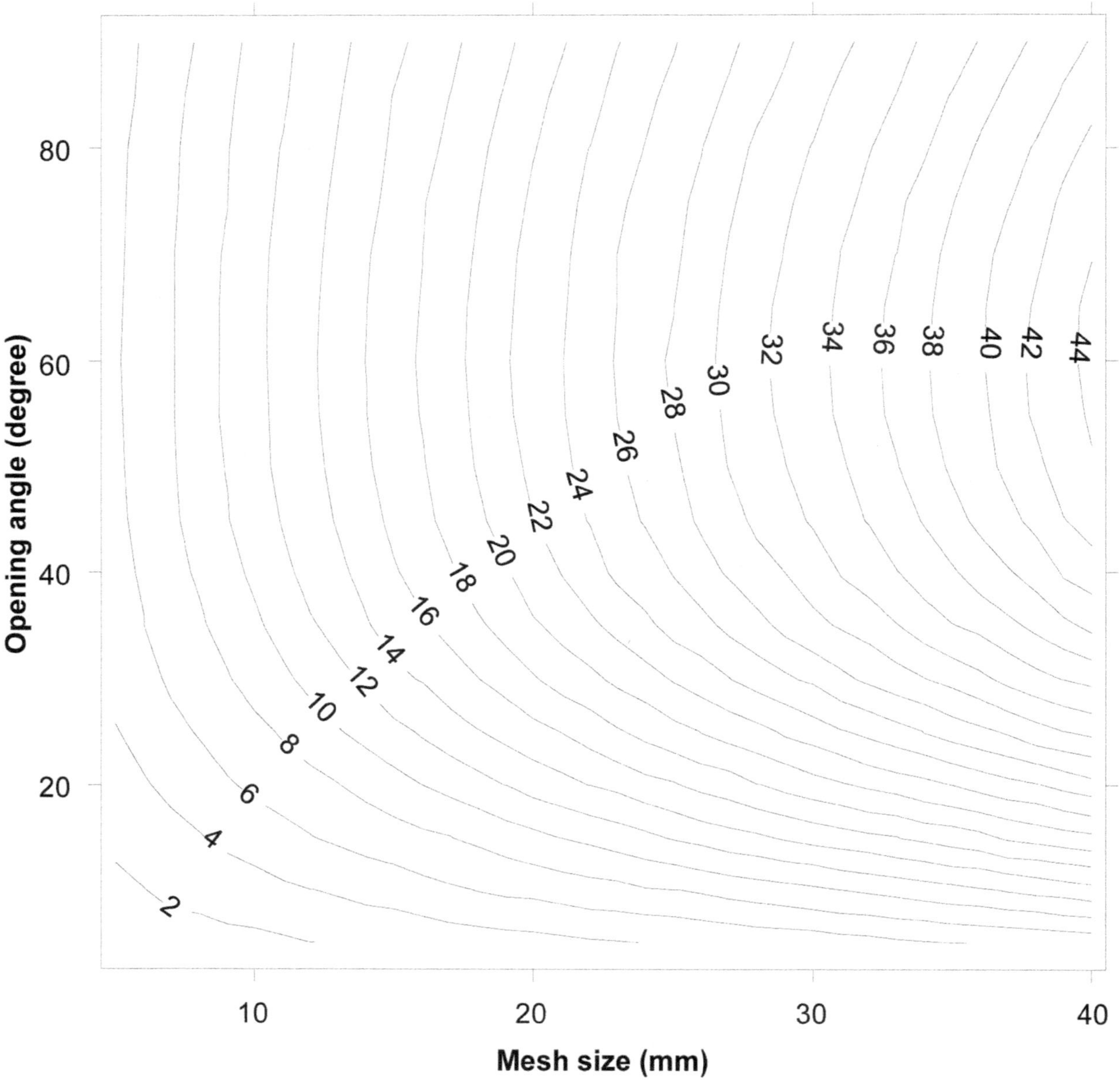


**Figure 7. Diamond mesh design guide for CS3.** The plot gives iso-L50 curves as a function of mesh size (mm) and mesh opening angle (oa).

assuming contributions of different mesh oa values. This kind of netting contact corresponds to what has been assumed when using FISHSELECT to investigate size selection for different species of fish. Using the observed oa range derived from underwater video recordings during commercial fishing, we explored, by combining the contribution of meshes with oa values of a slightly wider range than observed, the possibility to obtain a size selection curve similar to the experimentally obtained selectivity curve for the commercial netting. For this, we used simulated data for the specific commercial mesh size with oa values of 15, 20, 25, 30, 35, 40, 45 and 50° using the method described in Herrmann et al. [13]. We divided the L05 to L95 values from the experimental data into 5% steps and investigated whether it, by combining the same oa-value as observed on underwater recordings, was possible by simulation to obtain a similar size selection curve as experimentally observed. Finally, we used the estimated distribution of oa values together with the FISHSELECT results for other mesh sizes to predict the size selectivity of krill for trawls with these different mesh sizes following the procedure described in [13].

## Results

### Mesh size and shape description

The shape of the 15.4 mm meshes appeared relative stable during commercial fishing based on underwater observation (Fig. 4, Table 2). Pictures from the underwater video were captured and five individual meshes were identified and digitized (Table 2). Table 2 also provides the model fit statistics for a diamond mesh description of the commercial netting. Based on its high $R^2$ value, the meshes subsequently were described using the diamond mesh description. The five meshes listed in Table 2 covered the oa range in the captured picture frames, and the most open and closed meshes were selected for analysis. Table 2 indicates that 25–45° is the realistic oa range during commercial

**Table 6.** Values for the length of krill (L5-95) with different fixed retention likelihoods based on selectivity data from experimental fishing.

| L | Value (mm) |
|---|---|
| 5 | 26.23 (23.80–28.54) |
| 10 | 27.87 (25.80–29.83) |
| 15 | 28.90 (27.04–30.59) |
| 20 | 29.66 (27.92–31.24) |
| 25 | 30.30 (28.69–31.79) |
| 30 | 30.85 (29.33–32.27) |
| 35 | 31.36 (29.84–32.74) |
| 40 | 31.83 (30.27–33.19) |
| 45 | 32.28 (30.71–33.68) |
| 50 | 32.72 (31.09–34.12) |
| 55 | 33.16 (31.53–34.58) |
| 60 | 33.62 (31.92–35.10) |
| 65 | 34.09 (32.32–35.64) |
| 70 | 34.59 (32.73–36.28) |
| 75 | 35.15 (33.20–36.93) |
| 80 | 35.78 (33.66–37.78) |
| 85 | 36.55 (34.13–38.75) |
| 90 | 37.57 (34.96–40.03) |
| 95 | 39.22 (36.19–42.21) |

95% confidence limits in indicated in brackets.

fishing for krill in the tapered part of the trawl design used in this study.

## Use of preserved krill for morphological measurements

No significant difference in morphology was found between fresh krill collected from the trawl catch and those stored in borax-buffered 4% formalin for 10 months (total body length: $F = 0.1145$, $P = 0.9$; carapace width: $F = 0.1266$, $P = 0.9$). The minor differences in these parameters (Table 3) likely are due to slight differences in the locations at which measurements were taken and in interpretation of the caliper readings.

**Table 7.** Selectivity estimates for the commercial 15.4 mm diamond mesh trawl including fit-statistics based on experimental fishing.

| | |
|---|---|
| L50 (mm) | 32.72 (±1.74) |
| SR (mm) | 4.85 (±2.10) |
| SP | 0.55 (±0.04) |
| p-value | 0.26 |
| Deviance | 36.80 |
| DF | 32 |
| Total number in test | 393 |
| Total number in control | 416 |

95% confidence limits in indicated in brackets.

## Morphological measures and penetration models

The three measured cross sections had different cross section shapes. CS1 was best described by a flexelipse1, CS2 was best described by a flexelipse3, and CS3 was best described by a flexdrope2 (Table 4, Appendix S1). These models were chosen for use in the subsequent analyses because they displayed the highest $R^2$ value and the lowest AIC value. Figure 2 shows both the actual and the modeled cross section shapes and that they were in agreement. The model description of CS2 (lower part) did not include parts of the legs that were present in the cutting zone (Fig. 2). Krill legs are expected to have little effect on the optimal orientation, as they should fold up against the ventral side of the animal during mesh penetration with optimal orientation (head or tail first). Table 5 shows the length-based regression parameters for body length versus width and height in the cross sections.

## Comparison of FISHSELECT-based and experimental selectivity estimates

The design guide based on CS1_CS2 was used to predict the basic selective properties for krill for all relevant sizes of diamond mesh when the individuals meet the meshes at the optimal orientation (Fig. 6). The design guide covers the mesh size range from 5 to 40 mm and the opening angle range from 10 to 90°. The size selectivity of krill depends greatly on the mesh opening angles. This is especially true for meshes with oa values ranging from 10–45° where meshes with larger oa values have less effect on the predicted L50. The predicted L50 for a given mesh size is however increasing towards a mesh opening angle of 90° (which equals a square mesh). The design guide further indicates that even the small meshes used in some survey trawls (~7 mm) can be selective if the meshes are sufficiently open; if true, such surveys may underestimate the density of juvenile krill.

Compared to the contact mode CS1_CS2, little escapement occurs for the larger CS3 cross section, even for rather large meshes (Fig. 7). When the meshes of the commercially used trawl (15.4 mm) are open optimally, krill smaller than 18 mm length can potentially escape (Fig. 7). The experimentally obtained results show that L5 (5% retention likelihood) is above 26 mm (Table 6), which means that this type of contact plays no role in defining the size selection for krill in this type of gear. In reality, we retained only 50% of the individuals with a body length of 33 mm (L50) (Table 6). The experimental L50 of krill with the 15.4 mm commercial trawl was estimated to be 32.72 mm with an SR (L75–L25) value of 4.85 mm (Table 7). The selectivity curve shown in Figure 8 demonstrates that size selectivity occurs for individuals smaller than 40 mm in the commercial trawl. Based on the fit statistics in Table 7 it is demonstrated that the applied model (2) in Fig. 8 is able to describe the experimental data sufficiently (p-value>0.05). The experimental selectivity results also show that fewer than 5% of the krill smaller than 26 mm length that enters the trawl will be retained (Table 6). Based on this result, any contribution to the size selectivity of krill from contact modes with L50 less than 26 mm will be very limited. Furthermore, the majority of individuals seem to be able to meet the meshes with a far more optimal body orientation (CS1_C2), at least for their decisive (last) contact with the netting. This is clarified in Figure 9, in which the potential selectivity based on CS1_CS2 and CS3 is compared to the observed selective range in the commercial netting (15.4 mm), the observed oa range during commercial fishing, and the experimental L50 value. Selectivity based on CS3 does not reach the experimentally observed selective range; in contrast, the estimated selectivity based on CS1_CS2 has reasonable overlap with both the expected experimental oa range and the selective range for the 15.4 mm mesh size (Fig. 9).

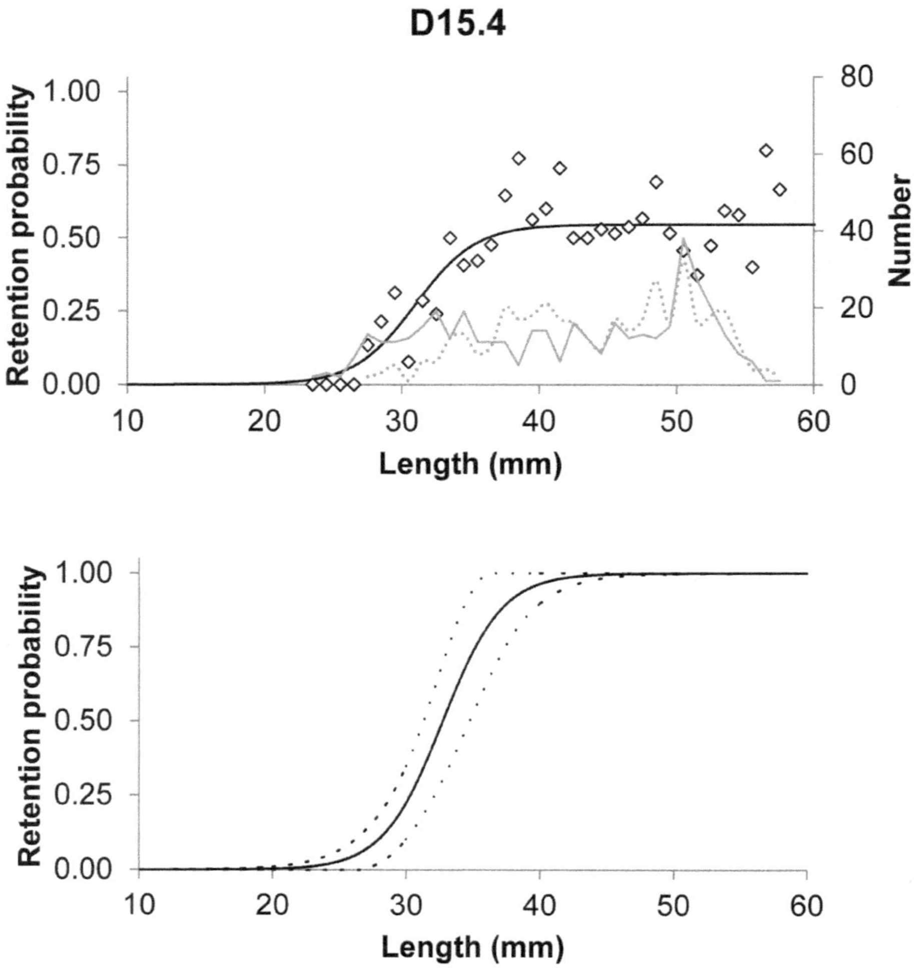


**Figure 8. Top plot show paired gear fit of the experimental data.** The population structure of the measured individuals are indicated (solid line = 7 mm survey trawl (*Macroplankton trawl*); broken line 15.4 mm commercial trawl). The lower plot is the size selection curve including 95% confidence limits.

However, Figure 9 shows that results could be slightly biased towards smaller oa values. Such difference may result from effect of attack angle with netting or none optimal rotation for krill during contact with the netting. This potential effect is investigated in detail in the next section.

## Effect of cross section orientation and attack angle

When examining the effect of cross section orientation and attack angle, we used only CS1_CS2, as CS3 was found to have no effect on the size selection of krill in the current trawl design. Figure 10 shows the potential effect of orientation of CS1_CS2 in the mesh for the relevant oa range rotated from a dorso-ventral orientation (see CS1 and CS2 in Fig. 2) (i.e., 0 to 90°). For the rotation range from 0 to 40°, the L50 value was nearly constant, indicating little effect of orientation over this range; it was only about 15–20% smaller than the maximal value obtained at 90° (Fig. 10). This relatively limited effect of rotation is also visible in Figure 5, but shows that individuals rotated 70–90° contrary to the other orientations will be retained by the mesh. Overall, the effect of cross section orientation in the mesh opening is relatively weak for a large range of rotation angles. This is due to the cross section shape of CS1_CS2, which is reasonably round shaped.

A more dramatic effect on the estimated L50 value was predicted for low attack angles (0–30°) but little effect was detected for large attack angles (Fig. 11). The very low tapering present in the commercial trawls targeting krill results in attack angles <5°. If the angle of attack had an important effect on the size selection of krill in the trawl designs tested, we would expect L50 values of around 10 mm for the experimental results, which was not is the case (Fig. 8).

The underwater recordings of escaping krill seem to indicate that they escape at the optimal attack angle perpendicular to the netting and head first (Fig. 12). This result demonstrates that the optimal FISHSELECT mode (i.e., CS1_CS2 with optimal orientation and attack angle) is a good approximation of the escape process. We therefore based our predictions of size selectivity of krill in trawls with other mesh sizes on this approximation and used only the optimal orientation (CS1_CS2).

## Simulation-based predictions and comparison to sea trail results

In Figure 13, the experimentally obtained selectivity results are indicated by the black curve, and the dashed lines show the 95% confidence limits. The thick grey curve shows the optimal FISHSELECT-based predictions for CS1_CS2 at optimal orien-

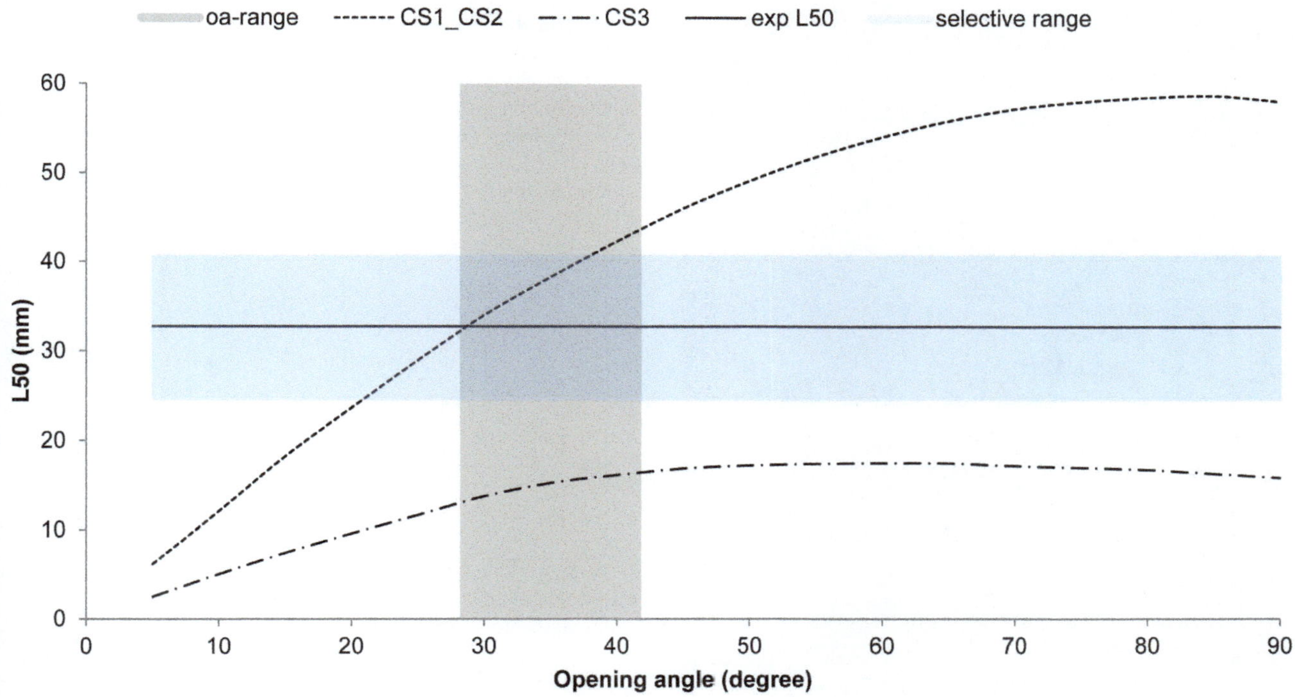


**Figure 9. The experimentally obtained L50 value is indicated with the solid line (exp L50).** L50 predictions for CS1_CS2 and for CS3 are indicated for the commercial mesh size (15.4 mm). The realistic mesh opening angles during commercial fishing is indicated with the vertical gray interval. The horizontal interval indicated the selective range for the 15.4 mm commercial mesh size.

tation and attack angle using the relative contributions of oa values according to Table 8. The similarity of the two curves indicates that it is possible to obtain a modeled size selection curve that is very similar to the one obtained experimentally by using realistic oa values. Table 8 also shows that this curve is reproduced nearly exclusively by contributions of meshes with oa values of 25, 30, and 35°, with contributions of 39.9, 45.33, and 14.74%, respectively. These results also indicate that less open meshes are more common than what would be expected based on the underwater recordings. This might be due to the effect of non-optimal rotation and/or the effect of attack angle. However, the size selection for krill seems to be well approximated by the FISHSELECT optimal mode, as was previously found to be the case for a number of fish species. Thus, it makes sense to make predictions based on the FISHSELECT optimal mode. This premise is validated by the similarity between the experimental and predicted selectivity curves shown in Figure 13. Figure 14 and Table 9 shows the predictions of size selectivity for krill using the oa distribution in Table 8 for the optimal orientation of CS1_CS2 for the range of mesh sizes from 6 to 28 mm. This figure shows the size selectivity consequences of using different mesh sizes in the krill trawl fishery, and it is valid under the assumption that trawls with these mesh sizes have a similar distribution of oas during fishing.

## Discussion

We identified, measured, and parameterized the morphology that determines size selectivity of krill in towed gears by applying

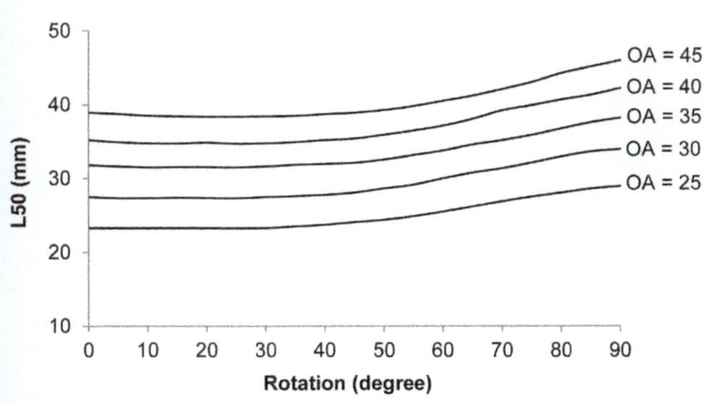


**Figure 10. The effect of krill encountering meshes with different orientation.** The penetration model (CS1_CS2) is rotated, at optimal attack angle perpendicular to the netting from 0–90°. 0° is a dorso ventral orientation equal to normal swimming orientation.

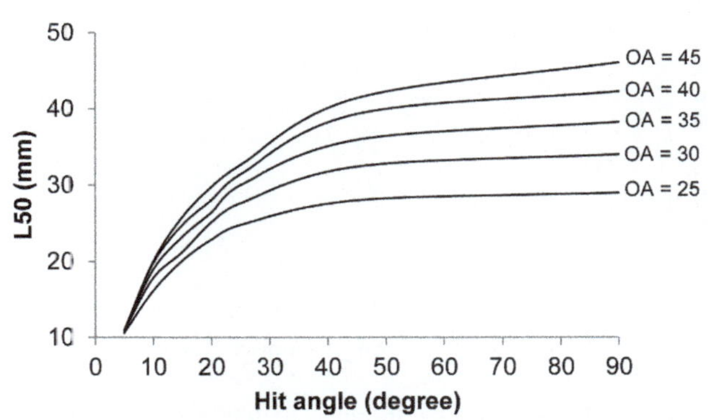


**Figure 11. The effect of varying attack angles in the penetration model for CS1-CS2 in the range of oa-values that were found relevant in the 15.4 mm trawl.**

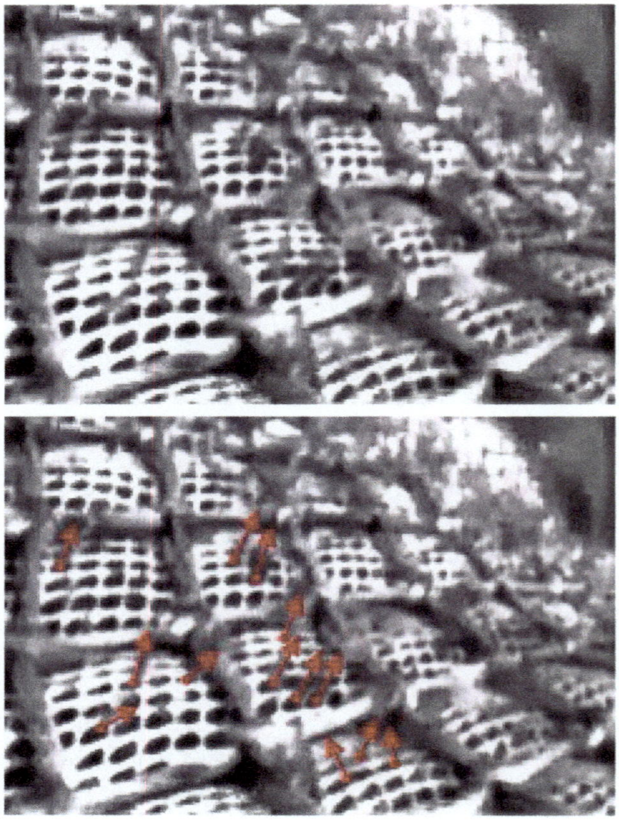

**Figure 12. Underwater images captured during fishing indicating escaping krill in the 15.4 mm mesh size.** The escapees are marked with red arrows in the lower photo and clearly demonstrate an optimal orientation of the krill escaping. The photo is taken 10 meters in front of the cod line during commercial fishing.

the FISHSELECT method with the modifications necessary for analyzing krill. Using the combination of this morphological description, realistic oa values for the meshes based on underwater observations made during commercial fishing operations, and experimental selectivity data, we were able to predict the selectivity for krill for different mesh sizes. This is the first time this has been attempted for krill, and it will be a useful tool for predicting the size selectivity of existing netting configurations and optimizing the size selectivity of future trawls designs based on management specifications in relation to the size selection. Such a predictive tool is especially valuable in expanding fisheries, such as the Antarctic trawl fishery for which few experimental data are available.

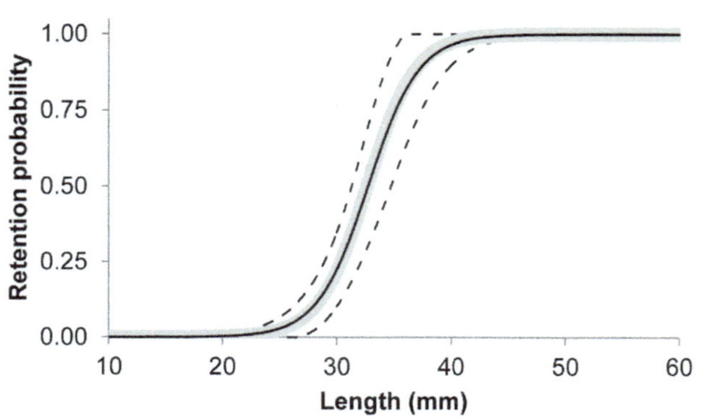


**Figure 13. Experimentally obtained data (black line) with 95% confidence limits (broken line).** Thick line (gray) is the predicted selectivity curve based on morphological based measurements (FISHSELECT) and the distribution of opening angle (oa)-values given in table 8.

**Table 8.** Simulated distribution of opening angles (oa) grouped in 5 cm. intervals that will result in identical selectivity curves for krill between simulated and experimental data.

| oa (degree) | Contribution (0–100%) |
|---|---|
| 15 | 0.00 |
| 20 | 0.01 |
| 25 | 39.90 |
| 30 | 45.33 |
| 35 | 14.74 |
| 40 | 0.01 |
| 45 | 0.01 |
| 50 | 0.00 |

We expected to find more random selectivity for krill than that generally observed for fish in trawls due to the relatively high towing speed used to fish for these small individuals. The tapering in commercial krill trawls is very low compared to that of traditional fish trawls, which results in a low attack angle relative to the flow direction during towing. Theoretically, this could dramatically reduce the estimated L50 value. We found that the selectivity of krill in the commercial trawl can be explained by assuming that individuals of all lengths meet the meshes in the optimal contact mode (CS1_CS2) at a more or less optimal attack angle. Underwater recordings made during commercial trawling show that krill escape the trawl head first and relatively perpendicular to the netting wall. This suggests that individual krill are capable of orientating themselves in relation to the trawl netting and meeting the meshes at an optimal orientation and attack angle. An alternative explanation for the observed size selectivity of krill is that selection is a more random process but the size of the commercial trawl (about 200 m long) provides so many contacts with the netting during passage to the codend that the

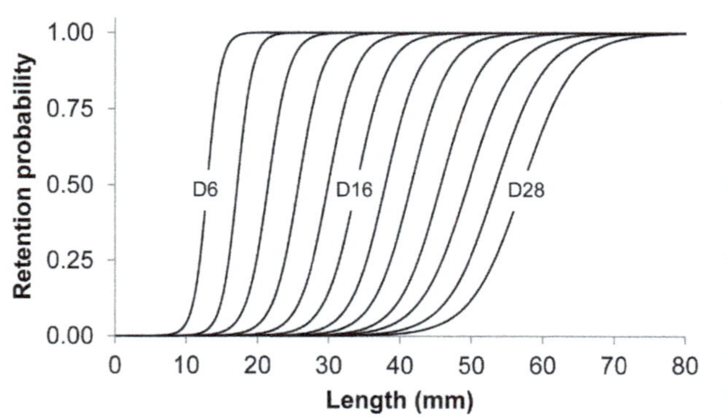


**Figure 14. Predicted selectivity of krill in different mesh sizes based on the weight factors of the different opening angle (oa)-values for optimal CS1_CS2.** Predictions are made from 6 to 28 mm meshes in steps of 2 mm assuming a similar distribution of oa.

**Table 9.** Predicted selectivity parameters (L50 and SR) for diamond mesh size ranging from 6 mm to 28 mm.

| Mesh size (mm) | L50 (mm) | SR (mm) |
|---|---|---|
| 6 | 12.96 | 2.06 |
| 8 | 17.17 | 2.39 |
| 10 | 21.48 | 3.25 |
| 12 | 25.72 | 3.81 |
| 14 | 29.93 | 4.34 |
| 16 | 33.91 | 4.83 |
| 18 | 37.93 | 5.24 |
| 20 | 41.55 | 5.77 |
| 22 | 45.68 | 6.35 |
| 24 | 49.49 | 7.22 |
| 26 | 53.44 | 7.56 |
| 28 | 57.20 | 7.97 |

catch loses the signature of a random process, as most individuals meet the mesh opening optimally at some point during transport through the trawl. Further studies are needed to determine which of these two processes is actually occurring.

We estimated an L50 of 32.72 mm, which indicates that there is substantial size selectivity in the fisheries that use the 15.4 mm mesh size. The SR value for krill is small compared to the experimentally observed SR values for other crustaceans such as *Nephrops* [14]. Possible explanations for the difference are that krill do not have claws and that the length of *Nephrops* trawls typically are much shorter than e.g. krill trawl which reduces the number of netting contacts during the catching process. The observed stability of the mesh opening (oa values) in the krill trawl during fishing may also result in a lower SR value.

Because a relatively large proportion of the length classes of krill (24–42 mm) potentially can escape through the commonly used mesh size, it is important to estimate the survival of escapees in such fishing gears. Siegel [5] estimated mortality rates of 5–25% of krill individuals escaping through the trawl netting. This estimate was based on the assumption that the mortality rate of the individuals passing through the net meshes equals the rate of lethally damaged individuals observed in the codend of the commercial trawl. If this is correct, the total mortality caused by the commercial fishery might be considerably higher than catch values that are reported to the CCAMLR. However, several substantially different trawl designs, using different mesh sizes, are used in the commercial krill fishery. The potential mortality and survival rate of escapees likely depend on the different gear designs used. If the survival rate of escapees in krill trawls is low and differs between trawl designs, it is important to apply gears that are the most sustainable.

## Acknowledgments

Thanks are given to AKER BioMarine ASA and Olympic ASA for providing vessels and crew free of charge for collection of technical and biological data. Further thanks are given to AKER BioMarine ASA for providing the underwater video images used in this study. Thanks are also given to Helle Andersen (DTU Aqua) who helped us digitize the scanned images of krill taken during the laboratory trials.

## Author Contributions

Conceived and designed the experiments: LAK BH BAK. Performed the experiments: LAK BAK BH. Analyzed the data: LAK BH BAK. Contributed reagents/materials/analysis tools: LAK BH BAK SAI SN AE. Wrote the paper: LAK BH BAK.

## References


1. Nicol S, Endo Y (1999) Krill fisheries – their development, management and ecosystem implications. Aquatic Living Resources 12: 105–120.
2. Hewitt RP, Watkins J, Naganobu M, Sushin V, Brierly AS, et al. (2004) Biomass of Antarctic krill in the Scotia Sea in January/February 2000 and its use in revising an estimate of precautionary yield. Deep-Sea Research Part II–Topical Studies in Oceanography 51: 1215–1236.
3. SC-CAMLR (2010) Report of the twenty-ninth meeting of the scientific committee (SC-CAMLR-XXVI). CCAMLR, Hobart, http://www.ccamlr.org/pu/e/e_pubs/sr/10/toc. htm
4. Pshenichnov L (2009) On incidental mortality of Antarctic krill at krill fishery. CCAMLR WG-EMM-09/11. 6 pp.
5. Siegel V (2007) Size selectivity of the RMT8 plankton net and a commercial trawl for Antarctic krill. CCAMLR WG-EMM-07/28. 9pp.
6. Atkinson A, Siegel V, Pakhomov EA, Rothery P, Loeb V, et al. (2008) Oceanic circumpolar habitats of Antarctic krill. Marine Ecology Progress Series 362: 1–23.
7. CCAMLR (2009a) Report of the twenty-eighth meeting of the Commission. Hobart Australia 26. October-6. November 2009. 201 pp.
8. CCAMLR (2009b) Report of the twenty-eighth meeting of the Scientific Committee. Hobart Australia 20–30. October 2009. 584 pp.
9. Herrmann B, Krag LA, Frandsen RP, Madsen N, Lundgren B, et al. (2009) Prediction of selectivity from morphological conditions: Methodology and a case study on cod (*Gadus morhua*). Fisheries Research 97: 59–71.
10. Krag LA, Herrmann B, Madsen N, Frandsen RP (2011) Size selection of haddock (*Melanogrammus aeglefinus*) in square mesh codends: A study based on assessment of decisive morphology for mesh penetration. Fisheries Research 110: 225–235.
11. Sistiaga M, Herrmann B, Nielsen KN, Larsen RB (2011) Understanding limits to cod and haddock separation using size selectivity in a multispecies trawl fishery: an application of FISHSELECT. Canadian Journal Fisheries and Aquatic Sciences 68: 927–940.
12. Herrmann B, Sistiaga M, Nielsen KN, Larsen RB (2012) Understanding the size selectivity of redfish (*Sebastes* spp.) in North Atlantic trawl codends. Journal of Northwest Atlantic Fishery Science 44: 1–13.
13. Herrmann B, Sistiaga M, Larsen RB, Nielsen KN, Grimaldo E (2013) Understanding sorting grid and codend size selectivity of Greenland halibut (*Reinhardtius hippoglossoides*). Fisheries Research 146: 59–73.
14. Frandsen RP, Herrmann B, Madsen N (2010) A simulation-based attempt to quantify the morphological component of size selection of *Nephrops norvegicus* in trawl codends. Fisheries Research 101: 156–167.
15. Wardle CS (1993) Fish behaviour and fishing gear. In: Behaviour of teleost fishes, 2nd edition. Pitcher TJ (Ed). Chapman and Hall, London. pp 609–644.
16. Lochhead JH (1961) Locomotion. In: Waterman T.H. (Ed), The physiology of the Crustacea. Academic Press, New York, pp 313–356.
17. Newland PL, Chapman CJ (1989) The swimming and orientation behaviour of the Norway lobster, *Nephrops norvegicus* (L), in relation to trawling. Fisheries Research 8, 63–80.
18. Krafft BA, Melle W, Knutsen T, Bagøien E, Broms C, et al. (2010) Distribution and demography of Antarctic krill in the Southeast Atlantic sector of the Southern Ocean during the austral summer 2008. Polar Biology 33: 957–968.
19. Makorov RR, Denys CJI (1981) Stages of sexual maturity of Euphausia *superba*. BIOMASS Handbook No. 11, pp 1–13.
20. Marr J (1962) The natural history and geography of the Antarctic krill *Euphausia superba*. Discov Rep 32: 33–464.
21. Akaike H (1973) Information theory and an extension of the maximum likelihood principle. Second International Symposium on Information Theory, (BN Petrov and F. Csaki, Eds) Budapest. Akademia Kaido, 267–281.
22. Fonteyne R (2005) Protocol for the use of an objective mesh gauge for scientific purposes. ICES Cooperative Research Report No. 279, 8 pp. ISBN 87-7482-044-3.
23. Wileman DA, Ferro RST, Fonteyne R, Millar RB (Eds) (1996) Manual of methods of measuring the selectivity of towed fishing gears. ICES Cooperative Research Report No. 215.

# 5

# Experimental Parameterisation of Principal Physics in Buoyancy Variations of Marine Teleost Eggs

**Kyung-Mi Jung[1,2*¤], Arild Folkvord[1,2], Olav Sigurd Kjesbu[1,3], Svein Sundby[1]**

**1** Institute of Marine Research (IMR) and Hjort Centre for Marine Ecosystem Dynamics, Bergen, Norway, **2** Department of Biology, University of Bergen, Bergen, Norway, **3** Centre for Ecological and Evolutionary Synthesis (CEES), Department of Biology, University of Oslo, Oslo, Norway

## Abstract

It is generally accepted that the high buoyancy of pelagic marine eggs is due to substantial influx of water across the cell membrane just before ovulation. Here we further develop the theoretical basis by applying laboratory observations of the various components of the fertilized egg in first-principle equations for egg specific gravity ($\rho_{egg}$) followed by statistical validation. We selected Atlantic cod as a model animal due to the affluent amount of literature on this species, but also undertook additional dedicated experimental works. We found that specific gravity of yolk plus embryo is central in influencing $\rho_{egg}$ and thereby the buoyancy. However, our established framework documents the effect on $\rho_{egg}$ of the initial deposition of the heavy chorion material in the gonad prior to spawning. Thereafter, we describe the temporal changes in $\rho_{egg}$ during incubation: Generally, the eggs showed a slight rise in $\rho_{egg}$ from fertilization to mid-gastrulation followed by a gradual decrease until full development of main embryonic organs just before hatching. Ontogenetic changes in $\rho_{egg}$ were significantly associated with volume and mass changes of yolk plus embryo. The initial $\rho_{egg}$ at fertilization appeared significantly influenced by the chorion volume fraction which is determined by the combination of the final chorion volume of the oocyte and of the degree of swelling (hydrolyzation) prior to spawning. The outlined principles and algorithms are universal in nature and should therefore be applicable to fish eggs in general.

**Editor:** Brian R. MacKenzie, Technical University of Denmark, Denmark

**Funding:** This study was funded by the project "Population differentiation of Atlantic cod" under the Programme on Biological mechanisms and by the Programme on Climate and Fish at the Institute of Marine Research in Norway. The funders had no role in study design, data collection and analysis, decision to publish, or preparation of the manuscript.

**Competing Interests:** The authors have declared that no competing interests exist.

* Email: kyungmi-toy@hanmail.net

¤ Current address: Fisheries Resources Management Division, National Fisheries Research and Development Institute, Busan, Republic of Korea

## Introduction

Buoyancy of fish eggs has been studied in a wealth of articles over the last decades [1], although the recent focus has shifted from actual measurements to understanding the molecular and physiological basis, in particular the active role of aquaporins during oocytic water uptake [2,3]. In line with earlier studies [4,5,6], Govoni & Forward [1] presents in their review a formulation of Archimedes principles where egg buoyancy is set as a function of the density of ambient water and of the various egg components. Although this principle is well adopted in science, the terminology and thereby the construction of this commonly seen formula deviates from physics of today, e.g. 'force' is defined in grams. Modern principal equations were presented in Kjesbu *et al.* [7]. However, as only data on eggs in the gastrulation stage were presented, their study does not address explanatory factors behind the noticed change in egg buoyancy from fertilization to hatching, i.e. 'incubation' (see Jung *et al.* [8,9] and references therein), the present topic of interest using Atlantic cod (*Gadus morhua*) as an object of study.

The introduction of the density gradient column by Coombs [10] for measuring specific gravity in fish eggs accelerated interest in buoyancy studies as it enabled recordings on live individual eggs at a precision and an accuracy far beyond any earlier methods. This high accuracy was a needed basis to model the relation between egg buoyancy and vertical distribution in the field [11], both for pelagic and mesopelagic fish eggs [12]. The first observations on specific gravity were focussing on differences across individual eggs (e.g. Solemdal & Sundby [13]) without taking into account possible changes in egg specific gravity throughout incubation. These individual differences were included in the model on vertical distribution of eggs by embedding a Gaussian distribution function on egg buoyancy balanced by the turbulent mixing [11]. Later it became evident that individual egg specific gravity also could vary systematically throughout incubation as exemplified on blue whiting eggs off the British Isles [14], Cape hake eggs off Namibia [15], and Atlantic mackerel and horse mackerel eggs off the British Isles [16] although these temporal changes in specific gravity was exceeded by the individual egg differences. The general feature of temporal change in the specific gravity throughout incubation, as demonstrated by Sundby *et al.* [15], was an initial increase after fertilization, and thereafter a decrease towards hatching. This development impacts the way the Cape hake eggs ascend from the spawning depth towards upper layers. In order to model with high accuracy how the vertical distribution of pelagic eggs is modified through incubation by the changes in the egg buoyancy in the presence of ambient turbulence, we need to both understand how the eggs are changing and how much. The present understanding of these changes throughout incubation, at least at the conceptual level [5,7], is incomplete to explain systematic trends in buoyancy as noticed in the laboratory during the period of egg incubation [8,17–19]. Craik & Harvey [4] showed that changes in water content of the egg influences the buoyancy, and such changes together with changes in composition of fat and proteins and changes through respiration could all act as factors to modify egg buoyancy onto the hatching (see Govoni & Forward [1] and references therein).

In the establishment of a true $\rho_{egg}$ model, first-principle equations should be applied and these algorithms should reflect the fact that a fertilized fish egg typically consists of three parts: chorion, perivitelline space (PVS) and yolk plus embryo (Appendix: Eq. 1). Thus, the total egg mass is the sum of masses of these components, which can be expressed by their separate volumes (V's) and specific gravities ($\rho$'s):

$$\rho_{egg} = \left(\rho_{cho}V_{cho} + \rho_{pvs}V_{pvs} + \rho_{yolk+emb}V_{yolk+emb}\right)/V_{egg} \quad \text{(Appendix : Eq.2)}$$

Here the contribution of the chorion requires some special attention. This structure consists of proteins [20], which harden following egg fertilization [21]. While the chorion volume ($V_{cho}$) may differ both among [22] and within individuals [7], the specific gravity of the wet chorionic material ($\rho_{cho}$) is roughly constant, even between different species, but appears to be exceedingly high: e.g. 1.20 g $cm^{-3}$ for Atlantic cod [7] and 1.18 g $cm^{-3}$ for Atlantic halibut (*Hippoglossus hippoglossus*) [23]. In general, egg buoyancy (i.e. $\Delta\rho$; $\Delta\rho = \rho_{water} - \rho_{egg}$) in pelagic fish eggs is less than 0.003 g $cm^{-3}$ [24], while $\rho_{water}$ ranges from about 1.010 (brackish) to 1.028 g $cm^{-3}$ (marine). As the chorion is permeable to the ambient seawater it is assumed that the specific gravity of the perivitelline space ($\rho_{pvs}$) is identical to $\rho_{water}$. Evidently, $\rho_{yolk+emb}$ varies intraspecifically; the reported values are 1.017 g $cm^{-3}$ and 1.008 g $cm^{-3}$, respectively, for marine and brackish cod eggs [7]. In the latter case identical figures were found in another, independent study: 1.007–1.009 g $cm^{-3}$ [25]. Thereby, the low $\rho_{yolk+emb}$ counteracts the high $\rho_{cho}$ to create the necessary hydrostatic lift of the egg. Both egg diameter and thereby egg volume ($V_{egg}$) and chorion thickness and chorion volume ($V_{cho}$) are considered constant throughout embryonic development, i.e., after the aforementioned initial short period of swelling and eggshell hardening that settles the volume of the egg. Therefore, also the sum of the volumes of the perivitelline space and the yolk plus embryo must be constant, i.e. $V_{pvs}+V_{yolk+emb}$ = constant, and hence changes in $V_{yolk+emb}$ during development must result in compensating changes in $V_{pvs}$ according to Eq. 3 (Appendix). Consequently, variability in $\rho_{egg}$ during incubation, i.e. from fertilization to hatching, must be caused by changes in $\rho_{yolk+emb}$. In regards to egg metabolism and physiological development, these changes in $\rho_{yolk+emb}$ can be caused by two factors: (1) changes in small molecules including ions (free amino acids, $NH_4^+$, $Cl^-$, $K^+$ and $Na^+$) and (2) changes in macromolecules (carbohydrates, proteins and lipids) [5].

During the work we encountered two hurdles: i) lack of sufficiently detailed algorithms of basic relationships and ii) missing or incomplete observations throughout incubation on key input variables for these equations. Hence we develop both the theoretical side (cf. Appendix) and a data collation programme undertaken under controlled conditions in the laboratory to provide values for the equations. The experimental study was performed to explore what *de facto* determines $\rho_{egg}$, both at fertilization and during incubation, by including associated measurements of a series of relevant variables (cf. Appendix). These formulations show that $V_{yolk+emb}$, $\rho_{yolk+emb}$ and $V_{cho}$ play important roles in defining $\rho_{egg}$. We hypothesize that (1) initial variation in $\rho_{egg}$ among individual newly spawned eggs from a single female, is due to differences in $V_{cho}$, and that (2) trends in $\rho_{egg}$ during embryonic development are caused by changes in $\rho_{yolk+emb}$ and $V_{yolk+emb}$. The value of $\rho_{yolk+emb}$ (Appendix: Eq. 4) was achieved by consulting the new measurements. Hence, the present series of theoretical and practical approaches confirmed the role of $\rho_{yolk+emb}$ in defining $\rho_{egg}$.

## Materials and Methods

### Ethics statement

All experiments followed the 'European Convention for the Protection of Vertebrate Animals used for Experimental and Other Scientific Purpose' (ETS123: http://conventions.coe.int/treaty/en/treaties/html/123.htm). Key experimenters were properly certified in laboratory animal science (FELASA category C researchers) (http://www.felasa.eu/accreditation-boards/accreditation-board-for-education-and-training1/). Permissions for collecting eggs were granted by the commercial hatchery of the Marine Harvest Cod (MHC) and the governmental hatchery at Austevoll Research Station, the Institute of Marine Research, Norway. Both hatcheries were properly approved and registered by Norwegian authorities (The Norwegian Directorate of Fisheries; http://www.fiskeridir.no) and followed strict veterinary and husbandry guidelines. Our study was restricted to the use of fertilized eggs only. An initial document search clarified that such early life stages are not classified as 'animals' (ETS123: Part1-General Principles, Article 1,2a). Hence, in the present case no specific permission was needed from the Norwegian Animal Research Authority.

### Biological material

Experiments were performed on eggs provided by the commercial hatchery Marine Harvest Cod (MHC) for one winter ($20^{th}$ Jan.–$16^{th}$ Feb. 2009) and spring season ($6^{th}$–$31^{st}$ May 2009) and by the hatchery at the Austevoll Research Station, Institute of Marine Research (IMR), for one fall season ($7^{th}$–$31^{st}$ Oct. 2009). Female and male cod were stocked together in single spawning tanks (winter: 45 females and 50 males; spring: 9 females and 20 males; fall: 27 females and 15 males) and induced to spawn naturally by artificial day light rhythms. MHC broodstock cod were originally captured near the entrance of Sognefjord (50 km north-east from Bergen), and female cod age and body weight were on average 9 years and 8–15 kg for the winter experiment and 12–13 years and 15–25 kg for the spring experiment. The cod at IMR Austevoll were born and raised in captivity from wild specimens captured in Øygarden (20 km west from Bergen), and were 5 years in the present fall experiment. Data on the female body weight was not available. Sea water of spawning tanks in the two hatcheries was maintained at temperatures of 6–7°C and salinities of 34–35.

Eggs were collected three times with an interval of one week during one month each season (Batch1-Batch9; Table S1), and then transported from the hatcheries to the IMR laboratory in Bergen. Immediately after arrival the eggs were gently poured into small flow-through aquaria set up in a refrigerated room at 6°C and incubated until hatching. Sea water was pumped from the neighbouring fjord (~100 m depth) and UV-filtered before entering the aquaria. Aquaria temperature varied between 7.6 and 8.3°C for winter, between 7.4 and 8.5°C for spring, and between 6.5 and 7.0°C for fall. Aquaria salinity was kept at 34–35 in the three seasons. The age of the egg was expressed in days post-fertilization (dpf) and specific description of egg developmental stages were according to Fridgeirsson [26]. Dead eggs sunken to the bottom were siphoned out every second day.

## Egg specific gravity

Egg specific gravity ($\rho_{egg}$) was determined using a density-gradient column (Martin Instruments Co. Ltd, U.K.) set up in the refrigerated room (6°C). All preparation of density gradient, calibration, accuracy, stability and operation have in general been described by Coombs [10] and specifically for the present work by Jung *et al.* [8]. UV-filtered sea water, distilled water, and NaCl were used to make the low and high salinity stock solutions, which is needed to develop the density gradient in the column. Each of three columns of 80 cm height was filled by continuously graded sea water solutions at 7°C, and the completed density gradients ranged from 1.0125 g $cm^{-3}$ at the surface of the column to 1.0325 g $cm^{-3}$ on the bottom. At each experimental trial, five glass floats of known and calibrated densities ($\rho_{float}$) were first introduced into the column from the top. The densities of the floats were originally reported at 23°C (or 18°C) with certified instruments in the Martin Instrument Co. (Three Crowns House, 160 Darlaston Road, Wednesbury WS10 7TA, UK). A density correction at 7°C was obtained according to the equation based on thermal expansion: $\rho_{float}$ at 7°C = $\rho_{float}$ at 23°C+(23−7)×$\rho_{float}$ at 23°C×0.000028. After 1 h the glass floats had positioned at their levels of neutral buoyancy in the column and could be used as reference levels of specific gravity. Prepared eggs were gently inserted from the top of the column and located at their neutral buoyancies. The vertical centre position of the glass floats were read with a precision of 1 mm and formed the basis of linear regressions between column position and specific gravity (always $r^2>0.99$). Due to the large number of eggs in the column and because the density gradient was set up to observe very small differences in egg specific gravity, we found it sufficiently accurate to count the eggs at every 10 mm vertical intervals. The eggs were first counted 30 min after the introduction into the column. The egg specific gravity, $\rho_{egg}$, was derived using the regression equation (specific gravity precision between 10-mm intervals: $\pm 2\times10^{-4}$ g $cm^{-3}$; specific gravity resolution between 10-mm intervals: $2.5\times10^{-4}$ g $cm^{-3}$; certified specific gravity precision of the calibrated floats: $\pm 2\times10^{-4}$ g $cm^{-3}$).

The measurements of $\rho_{egg}$ were carried out in two different ways; one was 'continuous measurements' over time and the other was 'time point measurements' at a given stage of egg development: (1) The purpose of the former measurements was to track daily changes in specific gravity of the same eggs. About 50 eggs at early developmental stages of 2–3 dpf were submerged in a single column and continuously monitored for two weeks until hatching. $\rho_{egg}$ was noted on a daily basis. Moribund and dead eggs in the column were counted every day for calculations of egg mortality rate and hatching rate from the initial number of eggs on the first day of the experiments. Moribund and dead eggs were apparent in that they were not transparent and with no sign of onward development from the previous day, or continuously sinking to the column bottom. Egg mortality rate was about 2–10% of the initial number of eggs. Hatching rate was about 2–8% at 12–13 dpf. (2) The latter measurements were aimed at studying co-variation of egg traits (see below) during development. About 100–300 eggs were introduced in another single column every second day. Subsets of these were recollected using a glass tube at assigned buoyancy layers (see below) for further scrutiny at different egg ages. Since the density gradient columns were newly made at each time point measurement occasion, the relatively large number of eggs did not affect the local oxygen conditions in the column. Due to the communal conditions of females and males in the same spawning tank, several egg batches were collected from different females on the same day. Generally, the individual differences in $\rho_{egg}$ are considerably larger than the observed temporal changes of one single egg throughout incubation. Therefore, in measuring the temporal changes in $\rho_{egg}$ we minimized the noise from the individual variations by selecting eggs at 2 dpf from a very narrow distribution (standard deviation: 0.0002 g $cm^{-3}$) in the columns. Thus, egg specific gravity layers of different egg batches at 2–3 dpf were carefully selected based on the narrow distributions in the column (see standard deviations measured in each specific gravity layer, Table S1). During development one specific gravity layer was selected from the upper part and the other was from the lower part of the whole distribution.

## Egg wet weight, dry weight, and water content

Images of eggs sampled from defined buoyancy layers were taken by a digital camera (Olympus, SZX10) equipped on a stereomicroscope at 10×. For the measurements of egg wet weight, about 50 eggs were sub-sampled from the photographed eggs, rinsed carefully in distilled water for 60 sec to remove sea water ions from the perivitelline space [17], dried on a paper (15–20 sec), placed equally in five pre-weighed Nunc cryotubes (i.e. 10 eggs in each cryotube), and weighed on Cahn 25 Automatic Electrobalance (accuracy: ±1 μg). Immediately, the five groups of 50 eggs were frozen at −80°C for dry weight measurements. A few of the Batch5 and Batch6 samples were lost after being frozen. Egg dry weight was determined by a freeze drier (Christ Alpha 1–4 Loc-1m) operating at −50°C for 24 hrs. Egg wet weights and dry weights were calculated by dividing by the relevant number of eggs in each cryotube. Fraction of water content was determined as the difference between wet weight and dry weight divided by the wet weight. Eggs were separately preserved in 4% formaldehyde for the measurement of chorion thickness.

## Chorion thickness

Three eggs representing each specific gravity layer were treated individually for this procedure. First, the diameter of an egg was measured under a fluorescent microscope (Nikon, AZ-100) at 25×, and placed between two microscope slides on a soft cutting-off plate. The two glasses were placed side by side with a distance of an egg diameter, preventing the egg from moving during cutting process. The egg was sectioned into two halves under a stereomicroscope by a scalpel, and placed in a water-filled petri dish. Each half was observed under a fluorescent microscope at 400× during which the background light intensity was set at a grey level of 150–160. Chorion thickness (t) and egg diameter (2r) were measured in ImageJ and used for the calculation of chorion volume ($V_{cho}$).

## Volume of yolk plus embryo and perivitelline space

With the captured egg images in ImageJ, individual yolk and embryo volume, $V_{yolk+emb}$, was determined by an equation for an ellipsoid which is approximately closest to its actual form. After mid-gastrulation (~4 dpf) the form of the embryo is starting to deviate from the ellipsoid shape. Hence, $V_{yolk+emb}$ was not measured after ~4 dpf. Perivitelline space volume ($V_{pvs}$) was calculated as: $V_{pvs} = V_{egg} - V_{cho} - V_{yolk+emb}$.

## Statistics

Statistical analyses were carried out using R version 2.10.0 and Statistica version 12 [27,28]. The instrumental set-up of the density gradient column with a relatively high number of eggs within small vertical distances in the column did not allow for tracking and retrieving individual eggs. The measured egg characteristics were thus averaged across layers in the following statistical analysis. As stated above, specific gravity resolution between 10-mm intervals was $2.5\times10^{-4}$ g cm$^{-3}$, while the certified specific gravity precisions of the calibrated floats are $\pm2\times10^{-4}$ g cm$^{-3}$ implying that recording of vertical level of eggs within 10-mm intervals was near the possible specified precision of the method.

Analysis of covariance (ANCOVA) was used to compare slopes of regressions between continuous and time point measurements for trends in $\rho_{egg}$ after the mid-gastrulation stage. Pearson correlation coefficient ($r$) was used to determine the strength of relationships between variables. In order to generate a predictive model of egg specific gravity we initially carried out a stepwise regression procedure with egg weight measures and different egg volume contributions ($V_{cho}$, $V_{pvs}$, and $V_{yolk+emb}$) as independent variables to determine which of these variables significantly contributed to the determination of observed specific gravity. Only cases with no missing values were used. Subsequently a re-analysis was carried out with extra data left out at the initial analysis and a final multiple regression analysis was carried out with significant independent variables.

## Results

### Trends in egg specific gravity during incubation

A similar pattern of temporal changes in egg specific gravity ($\rho_{egg}$) was observed among the three seasons (Figure 1A (winter), Figure 1B (spring), Figure 1C (fall) and Figures S1–S3). In general terms, fertilized eggs showed a slight increase in $\rho_{egg}$ until mid-gastrulation (3–4 dpf). Thereafter, the specific gravity gradually decreased until full development of main organs (9–10 dpf). Subsequently, $\rho_{egg}$ suddenly increased just prior to hatching (11–13 dpf). The egg total incubation period was 12 days for winter and spring and 13 days for fall at 7°C. The egg distribution in a column was relatively narrow at early stages, but widened at later stages. Within the single egg batches the standard deviations were not significantly different from those of egg volume, chorion volume, and yolk volume measured from a single batch. Also such variations tended to be a bit higher at later stages of 10 and 12 dpf, but it does not hamper understanding of the general trend during development.

The decreasing trend in $\rho_{egg}$ was similar between the two types of measurements (continuous *vs*. time point) for winter and spring, but different for fall (Table S2). Assuming that the environmental conditions of aquaria used for time point measurements were optimal for egg development, the lack of difference indicates that the eggs developed properly in the column in the winter and spring experiments.

### Initial changes in egg specific gravity as a function of perivitelline space and yolk volume dynamics

A slight increase in $\rho_{egg}$ from 2 to 4 dpf was generally found in all samples. For examination of causes for this, specific gravity and volume changes in each component were tracked, including the chorion, by setting the mean $\rho_{cho}$ to 1.20 g cm$^{-3}$ and $\rho_{pvs}$ identical to the ambient seawater (see Introduction). The specific gravity of yolk plus embryo ($\rho_{yolk+emb}$) was estimated using Eq. 4 (Appendix) and appeared to be similar between 2 and 4 dpf. In terms of volume changes, $V_{pvs}$ increased and $V_{yolk+emb}$ decreased from 2 to 4 dpf while $V_{cho}$ apparently was constant (Figure S4). Therefore, the noted increase in $V_{pvs}$ and the comparable decrease in $V_{yolk+emb}$ are presumed to cause the slight increase in the total $\rho_{egg}$ from 2 to 4 dpf.

### Changes in egg characteristics during development

Egg dry weight/egg volume did not show a systematic trend during development among the studied samples (Figure 2A). Only Batch4-b showed a significant negative correlation with age ($r = -0.4$, $p = 0.043$). Water content was positively correlated with age in two samples (Figure 2B), i.e. Batch2-b ($r = 0.48$, $p = 0.015$) and Batch4-b ($r = 0.41$, $p = 0.040$). Given the needed high precision of the measurements of dry weight and water content it is however, not surprising that trends are different for these measurements. $V_{yolk}/V_{egg}$ (where $V_{yolk}$ in this figure is the yolk excluding the embryo volume) consistently showed a decreasing trend (all occasions, $p<0.001$) (Figure 2C). $V_{cho}/V_{egg}$ also revealed a negative correlation with age for Batch2-b ($r = -0.55$, $p = 0.002$), Batch4-b ($r = -0.52$, $p = 0.001$), Batch5-a ($r = -0.40$, $p = 0.039$) and Batch7-a ($r = -0.61$, $p = 0.002$) (Figure 2D). However, chorion itself became slightly thinner at 12 dpf compared to the initial value at 2 dpf for all samples (Batch2-b: 6.7 to 6.4 μm; Batch3-b: 4.7 to 4.3 μm; Batch4-b: 7.5 to 6.6 μm; Batch5-a: 7.2 to 6.6 μm; Batch7-a: 6.8 to 6.1 μm from 2 dpf to 12 dpf), except for Batch8-a where chorion thickness (t) was measured to become slightly thicker (5.5 to 5.8 μm). Since the data material in all was limited, no statistical test was done. If, however, the chorion really became thinner, a speculation of the cause could be associated with enzymatic dissolution of the chorion from the inside related to the hatching process.

### Statistical relationships between initial egg specific gravity and egg components

Only the $V_{yolk+emb}$ and $V_{cho}$ contributed significantly to the specific gravity of the eggs (p<0.005, stepwise regression), while the $V_{pvs}$ and wet- and dry weights did not. Based on the initial analysis, a re-analysis was carried out with cases that had been excluded due to missing values in non-significant independent variables. The residuals of the extra cases from the original relationship was not significantly different from 0 (t-test, p>0.05), and the re-analysis with all available data provided the following relationship based on the multiple regression: $\rho_{egg} = 1.02697 - 0.01253\times V_{yolk+emb} + 0.000278\times V_{cho}$ (p<0.001, multiple regression, n = 14, $r^2 = 0.91$) (Figure 3).

### Initial specific gravity of yolk plus embryo

By inserting our measured values of egg specific gravity ($\rho_{egg}$), chorion thickness (t), egg diameter (2r), and yolk plus embryo volume ($V_{yolk+emb}$) into Eq. 4, we calculated $\rho_{yolk+emb}$ at 2–3 dpf. The relationship between $\rho_{egg}$ and $\rho_{yolk+emb}$ was highly significant. The linear regression was $\rho_{egg} = 1.6326\times\rho_{yolk+emb} - 0.6376$ ($r^2 = 0.94$) (Figure 4).

## Discussion

This study presents the first coherent explanation of the developmental and individual variability of egg specific gravity ($\rho_{egg}$) for marine pelagic fish eggs based on first principles. It shows the relative contribution from the various components of the egg to the resulting egg specific gravity. Chronologically, the critical events in determining the changes in egg specific gravity are as follows: The first determinant starts in the ovary with the deposition of the heavy chorion material ($M_{cho}$) associated with each oocyte [7]. The second determinant is associated with proteolytic cleavage and associated hydration of the yolk and the strong decrease in $\rho_{egg}$ prior to spawning as addressed in the work

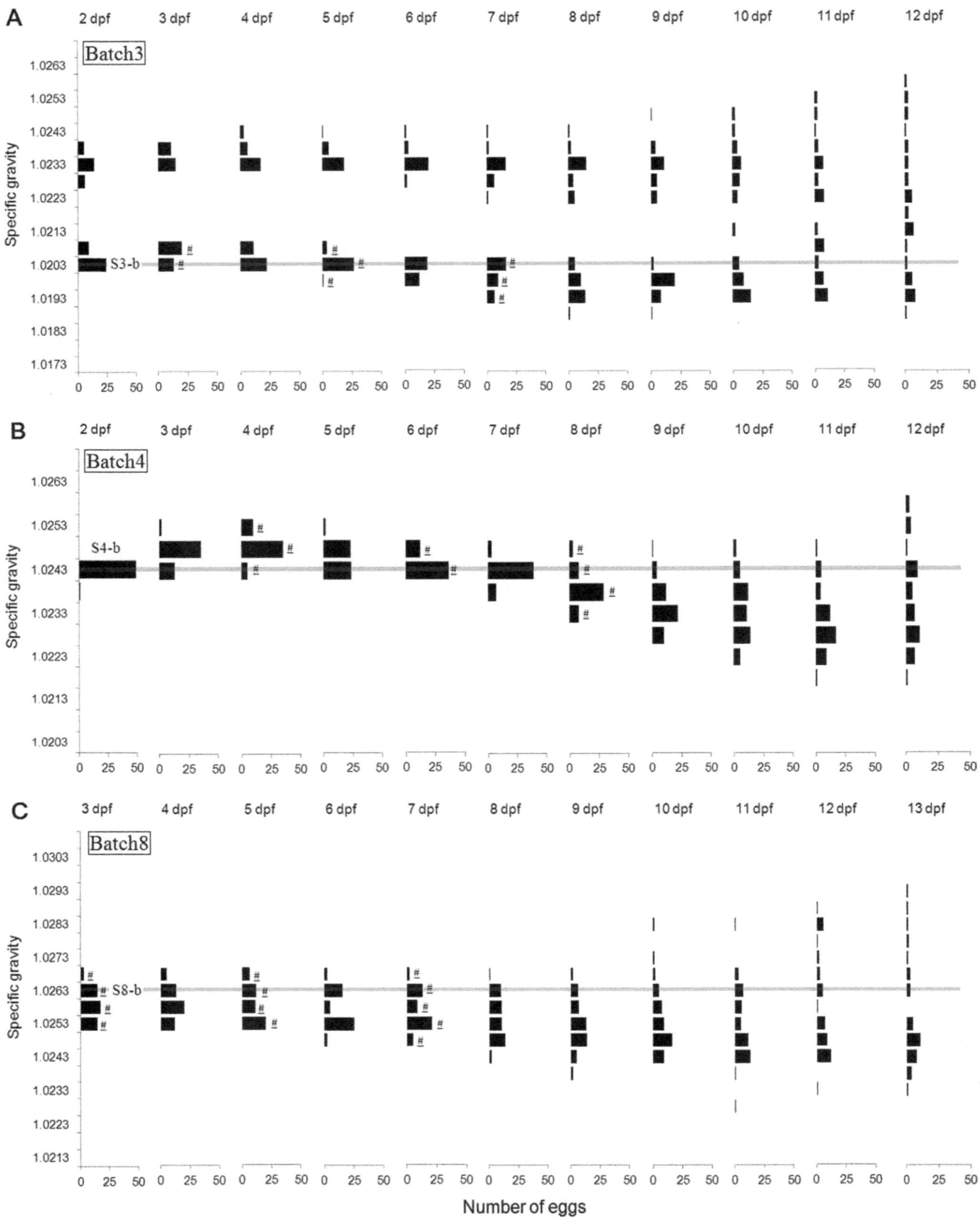


**Figure 1. Ontogenetic changes of three seasons from 2–3 dpf until hatching (continuous egg specific gravity measurements).** (A) Batch3, (B) Batch4, and (C) Batch8 represent winter, spring, and fall, respectively. Age of egg development is expressed in days post-fertilization (dpf). Horizontal lines refer to initial specific gravity of the selected layers (Batch3-b, Batch4-b and Batch8-b) at 2–3 dpf. Symbol of underlined # indicates the level of eggs used for slope tests in Table S2.

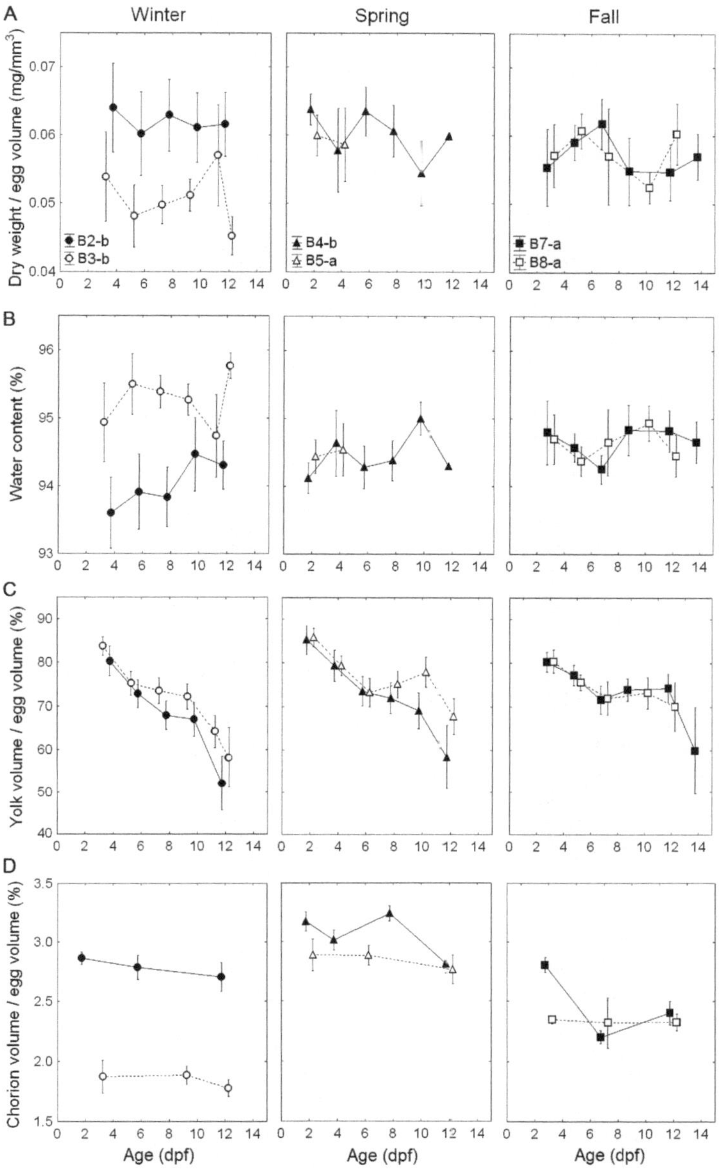


**Figure 2. Ontogenetic changes in (A) egg dry weight/egg volume, (B) water content, (C) yolk volume/egg volume, and (D) chorion volume/egg volume.** Selected buoyancy layers were Batch2-b and Batch3-b in winter, Batch4-b and Batch5-a in spring, and Batch7-a and Batch8-a in fall. As egg distribution overlapped with other layers of eggs, in particular after the latter half incubation, each egg component was divided by mean egg volume in order to avoid egg size effects except for the water content. Points refer to mean and one standard deviation.

by Craik & Harvey [4]. The third determinant is associated with the attained volume fraction of the neutrally buoyant perivitelline space ($V_{pvs}/V_{egg}$) as it manifests just after fertilization [4]. The fourth determinant is associated with the yolk plus embryo's capability in osmoregulation that causes the egg initially to lose water and become heavier. Thereafter, with the increasing capability in osmoregulation, caused by the development of the embryo, the water content is restored and the egg becomes lighter again [29]. The fifth determinant is associated with the increasing specific gravity prior to hatching. The cause of this final egg specific gravity increase is not yet fully clear but a possible mechanism is considered at the end of the discussion. The sixth and final determinant is the hatching process that ends with the newly hatched larvae get rid of the heavy chorion and becomes considerably lighter.

As the early notable feature of ontogenetic changes, eggs increase their specific gravity up to 0.03% of the initial value from fertilization until mid-gastrulation (~4 dpf). The slight increase in $\rho_{egg}$ is attributable to the volume decrease in yolk plus embryo and the corresponding volume increase in PVS (Figure S4). The shrinkage of yolk plus embryo ($V_{yolk+emb}$) without expansion of the chorion was also detected by Davenport *et al.* [30]. In addition, since egg osmoregulation is not effective until complete gastrulation [31], the passive water loss from the embryo to the hyper-osmotic ambient seawater is at maximum during the first four days [29,32]. Thus, the decline in $V_{yolk+emb}$, presumably resulting from osmotic water loss, may cause eggs to become slightly heavier at mid-gastrulation.

After 4 dpf, eggs generally show a gradual decline in $\rho_{egg}$ until 10–11 dpf, but with a rapid increase at 12–13 dpf just before hatching. The rate of decrease in $\rho_{egg}$ is 0.0002 g cm$^{-3}$ day$^{-1}$

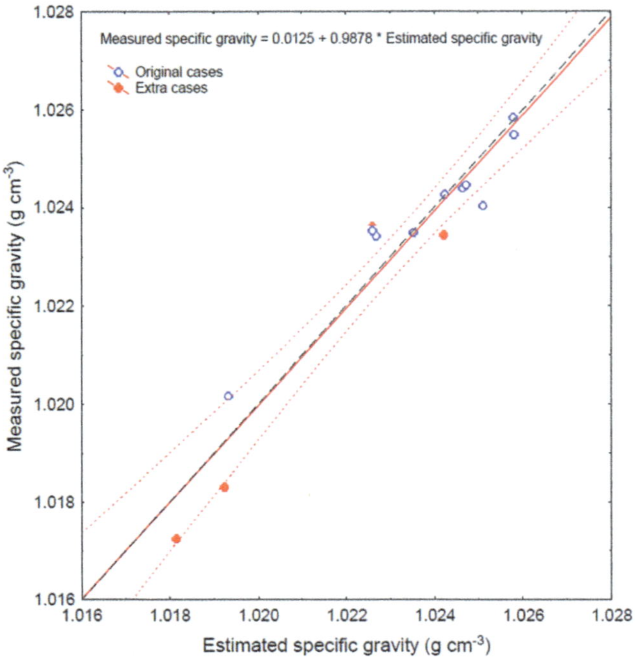


**Figure 3. Estimated specific gravity (g cm$^{-3}$) of cod eggs based on equation from multiple regression versus observed specific gravity from the same egg batches (solid line).** Open symbols for original cases without missing variable values and filled symbols for extra data points included after the initial analysis. Dashed line represents 1:1 line, and dotted lines represent 95% confidence regression bands.

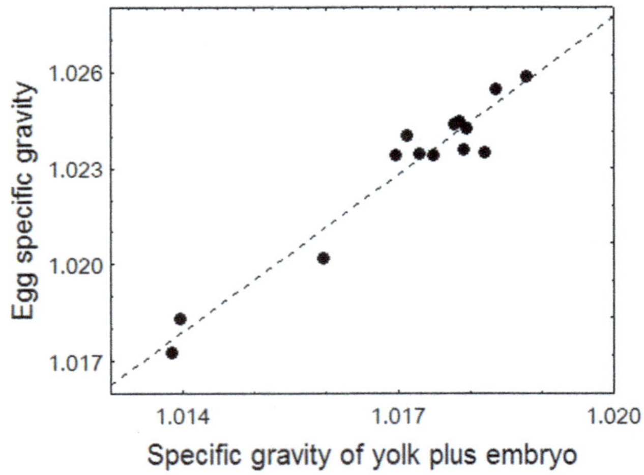


**Figure 4. Relationship between egg specific gravity and specific gravity of yolk plus embryo at 2–3 dpf.** The specific gravity of yolk plus embryo was estimated by Eq. 4 (Appendix).

(Table S2), implying that $\rho_{egg}$ changes by 0.0014 g cm$^{-3}$ from 4 to 11 dpf. During the same developmental stages, yolk volume fractions decrease with egg stages until hatching, while dry weight/ egg volume, water content, and chorion volume fraction do not show consistent trends during development among the egg samples. Finn *et al.* [33] reported a similar trend of smoothly decreasing $V_{yolk}$ in a single egg batch of Atlantic cod. They also found there were no significant changes in egg dry weight and water content throughout development. In a similar study for plaice (*Pleuronectes platessa*) eggs, wet weight, dry weight and water content were approximately constant until hatching [34]. Therefore, it seems that the changes in dry weight, water content, and chorion volume ($V_{cho}$) do not properly explain temporal variability in $\rho_{egg}$ during development.

Only changes in yolk plus embryo volume ($V_{yolk+emb}$) and specific gravity ($\rho_{yolk+emb}$) can cause the gradual decline in $\rho_{egg}$, since egg diameter (2r), chorion mass ($M_{cho}$), and chorion specific gravity ($\rho_{cho}$) are approximately constant throughout development. Although a slight decrease in chorion thickness (t) prior to hatching was observed, resultant changes in $M_{cho}$ are too small to significantly affect the $\rho_{egg}$. The change in $\rho_{yolk+emb}$ during development can, in turn, have two causes: (1) volume and mass changes in the yolk-sac linked to the water balance. Riis-Vestergaard [29] observed and calculated the change in yolk osmolarity in cod eggs raised in salinity of 34. He found a relatively rapid increase in yolk osmolarity from 340 at fertilization to around 410 at 3 dpf. Thereafter, the yolk osmolarity decreased to around 380 at 4 dpf and 320 at 10–12 dpf. The decrease in osmolarity from 380 to 320 corresponds to the decrease in $\rho_{egg}$ of about 0.00142 g cm$^{-3}$, i.e. very similar to our observation of 0.0014 g cm$^{-3}$ calculated by the reduction rate. (2) The other is caused by volume and mass changes in embryo structural composition such as lipids, proteins, muscles and cartilage, as the embryo develops. Finn *et al.* [33] studied the respiration of cod eggs. The oxygen consumption increased by a factor of 10 from fertilisation to hatching and the ammonia excretion by a factor of 6. However, there was no clear trend in the development of the wet weight of the eggs although a reduction in the dry weight seemed to occur. Consumption of lipids will increase wet weight as water must replace this consumption, while consumption of proteins will decrease the total wet weight of the egg. In the subsequent calculation we, therefore, consider the loss of mass from the yolk plus embryo due to respiration and metabolism to be largely compensated by the water. It implies that $\rho_{yolk+emb} \times V_{yolk+emb} = K$, where K is constant. This means that a decrease in $\rho_{yolk+emb}$ must be compensated by a corresponding increase in $V_{yolk+emb}$. Under these assumptions, we can calculate the change in $V_{pvs}$ (Appendix 5–8) derived from:

$$V_{pvs} = V_{egg} - V_{cho} - \left[K + (\rho_{cho} - \rho_{egg}) \times V_{cho}\right] / \rho_{egg} \quad (8)$$

By inserting the observed reduction of $\rho_{egg}$ (0.0014 g cm$^{-3}$) throughout incubation into the equation, the calculated changes in $V_{pvs}$ are only from 13.3 to 13.4% of the $V_{egg}$ for an average egg (i.e. mean values of 2r, t, $\rho_{egg}$, $V_{pvs}$). This is in contrast to the much larger variation in $V_{pvs}$ which varies from 17 to 22% of total egg volume during development [17]. Hence, we can conclude that the changes in $\rho_{egg}$ and $V_{pvs}$ throughout the first part of the incubation period onto 11 dpf are caused significantly by the changes in yolk osmolarity rather than the changes in embryo composition of proteins, bones, and muscles. Accordingly, the increase in osmolarity of yolk plus embryo after spawning results in loss of water and, consequently, reduced $V_{yolk+emb}$, increased $\rho_{yolk+emb}$, and subsequently increased $V_{pvs}$ from fertilization to 4 dpf. Thereafter, the $\rho_{yolk+emb}$ decreases since the decreasing yolk and the increasing embryo begin to actively take up water resulting in the increase in $V_{yolk+emb}$ and the corresponding decrease in $V_{pvs}$.

Our explanation for developmental variability in $\rho_{egg}$ is consistent with egg osmoregulatory work [29]. Precursors of chloride cells are present at morula stage (~2 dpf in the current

**Table 1.** Calculated influence of three egg components on egg specific gravity ($\rho_{egg}$) for Atlantic cod based on observed ranges of the components.

| | Difference in $\rho_{egg}$ (g cm$^{-3}$) | Range of $\rho_{egg}$ (g cm$^{-3}$) | D (mm) | t (μm) | $V_{cho}/V_{egg}\times 100$ (%) | $V_{pvs}/V_{egg}\times 100$ (%) | $\rho_{yolk+emb}$ (g cm$^{-3}$) |
|---|---|---|---|---|---|---|---|
| I | 0.00626 | 1.02647 | 1.20 | 9.0 | 4.43 | 13.5 | 1.0171 |
| | | 1.02021 | 1.50 | 3.7 | 1.47 | 13.5 | 1.0171 |
| II | 0.00063 | 1.02348 | 1.35 | 6.5 | 2.86 | 18.0 | 1.0171 |
| | | 1.02285 | 1.35 | 6.5 | 2.86 | 9.0 | 1.0171 |
| III | 0.00464 | 1.02479 | 1.35 | 6.5 | 2.86 | 13.5 | 1.0188 |
| | | 1.02015 | 1.35 | 6.5 | 2.86 | 13.5 | 1.0140 |

I: Influence of fraction of chorion volume in percent of total egg volume, $(V_{cho}/V_{egg})\times 100$, by keeping the two other components constant and at average values. The observed upper value is based on smallest observed egg (D = 1.20 mm) [13] combined with the thickest observed chorion (t = 9.0 μm) [30] and the lower value is based on the largest observed egg (D = 1.50 mm) [13] and the thinnest observed chorion (t = 3.7 μm) (present study). II: Influence of the fraction of perivitelline space in percent of total egg volume, $(V_{pvs}/V_{egg})\times 100$, by keeping the two other components at constant and at average values. III: Influence of specific gravity of yolk plus embryo, $\rho_{yolk+emb}$, by keeping the two other components constant and at average values. The observed ranges of egg diameter, D, and chorion thickness, t, are based on the literature. All other ranges are based on the present data.

study) and apparently functional after mid-gastrulation (~4 dpf). Hence, the eggs regulate water loss and salt load by means of the chloride cells [31]. Restricted water loss has been found for cod [29,32], and there was a positive relationship between buoyant eggs and sodium-potassium ATPase pump [9]. Mangor-Jensen [17] demonstrated that the mechanism for providing hydrostatic lift is the large volume of diluted tissue water located in the yolk and subdermal spaces. Accordingly, osmoregulatory ability of developing embryos is highly correlated with keeping proper egg buoyancy. Although the in-depth analyses of the changes in eggs throughout incubation are mainly demonstrated in cod, similar observations on changes in $\rho_{egg}$ have been reported for other species where continuous measurements of egg specific gravity have been conducted. These include anchovy (*Engraulis encrasicolus*) [35], cape hake (*Merluccius capensis*) [15] and blue whiting (*Micromesistius poutassou*) [14].

Moreover, developmental changes in $\rho_{egg}$ are also associated with catabolism of various energy substrates (carbohydrates, free amino acids (FAA), and lipids). For Atlantic cod eggs, FAA, which is markedly denser than ambient seawater, contributes 75% to aerobic catabolism [33]. The FAA is consumed for the purpose of respiration and protein synthesis, and 50% of the initial amount of FAA disappears before hatching [36]. Carbohydrates and lipids are present in low amounts during development as 0.3–0.7% and 10–13% of egg dry weight, respectively [33]. As a consequence, the substantial loss in FAA is likely to contribute to the decline in the yolk osmolarity throughout development.

Of egg structural materials, carbohydrates and proteins are denser than sea water and provide downwardly directed forces on the egg. In contrast, lipid and pure water are less dense and provide upward forces [4]. As such, egg specific gravity is initially determined by the balance of those materials in which proteins and water are regarded as major factors and carbohydrates and lipid as minor factors [4,5,6]. In the present study, chorion volume ($V_{cho}$) at early egg stages consists from 2 (thickness: 3.8 μm) to 3.6% (thickness: 7.8 μm) of total egg volume ($V_{egg}$) and significantly associates with $\rho_{egg}$ ranging from 1.017 to 1.026 g cm$^{-3}$. Similar positive relationship was observed for Baltic cod eggs in brackish water [37]. In their study, chorion thickness varied from 3 to 5 μm and positively correlated with $\rho_{egg}$ ranging from 1.010 to 1.014 g cm$^{-3}$. Kjesbu *et al.* [7] pointed out that two eggs with the same chorion volume ($V_{cho}$) would have different egg specific gravity depending on egg diameter, because the ratio $V_{cho}/V_{egg}$ would change. Hence, a larger egg would have lower specific gravity than a smaller egg. Our results confirm the significant contribution of chorion volume to determining the initial egg specific gravity.

The $V_{pvs}$ has been earlier reported to be about 20% of total egg volume [17,30]. In the present study, the $V_{pvs}$ largely varies from 9 to 18% at 2–3 dpf. Also the $\rho_{yolk+emb}$ estimated in here ranges largely from 1.014 to 1.019 g cm$^{-3}$, but mean $\rho_{yolk+emb}$ is surprisingly similar to the previous measurement of Kjesbu *et al.* [7], 1.0170 *vs.* our estimate of 1.0171 g cm$^{-3}$. Measurements on specific gravity on the hardened chorion material, $\rho_{cho}$, from cod eggs Kjesbu *et al.* [7] show relatively constant values around 1.20 g cm$^{-3}$. Also independent measurement on chorions from six Atlantic halibut eggs showed similar average value, i.e. 1.1842 g cm$^{-3}$ [23]. Table 1 shows sensitivity of $\rho_{egg}$ to the presently observed ranges of egg size, $V_{cho}$, $V_{pvs}$, and $\rho_{yolk+emb}$. It appears from the table that the specific gravity of the cod eggs changes by 0.0063 g cm$^{-3}$ across the range of egg size and $V_{cho}$, i.e. the smallest eggs with thickest chorions are 0.0063 g cm$^{-3}$ heavier than the extreme large eggs with thin chorions. In comparison, the extreme changes in $\rho_{yolk+emb}$ cause $\rho_{egg}$ to change by 0.0046 g cm$^{-3}$, which is about 70% of the variation caused by the chorion volume fraction. It should be emphasized that total range of $\rho_{yolk+emb}$ among individual eggs (i.e. 0.0046 g cm$^{-3}$) is 3.3 times larger than the individual change in $\rho_{egg}$ throughout incubation (i.e. 0.0014 g cm$^{-3}$). We have not measured the initial degree of hydrolyzation of the proteins in the yolk of the present study. However, noting that the large difference in $\rho_{egg}$ between Baltic cod and Norwegian Atlantic cod can partly be ascribed the variation in hydrolyzation of the yolk [17,19] it is likely that the present observed (and smaller) differences in $\rho_{yolk+emb}$ among individual eggs in Norwegian Atlantic cod could be explained by moderate individual differences in yolk hydrolyzation and, hence, the degree of egg swelling prior to spawning. While the chorion volume ($V_{cho}$) is determined during the oocyte stage [38], the chorion volume fraction ($V_{cho}/V_{egg}$) is determined by subsequent hydrolyzation prior to spawning. The present findings on the variation range of the $V_{pvs}$ are larger than earlier findings in the literature; 9–18% compared to 18–22% in Kjesbu *et al.* [7]. However, still with this larger range it has relatively small effects on the resulting changes in $\rho_{egg}$: 0.0006 g cm$^{-3}$, which is less than

10% of the variation in $\rho_{egg}$ caused by the chorion volume fractions.

Some of the changes in $\rho_{egg}$ during development may also be related to differences in the proximal composition of the yolk and embryo component, resulting in slight changes in $\rho_{yolk+emb}$ from fertilization to hatching. For example, relatively higher free amino acid (FAA) content is expected in the yolk compared to the embryo [5,36], with correspondingly higher relative protein content in the embryo as a result. Preliminary estimations of the $\rho_{yolk}$ versus $\rho_{emb}$ based on unfed newly hatched cod larvae, suggest that these differ by approx. 0.014 g cm$^{-3}$ (own unpubl. results). During the initial few days after fertilization this difference is not expected to be of importance in the overall $\rho_{yolk+emb}$ due to the relatively high mass of yolk relative to mass of embryo. However, as the volume of the embryo increases rapidly towards the end of incubation it is expected to contribute more to increasing the overall $\rho_{egg}$.

In conclusion, based on our working hypotheses, the present study confirms that the initial $\rho_{egg}$ is largely influenced by chorion volume fractions, and that the ontogenetic variability in $\rho_{egg}$ during incubation is attributable to the volume and mass changes of yolk plus embryo. With the previous findings of water balance and metabolism during egg stages, we demonstrate that swelling of embryos by uptake of water and drop in yolk osmolarity are main causes of decreasing $\rho_{egg}$ from mid-gastrulation to full development of main organs.

## Supporting Information

**Figure S1 Time point egg specific gravity measured in winter every second day during development.** Horizontal lines refer to initial specific gravity of the selected layers (Batch2-a and b, Batch3-a, b and c) at 2–3 dpf. Symbol of # indicates sampled levels for further analyses on egg composition shown in Figure 2 under the assumption that eggs initially having the same specific gravity as the Batch2-b and Batch3-b at 2–3 dpf would fall into the sampled levels with age. The underlined # in Batch3 indicates the level of eggs used for slope tests in Table S2.

**Figure S2 Time point egg specific gravity measured in spring every second day during development.** Horizontal lines refer to initial specific gravity of the selected layers (Batch4-a and b, Batch5-a and b) at 2 dpf. Symbol of # indicates pick-up levels for further analyses on egg composition shown in Figure 2 under the assumption that eggs initially having the same specific gravity as the Batch4-b and Batch5-a at 2 dpf would fall into the sampled levels with age. The underlined # in Batch4 indicates the level of eggs used for slope tests in Table S2.

**Figure S3 Time point egg specific gravity measured in fall every second day during development.** Horizontal lines refer to initial specific gravity of the selected layers (Batch7-a and b, Batch8-a and b) at 3 dpf. Symbol of # indicates pick-up levels for further analyses on egg composition shown in Figure 2 under the assumption that eggs initially having the same specific gravity as the Batch7-a and Batch8-a at 3 dpf would fall into the sampled levels with age. The underlined # in Batch8 indicates the level of eggs used for slope tests in Table S2.

**Figure S4 Initial changes in egg specific gravity caused by perivitelline space and yolk volume dynamics.** Left panel is changes in mean egg specific gravity and right panel is changes in mean volume fractions of yolk plus embryo, perivitelline space and chorion from 2 to 4 dpf, represented by spring time point measurements of (A) Batch4-b, (B) Batch5-a and (C) Batch6-a. Points refer to mean and one standard deviation.

**Table S1 Egg characteristics at 2–3 days post-fertilization (dpf) from time point measurements.**

**Table S2 Comparison of two types of measurements (continuous *vs.* time point).**

**Appendix S1 Theoretical equations for egg specific gravity containing (1) list of symbols in equations and (2) algorithms of specific gravity at fertilization and during development.**

## Acknowledgments

We thank the commercial hatchery Marine Harvest Cod and Austevoll hatchery of IMR for the safe delivery of cod eggs to the laboratory and useful information of the broodfish.

## Author Contributions

Conceived and designed the experiments: KMJ AF OSK SS. Performed the experiments: KMJ SS. Analyzed the data: KMJ AF OSK SS. Contributed reagents/materials/analysis tools: KMJ AF OSK SS. Wrote the paper: KMJ AF OSK SS.

## References


1. Govoni JJ, Forward RB Jr (2008) Buoyancy. In: Finn RN, Kapoor BG, editors. Fish Larval Physiology. Enfield, NH: Science Publishers. pp. 495–521.
2. Cerdà J (2009) Molecular pathways during marine fish egg hydration: the role of aquaporins. J Fish Biol 75: 2175–2196.
3. Cerdà J, Finn RN (2010) Piscine aquaporins: an overview of recent advances. J Exp Zool 313A: 623–650.
4. Craik JCA, Harvey SM (1987) The causes of buoyancy in eggs of marine teleosts. J Mar Biol Ass UK 67: 169–182.
5. Thorsen A, Kjesbu OS, Fyhn HJ, Solemdal P (1996) Physiological mechanisms of buoyancy in eggs from brackish water cod. J Fish Biol 48: 457–477.
6. Riis-Vestergaard J (2002) Energy density of marine pelagic fish eggs. J Fish Biol 60: 1511–1528.
7. Kjesbu OS, Kryvi H, Sundby S, Solemdal P (1992) Buoyancy variations in eggs of Atlantic cod (*Gadus morhua* L.) in relation to chorion thickness and egg size: theory and observations. J Fish Biol 41: 581–599.
8. Jung K, Folkvord A, Kjesbu OS, Agnalt AL, Thorsen A, et al. (2012a) Egg buoyancy variability in local populations of Atlantic cod (*Gadus morhua*). Mar Biol 159: 1969–1980.
9. Jung K, Svardal AM, Eide T, Thorsen A, Kjesbu OS (2012b) Seasonal trends in adenylate nucleotide content in eggs of recruit and repeat spawning Atlantic cod (*Gadus morhua* L.) and implications for egg quality and buoyancy. J Sea Res 73: 63–73.
10. Coombs SH (1981) A density-gradient column for determining the specific gravity of fish eggs, with particular reference to eggs of the mackerel *Scomber scombrus*. Mar Biol 63: 101–106.
11. Sundby S (1983) A one-dimensional model for the vertical distribution of pelagic fish eggs in the mixed layer. Deep-Sea Res 30: 645–661.
12. Sundby S (1991) Factors affecting the vertical distribution of eggs. ICES Mar Sci Symp 192: 33–38.
13. Solemdal P, Sundby S (1981) Vertical distribution of pelagic fish eggs in relation to species, spawning behavior and wind conditions. ICES CM 1981/G: 77. 1–27.
14. Ådlandsvik B, Coombs S, Sundby S, Temple G (2001) Buoyancy and vertical distribution of eggs and larvae of blue whiting (*Micromesistius poutassou*): observations and modeling. Fish Res 50: 59–72.
15. Sundby S, Boyd AJ, Hutchings L, O'Toole MJ, Thorisson K, et al. (2001) Interaction between Cape hake spawning and the circulation in the Northern Benguela upwelling ecosystem. S Afr J Mar Sci 23: 317–336.
16. Coombs SH, Morgans D, Halliday NC (2001) Seasonal and ontogenetic changes in the vertical distribution of eggs and larvae of mackerel (*Scomber scombrus* L.) and horse mackerel (*Trachurus trachurus* L.) Fish Res 50: 27–40.

17. Mangor-Jensen A (1987) Water balance in developing eggs of the cod *Gadus morhua*. Fish Phys Biochem 3: 17–24.
18. Anderson J, deYoung B (1994) Stage-dependent density of cod eggs and larvae (*Gadus morhua* L.) in Newfoundland waters. ICES Mar Sci Symp 198: 654–665.
19. Nissling A, Vallin L (1996) The ability of Baltic cod eggs to maintain neutral buoyancy and the opportunity for survival in fluctuation conditions in the Baltic Sea. J Fish Biol 48: 217–227.
20. Oppen-Berntsen DO, Helvik JV, Walther BT (1990) The major structural proteins of cod (*Gadus morhua*) eggshells and protein crosslinking during teleost egg hardening. Dev Biol 137: 258–265.
21. Kjørsvik E, Lønning S (1983) Effects of egg quality on normal fertilization and early development of the cod, *Gadus morhua* L. J Fish Biol 23: 1–12.
22. Lønning S, Kjørsvik E, Falk-Petersen IB (1988) A comparative study of pelagic and demersal eggs from common marine fishes in northern Norway. Sarsia 73: 49–60.
23. Mangor-Jensen A, Huse I (1991) On the changes in buoyancy of halibut, *Hippoglossus hippoglossus* (L.), larvae caused by hatching - a theoretical view. J Fish Biol 39: 133–135.
24. Sundby S (1997) Turbulence and ichthyoplankton: influence on vertical distributions and encounter rates. Sci Mar 61: 159–176.
25. Westin L, Nissling A (1991) Effects of salinity on spermatozoa motility, percentage of fertilized eggs and egg development of Baltic cod (*Gadus morhua*), and implications for cod stock fluctuations in the Baltic. Mar Biol 108: 5–9.
26. Fridgeirsson E (1978) Embryonic development of five species of gadoid fishes in Icelandic waters. Rit Fisk 5: 1–68.
27. R Development Core Team (2010) R: A language and environment for statistical computing. R Foundation for Statistical Computing, Vienna Austria ISBN 3-900051-07-0. URL http://www.R-project.org/.
28. StatSoft Inc. (2013) STATISTICA (data analysis software system), version 12. www.statsoft.com.
29. Riis-Vestergaard J (1984) Water balance in cod eggs. In: Dahl E, Danielsse DS, Molesness E, Solemdal P, editors. The propagation of cod, Gadus morhua. Part 1. Flødevigen rapportser 1. pp. 87–103.
30. Davenport J, Lønning S, Kjørsvik E (1981) Osmotic and structural changes during early development of eggs and larvae of the cod, *Gadus morhua* L. J Fish Biol 19: 317–331.
31. Adoff GR (1986) Anatomical studies of developing eggs and larvae of cod (Gadus morhua L.); Basic studies and effects of oil exposure. In: Fyhn HJ, editors. Fish larval physiology and anatomy. Bergen: University of Bergen, pp. 51–115.
32. Mangor-Jensen A (1986) Osmoregulation in eggs and larvae of the cod (Gadus morhua L.); Basic studies and effects of oil exposure. In: Fyhn HJ, editors. Fish larval physiology and anatomy. Bergen: University of Bergen. pp. 117–164.
33. Finn RN, Fyhn HJ, Evjen MS (1995) Physiological energetic of developing embryos and yolk-sac larvae of Atlantic cod (*Gadus morhua*). I. Respiration and nitrogen metabolism. Mar Biol 124: 355–369.
34. Skjærven KH, Finn RN, Kryvi H, Fyhn HJ (2003) Yolk resorption in developing plaice (Pleuronectes platessa). In: Browman HI, Skiftesvik AB, editors. The big fish bang. Proceedings of the 26th annual larval fish conference. Bergen: Institute of Marine Research. pp. 193–209.
35. Ospina-Álvarez A, Palomera I, Parada C (2012) Changes in egg buoyancy during development and its effects on the vertical distribution of anchovy eggs. Fish Res 117–118: 86–95.
36. Fyhn HJ, Serigstad B (1987) Free amino acids as energy substrate in developing eggs and larvae of the cod *Gadus morhua*. Mar Biol 96: 335–341.
37. Nissling A, Kryvi H, Vallin L (1994) Variation in egg buoyancy of Baltic cod *Gadus morhua* and its implications for egg survival in the Baltic Sea. Mar Ecol Prog Ser 110: 67–74.
38. Kjesbu OS, Kryvi H, Norberg B (1996) Oocyte size and structure in relation to blood plasma steroid hormones in individually monitored, spawning Atlantic cod. J Fish Biol 49: 1197–1215.

6

# Contextualising the Last Survivors: Population Structure of Marine Turtles in the Dominican Republic

**Carlos Carreras[1]*, Brendan J. Godley[1], Yolanda M. León[2,3], Lucy A. Hawkes[1], Ohiana Revuelta[4], Juan A. Raga[4], Jesús Tomás[1,4]**

**1** Centre for Ecology and Conservation, University of Exeter, Penryn, United Kingdom, **2** Grupo Jaragua, Santo Domingo, Dominican Republic, **3** Instituto Tecnológico de Santo Domingo, Santo Domingo, Dominican Republic, **4** Cavanilles Institute of Biodiversity and Evolutionary Biology (Scientific Park), University of Valencia, Valencia, Spain

**Abstract**

Nesting by three species of marine turtles persists in the Dominican Republic, despite historic threats and long-term population decline. We conducted a genetic survey of marine turtles in the Dominican Republic in order to link them with other rookeries around the Caribbean. We sequenced a $740_{bp}$ fragment of the control region of the mitochondrial DNA of 92 samples from three marine turtle species [hawksbill (n = 48), green (n = 2) and leatherback (n = 42)], and incorporated published data from other nesting populations and foraging grounds. The leatherback turtle (*Dermochelys coriacea*) in the Dominican Republic appeared to be isolated from Awala-Yalimapo, Cayenne, Trinidad and St. Croix but connected with other Caribbean populations. Two distinct nesting populations of hawksbill turtles (*Eremochelys imbricata*) were detected in the Dominican Republic and exhibited interesting patterns of connectivity with other nesting sites and juvenile and adult male foraging aggregations. The green sea turtle (*Chelonia mydas*) has almost been extirpated from the Dominican Republic and limited inference could be made from our samples. Finally, results were compared with Lagrangian drifting buoys and published Lagrangian virtual particles that travelled through the Dominican Republic and Caribbean waters. Conservation implications of sink-source effects or genetic isolation derived from these complex inter-connections are discussed for each species and population.

**Editor:** Graeme Clive Hays, University of Wales Swansea, United Kingdom

**Funding:** The present study is included as part of a conservation project funded by: the Spanish International Cooperation Agency (AECI, projects: A/2991/05 and A/5641/06), the Spanish Ministry of Education and Sciences (CGL2006-02936-BOS), the General Foundation of the University of Valencia, and the European Union (Marie Curie grants, FP6 & 7). BJG is supported by NERC and the Darwin Initiative. LAH is supported by the Biotechnology and Biological Sciences Research Council of the UK. JAR and JT are also supported by project Prometeo/2011/40 of 'Conselleria de Educació' (Generalitat Valenciana) and project CGL2011-30413 of the Spanish Ministry of Sciences and Innovation. The funders had no role in study design, data collection and analysis, decision to publish, or preparation of the manuscript.

**Competing Interests:** The authors have declared that no competing interests exist.

* E-mail: carlos_c_h@hotmail.com

## Introduction

Many marine vertebrates have complex life cycles requiring the use of different habitats that are often spread over wide spatio-temporal scales and usually result in a network of connections among different populations and between populations and distant feeding grounds [1–7]. This complex structure is especially relevant to endangered species, in which threat may be highly localized, yet have a potentially profound effect on distant areas. Thus, the knowledge of migratory pathways, population structure and connectivity of the most threatened populations in comparison to other populations and feeding grounds is crucial for an effective application of conservation actions. Marine turtles are one of the best examples of this complex structuring [8–9] and highly migratory behaviour [10], and have been subject to increasing focus for molecular research and conservation over the last few decades. This is especially relevant for management, as all marine turtle species are of conservation concern, and action plans often need to be international in scope [11]. As the Caribbean hosts numerous populations of several marine turtle species [12], gaps in information may lead to undiagnosed population sinks to otherwise protected stocks [11,13].

The Dominican Republic has, for many years, presented a gap of knowledge in marine turtle biology and conservation. Although it has been suggested as an important nesting area for several marine turtle species [14–15] there has been a paucity of monitoring data. Furthermore, there is clear evidence of a long history of harvesting and exploitation of marine turtle meat, eggs and shell as an important resource for local communities [14–18]. In addition to these threats, turtles are also incidentally captured at sea [19–20]. Although marine turtles have been legally protected by law since 1966 in the Dominican Republic [21–22] it was estimated that there was an annual capture of between 1000 and 2000 green (*Chelonia mydas*), loggerhead (*Caretta caretta*) and leatherback (*Dermochelys coriacea*) turtles during the 1980s [19]. In addition, a total of 4366 kg of hawksbill (*Eretmochelys imbricata*) shell was exported to Japan between 1970 and 1986 [23]. Turtle shell exploitation was also detected more recently [17–18,24–25]. The country also receives four million tourists annually that result in a significant degradation of coastal habitats [26–27].

The consequences of the accumulation of these threats to the Dominican Republic nesting populations were not properly addressed until a recent study [22], which suggested that the Dominican Republic is an important nesting area for the hawksbill

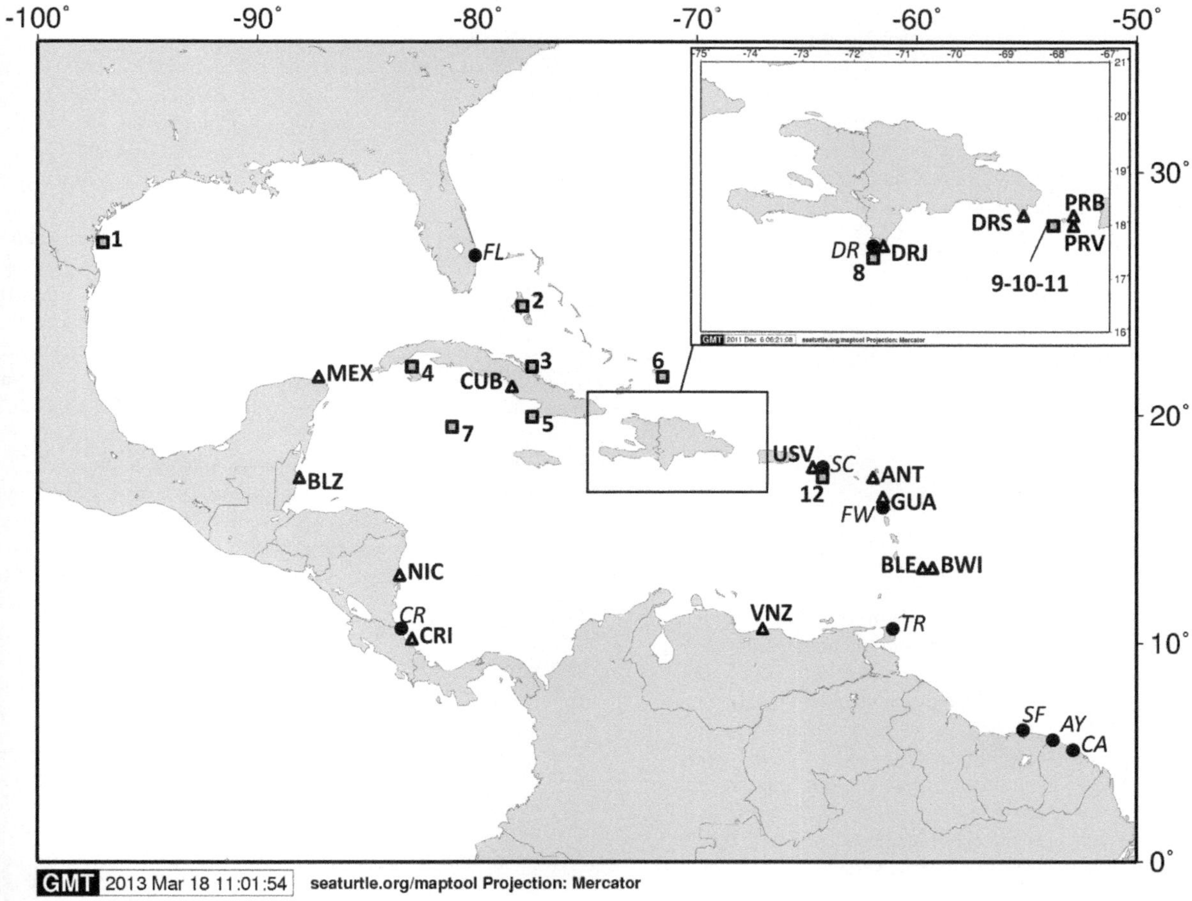


**Figure 1. Marine turtle nesting sites and feeding grounds considered in this study.** Leatherback nesting sites are represented by black circles labelled in italics. AY: Awala-Yalimapo; CA: Cayenne; CR Costa Rica (Atlantic); FL: Florida; FW: French West Indies; SC: Saint Croix; SF: Suriname/FG; TR: Trinidad; DR: Dominican Republic. Hawksbill nesting sites are represented by white triangles labelled in bold. ANT: Antigua; BLE: Barbados Leeward; BWI: Barbados Windward; BLZ: Belize; CRI: Costa Rica; CUB: Cuba; GUA: Guadeloupe; MEX: Mexico; NIC: Nicaragua; PRV: Puerto Rico [35]; PRB: Puerto Rico [64]; USV: US Virgin Islands; VNZ: Venezuela; DRJ: Dominican Republic-Jaragua; DRS: Dominican Republic-Saona. The hawksbill population of BRZ: Brazil and the leatherback populations of BZ: Brazil and SA: South Africa are not included in the map. Hawksbill feeding grounds are represented by numbered grey squares. 1. Texas; 2. Bahamas; 3. Cuba D; 4. Cuba B; 5. Cuba A; 6. Turk and Caicos; 7. Caiman Islands; 8. Dominican Republic; 9. Puerto Rico residents; 10. Puerto Rico recruits; 11. Puerto Rico pooled; 12. US Virgin Islands. Sampling sites of the present study were the Jaragua National Park (DRJ or DR) and Saona Island (DRS). Map created using MAPTOOL [106].

and leatherback turtle and includes some sporadic nesting by green turtles. However, the persistence of threats appears to have led to population reduction and a significant contraction of nesting habitats, with nesting largely restricted to protected areas [14–15].

Genetic markers have been widely used to establish links between populations and feeding grounds and to infer relative exposure to threats [8–9,28]. Maternally inherited mitochondrial DNA (mtDNA) is particularly well suited to assess such links as nesting female turtles exhibit marked site fidelity [8–9] permitting the definition of isolated Management Units (MUs) [29]. Juvenile turtles, however, may be widely dispersed and occupy foraging areas comprised of turtles of differing provenance [28,30–31]. For some species it has been shown that this mixing still remains after sexual maturity and could lead to male mediated gene flow between populations [32–33] although not in all cases [34]. For this reason, the concept of Regional Management Units (RMUs) has been proposed [13]. These would function as higher level conservation units that include all isolated populations (MUs) linked by the use of common feeding grounds, since mortality occurring in shared feeding areas will affect contributing nesting populations [13]. Mixed Stock Analysis (MSA) has provided the tool with which to link feeding grounds with nesting habitats and hence establish relationships at the RMU level for many sea turtle species testing what proportion of turtles from a feeding ground come from each nesting population from a *mixed stock* point of view [30,35–42]. The recent development of the 'many to many' analysis [43] allowed the testing of how many individuals from a nesting population use each feeding ground from a *nesting population* point of view [44–45].

Other approaches have involved the study of ocean currents, and have demonstrated that they may play a crucial role in defining hatchling and juvenile marine turtle dispersion [30,41,46–48]. To understand population connectivity, therefore, empirical data describing sea surface currents may be informative [49]. For example, data describing the tracks of Lagrangian drifting buoys [44,50], or biophysical modelling of oceanic dispersal [51–53] could be used to simulate the movements of a passively drifting hatchling turtle to complement the tagging of individuals [51,54–56]. Thus, a multidisciplinary approach has been recommended to assess both the connectivity among populations [57] and the definition of RMUs [13].

Here we combine genetic data from the Dominican Republic turtle nesting population with published genetic data from other nesting populations and feeding grounds, as well as tracks of passive drifter buoys and satellite tracking of adult turtles to: 1) determine if Dominican Republic turtle nesting populations are isolated from others in the region; 2) determine if mtDNA sequences reflect population size reductions; and 3) determine which feeding grounds are used by Dominican Republic marine turtles.

## Materials and Methods

During the 2007 and 2008 nesting seasons (from March to October), samples from three nesting marine turtle species were collected within an extensive monitoring project to assess nesting marine turtle abundance and distribution in the Dominican Republic [22] (Figure 1, Table 1). Samples from green turtles (n = 2; DRS), hawksbill turtles (n = 48; DRS n = 33, DRJ n = 15) and leatherback turtles (n = 42; DRJ) were collected from the Jaragua National Park (DRJ/DR) and Saona Island (DRS), with the permission and support of the office of Protected Areas of the Ministry of Environment and Natural Resources of the Dominican Republic Government. Muscle and skin samples were collected from dead hatchlings found in nests after emergence, so sampling had no impact on living animals and thus the study did not require the approbation of any ethical animal committee. Only one hatchling per nest and female was sampled in order to ensure independence of samples. Due to the lack of resources in the study area and the low density of nesting events in some places and species, it was not possible to identify and tag all females while nesting. However, a filtering method applied in other marine turtle species that combine remigration interval, sample location and haplotype found [58] was applied in order to avoid the risk of pseudoreplication. Samples were stored in 95% ethanol (n = 1 hatchling sample per nest). Finally, samples were transported for analysis following international CITES regulations.

DNA was extracted using the QIAamp extraction kit (QIAGEN®) following the manufacturer's instructions. We amplified a ~740$_{bp}$ fragment of the mtDNA control region using the primers LCM15382 (5′-GCTTAACCCTAAAGCATTGG-3′) and H950 (5′GTCTCGGATTTAGGGGTTT-3′) [59]. This fragment includes the region historically surveyed for several marine turtle species in previous studies having lengths of 496$_{bp}$ (leatherback sea turtle), 491$_{bp}$ (green sea turtle) or 384$_{bp}$/480$_{bp}$ (hawksbill sea turtle). Our 25 µL polymerase chain reaction (PCR) included the following: genomic DNA, 1× PCR Buffer, 2 mM $MgCl_2$, 0.12 mM dNTP, 0.2, µm of each primer and 0.04 U/µL of Taq polymerase. After an initial 5 min denaturing step (94°C), our PCR protocol consisted of 35 cycles of the following temperature regime: 1 min at 94°C (denaturing), 1 min at 52°C (annealing) and 90 s at 72°C (extension). In addition, we included a final extension step of 10 min at 72°C. Following PCR, we removed single-stranded DNA by digesting 5 µL of PCR product with 2 µL of a combined Exonuclease I and Shrimp Alkaline Phosphatase solution (ExoSAP-IT®). The reaction mixture was incubated for 15 min at 37°C, followed by other 15 min incubation at 80°C to inactivate the two enzymes. We sequenced both forward and reverse strands using the BigDyeTM Primer Cycle Sequencing Kit (Applied Biosystems) run on an automated DNA sequencer (ABI PRISM 3100). For each sequencing reaction, we used 2 µL of our PCR product in a 10 µL reaction mix under the following conditions: 1 m denaturing step at 96$^{a}$ followed by 25 cycles consisting of an initial denaturing of 10 s at 96°C, 5 s at 50°C (annealing) and 4 m at 60°C (extension). Products were purified by ethanol precipitation before enter the sequencer.

Sequences were aligned by eye using the program BioEdit 5.0.9 [60] and compared with the short (<500$_{bp}$) haplotypes previously described for the leatherback turtle [61–63] and the hawksbill turtle [35,38,40,64–67]. Additionally, the whole fragment was compared to known long (>500$_{bp}$) haplotypes described in those manuscripts that used the same or similar primers in the leatherback turtle [62–63] and the hawksbill turtle [35,40–42,67]. Green turtle sequences were compared with the haplotypes found in the database maintained by the Archie Carr Center for Sea Turtle Research (http://accstr.ufl.edu/) that includes all published haplotypes. Posterior statistical analyses were carried out with short, long sequences or both depending on the data found in the published literature.

### Population Structure

Differences in haplotype frequencies of samples from the same species at different locations within Dominican Republic were assessed using a Chi-square test. Values were compared to the

**Table 1.** Haplotype frequencies found in Dominican Republic marine turtle populations.

| **Hawksbill (*E. imbricata*)** | | | | **Leatherback (*D. coriacea*)** | | | **Green (*C. mydas*)** | | |
|---|---|---|---|---|---|---|---|---|---|
| **Haplotype** | **GAN** | **DRJ** | **DRS** | **Haplotype** | **GAN** | **DR** | **Haplotype** | **GAN** | **DRS** |
| Ei-A01 (A/CU1) | EF210779 | 1 | 3 | Dc_A1 (A) | EF513272 | 38 | CM-A5 | Z50127 | 2 |
| Ei-A09 (F/c) | EF210783 | 2 | – | Dc_C (C) | EF513272 | 4 | | | |
| Ei-A11 (F/PR1) | EF210784 | 3 | 22 | | | | | | |
| Ei-A18 (L/PR3) | EF210786 | – | 2 | | | | | | |
| Ei-A20 (N/PR2) | EF210788 | – | 6 | | | | | | |
| Ei-A23 (Q/MX1) | EF210791 | 4 | – | | | | | | |
| Ei-A43 (Q/MX2) | EF210794 | 4 | – | | | | | | |
| Ei-A47 (L/PR3) | EF210787 | 1 | – | | | | | | |
| **TOTAL** | | **9** | **30** | | | **42** | | | **2** |

Hawksbill frequencies are given for 740$_{bp}$ haplotypes and the equivalences for 380$_{bp}$ and 480$_{bp}$ haplotypes are given in brackets respectively (380$_{bp}$/480$_{bp}$). Leatherback frequencies are given for 711$_{bp}$ haplotypes and the equivalences for 496$_{bp}$ are given in brackets. DRJ/DR: Jaragua National Park; DRS: Saona Island. GAN: GenBank Accession Number of each haplotype. Hawksbill haplotypes defined in [35]. Leatherback haplotypes defined in [62]. Green haplotype described in [84] and compiled in http://accstr.ufl.edu/.

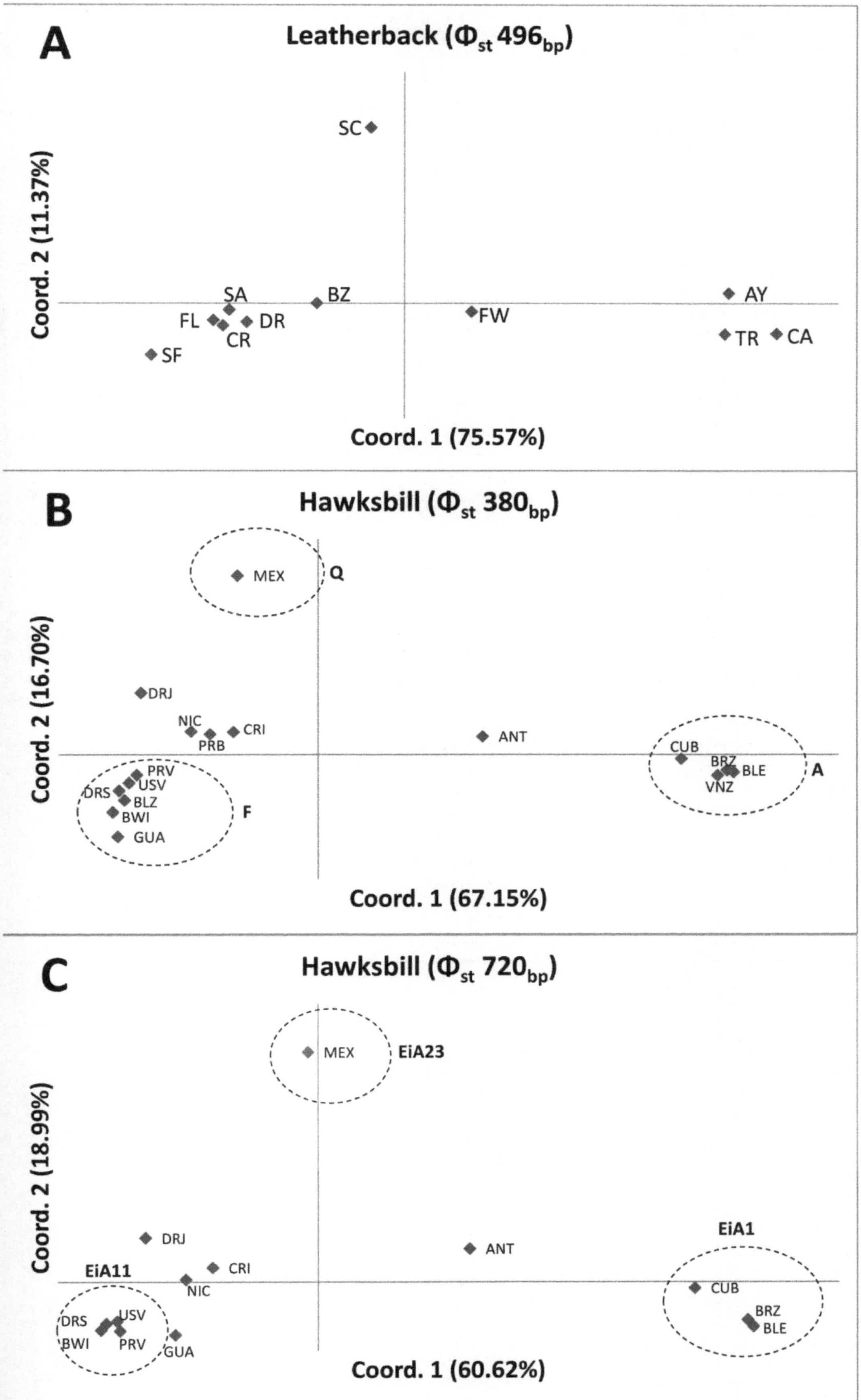
A
Leatherback (Φst 496bp)
SC
SA
BZ
FL
DR
CR
SF
FW
AY
TR
CA
Coord. 2 (11.37%)
Coord. 1 (75.57%)
B
Hawksbill (Φst 380bp)
MEX
Q
DRJ
NIC
CRI
PRB
ANT
CUB
BRZ
BLE
VNZ
A
PRV
USV
DRS
BLZ
BWI
GUA
F
Coord. 2 (16.70%)
Coord. 1 (67.15%)
C
Hawksbill (Φst 720bp)
MEX
EiA23
DRJ
ANT
EiA1
CRI
EiA11
NIC
CUB
USV
DRS
BWI
PRV
GUA
BRZ
BLE
Coord. 2 (18.99%)
Coord. 1 (60.62%)

**Figure 2. PCA including Atlantic leatherback and hawksbill populations using $\Phi_{st}$.** The percentage of the variability explained by each coordinate is shown in brackets. For the leatherback turtle(A) PCA was done using the 496bp fragment. For the hawksbill turtles PCA was done either considering the $380_{bp}$ fragment (B) and the $720_{bp}$ fragment (C). Leatherback nesting populations: AY: Awala-Yalimapo; BZ: Brazil; CA: Cayenne; CR Costa Rica (Atlanic); FL: Florida; FW: French West Indies; SA: South Africa; SC: St. Croix; SF: Suriname/FG; TR: Trinidad; DR: Dominican Republic. Hawksbill nesting populations: ANT: Antigua; BLE: Barbados Leeward; BWI: Barbados Windward; BLZ: Belize; BRZ: Brazil; CRI: Costa Rica; CUB: Cuba; GUA: Guadeloupe; MEX: Mexico; NIC: Nicaragua; PRV: Puerto Rico [35]; PRB: Puerto Rico [64]; USV: US Virgin Islands; VNZ: Venezuela; DRJ: Dominican Republic-Jaragua; DRS: Dominican Republic-Saona. Dashed circles represent the groups detected in previous studies [40] indicating the haplotype found at higher frequency.

distributions observed by randomizing individuals among populations using Monte-Carlo resampling [68] as implemented in the program CHIRXC [69]. Additionally we computed the exact test based on haplotypes frequencies [70] in Arlequin 3.0. Both analyses were used to test if samples from different locations could be grouped or should be considered separately.

In order to assess the genetic diversity compared to the other Atlantic populations we calculated haplotype diversity ($h$) and nucleotide diversity ($\pi$) [71] of each population and species using the program Arlequin 3.0 [72]. Fu's Fs neutrality test for the detection of population growth [73] was undertaken with the DnaSP 5.0 software package [74] for each nesting population of the Dominican Republic. Fs tends to be negative under an excess of recent mutations and a significant negative value was taken as evidence of recent population expansion. Differentiation among population pairs was assessed considering frequency based genetic distances ($\Phi_{st}$) using Arlequin 3.0 [72]. Significance of differentiation was tested using a Chi-square test and computing the exact test based on haplotypes frequencies as explained above. Genetic distances were used to perform a Principal Coordinate Analysis (PCA) with the package GenAlEX 6.2 [75] in order to distribute, in a two-dimension space, the genetic variability found. A sequential Bonferroni correction was not applied for multiple pair-wise comparisons, since they dramatically increase the probability for type II error ($\beta$: assume no differentiation when it does exist), an effect that becomes worse as many P-values are discarded [76–77]. In substitution, we applied the False Discovered Rate (FDR) correction that calculates the most appropriate threshold for the P-value significance considering the multiple comparisons involved in the analysis under an expected original threshold of $P<0.05$ [78].

## Hawksbill Juvenile and Male Dispersion

Considering our samples and published information available, a mixed stock analysis was performed in order to test the dispersion of hawksbill juveniles originating in Dominican Republic nesting areas. We used the *rookery centred* approach of the 'many-to-many' analysis [43] to test how juveniles hatched in the Dominican Republic disperse to all known Caribbean feeding grounds and a *mixed stock centred* approach to test the relative importance of Dominican Republic to the individual feeding grounds. For the nesting population baseline we used the haplotype frequencies previously described in the literature [35,40,67] but incorporating the frequencies of the Dominican Republic Jaragua and Saona (*present study*) (Table S1). Additionally, we tested the contribution of the Caribbean nesting populations to the juvenile feeding ground located south-west of the Dominican Republic [38]. All 'many to many' mixed stock analyses were conducted using only the short ($380_{bp}$) haplotype frequencies, as this analysis requires detailed information of haplotype frequencies of nesting populations and feeding areas and the dataset regarding putative feeding grounds available in the literature is most extensive for this fragment (Table S2). Nesting populations and feeding grounds located in the eastern Atlantic [79–80] were not included in the analysis as they have been shown to be highly isolated from all Caribbean populations and their contribution to Caribbean feeding grounds has been shown to be negligible [79]. Finally, a 'one-to-many' analysis was conducted to test the possible contribution of Dominican Republic turtles to the adult male aggregation found

**Table 2.** Genetic distances ($\Phi_{st}$) between Atlantic/Indic leatherback nesting populations.

| | AY | BZ | CA | CR | FL | FW | SA | SC | SF | TR | DR | Reference |
|---|---|---|---|---|---|---|---|---|---|---|---|---|
| AY | – | 0.168* | 0.037* | N | N | 0.108* | N | N | N | N | 0.277* | [63] |
| BZ | 0.184* | – | 0.193 | N | N | −0.019 | N | N | N | N | −0.061 | [62] |
| CA | 0.065* | 0.202 | – | N | N | 0.095 | N | N | N | N | **0.285*** | [63] |
| CR | **0.301*** | −0.065 | **0.306*** | – | N | N | N | N | N | N | N | [61] |
| FL | **0.299*** | −0.009 | 0.331 | −0.019 | – | N | N | N | N | N | N | [61] |
| FW | 0.126* | −0.018 | 0.097 | 0.040 | 0.090 | – | N | N | N | N | 0.022 | [63] |
| SA | **0.268** | −0.032 | 0.308 | −0.035 | 0.000 | 0.072 | – | N | N | N | N | [61] |
| SC | **0.251*** | 0.057 | **0.265*** | 0.108* | 0.113 | 0.095* | 0.093 | – | N | N | N | [61] |
| SF | **0.406*** | 0.058 | **0.415*** | 0.020 | 0.000 | 0.145 | 0.000 | **0.180*** | – | N | N | [61] |
| TR | 0.061* | 0.152 | −0.038 | **0.243*** | 0.274 | 0.057 | 0.253 | **0.222*** | **0.346*** | – | N | [61] |
| DR | **0.292*** | −0.061 | **0.289*** | −0.027 | 0.001 | 0.022 | −0.014 | 0.106* | 0.037 | **0.226*** | – | PS |

Genetic distances based on $496_{bp}$ and $711_{bp}$ sequences. Significant values given by the exact test and after FDR correction are marked with (*) for the exact text and in bold for the Chi-square test (for a $P<0.05$ $FDR_{496bp}=0.0109$). AY: Awala-Yalimapo; BZ: Brazil; CA: Cayenne; CR Costa Rica (Atlanic); FL: Florida; FW: French West Indies; SA: South Africa; SC: St. Croix; SF: Suriname/FG; TR: Trinidad; DR: Dominican Republic. Pacific populations [61] were included in the analysis but have not been included in the table as all were significantly different from all Atlantic/Indic populations. N: pairwise comparison not possible as $711_{bp}$ sequences were not available for some of the populations. PS: Present Study.

**Table 3.** Genetic distances ($\Phi_{st}$) between Caribbean hawksbill nesting populations.

| | ANT | BLE | BWI | BLZ | BRZ | CRI | CUB | GUA | MEX | NIC | PRV | PRB | USV | VNZ | DRJ | DRS | *Reference* |
|---|---|---|---|---|---|---|---|---|---|---|---|---|---|---|---|---|---|
| ANT | – | **0.373*** | **0.407*** | N | **0.381*** | **0.203*** | **0.206*** | **0.563*** | **0.514*** | **0.284*** | **0.495*** | N | **0.414*** | N | **0.345*** | **0.418*** | [67] |
| BLE | **0.372*** | – | **0.913*** | N | **0.076*** | **0.661*** | **0.076** | **0.951*** | **0.978*** | **0.678*** | **0.854*** | N | **0.849*** | N | **0.912*** | **0.902*** | [40] |
| BWI | **0.460*** | **0.925*** | – | N | **0.896*** | 0.070* | **0.740*** | **0.287*** | **0.679*** | 0.032* | 0.041* | N | −0.012 | N | 0.132* | 0.014* | [40] |
| BLZ | **0.381*** | **0.916*** | −0.027 | – | N | N | N | N | N | N | N | N | N | N | N | N | [107] |
| BRZ | **0.380*** | **0.099*** | **0.906*** | **0.892*** | – | **0.667*** | 0.085* | **0.936*** | **0.942*** | **0.684*** | **0.851*** | N | **0.843*** | N | **0.886*** | **0.888*** | [67] |
| CRI | **0.231*** | **0.644*** | 0.085* | 0.025 | **0.651*** | – | **0.517*** | **0.279*** | **0.412*** | 0.010* | **0.134*** | N | **0.070*** | N | **0.064*** | 0.078* | [67] |
| CUB | **0.210*** | 0.078 | **0.762*** | **0.716*** | **0.090*** | **0.511*** | – | **0.824*** | **0.758*** | **0.568*** | **0.755*** | N | **0.720*** | N | **0.688*** | **0.740*** | [67] |
| GUA | **0.599*** | **0.950*** | 0.008 | 0.070 | **0.937*** | **0.203*** | **0.839*** | – | **0.811*** | **0.240*** | **0.347*** | N | **0.308*** | N | **0.415*** | **0.405*** | [67] |
| MEX | **0.618*** | **0.997*** | **0.636*** | **0.629*** | **0.973*** | **0.311*** | **0.850*** | **0.723*** | – | **0.420** | **0.649** | N | **0.613** | N | **0.438** | **0.661*** | [35,67] |
| NIC | **0.329*** | **0.707*** | 0.048* | 0.012* | **0.711*** | 0.005* | **0.594*** | **0.136*** | **0.270*** | – | **0.085*** | N | 0.030* | N | 0.055* | 0.040* | [67] |
| PRV | **0.519*** | **0.840*** | **0.045*** | **0.048*** | **0.838*** | **0.138*** | **0.749*** | **0.081*** | **0.443*** | **0.111*** | – | N | 0.025* | N | **0.194*** | −0.012 | [35] |
| PRB | **0.316*** | **0.848*** | **0.212*** | **0.111*** | **0.828*** | 0.043* | **0.625*** | **0.428*** | **0.536*** | 0.093* | **0.167*** | – | N | N | N | N | [48] |
| USV | **0.453*** | **0.852*** | −0.019 | −0.023 | **0.846*** | **0.078*** | **0.730*** | 0.035* | **0.466*** | 0.044* | 0.036* | **0.182*** | – | N | 0.132* | −0.004 | [67] |
| VNZ | 0.231 | 0.000 | **0.838*** | **0.753*** | −0.003 | **0.495*** | −0.017 | **0.923*** | **0.994*** | **0.612*** | **0.786*** | **0.611*** | **0.773*** | – | N | N | [38] |
| DRJ | **0.433*** | **0.948*** | **0.165*** | **0.102*** | **0.919*** | 0.079* | **0.742*** | **0.343*** | **0.281*** | 0.053* | **0.157*** | 0.114* | 0.127* | **0.842*** | – | **0.136*** | PS |
| DRS | **0.461*** | **0.901*** | 0.010 | 0.005 | **0.887*** | 0.085* | **0.750*** | **0.068*** | **0.552*** | 0.064* | −0.014 | **0.131*** | 0.002 | **0.800*** | **0.123*** | – | PS |

Below the diagonal distances based on $380_{bp}$ traditional sequence, above diagonal distances based on the $720_{bp}$ sequence. Significant values given by the exact test and after FDR correction are marked with (*) for the exact text and in bold for the Chi-square test (for a $P<0.05$ $FDR_{380bp} = 0.0093$; $FDR_{740bp} = 0.0101$). ANT: Antigua; BLE: Barbados Leeward; BWI: Barbados Windward; BLZ: Belize; BRZ: Brazil; CRI: Costa Rica; CUB: Cuba; GUA: Guadeloupe; MEX: Mexico; NIC: Nicaragua; PRV: Puerto Rico [35]; PRB: Puerto Rico [64]; USV: US Virgin Islands; VNZ: Venezuela; DRJ: Dominican Republic-Jaragua; DRS: Dominican Republic-Saona. N: pairwise comparison not possible as $720_{bp}$ sequences were not available for some of the populations. PS: Present Study.

in Puerto Rico [35]. This analysis was conducted using either short (Table S1) or long sequences (Table S3). As we compared the nesting population haplotype frequencies to a single foraging area, the haplotype frequencies of other feeding grounds were not needed. All mixed stock analysis included population size as a weighting factor as several studies have proved that the inclusion of this factor improved the accuracy of results [41,81]. Source population sizes (as mean number of nesting females) were taken from the literature [22,67,82] (Tables S1 and S3).

## Lagrangian Buoys Dispersion

In order to simulate hatchling turtle dispersal, data describing the tracks of satellite-tracked surface drifter buoys were obtained from the Global Drifter Program of the National Oceanic and Atmospheric Administration (NOAA, USA). These Lagrangian buoys are periodically released throughout the year at varying locations and are tracked by satellite (RAMS, Argos, EOLE), providing several positional fixes per day (accuracy 0.1–2.0 km; http://www.aoml.noaa.gov/). An initial query within the locations database was undertaken in order to select all locations plotted near (<50 km) either the Jaragua National Park (DRJ/DR) or Saona Island (DRS) nesting populations. Then, all buoys with at least one location obtained near one of these Dominican Republic nesting areas were selected as an evidence of passive arrival or departure from the area.

# Results

The two green turtle samples were collected in Saona Island (DRS) and exhibited the haplotype CM-A5 (Table 1) which is found at high frequency in Suriname [83], Aves Island and Venezuela [84] but is also present at very low frequencies in other nesting populations of the Atlantic, including Mexico [83], Costa Rica [85], and São Tomé and Príncipe [86]. Due to this low sample size, no further statistical analyses were performed for this species.

## Leatherback Population Structure

Two haplotypes were found among leatherback samples when considering either the $496_{bp}$ fragment or the $711_{bp}$ fragment (Table 1). Genetic variability of the Dominican Republic nesting population was similar or higher than the other Atlantic populations [DR: $h$ (SD) = 0.176 (0.074), $\pi$ (SD) = 0.0014 (0.0012)] with the exceptions of some populations outside the Caribbean like Awala-Yalimapo [AY: $h$ (SD) = 0.780 (0.061), $\pi$ (SD) = 0.0051 (0.0033)], Cayenne [CA: $h$ (SD) = 0.519 (0.030), $\pi$ (SD) = 0.0042 (0.0027)], French West Indies [FW: $h$ (SD) = 0.340 (0.090), $\pi$ (SD) = 0.0027 (0.0020)], Saint Croix [SC: $h$ (SD) = 0.589 (0.067), $\pi$ (SD) = 0.0024 (0.0018)] and Trinidad [TR: $h$ (SD) = 0.501 (0.043), $\pi$ (SD) = 0.0040 (0.0026)]. No recent expansion was suggested for Dominican Republic leatherback population, independently of the length of the marker used (Fu's Fs neutrality test: $496_{bp}$: Fs = 2.664, P = 0.198; $711_{bp}$: Fs = 3.307, P = 0.134). As previous studies with the $496_{bp}$ fragment indicated [61], pairwise population analysis showed very deep differentiation between all Atlantic/Indic and all Pacific populations (data not shown) so the latter group of populations were not considered for future analysis. The Dominican Republic nesting population exhibited moderate levels of differentiation with other Atlantic/Indian ocean nesting populations (Table 2) being significantly different from Awala-Yalimapo, Cayenne, Saint Croix and Trinidad. The separation of these four populations from the others was corroborated by the PCA analysis, with their two principal coordinates explaining an accumulated 86.9% of the genetic variability found in all populations (Figure 2A). The comparisons using the $711_{bp}$ fragment yielded similar results (Table 2).

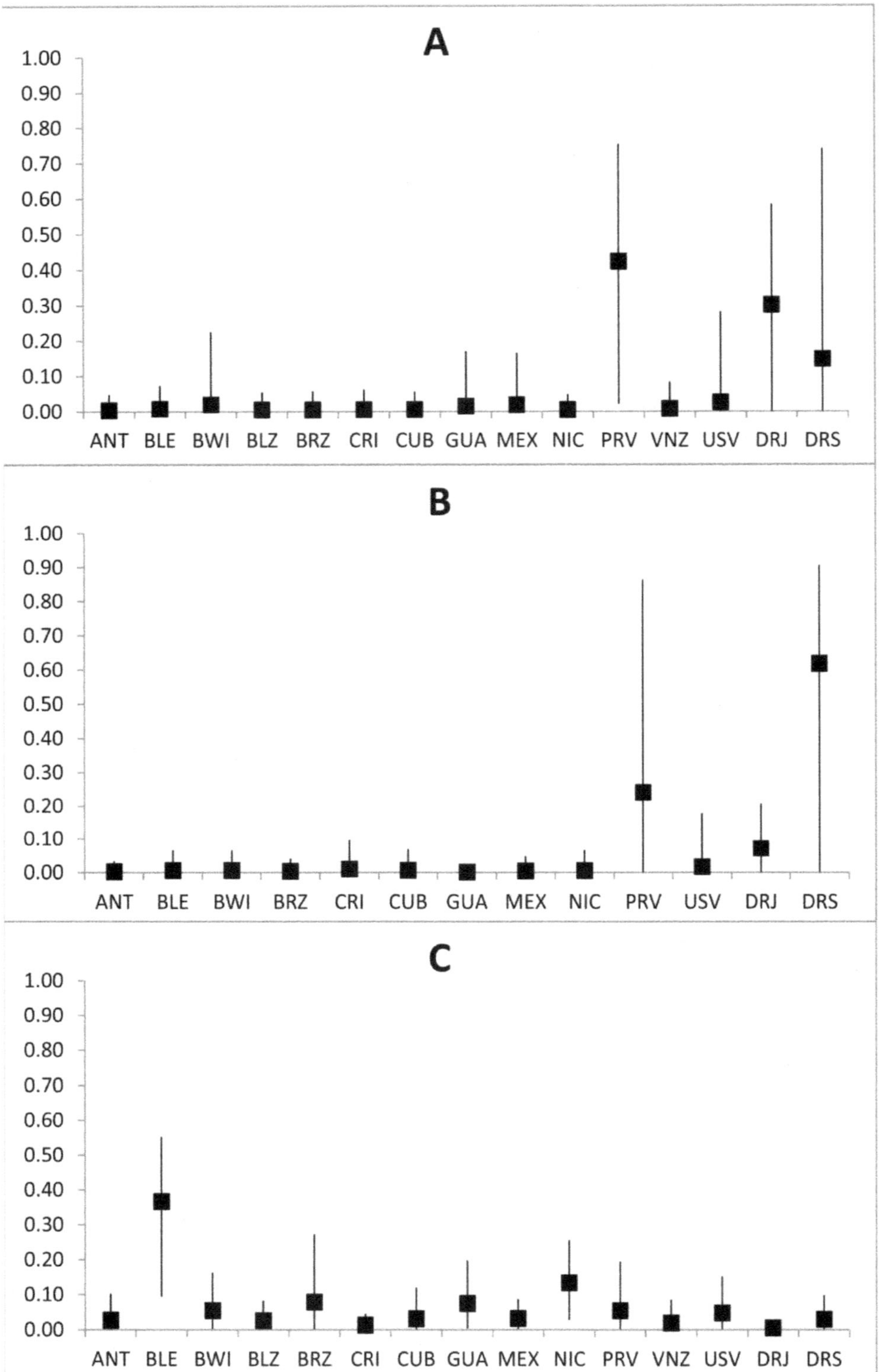


**Figure 3. Mixed Stock analysis.** Relative contribution and 95% confidence interval of each hawksbill turtle nesting area to the male aggregation of Mona Island (Puerto Rico) using A) the short ($380_{bp}$) mtDNA fragment, B) the long ($720_{bp}$) mtDNA fragment and C) to the SW Dominican Republic feeding ground using the short ($380_{bp}$) mtDNA fragment. ANT: Antigua; BLE: Barbados Leeward; BWI: Barbados Windward; BLZ: Belize; BRZ: Brazil; CRI: Costa Rica; CUB: Cuba; GUA: Guadeloupe; MEX: Mexico; NIC: Nicaragua; PRV: Puerto Rico [35]; PRB: Puerto Rico [64]; USV: US Virgin Islands; VNZ: Venezuela; DRJ: Dominican Republic-Jaragua; DRS: Dominican Republic-Saona.

**Table 4.** 'Many to many' mixed stock analysis results.

| | *Rookery centred* | | *Mixed stock centred* | | |
|---|---|---|---|---|---|
| **Mixed stock** | **DRJ** | **DRS** | **DRJ** | **DRS** | *Reference* |
| 1. Texas | 6.9 (0.2–23.3) | 2.9 (0.1–10.9) | 0.8 (0.0–3.0) | 2.2 (0.1–8.1) | [38] |
| 2. Bahamas | 7.5 (0.2–25.8) | 5.1 (0.2–18.7) | 0.6 (0.0–2.3) | 2.6 (0.1–8.9) | [38] |
| 3. Cuba D | 8.8 (0.3–30.4) | 9.4 (0.4–28.3) | 0.7 (0.0–2.8) | 5.2 (0.2–16.1) | [65] |
| 4. Cuba B | 7.6 (0.2–26.7) | 8.2 (0.3–26.3) | 0.4 (0.0–1.4) | 2.6 (0.1–9.4) | [65] |
| 5. Cuba A | 7.6 (0.2–25.5) | 6.3 (0.2–23.9) | 0.3 (0.0–1.1) | 1.5 (0.0–5.7) | [65] |
| 6. Turk and Caicos | 7.4 (0.2–23.1) | 6.7 (0.2–22.5) | 0.5 (0.0–1.9) | 2.8 (0.1–9.4) | [42] |
| 7. Cayman Islands | 7.8 (0.2–26.7) | 8.5 (0.2–27.8) | 0.4 (0.0–1.7) | 3.1 (0.1–10.0) | [41] |
| 8. Dominican Republic | 8.0 (0.3–27.0) | 7.2 (0.2–23.0) | 0.5 (0.0–2.0) | 3.0 (0.1–9.8) | [38] |
| 9. Puerto Rico residents | 7.1 (0.2–25.2) | 7.4 (0.2–26.5) | 0.4 (0.0–1.7) | 3.0 (0.1–11.2) | [35] |
| 10. Puerto Rico recruits | 7.8 (0.2–27.2) | 11.6 (0.4–33.5) | 0.3 (0.0–1.3) | 3.5 (0.1–12.4) | [35] |
| 11. Puerto Rico pooled | 8.7 (0.2–27.8) | 11.6 (0.4–31.9) | 0.6 (0.0–2.1) | 5.3 (0.2–15.3 | [38,65] |
| 12. US Virgin Islands | 7.4 (0.2–25.2) | 7.0 (0.2–24.7) | 0.3 (0.0–1.2) | 2.1 (0.1–7.9) | [38] |
| 13. Unknown | 7.3 (0.2–26.0) | 8.0 (0.3–28.2) | NA | NA | |

*Rookery centred* analysis includes the percentage of Dominican Republic juveniles that use a mixed stock and an esteem of juveniles that disperse to *unknown* juvenile feeding grounds (last line). *Mixed stock centred* analysis includes the percentage of turtles from the mixed stock coming from Dominican Republic nesting populations. 95% confidence intervals are shown in brackets. NA: Not applicable.

## Hawksbill Population Structure

A total of 8 haplotypes were found in the Dominican Republic hawksbill samples using the whole $720_{bp}$ sequence (Table 1). When we truncated the sequence for comparisons with previous studies, we found 5 haplotypes considering the $380_{bp}$ and 6 considering the $480_{bp}$ fragment (Table 1). The two sample sites in the Dominican Republic were genetically different considering both short and long fragments ($380_{bp}$ fragment: $\Phi_{st} = 0.123$; $720_{bp}$ fragment: $\Phi_{st} = 0.223$; $P<0.001$ in all cases, both for the Chi-square and exact test) and hence were treated as two independent units for all posterior analysis. Genetic variability of the two Dominican Republic populations was similar or higher than other nesting populations [$380_{bp}$: DRJ: $h$ (SD) = 0.638 (0.093), $\pi$ (SD) = 0.0037 (0.0027); DRS: $h$ (SD) = 0.526 (0.089), $\pi$ (SD) = 0.0044 (0.0029); $720_{bp}$: DRJ: $h$ (SD) = 0.848 (0.054), $\pi$ (SD) = 0.0047 (0.0029); DRS: $h$ (SD) = 0.527 (0.089), $\pi$ (SD) = 0.0035 (0.0021)]. No recent expansion was suggested for either populations irrespective of the length of the marker used (Fu's Fs neutrality test: $380_{bp}$: DRJ: Fs = 0.440, P = 0.269; DRS: Fs = 1.957, P = 0.167; $720_{bp}$: DRJ: Fs = 0.286, P = 0.233; DRS: Fs = 3.581, P = 0.062). Analysis of genetic structuring of Dominican Republic nesting beaches in relation to other populations in the Caribbean showed deep levels of differentiation with the exceptions of the DRS population that showed no significant differences to the proximate nesting aggregation of Mona Island in Puerto Rico (PRV), the populations of Belize (BLZ), Barbados Windward (BWI) and U.S. Virgin Islands (USV) (Table 3). The similarity between DRS and PRV, and between DRS and USV was confirmed also for the $720_{bp}$ fragment while BWI yielded significant differentiation with DRS by means of the exact test, but not for the Chi-square test (Table 3). The lack of differentiation between DRS and BLZ was not confirmed as no long sequences were available for the latter. The accumulated first two coordinates of the PCA explained a high percentage of the genetic variability found both for the $380_{bp}$ (83.9%; Figure 2B) and the $720_{bp}$ (79.61%; Figure 2C) fragments and confirmed the complete isolation of DRJ and the relative proximity of DRS to PRV, USV and BWI (Figure 2C).

## Hawksbill Juvenile and Male Dispersion

The *rookery centred* 'many-to-many' mixed stock analysis for Dominican Republic juveniles suggested that turtles originating from the two nesting areas are likely to have been distributed in foraging grounds across the Caribbean, although the 95% confidence intervals were very high (Table 4). The *mixed stock centred* analysis showed that the proportion of turtles in the genotyped Caribbean juvenile feeding grounds coming from Dominican Republic nesting populations is very low, being less than 0.01 from Jaragua National Park and less than 0.06 from Saona Island in all cases (Table 4). Despite these results, the mixed stock analysis with the short fragment showed that both Jaragua National Park (DRJ) and Saona Island (DRS) significantly contributed to the adult male aggregation in Puerto Rico, along with the breeding stocks from Puerto Rico itself, (DRJ: 0.30 (0.00–0.58), DRS: 0.15 (0.00–0.74), Figure 3A). However, the contribution of Saona Island exceeded that for Jaragua National Park when the new long fragment was considered (DRJ: 0.07 (0.00–0.21), DRS: 0.62 (0.00–0.91), Figure 3B). The population of Barbados leeward has the highest contribution of juveniles to the SW Dominican Republic feeding aggregation (Figure 3B).

## Lagrangian Buoys Dispersion

A total of 22 passive drifter buoys approached and/or left the Jaragua National Park and Saona Island areas (DRJ: n = 10; DRS: n = 11; both: n = 1) (Figure 4) between 1996 and 2010. All buoys that arrived at the Dominican Republic did so from the south-east of the island. They arrived year round, with 72.7% being recorded during the sampling period and with 45.5% arriving during the summer. However, buoys that left the Dominican Republic travelled in different directions: all those from Jaragua travelled south-west into the Caribbean (Figure 4A) while some of those from Saona Island (25%) travelled north into the open Atlantic

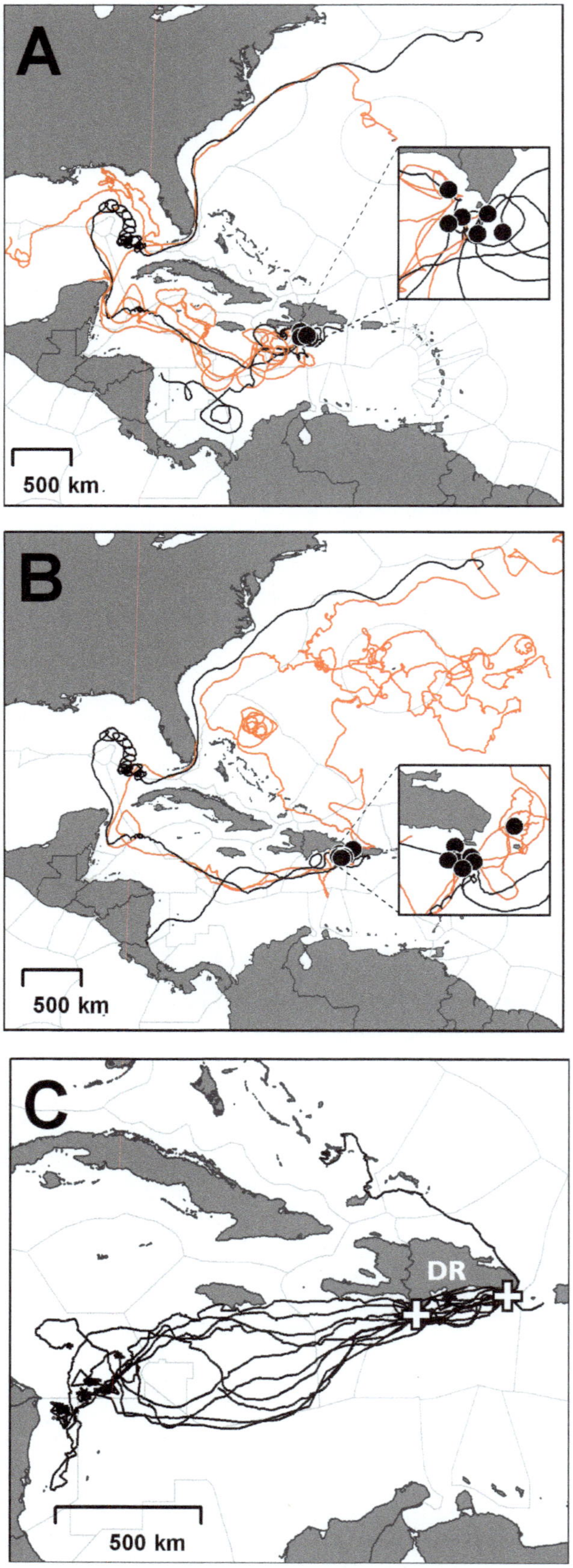


**Figure 4. Tracks of passive drifter buoys and satellite traked turtles.** (A, B) Tracks of passive drifter buoys on departure from the (A) Jaragua National Park or (B) Saona Island (starting point indicated with black circles). Red lines show tracks on departure during the nesting season. Figure (C) shows the migrations of ten satellite tracked adult female hawksbill turtles [102]. White crosses show deployment locations for satellite tracked turtles in Jaragua and Saona Island. Pale grey lines show the extent of the Exclusive Economic Zone for each Caribbean country. Note part (C) is to a different scale.

(Figure 4B). These differences in the trajectories remained when only tracks during the nesting season were considered (Figure 4). Mean drifting time was 307 days but was highly variable between buoys (7–941 days). Mean drifting time of the buoys following departure from study area (Figure 4) was 265 days (10–883 days).

## Discussion

Conservation planning for mobile species such as marine turtles, depends on robust spatio-temporal information about the RMU of interest [13]. Since the MUs [29], which compose a RMU, may be sharing common reproductive or feeding areas (even if they are genetically isolated), the lack of information about many MUs may lead to poor or biased management effort. Such information gaps are particularly pertinent where threats vary among MUs [87–88].

### Green Turtle

Green turtles in the Dominican Republic [22] declined from *ca.* 260 green turtles nesting per year in the 1980s [14] to near extirpation at present [22], and precluded robust analysis of population structuring or connectivity. However, they appear to share a haplotype very common in Suriname and Aves Island, hence suggesting that population recovery through immigration could be possible in the future if these populations are still connected.

### Leatherback Turtle Population Structure

The leatherback turtle is thought to be the least philopatric of the marine turtles [89–91], and consequently genetic markers in the present study and others [61] have demonstrated high levels of population connectivity. The nesting populations of Awala-Yamalipo, Cayenne and Trinidad emerged as a distinct MU, separated from St Croix MU. The Dominican Republic nesting population from the present study was similar to the other Caribbean nesting populations forming part of a third MU in the region. The use of $711_{bp}$ sequences did not change the number of haplotypes in the Dominican Republic in contrast to previous studies, where some $496_{bp}$ haplotypes were subdivided into multiple $711_{bp}$ haplotypes [62–63]. Future work expanding the use of $711_{bp}$ sequences to other nesting and feeding areas is needed in order to test if longer sequences improve resolution of population structuring as they have for the loggerhead turtle [44].

The connection between the Dominican Republic leatherback population with several larger populations within the Atlantic would favour putative recovery through immigration [89–91]. However, unless ameliorated, the anthropogenic stressors that have contributed to population decline would likely affect immigrant turtles equally and hence our study site may act as a sink. The Costa Rica nesting aggregation, for example, is one of the largest populations in the Caribbean, but is thought to be declining despite some of the most heavily resourced conservation efforts in the world [90]. Sink areas like the Dominican Republic may have been partially responsible for such a decline and is worthy of future investigation. Finally, the genetic signature of important nesting sites in Colombia-Panama [92] still remains unknown, and hence their potential role as source populations within the Caribbean remains to be elucidated.

## Hawksbill Turtle Population Structure

Previous studies in the Caribbean have shown that the hawksbill turtle exhibits a high degree of philopatry resulting in fine scale population structure [38,65–66,93], sometimes at surprisingly small geographic scales [40]. The present study offers striking support for this, demonstrating that in the Dominican Republic, the two hawksbill populations were genetically distinct despite being separated by only 300 km straight-line distance. Hence, we strongly recommend finer scale sampling efforts in order to detect the genetic richness of a given territory as different MUs may be separated by much smaller distances [40] than classically thought [93].

Beyond the fine scale structuring detected within the Dominican Republic, the two populations are clearly isolated from almost all other Caribbean nesting populations with the exception of DRS with Puerto Rico, US Virgin Islands and Belize in terms of maternally inherited DNA (mtDNA). The nesting aggregation in Mona Island, Puerto Rico [35,82] is only 68 km straight distance from Saona Island, in the Dominican Republic, while the US Virgin Islands are 410 km. The connection with Belize is more difficult to explain but this population is located at the same latitude as the Dominican Republic. A study of the loggerhead marine turtles in Florida [94] noted that haplotype frequencies were similar for nesting populations at similar magnetic fields, a putative homing cue for nesting females. However, the statistical significance between Belize and DRS disappeared only after FDR correction for one of the tests ($P = 0.042$ for the exact test) and the Belizean sample size was very small, and thus potentially misleading. Future analyses of the Belizean nesting population including more samples are needed to test if this connection is real or an artefact of sample size. The populations of Puerto Rico, US Virgin Islands and DRS exhibited some levels of connection with Barbados (windward), depending on the length of the marker and the statistical test used. This upper level of structuring was clearly detected by the PCA analysis and supports the AMOVA analyses conducted in previous studies [40]. In both cases, three clusters were clearly defined and the haplotype composition suggests an evolutionary origin for such structuring. The first group is characterized by the high frequency of the Q ($380_{bp}$)/EiA23 ($720_{bp}$) haplotype, the second by the high frequency of the F ($380_{bp}$)/EiA11 ($720_{bp}$) haplotype and the last by the high frequency of the A ($380_{bp}$)/EiA1 ($720_{bp}$) haplotype (Figure 2B, 2C). Based on data gathered using multiple techniques, the whole Caribbean has been proposed as a unique RMU for the hawksbill sea turtle [13], comprised of several different isolated MUs defined using genetic markers [35,38,40,93]. Here we highlight the existence of an intermediate level of structuring associated with haplotype composition, as detected in previous studies [40]. Whether this intermediate level corresponds to a RMU or not needs to be tested in the future using other markers (such as microsatellites or SNPs) and combining the results with all available information of the species in the area.

The structuring of the Caribbean stocks has strong implications for the conservation of the hawksbill nesting populations of the Dominican Republic. The DRJ population is completely isolated, suggesting it is unlikely to receive females by immigration. The small population size [22] could favour the loss of the maternally inherited genetic diversity through mechanisms such as genetic drift or inbreeding. On the other hand, DRS population fate would appear to be linked to other nesting populations, especially Puerto Rico and US Virgin Islands (but also Barbados and perhaps Belize). However, this also means that the threats to the species detected in the area of Saona have potential impacts for these linked nesting aggregations. Fortunately, populations of hawksbill turtles nesting in Puerto Rico [82] and Barbados [82,95] have increased recently. This increase may have a future positive effect on DRS nesting population through immigration of non-philopatric females from these two areas.

Regardless of female philopatry, hawksbill turtles appear to be highly mobile during the juvenile developmental phase. Blumenthal et al. [41] modelled the passive dispersal of virtual particles released in known hawksbill nesting areas during the hatchling dispersal phase and compared it with genetic markers to support the hypothesis that juvenile dispersion is highly dependent on current patterns. For the Greater Antilles ecoregion, the model predicted that the particles would divide in two well differentiated branches when arriving the Dominican Republic, one heading south-west and entering the Caribbean and the other heading north and north-west and entering the open Atlantic. Lagrangian particle modelling is an excellent approach to describe the general pattern of passive dispersal as it easily provides a statistically robust sample size, has a wide spatio-temporal coverage [49,96] and it is possible to add behavioural components to the modelled particles [53,97–98]. However, this approach is sensitive to model resolution [96] and sometimes fails in the detection of small scale current variations [49]. On the other hand, Lagrangian drifting buoys reflect high resolution drifting trajectories, but usually have a limited spatio-temporal coverage and sample sizes are much lower than modelling, so biases can be also found [49]. For these reasons it is desirable to contextualise findings with both techniques to study the effect of passive drifting in dispersal [49,96]. The Lagrangian drifting buoys that departed from the Dominican Republic (present study) followed similar patterns than described in the Lagrangian particles model of Blumenthal et al. [41] but also showed that, depending on the release point, the buoys may take only the southern branch into the Caribbean (Jaragua National park, Figure 4A) or also enter the open Atlantic (Saona Island, Figure 4B). This subtle small scale variation on the current system may produce a differential dispersion of hatchlings from both Dominican Republic nesting populations, as hatchlings are known to be highly influenced by the dominant currents. However, marine turtle hatchlings and juveniles are also known to be able to contribute substantially to the net movement by weak directed swimming thus escaping from cold or highly predated areas [49,53,56,97–98]. Thus, directed swimming of hatchlings from these areas could greatly affect the proportion of individuals entering the open Atlantic or the inner Caribbean sea. Recent papers proposed the 'learned migration goal' hypothesis (LMG) [51,56] to explain how adult marine turtles can be influenced by the currents system despite being able to swim against them. Under the LMG theory, adult turtles would follow the preferred route from those learned as hatchlings and juveniles and fixed in a magnetic map [99–101]. Thus, the subtle oceanic differences detected within the vicinity of Dominican Republic nesting beaches would be reflected in females' behaviour and could explain the significant genetic differences observed between Saona Island and Jaragua National Park nesting populations, irrespective of their proximity to one another. A recent telemetry study [102] showed that the nesting females that left the country (the majority from Saona Island) took the southwestern route after the nesting season, heading into the Caribbean towards the western Caribbean basin (Figure 4C). The only satellite tracked adult female to head northwards was deployed from Saona Island, supporting the idea that adult turtle migration may be still partially influenced by the oceanographic currents that affected them as hatchlings and juveniles [51,56].

## Hawksbill Juvenile and Male Dispersion

Considering all these results and the putative dispersal mechanisms of the species [41], one may expect that DRJ juveniles would disperse mainly southwards towards inner Caribbean feeding grounds while DRS juveniles may disperse north and eastwards into the open Atlantic, Bahamas, Cuba (foraging ground D) or the Turks and Caicos Islands. Surprisingly, the *rookery centred* mixed stock analysis, did not support this hypothesis, as it predicted an homogeneous distribution of juveniles originating in both Dominican Republic nesting populations. Such mixed stock analysis usually yields wide confidence intervals in marine turtles due to the existence of common haplotypes and results may be taken with caution. Furthermore, the population sizes of the two Dominican Republic nesting populations [22] are one or two orders of magnitude lower than the most abundant nesting areas [82] and hence the production of hatchlings is likely much lower. As a consequence, the contribution of Dominican Republic populations to all the juvenile feeding grounds would be necessarily very low, as indicated by the *mixed stock centred* analysis. For instance, the juvenile feeding ground located near DRJ [38,103] receives turtles mainly from Barbados, [40] in agreement with the particle dispersal model [41] and the drifter buoys that arrived in the area (*present study*), but it also receives turtles from Cuba. This means that any extreme mortality of this juvenile aggregation would have an impact on these populations.

The contribution of both Dominican Republic nesting populations to the Puerto Rico male aggregation is very clear. Satellite tracks of adult males and females in Puerto Rico showed that adults move from Puerto Rico and arrive to the vicinity of DRJ and DRS and remained there, possibly to breed [104]. The present study has shown that adult individuals from these three nesting aggregations share the same foraging areas. Given that DRJ is clearly a different genetic unit, the fact that adult males are using the same foraging areas may indicate the existence of common mating areas and opens the possibility of male mediated gene flow, as found for other sea turtle species [32–33,105]. The use of biparentally inherited markers, such as microsatellites or SNPs, is needed to test the existence of male mediated gene flow in these areas.

## Genetic Variability of Leatherback and Hawksbill Nesting Populations

Historical measures of the genetic variability would be desirable in order to detect a recent loss of the genetic variability in the Dominican Republic nesting populations caused by the recent population decline [22]. Unfortunately, only recent measures from other populations can be obtained using current haplotype frequencies, and all populations have been affected to some extent by human activities. However, the conservation status and population sizes of marine turtle populations within the Caribbean are highly variable, so a comparison of the genetic variability among them would provide a relative measure of the genetic health of the studied populations. Thus, it is reasonable to conclude that the extreme recent reduction of the leatherback and hawksbill Dominican Republic nesting populations has not yet been reflected in a substantial loss of genetic variability, as they had values similar or higher to almost all other populations in the same area that have high population sizes or are increasing due to conservation efforts. The addition of measures of variability obtained from biparentally inherited markers would provide greater insight about how population reduction impacts in the genetic variability.

## Conclusions

One of the milestones in the conservation of endangered species is the detection and quantification of threats affecting declining populations. However, the contextualisation of these potential conservation sinks in a wider regional area is necessary when complex life histories and complex populations structures are present, such as in sea turtles [8–9]. The present study provides one such case study and goes beyond the detection of the threats of the Dominican Republic and the local decline [22]. The detection of fine scale structuring within Dominican Republic hawksbill populations, the isolated nature of the Jaragua National Park hawksbill nesting population, the establishment of migratory pathways involving the threatened Dominican Republic marine turtles and the use of common feeding grounds at different life stages has been crucial in understanding which populations might be affected by a local sink and which healthy populations might act as a source of individuals to help the recovery of threatened populations.

## Supporting Information

**Table S1 Haplotype frequencies of Caribbean Hawksbill marine turtle nesting populations.** Using the short (380 bp) fragment and including nesting population size (nests/year) used as a baseline for the Mixed Stock Analysis. Puerto Rico data from PRB [1] was not included in the baseline as PRV [2] included a larger sampling set from the same location. PS: Present Study.

**Table S2 Haplotype frequencies of Caribbean Hawksbill marine turtle feeding grounds.** Using the short (380 bp) fragment.

**Table S3 Haplotype frequencies of Caribbean Hawksbill marine turtle nesting populations.** Using the long (720 bp) fragment and including nesting population size (nests/year) used as a baseline for the Mixed Stock Analysis. PS: Present Study.

## Acknowledgments

We specially thank the support of the Dominican NGO Grupo Jaragua and of The Nature Conservancy (TNC). The project counted on the support of the Dominican Environmental authorities. We are very grateful to rangers of the JNP and Del Este NP, especially to Pelagio Paulino 'Negro'. The authors wish also to acknowledge use of the Maptool program for analysis and graphics in this paper. Maptool is a product of SEATURTLE.ORG. (Information is available at www.seaturtle.org).

## Author Contributions

Conceived and designed the experiments: JT BJG CC. Performed the experiments: CC. Analyzed the data: CC LAH. Contributed reagents/materials/analysis tools: OR BJG JT JAR YML. Wrote the paper: CC BJG JT. Reviewed the manuscript: YML LAH OR JAR.

## References

1. Schmidt J (1923) Breeding places and migrations of the eel. Nature 111: 51–54.
2. Campagna C, Piola AR, Marin MR, Lewis M, Fernandez T (2006) Southern elephant seal trajectories, fronts and eddies in the Brazil/Malvinas Confluence. Deep-Sea Res Part I-Oceanogr Res Pap 53: 1907–1924.
3. Cotte C, Park YH, Guinet C, Bost CA (2007) Movements of foraging king penguins through marine mesoscale eddies. P Roy Soc B-Biol Sci 274: 2385–2391.

4. Cotte C, d'Ovidio F, Chaigneau A, Levy M, Taupier-Letage I, et al. (2011) Scale-dependent interactions of Mediterranean whales with marine dynamics. Limnol Oceanogr 56: 219–232.
5. Horton TW, Holdaway RN, Zerbini AN, Hauser N, Garrigue C, et al. (2011) Straight as an arrow: humpback whales swim constant course tracks during long-distance migration. Biol Lett 7: 674–679.
6. Sleeman JC, Meekan MG, Wilson SG, Polovina JJ, Stevens JD, et al. (2010) To go or not to go with the flow: Environmental influences on whale shark movement patterns. J Exp Mar Biol Ecol 390: 84–98.
7. Lohmann KJ, Witherington BE, Lohmann CMF, Salmon M (1997) Orientation, navigation and natal beach homing in sea turtles. In: Lutz PL, Musick JA, editors. The biology of sea turtles. Boca Raton: CRC Press. 107–136.
8. Bowen BW, Karl SA (2007) Population genetics and phylogeography of sea turtles. Mol Ecol 16: 4886–4907.
9. Lee PLM (2008) Molecular ecology of marine turtles: New approaches and future directions. J Exp Mar Biol Ecol 356: 25–42.
10. Hays GC, Scott R (2013) Global patterns for upper ceilings on migration distance in sea turtles and comparisons with fish, birds and mammals. Funct Ecol DOI: 10.1111/1365-2435.12073.
11. Hamann M, Godfrey MH, Seminoff JA, Arthur K, Barata PCR, et al. (2010) Global research priorities for sea turtles: informing management and conservation in the 21st century. Endang Species Res 11: 245–269.
12. DiMatteo AD, Fujioka E, Wallace BP, Hutchinson BJ, Cleary J, et al. (2009) SWOT Database Online. Available: http://seamap.env.duke.edu/swot.
13. Wallace BP, DiMatteo AD, Hurley BJ, Finkbeiner EM, Bolten AB, et al. (2010) Regional Management Units for Marine Turtles: A Novel Framework for Prioritizing Conservation and Research across Multiple Scales. Plos One 5 (12): e15465. doi:10.1371/journal.pone.0015465.
14. Ottenwalder JA (1981) Estudio preliminar sobre el estado, distribución, y biología reproductiva de las tortugas marinas en la República Dominicana. Tesis de Licenciatura, Universidad Autónoma de Santo domingo, Santo Domingo.
15. Ross JP, Ottenwalder JA (1983) The leatherback sea turtle, *Dermochelys coriacea*, nesting in the Dominican Republic. Advances in herpetology and evolutionary biology Essays in honor of Ernest E Williams. 706–713.
16. Dominguez TJ, Villalba AA (1994) Trade of hawksbill carapaces in Santo Domingo, Dominican Republic. In: Bjorndal A, Bolten A, Johnson DA. Eliazar PJ, editors. Proceedings of the fourteenth Annual Symposium on Sea turtle Biology and Conservation: 34–35.
17. Mota JM, Leon YM (2003) Beliefs and perceptions associated with sea turtle products in the Dominican Republic. In: Pilcher NJ, editor. Proceedings of the twenty-third Annual Symposium on Sea Turtle Biology and Conservation: 197–199.
18. Feliz P, Leon YM, Tomás J, Hierro K, Mateo A, et al. (2008) Tortoiseshell trade in Santo Domingo, Dominican Republic: discouraging news for Caribbean hawksbill. In: Dean K, López Castro MC, editors. Proceedings of the 28th Annual sea turtle symposium, Loreto, Baja California Sur, Mexico: 247.
19. Ottenwalder JA (1996) Conservation and Management of Sea Turtles in the Dominican Republic. CITES.
20. Aucoin S, Leon YM (2007) Preliminary data on hawksbill turtle (*Eretmochelys imbricata*) bycatch in an artisanal gillnet used near Jaragua National Park, Dominican Republic. Proceedings of the Gulf and Caribbean Fisheries Institute 60: 169–172.
21. Powell R, Inchaustegui SJ (2009) Conservation of the herpetofauna of the Dominican Republic. Appl Herpetol 6: 103–122.
22. Revuelta O, León YM, Feliz P, Godley BJ, Raga A, et al. (2012) Protected areas host important remnants of marine turtle nesting stocks in the Dominican Republic. Oryx 46: 348–358.
23. Fleming EH (2001) Swimming against the Tide: recent surveys of exploitation, trade and management of marine turtles in the Northern Caribbean. Washington DC, USA: TRAFFIC North america.
24. Stam S, Stam R (1992) Turtle trouble in the Dominican Republic. Marine Turtle Newsletter 57: 19.21.
25. Reuter A, Allan C (2006) Tourists, turtles and trinkets: a look at the trade in marine turtle products in the Dominican Republic and Colombia. Washington, DC, USA: TRAFFIC North America.
26. Leon YM (2004) Community impacts of coastal tourism in the Dominican Republic. PhD dissertation. Kingston, Rhode Island, USA: University of Rhode Island.
27. Wielgus J, Cooper E, Torres R, Burke L (2010) Coastal capital: Dominican Republic. Case studies on the economic value of coastal ecosystems in the Dominican Republic. Washington DC, USA: World Resources Institute.
28. Laurent L, Casale P, Bradai MN, Godley BJ, Gerosa G, et al. (1998) Molecular resolution of marine turtle stock composition in fishery bycatch: a case study in the Mediterranean. Mol Ecol 7: 1529–1542.
29. Moritz C (1994) Defining Evolutionarily Significant Units for Conservation. Trends Ecol Evol 9: 373–375.
30. Carreras C, Pont S, Maffucci F, Pascual M, Barcelo A, et al. (2006) Genetic structuring of immature loggerhead sea turtles (*Caretta caretta*) in the Mediterranean Sea reflects water circulation patterns. Mar Biol 149: 1269–1279.
31. Bowen BW, Abreu-Grobois A, Balazs GH, Kamezaki N, Limpus CJ, et al. (1995) Trans-Pacific migrations of the loggerhead turtle (*Caretta caretta*) demonstrated with mitochondrial DNA markers. Proc Natl Acad Sci U S A 92: 3731–3734.
32. Bowen BW, Bass AL, Soares L, Toonen RJ (2005) Conservation implications of complex population structure: lessons from the loggerhead turtle (*Caretta caretta*). Mol Ecol 14: 2389–2402.
33. FitzSimmons NN, Moritz C, Limpus CJ, Pope L, Prince R (1997) Geographic structure of mitochondrial and nuclear gene polymorphisms in Australian green turtle populations and male-biased gene flow. Genetics 147: 1843–1854.
34. Carreras C, Pascual M, Cardona L, Marco A, Bellido JJ, et al. (2011) Living Together but Remaining Apart: Atlantic and Mediterranean Loggerhead Sea Turtles (*Caretta caretta*) in Shared Feeding Grounds. J Hered 102: 666–677.
35. Velez-Zuazo X, Ramos WD, van Dam RP, Diez CE, Abreu-Grobois A, et al. (2008) Dispersal, recruitment and migratory behaviour in a hawksbill sea turtle aggregation. Mol Ecol 17: 839–853.
36. Bolten AB, Bjorndal KA, Martins HR, Dellinger T, Biscoito MJ, et al. (1998) Transatlantic developmental migrations of loggerhead sea turtles demonstrated by mtDNA sequence analysis. Ecol Appl 8: 1–7.
37. Bowen BW, Bass AL, Chow SM, Bostrom M, Bjorndal KA, et al. (2004) Natal homing in juvenile loggerhead turtles (*Caretta caretta*). Mol Ecol 13: 3797–3808.
38. Bowen BW, Grant WS, Hillis-Starr Z, Shaver DJ, Bjorndal A, et al. (2007) Mixed-stock analysis reveals the migrations of juvenile hawksbill turtles (*Eretmochelys imbricata*) in the Caribbean Sea. Mol Ecol 16: 49–60.
39. Monzon-Arguello C, Rico C, Carreras C, Calabuig P, Marco A, et al. (2009) Variation in spatial distribution of juvenile loggerhead turtles in the eastern Atlantic and western Mediterranean Sea. J Exp Mar Biol Ecol 373: 79–86.
40. Browne D, Horrocks J, Abreu-Grobois A (2010) Population subdivision in hawksbill turtles nesting on Barbados, west Indies, determined from mitochondrial DNA control region sequences. Conserv Genet 11: 1541–1546.
41. Blumenthal JM, Abreu-Grobois FA, Austin TJ, Broderick AC, Bruford MW, et al. (2009) Turtle groups or turtle soup: dispersal patterns of hawksbill turtles in the Caribbean. Molecular Ecology 18: 4841–4853.
42. Richardson PB, Bruford MW, Calosso MC, Campbell LM, Clerveaux W, et al. (2009) Marine Turtles in the Turks and Caicos Islands: Remnant Rookeries, Regionally Significant Foraging Stocks, and a Major Turtle Fishery. Chelonian Conserv Bi 8: 192–207.
43. Bolker BM, Okuyama T, Bjorndal KA, Bolten AB (2007) Incorporating multiple mixed stocks in mixed stock analysis: 'many-to-many' analyses. Mol Ecol 16: 685–695.
44. Monzon-Arguello C, Rico C, Naro-Maciel E, Varo-Cruz N, Lopez P, et al. (2010) Population structure and conservation implications for the loggerhead sea turtle of the Cape Verde Islands. Conserv Genet 11: 1871–1884.
45. Monzon-Arguello C, Lopez-Jurado LF, Rico C, Marco A, Lopez P, et al. (2010) Evidence from genetic and Lagrangian drifter data for transatlantic transport of small juvenile green turtles. J Biogeogr 37: 1752–1766.
46. Bass AL, Epperly SP, Braun-McNeill J (2006) Green turtle (*Chelonia mydas*) foraging and nesting aggregations in the Caribbean and Atlantic: Impact of currents and behavior on dispersal. J Hered 97: 346–354.
47. Luke K, Horrocks JA, LeRoux RA, Dutton PH (2004) Origins of green turtle (*Chelonia mydas*) feeding aggregations around Barbados, West Indies. Mar Biol 144: 799–805.
48. Okuyama T, Bolker BM (2005) Combining genetic and ecological data to estimate sea turtle origins. Ecol Appl 15: 315–325.
49. Fossette S, Putman NF, Lohmann KJ, Marsh R, Hays GC (2012) A biologist's guide to assessing ocean currents: a review. Mar Ecol Prog Ser 457: 285–301.
50. Hays GC, Marsh R (1997) Estimating the age of juvenile loggerhead sea turtles in the North Atlantic. Can J Zool-Rev Can Zool 75: 40–46.
51. Hays GC, Fossette S, Katselidis KA, Mariani P, Schofield G (2010) Ontogenetic development of migration: Lagrangian drift trajectories suggest a new paradigm for sea turtles. J R Soc Interface 7: 1319–1327.
52. Botsford LW, White JW, Coffroth MA, Paris CB, Planes S, et al. (2009) Connectivity and resilience of coral reef metapopulations in marine protected areas: matching empirical efforts to predictive needs. Coral Reefs 28: 303–305.
53. Scott R, Marsh R, Hays GC (2012) A little movement orientated to the geomagnetic field makes a big difference in strong flows. Mar Biol 159: 481–488.
54. Shillinger GL, Di Lorenzo E, Luo H, Bograd SJ, Hazen EL, et al. (2012) On the dispersal of leatherback turtle hatchlings from Mesoamerican nesting beaches. P Roy Soc B-Biol Sci 279: 2391–2395.
55. Shillinger GL, Hazen EL, Bailey H, Bograd SJ, Godley B, et al. (2012) Tagging through the stages: ontogeny in biologging. Mar Ecol Prog Ser 457: 163–301.
56. Gaspar P, Benson SR, Dutton PH, Reveillere A, Jacob G, et al. (2012) Oceanic dispersal of juvenile leatherback turtles: going beyond passive drift modeling. Mar Ecol Prog Ser 457: 265–284.
57. Godley BJ, Barbosa C, Bruford M, Broderick AC, Catry P, et al. (2010) Unravelling migratory connectivity in marine turtles using multiple methods. J Appl Ecol 47: 769–778.
58. Garofalo L, Mingozzi T, Mico A, Novelletto A (2009) Loggerhead turtle (*Caretta caretta*) matrilines in the Mediterranean: further evidence of genetic diversity and connectivity. Mar Biol 156: 2085–2095.
59. Abreu-Grobois A, Horrocks J, Formia A, Leroux R, Velez-Zuazo X, et al. (2006) New mtDNA dloop primers which work for a variety of marine turtle species may increase the resolution capacity of mixed stock analysis. In: Frick

MG, Panagopoulou A, Rees A, Williams KL, editors; Proceedings of the twenty-six Annual Symposium on Sea Turtle Biology and Conservation, Crete, Greece: 179.

60. Hall TA (1999) BioEdit: a user-friendly biological sequence alignment editor and analysis program for Windows 95/98/NT. Nucleic Acids Symposium Series 41: 95–98.
61. Dutton PH, Bowen BW, Owens DW, Barragan A, Davis SK (1999) Global phylogeography of the leatherback turtle (*Dermochelys coriacea*). J Zool 248: 397–409.
62. Vargas SM, Araujo FCF, Monteiro DS, Estima SC, Almeida AP, et al. (2008) Genetic diversity and origin of leatherback turtles (*Dermochelys coriacea*) from the Brazilian coast. J Hered 99: 215–220.
63. Molfetti E, Vilaça ST, Georges J-Y, Plot V, Delcroix E, et al. (2013) Recent Demographic History and Present Fine-Scale Structure in the Northwest Atlantic Leatherback (*Dermochelys coriacea*) Turtle Population. Plos One 8 (3): e58061. doi:58010.51371/journal.pone.0058061.
64. Bass AL, Good DA, Bjorndal KA, Richardson JI, Hillis ZM, et al. (1996) Testing models of female reproductive migratory behaviour and population structure in the Caribbean hawksbill turtle, *Eretmochelys imbricata*, with mtDNA sequences. Mol Ecol 5: 321–328.
65. Diaz-Fernandez R, Okayama T, Uchiyama T, Carrillo E, Espinosa G, et al. (1999) Genetic sourcing for the hawksbill turtle, *Eretmochelys imbricata*, in the northern Caribbean region. Chelonian Conserv Bi 3: 296–300.
66. Troeng S, Dutton PH, Evans D (2005) Migration of hawksbill turtles *Eretmochelys imbricata* from Tortuguero, Costa Rica. Ecography 28: 394–402.
67. Leroux RA, Dutton PH, Abreu-Grobois FA, Lagueux CJ, Campbell CL, et al. (2012) Re-examination of Population Structure and Phylogeography of Hawksbill Turtles in the Wider Caribbean Using Longer mtDNA Sequences. J Hered 103: 806–820.
68. Rolf DA, Bentzen P (1989) The statistical analysis of mitochondrial DNA polymorphisms: $X^2$ and the problem of small samples. Mol Biol Evol 6: 539–545.
69. Zaykin DV, Pudovkin AI (1993) Two programs to estimate the significance of $X^2$ values using pseudoprobability tests. J Hered 84: 152.
70. Goudet J, Raymond M, de Meeüs T (1996) Testing differentiation in diploid populations. Genetics 144: 1931–1938.
71. Nei M (1982) Evolution of Human Races at the Gene Level. In: Bonne-Tamir B, Cohen T, Goodman RM, editors. Human Genetics Part A: The Unfolding Genome;. New York, NY: Alan R. Liss. 167–181.
72. Excoffier L, Laval G, Schneider S (2005) Arlequin (version 3.0): An integrated software package for population genetics data analysis. Evol Bioinform 1: 47–50.
73. Fu YX (1997) Statistical tests of neutrality of mutations against population growth, hitchhiking and background selection. Genetics 147: 915–925.
74. Rozas J, Sanchez-DelBarrio JC, Messeguer X, Rozas R (2003) DnaSP, DNA polymorphism analyses by the coalescent and other methods. Bioinf 19: 2496–2497.
75. Peakall R, Smouse PE (2006) GENALEX 6: genetic analysis in Excel. Population genetic software for teaching and research. Mol Ecol Notes 6: 288–295.
76. Moran MD (2003) Arguments for rejecting the sequential Bonferroni in ecological studies. Oikos 100: 403–405.
77. Cabin RJ, Mitchell RJ (2000) To Bonferroni or not to bonferroni: when and how are the questions. Bull Ecol Soc Am 81: 246–248.
78. Narum SR (2006) Beyond Bonferroni: Less conservative analyses for conservation genetics. Conserv Genet 7: 783–787.
79. Monzon-Arguello C, Loureiro NS, Delgado C, Marco A, Lopes JM, et al. (2011) Principe island hawksbills: Genetic isolation of an eastern Atlantic stock. J Exp Mar Biol Ecol 407: 345–354.
80. Monzon-Arguello C, Rico C, Marco A, Lopez P, Felipe Lopez-Jurado L (2010) Genetic characterization of eastern Atlantic hawksbill turtles at a foraging group indicates major undiscovered nesting populations in the region. J Exp Mar Biol Ecol 387: 9–14.
81. Bass AL, Epperly SP, Braun-McNeill J (2004) Multi-year analysis of stock composition of a loggerhead turtle (*Caretta caretta*) foraging habitat using maximum likelihood and Bayesian methods. Conserv Genet 5: 783–796.
82. Mortimer JA, Donnelly M (2007) Marine Turtle Specialist Group 2007 IUCN Red List status assessment Hawksbill turtle (*Eretmochelys imbricata*). Available: http://www.iucn-mtsg.org/red_list/ei/index.shtml.
83. Encalada SE, Lahanas PN, Bjorndal KA, Bolten AB, Miyamoto MM, et al. (1996) Phylogeography and population structure of the Atlantic and Mediterranean green turtle *Chelonia mydas*: A mitochondrial DNA control region sequence assessment. Mol Ecol 5: 473–483.
84. Lahanas PN, Bjorndal KA, Bolten AB, Encalada SE, Miyamoto MM, et al. (1998) Genetic composition of a green turtle (*Chelonia mydas*) feeding ground population: evidence for multiple origins. Mar Biol 130: 345–352.
85. Bjorndal KA, Bolten AB, Troeng S (2005) Population structure and genetic diversity in green turtles nesting at Tortuguero, Costa Rica, based on mitochondrial DNA control region sequences. Mar Biol 147: 1449–1457.
86. Formia A, Godley BJ, Dontaine JF, Bruford MW (2006) Mitochondrial DNA diversity and phylogeography of endangered green turtle (*Chelonia mydas*) populations in Africa. Conserv Genet 7: 353–369.
87. Pulliam HR (1988) Sources, sinks, and population regulation. Am Nat 132: 652–661.
88. Pulliam HR, Danielson BJ (1991) Sources, sinks, and habitat selection - a landscape perspective on Population-Dynamics. Am Nat 137: S50–S66.
89. Ordonez C, Troeng S, Meylan A, Meylan P, Ruiz A (2007) Chiriqui Beach, Panama, the most important leatherback nesting beach in Central America. Chelonian Conserv Bi 6: 122–126.
90. Troeng S, Harrison E, Evans D, de Haro A, Vargas E (2007) Leatherback turtle nesting trends and threats at Tortuguero, Costa Rica. Chelonian Conserv Bi 6: 117–122.
91. Chacon-Chaverri D, Eckert KL (2007) Leatherback sea turtle nesting at Gandoca Beach in Caribbean Costa Rica: Management recommendations from fifteen years of conservation. Chelonian Conserv Bi 6: 101–110.
92. Patino-Martinez J, Marco A, Quinones L, Godley B (2008) Globally significant nesting of the leatherback turtle (*Dermochelys coriacea*) on the Caribbean coast of Colombia and Panama. Biol Conserv 141: 1982–1988.
93. Bass AL (1999) Genetic analysis to elucidate the natural history and behavior of hawksbill turtles (*Eretmochelys imbricata*) in the wider Caribbean: A review and re-analysis. Chelonian Conserv Bi 3: 195–199.
94. Shamblin BM, Dodd MG, Bagley DA, Ehrhart LM, Tucker AD, et al. (2011) Genetic structure of the southeastern United States loggerhead turtle nesting aggregation: evidence of additional structure within the peninsular Florida recovery unit. Mar Biol 158: 571–587.
95. Beggs J, Horrocks J, Krueger B (2007) Increase in hawksbill sea turtle *Eremochelys imbricata* nesting in Barbados, West Indies. Endang Species Res 3: 159–168.
96. Putman NF, He R (2013) Tracking the long-distance dispersal of marine organisms: sensitivity to ocean model resolution. J R Soc Interface 10: 20120979.
97. Putman NF, Verley P, Shay TJ, Lohmann KJ (2012) Simulating transoceanic migrations of young loggerhead sea turtles: merging magnetic navigation behavior with an ocean circulation model. J Exp Biol 215: 1863–1870.
98. Putman NF, Scott R, Verley P, Marsh R, Hays GC (2012) Natal site and offshore swimming influence fitness and long-distance ocean transport in young sea turtles. Mar Biol 159: 2117–2126.
99. Lohmann KJ, Hester JT, Lohmann CMF (1999) Long-distance navigation in sea turtles. Ethol Ecol Evol 11: 1–23.
100. Lohmann KJ, Putman NF, Lohmann CMF (2008) Geomagnetic imprinting: A unifying hypothesis of long-distance natal homing in salmon and sea turtles. Proc Natl Acad Sci U S A 105: 19096–19101.
101. Freake MJ, Muheim R, Phillips JB (2006) Magnetic maps in animals: A theory comes of age? Q Rev Biol 81: 327–347.
102. Hawkes LA, Tomás J, Revuelta O, Leon YM, Blumenthal JM, et al. (2012) Migratory patterns in hawksbill turtles described by satellite tracking. Mar Ecol Prog Ser 461: 223–232.
103. Leon YM, Diez CE (1999) Population structure of hawksbill turtles on a foraging ground in the Dominican Republic. Chelonian Conserv Bi 3: 230–236.
104. Van Dam RP, Diez CE, Balazs GH, Colon LAC, McMillan WO, et al. (2008) Sex-specific migration patterns of hawksbill turtles breeding at Mona Island, Puerto Rico. Endang Species Res 4: 85–94.
105. Carreras C, Pascual M, Cardona L, Aguilar A, Margaritoulis D, et al. (2007) The genetic structure of the loggerhead sea turtle (*Caretta caretta*) in the Mediterranean as revealed by nuclear and mitochondrial DNA and its conservation implications. Conserv Genet 8: 761–775.
106. SEATURTLE.ORG Maptool (2002) Available: http://www.seaturtle.org/maptool/.SEATURTLE.ORG, Inc. Accessed 2013 Mar 18.
107. Meylan AB, Donnelly M (1999) Status justification for listing the hawksbill turtle (*Eretmochelys imbricata*) as Critically Endangered on the 1996 IUCN Red List of Threatened Animals. Chelonian Conserv Bi 3: 200–224.

7

# Integrating Multibeam Backscatter Angular Response, Mosaic and Bathymetry Data for Benthic Habitat Mapping

**Rozaimi Che Hasan[1,2], Daniel Ierodiaconou[1]*, Laurie Laurenson[1], Alexandre Schimel[1]**

**1** Deakin University, Centre for Integrative Ecology, School of Life and Environmental Sciences, Faculty of Science and Built Environment, Warrnambool, Victoria, Australia, **2** UTM Razak School of Engineering and Advanced Technology, Universiti Teknologi Malaysia, Kuala Lumpur, Malaysia

## Abstract

Multibeam echosounders (MBES) are increasingly becoming the tool of choice for marine habitat mapping applications. In turn, the rapid expansion of habitat mapping studies has resulted in a need for automated classification techniques to efficiently map benthic habitats, assess confidence in model outputs, and evaluate the importance of variables driving the patterns observed. The benthic habitat characterisation process often involves the analysis of MBES bathymetry, backscatter mosaic or angular response with observation data providing ground truth. However, studies that make use of the full range of MBES outputs within a single classification process are limited. We present an approach that integrates backscatter angular response with MBES bathymetry, backscatter mosaic and their derivatives in a classification process using a Random Forests (RF) machine-learning algorithm to predict the distribution of benthic biological habitats. This approach includes a method of deriving statistical features from backscatter angular response curves created from MBES data collated within homogeneous regions of a backscatter mosaic. Using the RF algorithm we assess the relative importance of each variable in order to optimise the classification process and simplify models applied. The results showed that the inclusion of the angular response features in the classification process improved the accuracy of the final habitat maps from 88.5% to 93.6%. The RF algorithm identified bathymetry and the angular response mean as the two most important predictors. However, the highest classification rates were only obtained after incorporating additional features derived from bathymetry and the backscatter mosaic. The angular response features were found to be more important to the classification process compared to the backscatter mosaic features. This analysis indicates that integrating angular response information with bathymetry and the backscatter mosaic, along with their derivatives, constitutes an important improvement for studying the distribution of benthic habitats, which is necessary for effective marine spatial planning and resource management.

**Editor:** Simon Thrush, University of Auckland, New Zealand

**Funding:** This work was supported by the National Heritage Trust and Caring as part of the Victorian Marine Habitat Mapping Project with project partners Glenelg Hopkins Catchment Management Authority, Department of Sustainability and Environment, Parks Victoria, University of Western Australia and Fugro Survey. The funders had no role in study design, data collection and analysis, decision to publish, or preparation of the manuscript.

**Competing Interests:** The authors have declared that no competing interests exist.

* E-mail: iero@deakin.edu.au

## Introduction

Marine biodiversity worldwide is under pressure from a wide variety of anthropogenic activities [1,2]. The mapping of marine habitats is viewed as the first step in the process of studying, managing, protecting and ultimately conserving marine biodiversity [3]. Multibeam echo sounders (MBES) are now extensively used for this purpose, chiefly because they present technological capabilities (swath coverage, acquisition of high-resolution bathymetry, wide depth range) that all other existing systems, such as single-beam echo sounders, side-scan sonars or Light Detection And Ranging (LiDAR), fail to combine [4]. Various methods of classifying MBES data into habitat maps have been developed over the past two decades. These methods vary widely in terms of the classification algorithms that are implemented, but also in the data features used for classification. There are three types of MBES datasets commonly used as features and/or sources of derivative features for the classification process: backscatter mosaic, backscatter angular response and bathymetry.

A MBES backscatter mosaic is a georeferenced grey-level image representing the acoustic intensity scattered by the seabed, with different seabed types usually showing different intensity levels [5]. Since the acoustic intensity scattered by the seabed is varying with the angle of incidence of the acoustic signal at the seafloor at the time of data acquisition, a statistical normalization of this angular variation is required prior to forming the backscatter mosaic, so that the intensity variations in the image are due to geographical changes in seafloor-type only [6]. This normalization process implies that the quantitative aspect of the intensity level is lost, so that any analysis of the resulting backscatter mosaic requires some form of qualitative interpretation or ground-truthing [7]. The backscatter mosaic grey-level has been extensively used as a feature in many classification techniques [8–11] or as a source of derivative features describing, among other image characteristics, the grey-level statistics [12,13] or the texture [14].

The MBES backscatter angular response is the acoustic intensity scattered by the seabed as a function of the angle of incidence of the acoustic signal at the seafloor. Often represented as the mean angular curve, the backscatter angular response is characteristic of the type of seafloor that reflected the acoustic signal [7]. Since the angular response is not normalized like the backscatter mosaic, it potentially allows the extraction of quantitative seafloor characteristics [7]. Forming a useful mean angular response curve requires the collection of several data samples from the widest angular range possible. In practice, this is obtained by combining several consecutive pings over a full or half swath, which leads to a spatial resolution that is considerably coarser compared to that achieved in the backscatter mosaic format. Furthermore, the large area of seabed thus covered might not present a homogenous seabed type, and thus lead to errors in the angular response analysis. As a consequence, approaches based on exploiting features describing the backscatter angular response curves have remained relatively scarce to date in comparison to those exploiting the backscatter mosaic format [4]. However there has been a renewed interest in this type of analysis recently, with a number of studies testing a number of different features for their predictive power [15–19].

Bathymetry is the data type MBES were originally designed to record. Bathymetry is a major driver of species distributions in coastal waters as depth influences the amount of light reaching the seafloor and exposure to wave action and tide induced currents. In addition, full-coverage bathymetry allows the extraction of seascape metrics that may be used to estimate variations in environmental complexity, which might influence the area available for settlement, food and protection from predation [20]. The predictive power of MBES bathymetry data and their derivative metrics in revealing habitat spatial distribution patterns and the relationship between seabed type and benthic habitats has often been demonstrated [21–24].

The past decade has seen an increase in classification techniques developed to exploit features commonly derived from two of these three MBES data sources. Methodologies that integrate both bathymetry derivatives and backscatter mosaic features have become commonplace and have shown improvements in class separation and overall classification success [10,25–29]. In parallel, Fonseca et al. [30] suggested that backscatter mosaic and angular response should be combined to use both the fine spatial resolution of the former and the predictive power of the latter. Although promising, this suggestion has rarely been implemented [31,32]. Finally, there has been attempts at integrating features extracted from bathymetry and angular response curves although they have remained scarce to date [33]. However, to our knowledge, no benthic habitat classification methodology has been designed to integrate features derived from all three MBES data sources. Given the improvements in classification accuracy obtained by the more recent methodologies combining two datasets over the more traditional methodologies that only exploit one, it can be expected that the integration of features extracted from the three data sources could further improve the class-differentiation process.

Irrespective of whether existing habitat-mapping classification techniques focus on backscatter mosaic, backscatter angular response or bathymetry data (or their integration), those that exploit a set of several data features often fail to assess which of these features contribute the most to classification success. With an increasing number of classification approaches being available that use an increasing number of features derived from MBES data, future classification efforts should be accompanied with the identification of which features are the most relevant to classification success. This issue is becoming particularly pressing with the increasing volume of data and the growing demand for mapping products [4]. Random Forests (RF) may address this specific requirement, as they provide a measure of relative importance for each feature as a complement to their classification output. Typically, an RF algorithm works by training several decision trees, and combining their results through a voting process with the number of trees set by the user, and each tree voting for a particular class [34]. Contrary to standard decision tree algorithms that split nodes based on the best split amongst all variables, RF algorithms split nodes by using the best among a subset of predictors that are randomly chosen at each node [35]. The capability of an algorithm to estimate the relative importance of each feature stems from this random subset selection process. RF algorithms have repeatedly proven successful in predicting fish assemblages [36], in mapping near shore epi-macrobenthic species richness from airborne LiDAR data [37] and in mapping benthic habitats from Autonomous Underwater Vehicle (AUV) images [38].

Accordingly, the objectives of the present study are to integrate angular response features with standard products derived from both bathymetry and backscatter mosaic and assess whether this integration lead to increased classification accuracy, using the capability of the RF algorithm to estimate the relative importance of each feature.

## Methods

### Ethics statement

The remotely sensed techniques used in this study (both sonar and video observations) did not require a permit for their use in survey, thus no specific permission were sought for data collection in the study area location (GPS coordinates can be found on Figure 1). The field studies did not involve endangered or protected species.

### Study area

The study site encompassed a 42 $km^2$ area, with sea depths ranging from 11 to 80 m, located off Cape Duquesne in Discovery Bay, in the state of Victoria, south-eastern Australia (Figure 1). Shallow reefs in this area support diverse assemblages of red algae and kelps (dominated by *Ecklonia radiata*, *Phyllospora comosa* and *Durvillaea potatorum*), while deeper reefs are dominated by invertebrate communities with sponges, ascidians, bryozoans and gorgonian corals [39].

### Acoustic data acquisition

The acoustic data were acquired on the 6th and 7th of November 2005 using a hull-mounted Reson SeaBat 8101 MBES, integrated with a Fugro "Starfix.HP" Differential GPS system for positioning (±0.30 m accuracy) and an Applanix POS MV (Position and Orientation Systems for Marine Vessels) for heave, pitch, roll and yaw corrections (±0.02° accuracy). Real-time navigation, data-logging, quality control and display were provided by the Fugro Starfix suite 8.1 software. Daily sound velocity profiles were collected to correct the acoustic data for water-column sound speed variations.

### Bathymetry map and derivatives

Depth soundings were cleaned using the Fugro Starfix suite, reduced to the lowest astronomical tide datum using tidal observations, and gridded to produce a bathymetric grid at 2.5 m resolution (Figure 1). Six derivative layers were produced from the high-resolution bathymetry grid using various GIS software (Table 1); aspect, rugosity, maximum curvature, bathymetric position index (BPI), slope and complexity [20,24,40]. These six layers were selected based on their successful application

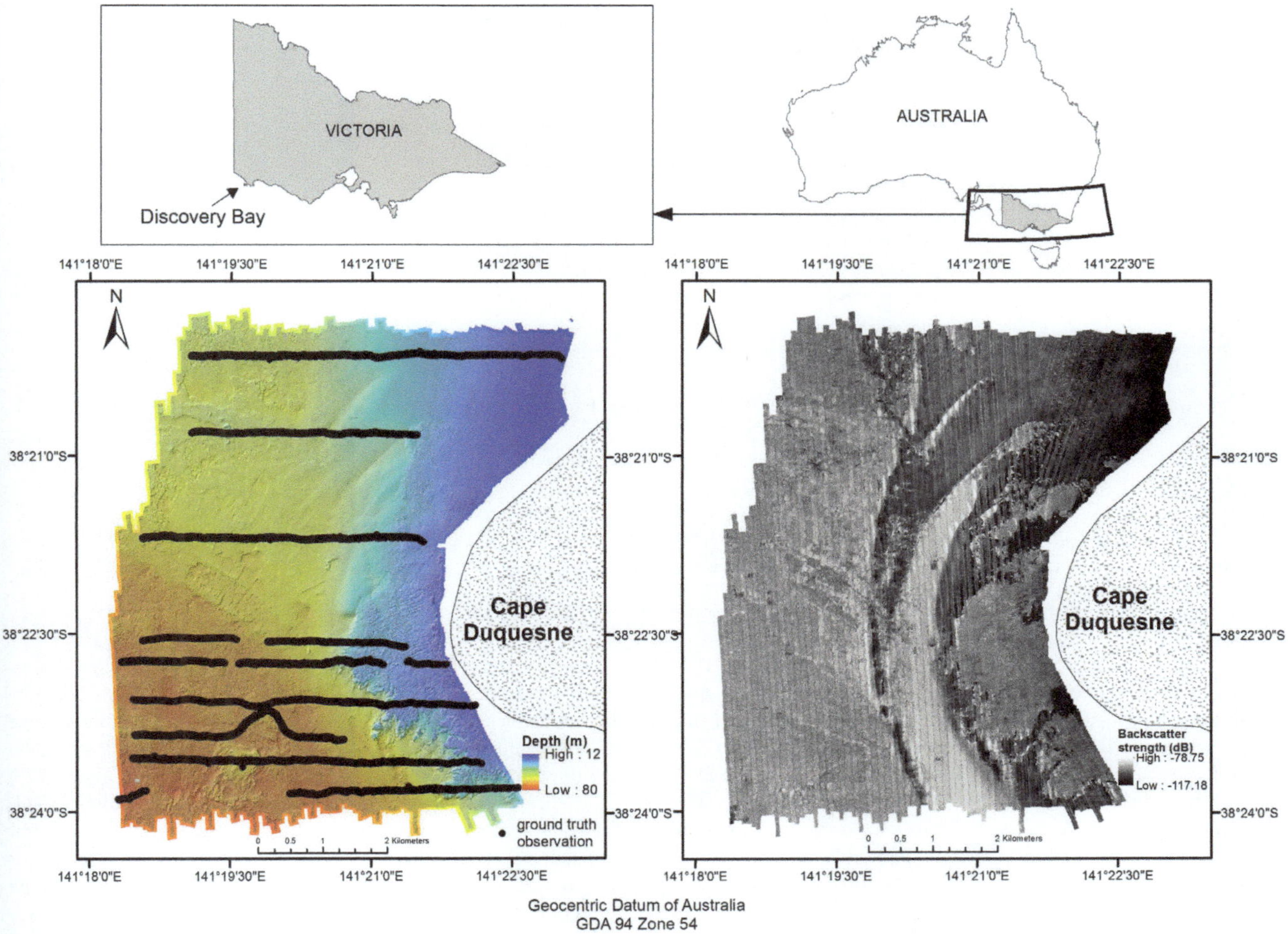


**Figure 1. Study site location (top panels), bathymetry grid with superimposed ground truth observation transects (bottom left panel) and backscatter mosaic (bottom right panel).**

in thematic benthic habitat map construction of previous studies [25–27].

## Backscatter mosaic and derivatives

The backscatter mosaic was obtained using the CMST MB Process software v10.04.04.2, developed by Curtin University's Centre for Marine Science and Technology [41,42]. First, vessel attitude data (i.e. roll, pitch, yaw, heave and heading) and sounding slant range were used to estimate the actual depth and location of measurements within each beam in every ping. Then, raw signal amplitude data were reduced to seafloor backscatter intensity using radiometric corrections, including TVG (Time-Varying Gain) removal, the estimation of spreading and absorption losses and the compensation for the beam footprint size. The angular dependence was then statistically compensated using a 'sliding window' of 25 consecutive pings and a reference angle of 30°. Finally, a backscatter mosaic was produced at 2.5 m resolution and exported for further analysis (Figure 1).

Six derivatives were produced from the backscatter mosaic using the ENVI 4.7 software (Table 1): specifically, Red, Green and Blue layers of Hue, Saturation and Intensity (HSI) [43], and the Haralick texture features Homogeneity, Entropy and Correlation, calculated from Grey Level Co-occurrence Matrices (GLCM) [44]. Like the bathymetry derivatives, the three HSI layers were selected based on their ability to produce accurate benthic habitat maps in previous studies [25–27]. Homogeneity, Entropy and Correlation were selected, among a wide range of other texture features available [44], based on their reported importance in previous texture-based habitat mapping efforts [14,28,45,46] and on their belonging to three different groups, so as to minimise risks of correlation [47]. The three texture features were obtained by calculating the GLCMs in the 0°, 45°, 90° and 135° directions over the 8-bit backscatter mosaic (with no greyscale normalisation applied), extracting the features from each GLCM direction, and averaging the results.

## Angular response derivatives

The angular response curves were obtained by combining the seafloor backscatter intensity samples produced by the MB Process software (prior to the statistical angular compensation that leads to the mosaic) and a segmentation of the mosaic. First, the mosaic was segmented into separate contiguous regions using a region-growing algorithm in Spring v5.1 software. By initially using each mosaic pixel as a distinct region ("seeds"), the algorithm recursively aggregated the neighbouring regions presenting the maximum grey-level similarity, as long as this maximum similarity fell under a user-defined similarity threshold that became increasingly less stringent as the algorithm progressed [48]. At

**Table 1.** Detailed explanations of all layers derived from bathymetry and the backscatter mosaic.

| Derivatives (original layer) | Details | Analysis window (pixel size: 2.5 m) | Software |
|---|---|---|---|
| Aspect (Bathymetry) | Describes the azimuthal direction of the steepest slope through the points in the analysis window [20] | 3×3 | Spatial Analyst (ArcGIS 9.3) |
| Rugosity (Bathymetry) | A measure of structural complexity represented by the ratio of surface area to planar area [24] | 3×3 | Benthic Terrain Modeller Tool for ArcGIS |
| Maximum curvature (Bathymetry) | Describes the steepest curve of either plan or profile convexity through a defined cell neighbourhood [40] | 3×3 | ENVI 4.7 |
| Bathymetric Position Index (Bathymetry) | Compares the elevation of each cell in a digital elevation model to the mean elevation of a specified neighbourhood around that cell [68] | Inner radius = 10, Outer radius = 10, Scale factor = 125 | Benthic Terrain Modeller Tool for ArcGIS |
| Slope (Bathymetry) | Describes the maximum change in elevation between each cell and cells in the analysis neighbourhood [20] | 3×3 | Spatial Analyst (ArcGIS 9.3) |
| Complexity (Bathymetry) | Describes the rate of change in the bathymetry slope [20] | 3×3 | ENVI 4.7 |
| Red, Green and Blue bands of Hue, Saturation and Intensity (Backscatter mosaic) | High and low frequency information of an image after application of high and low pass filters, producing three band images of Red, Green and Blue [43]. | 3×3 (high pass filter) and 11×11 (low pass filter) | ENVI 4.7 |
| Homogeneity, Entropy and Correlation texture features (Backscatter mosaic) | Texture features calculated from Grey Level Co-occurrence Matrices (GLCM) [44] | 7×7 | ENVI 4.7 |

the end of this recursive process, a second user-defined area threshold that specified a minimum region-size, allowed the smallest regions to be aggregated with larger adjacent regions. For this study, a similarity threshold of 1 and an area threshold of 2500 were used as parameters to produce the segmentation. The segmentation was then imported into ArcMap, in which all segments were vectorised as polygons (Figure 2a). Finally, using proprietary Matlab code (available in Supplementary S1), the seafloor backscatter intensity samples and their associated angle of incidence were compiled for all MBES data files over each polygon, and the mean intensity value (in dB space) for each angle was computed. This process resulted in a single backscatter angular response curve for each segment (Figure 2b and c).

Four derivatives were produced from the backscatter angular response curves using Matlab: mean, least square slope, skewness and kurtosis of the backscatter intensity within 30 to 50° incidence angles. The derivative values were then attributed to their respective polygons, and rasterised at a resolution of 2.5 m using ArcMap (Figure 3).

## Data layer correlation analysis

The 18 data layers that were obtained from the acoustic data (bathymetry + six bathymetry derivatives + mosaic grey-level + six mosaic derivatives + four angular response derivatives) were tested for correlation in ENVI 4.7, through the computation of Pearson's linear correlation coefficient ($R^2$).

## Towed video observations

A VideoRay PRO 3 GTO Remotely Operated Vehicle (ROV) was used to provide ground truth information for model building and evaluation (Figure 1). Underwater acoustic positioning of the towed video system was achieved by using a Tracklink 1500MA Ultra Short Base Line (USBL) acoustic tracking system, with vessel errors (roll, pitch and yaw) being corrected for by using a KVH Industries motion sensor. An Omnistar wide-area Differential Global Positioning System (DGPS) was used to fix the vessel location, and apply corrections for acoustically positioned video (±2.5 m accuracy). The recorded video data were classified according to the Victorian Towed Video Classification scheme to identify benthic biota classes. The classification scheme followed the guidelines published by the Interim Marine and Coastal Regionalisation for Australia [49]. Seven habitat classes were identified from the video observations; Mixed Brown algae (MB), Invertebrates (INV), Mixed Red algae and Invertebrates (MRI), Mixed Brown algae and Invertebrates (MBI), Mixed Brown algae and Mixed Red algae (MBMR), Mixed Green algae and Invertebrates (MGI) and No Visible Biota (NVB). All available reference data were randomly sampled for model development (70%) and for the accuracy assessment (30%) (Table 2).

## Random Forests

Supervised RF decision trees were implemented to train and subsequently model class predictions. A RF algorithm programmed in Matlab was used [50], combined with a proprietary Matlab routine (available in Supplementary S1) developed to read and process multilayer images in native ENVI format. For this study, the RF algorithm parameters *m* (number of predictors randomly chosen for each split) and *ntree* (number of trees generated) were set to the integral part of the square root of the total number of variables (default setting) and 200 (to minimise errors rates) respectively.

Initially, two RF models were generated to test for the relevance of the angular response derivatives (Table 3). The first model was limited to bathymetry and the backscatter mosaic and their derivatives (Model 1), while the second model also incorporated the angular response derivatives (Model 2). The contribution of each input layer to each of these two models was ranked by importance (scaled from 0 to 1). Additional RF models were then generated to test whether the success rate of these initial models could be achieved by using fewer input layers. First, an RF model was generated using only the input layers with an importance score of 1 (Model 3). Then, additional RF models were generated, in which the input layers of lesser importance were gradually added, based on two rules: (1) three or less variables should be added at a time, and (2) the differences in the importance score between the added variables should be less than 0.2 (Models 4 to 6). The accuracy of each model was assessed by forming an error matrix, and computing its overall accuracy and Kappa coefficient

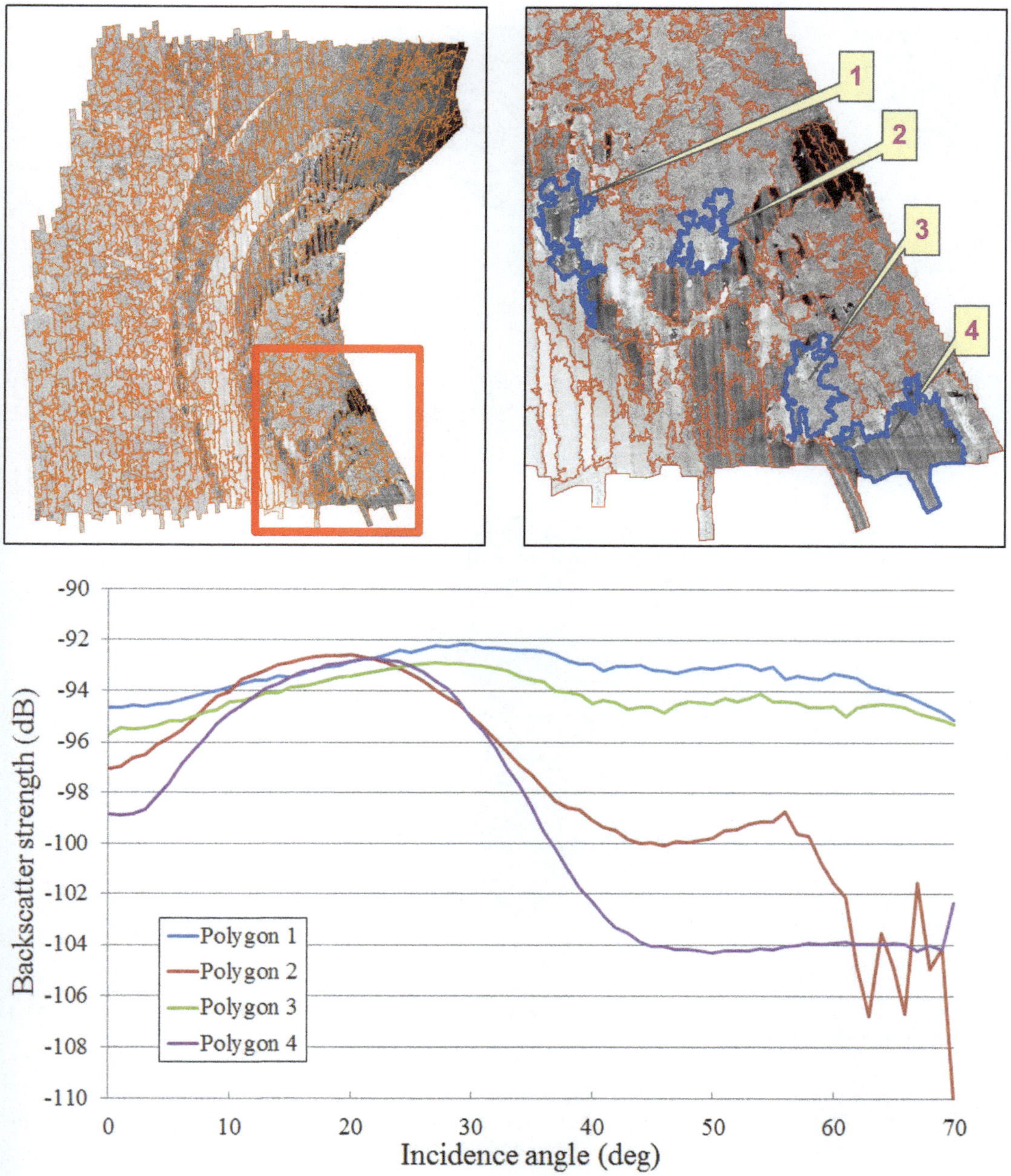


**Figure 2.** (a) Backscatter mosaic overlaid with the results of the region-growing segmentation. (b) Detailed view of the mosaic segmentation as indicated by the red box in (a). (c) Examples of four different angular response curves computed from polygons 1–4 as indicated by the blue sections in (b).

[51]. Z statistics were computed from pairwise combinations of the error matrices to compare the model outputs [52].

## Results

### Backscatter angular response derivatives

The angular response mean (Figure 3a) was visually very similar to the backscatter mosaic (Figure 1), but presented two peculiarities. First, it showed more contrast and discrimination between the low and high backscatter regions of the study site. Second, it did not present the along-track artefacts that were still highly visible in the mosaic, despite statistical compensation. Similar observations were made for the angular response slope, although the low/high backscatter discrimination was less distinct (Figure 3b). The angular response skewness and kurtosis maps (Figure 3c and 3d) were very similar to each other, but showed little resemblance to the backscatter mosaic, the angular response mean and the angular response slope maps. In particular, the skewness and kurtosis maps appeared to highlight areas in the south-west of the study site that were not evident in the backscatter mosaic.

### Correlation of the layers

Strong auto-correlation (>0.5) was found among several layers derived from backscatter data, but not among bathymetry products (Table 4). The highest $R^2$ was found between the GLCM Homogeneity and Entropy layers (0.98). Confirming the visual analysis in the previous section, a high correlation was found between the angular response mean and angular response slope

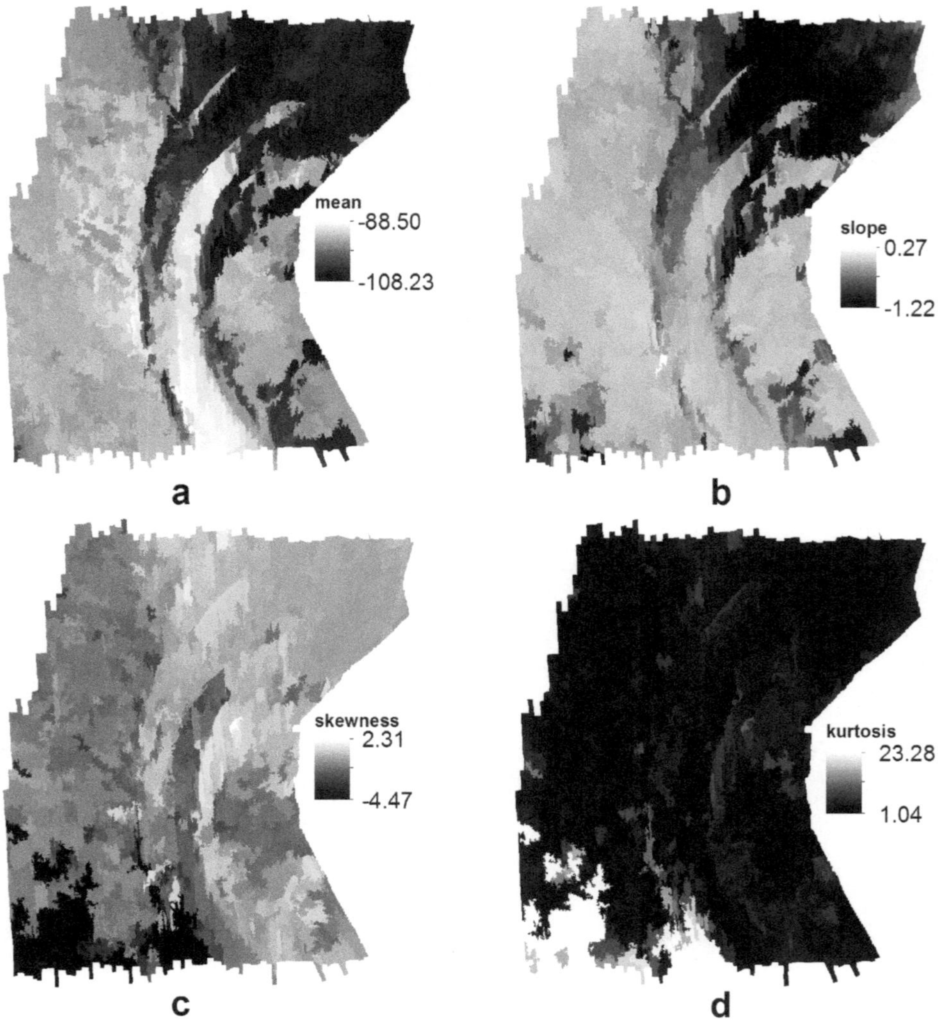


**Figure 3. Maps of the angular response features, derived from the backscatter angular response curves between 30° and 50°; (a) mean, (b) least square slope, (c) skewness and (d) kurtosis.**

(0.75), between the angular response skewness and angular response kurtosis (0.72) and between the mosaic and the angular response mean (0.53). In comparison, the maximum $R^2$ measured between two layers derived from the bathymetry was relatively low (0.43; between complexity and bathymetry slope). Relatively low correlation was also observed between the original backscatter mosaic and bathymetry (0.33).

## Variable importance and feature selection

Bathymetry appeared the most important variable in the first model, which included all layers, except angular response

**Table 2.** Number of samples used for model development and accuracy assessment, for each biota class.

| Biota class | Number of samples used for model development | Number of samples used for accuracy assessment |
|---|---|---|
| Mixed Brown algae (MB) | 1107 | 475 |
| Invertebrates (INV) | 11830 | 5070 |
| Mixed Red algae and Invertebrates (MRI) | 1391 | 596 |
| No Visible Biota (NVB) | 11915 | 5107 |
| Mixed Brown algae and Invertebrates (MBI) | 593 | 254 |
| Mixed Brown algae and Mixed Red algae (MBMR) | 62 | 26 |
| Mixed Green algae and Invertebrates (MGI) | 749 | 321 |

**Table 3.** Variable combinations and classification accuracy results for the different models.

| **Model number** | | | #1 | #2 | #3 | #4 | #5 | #6 |
|---|---|---|---|---|---|---|---|---|
| **Total number of variables in model** | | | **14** | **18** | **2** | **5** | **8** | **11** |
| Model variables | Bathymetry | | √ | √ | √ | √ | √ | √ |
| | Bathy derivatives | Complexity | √ | √ | | | | √ |
| | | Aspect | √ | √ | | | | |
| | | BPI | √ | √ | | | | |
| | | Slope | √ | √ | | | | |
| | | Maximum Curvature | √ | √ | | | | |
| | | Rugosity | √ | √ | | | | √ |
| | Mosaic | | √ | √ | | | | |
| | Mosaic derivatives | Red HSI | √ | √ | | | √ | √ |
| | | Green HSI | √ | √ | | | √ | √ |
| | | Blue HSI | √ | √ | | | √ | √ |
| | | GLCM Homogeneity | √ | √ | | | | |
| | | GLCM Entropy | √ | √ | | | | |
| | | GLCM Correlation | √ | √ | | | | √ |
| | AR derivatives | AR mean | | √ | √ | √ | √ | √ |
| | | AR slope | | √ | | √ | √ | √ |
| | | AR skewness | | √ | | √ | √ | √ |
| | | AR kurtosis | | √ | | √ | √ | √ |
| Model performance | Overall accuracy (%) | | 88.5 | 93.6 | 90.2 | 89.7 | 92.2 | 93.5 |
| | Kappa coefficient | | 0.81 | 0.90 | 0.84 | 0.83 | 0.88 | 0.90 |

The tick symbol (√) indicates the variable was used in a model.

**Table 4.** Correlation of determination ($R^2$) using the linear Pearson correlation measure between all variables used in this study.

| Variables | Bathy | Comp | Asp | BPI | Slo | Maxc | Rug | Mos | HSIR | HSIG | HSIB | Hom | Ent | Cor | ARmean | ARslo | ARsk | ARkur |
|---|---|---|---|---|---|---|---|---|---|---|---|---|---|---|---|---|---|---|
| Bathy | 1 | - | - | - | - | - | - | - | - | - | - | - | - | - | - | - | - | - |
| Comp | 0.00 | 1 | - | - | - | - | - | - | - | - | - | - | - | - | - | - | - | - |
| Asp | 0.06 | 0.02 | 1 | - | - | - | - | - | - | - | - | - | - | - | - | - | - | - |
| BPI | 0.00 | 0.03 | 0.00 | 1 | - | - | - | - | - | - | - | - | - | - | - | - | - | - |
| Slop | 0.00 | 0.43 | 0.00 | 0.03 | 1 | - | - | - | - | - | - | - | - | - | - | - | - | - |
| Maxc | 0.00 | 0.18 | 0.00 | 0.08 | 0.34 | 1 | - | - | - | - | - | - | - | - | - | - | - | - |
| Rug | 0.00 | 0.13 | 0.00 | 0.01 | 0.38 | 0.20 | 1 | - | - | - | - | - | - | - | - | - | - | - |
| Mos | 0.33 | 0.01 | 0.01 | 0.00 | 0.00 | 0.00 | 0.00 | 1 | - | - | - | - | - | - | - | - | - | - |
| HSIR | 0.24 | 0.00 | 0.01 | 0.00 | 0.01 | 0.00 | 0.00 | 0.18 | 1 | - | - | - | - | - | - | - | - | - |
| HSIG | 0.26 | 0.04 | 0.03 | 0.00 | 0.01 | 0.00 | 0.00 | 0.32 | 0.35 | 1 | - | - | - | - | - | - | - | - |
| HSIB | 0.06 | 0.05 | 0.02 | 0.01 | 0.02 | 0.01 | 0.00 | 0.10 | 0.08 | 0.08 | 1 | - | - | - | - | - | - | - |
| Hom | 0.10 | 0.01 | 0.01 | 0.00 | 0.00 | 0.00 | 0.00 | 0.33 | 0.07 | 0.00 | 0.28 | 1 | - | - | - | - | - | - |
| Ent | 0.11 | 0.01 | 0.01 | 0.00 | 0.00 | 0.00 | 0.00 | 0.36 | 0.04 | 0.00 | 0.23 | **0.98** | 1 | - | - | - | - | - |
| Cor | 0.05 | 0.01 | 0.00 | 0.00 | 0.01 | 0.00 | 0.00 | 0.09 | 0.12 | 0.05 | 0.07 | 0.00 | 0.00 | 1 | - | - | - | - |
| ARmean | 0.38 | 0.07 | 0.05 | 0.00 | 0.02 | 0.01 | 0.00 | **0.53** | 0.33 | 0.47 | 0.03 | 0.11 | 0.13 | 0.11 | 1 | - | - | - |
| ARslo | 0.26 | 0.11 | 0.05 | 0.01 | 0.04 | 0.01 | 0.00 | 0.24 | 0.22 | 0.36 | 0.02 | 0.02 | 0.02 | 0.04 | **0.75** | 1 | - | - |
| ARsk | 0.22 | 0.01 | 0.02 | 0.00 | 0.00 | 0.00 | 0.00 | 0.10 | 0.14 | 0.11 | 0.00 | 0.01 | 0.01 | 0.03 | 0.19 | 0.06 | 1 | - |
| ARkur | 0.10 | 0.00 | 0.00 | 0.00 | 0.00 | 0.00 | 0.00 | 0.01 | 0.03 | 0.01 | 0.00 | 0.00 | 0.00 | 0.00 | 0.01 | 0.00 | **0.72** | 1 |

Highlighted values indicate high correlation ($R^2$>0.5). Bathy = Bathymetry, Comp = Complexity, Asp = Aspect, BPI = Bathymetric Position Index, Slo = Bathymetry slope, Maxc = Maximum curvature, Rug = Rugosity, Mos = Backscatter mosaic, HSIR = Red layer of Hue Saturation and Intensity (HSI), HSIG = Green layer of HSI, HSIB = Blue layer of HSI, Hom = GLCM Homogeneity, Ent = GLCM Entropy, Cor = GLCM Correlation, ARmean = Mean of angular response, ARslo = Slope of angular response, ARsk = Skewness of angular response and ARkur = Kurtosis of angular response.

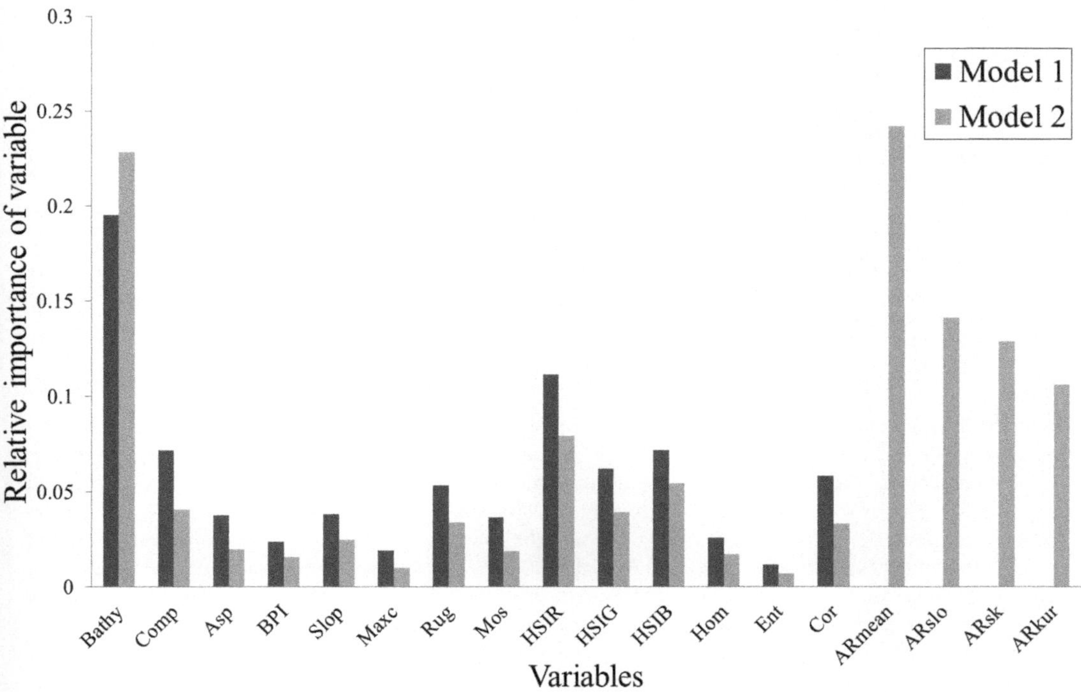


**Figure 4. Measures of the relative importance of the variables produced from Models 1 and 2.** Bathy = Bathymetry, Comp = Complexity, Asp = Aspect, BPI = Bathymetric Position Index, Slo = Bathymetry slope, Maxc = Maximum curvature, Rug = Rugosity, Mos = Backscatter mosaic, HSIR = Red layer of Hue Saturation and Intensity (HSI), HSIG = Green layer of HSI, HSIB = Blue layer of HSI, Hom = GLCM Homogeneity, Ent = GLCM Entropy, Cor = GLCM Correlation, ARmean = Mean of angular response, ARslo = Slope of angular response, ARsk = Skewness of angular response and ARkur = Kurtosis of angular response.

derivatives (Model 1, 14 variables Figure 4). The Red HSI layer ranked second in importance. All other variables including, interestingly, the mosaic itself, were found to be of very low importance.

After the angular response derivatives were added (Model 2, 18 variables, Figure 4), bathymetry and angular response mean ranked equally as the two most important variables, closely followed by the three other angular response features (slope, skewness and kurtosis). The other variables demonstrated similar levels of low importance, as described previously for Model 1.

This order of variable importance was used to construct the subsequent four models (Models 3 to 6, see Table 3). Model 3 contained the two most important variables; bathymetry and angular response mean. The other three angular response variables (slope, skewness and kurtosis) followed with moderate importance, and were, therefore, added to generate Model 4. The three HSI layers followed with relatively little importance, and were added to generate Model 5. Finally, complexity, rugosity and GLCM correlation showed slightly more importance compared to the rest of the variables, and were added to generate Model 6.

## Model performance

The two original models (Models 1 and 2) performed very well, obtaining overall high accuracy and kappa coefficients (Table 3). The inclusion of variables derived from the backscatter angular response from Model 1 into Model 2 increased the overall accuracy by 5.1% (88.5% to 93.6%) and the Kappa coefficient by 0.09 (0.81 to 0.90). The accuracy for all individual classes improved with inclusion of angular response derivatives, particularly for MBMR and MGI (Figure 5).

Interestingly, the simple models also achieved high accuracy, with Model 3 achieving 90.2% overall accuracy and 0.84 Kappa coefficient, despite being only driven by two variables (bathymetry and angular response mean, Table 3). The least parsimonious of the simple models (Model 6, Figure 6) performed as well as the full model (Model 2), with 93.5% overall accuracy (down by only 0.1%) and the same Kappa coefficient (0.90). Indeed, pairwise comparison of the error matrices from Models 2 and 6 indicated no significant difference between these matrices ($Z = 0.29$, Table 5).

## Discussion

Overall, all models derived in this study achieved good accuracies, scoring between 88.5% and 93.6% (Table 3). These scores were slightly above those reached by previous studies implementing different decision tree techniques, such as CART, Quick, Unbiased, Efficient Statistical Trees (QUEST) and Classification Rule with Unbiased Interaction Selection and Estimation (CRUISE), in comparable habitats of south west Victoria, Australia [84% in 22, and 87% in 25, 80% in 26, 83% in 27]. In addition, the unique capability of RF algorithms to assess the importance of the various predictors was used to build simpler models. In the present work, the optimal model was Model 6, because it only used 11 of the 18 features derived in this research to achieve accuracy levels that were equivalent to the full model that implemented the entire set of 18 variables (Model 2). Although RF algorithms may include many variables while remaining insensitive to over-fitting [53], the use of fewer variables in the classification process has very practical benefits, in terms of gain of computer processing time and effort.

These improved results were obtained using a novel approach to integrate features derived from MBES backscatter angular response curves – which are good predictors of sediment grain-size

**Table 5.** Pairwise comparison of error matrices between Model 2 and the four simpler models.

| Pairwise combination | | Z statistic | Significant/Not significant* |
|---|---|---|---|
| Model 2 | Model 3 | 9.54 | Significant |
| Model 2 | Model 4 | 11.22 | Significant |
| Model 2 | Model 5 | 3.91 | Significant |
| Model 2 | Model 6 | 0.29 | Not significant |

Model 2 included all variables, while the simpler models contained different combinations of the variables (see Table 3). The significant level indicates whether two error matrices (i.e. from two different models) are completely different (significant), or capable of producing similar results (not significant).
*Significant at the 95% confidence interval (critical value Z = 1.96).

[17] – with features describing the texture and patterns in the backscatter mosaic – which are good predictors of seafloor substrate types [54] – as well as bathymetry and its most common derived seascapes – which are good predictors of biological communities distribution [22]. To our knowledge, this constitutes the first benthic habitat mapping methodology exploiting the three main MBES data sources that are bathymetry, mosaic and angular response. In many fields linked to land mapping, improvements in classification accuracy have similarly followed from the availability of a large number of new features (i.e. spectral bands in hyperspectral remote sensing) [55,56]. In comparison, the number of variables available for mapping in the marine realm (primarily from MBES) are severely restricted. The present study contributes to the benthic habitat mapping field by increasing the range of available acoustic variables that can be combined to characterise benthic habitats.

The main result of this study is how statistical features describing the backscatter angular response curves considerably improved class differentiation. First, the classification accuracy and the individual class accuracies were greatly improved by adding angular response features (Model 2 compared to Model 1). Secondly, the angular response mean was found to be the most important of all backscatter data derivatives, out-performing even the backscatter mosaic. In fact, all of the angular response features were ranked as more important compared to the backscatter mosaic or any of its derivatives. Finally, a simple model using only the angular response mean and bathymetry yielded higher accuracy (Model 3; 90.2%) compared to a model using mosaic, 6 mosaic derivatives, bathymetry and 6 bathymetry derivatives (Model 1; 88.5%). These results suggest that the methodology presented here deriving statistical features from the backscatter angular response, successfully captured the characteristics of backscatter variation at moderate incidence angles, which are known to successfully discriminate between seabed types [18] or benthic communities [41]. The fact that the angular response mean has the appearance of a de-noised version of the backscatter mosaic (Figures 1 and 3) probably also contributed to the success of this feature. In effect, the speckle and nadir noise commonly displayed in backscatter mosaics are likely to be responsible for errors in the classification process, and, hence, hinder the predictive power of the backscatter mosaic. By effectively enhancing the meaningful backscatter contrasts between the small regions of the study site, while removing the noise and maintaining the mosaic spatial resolution, the methodology used to derive the angular response mean in this study may be viewed as creating an improved version of the backscatter mosaic.

The other three angular response features (slope, skewness and kurtosis) were found to be more important compared to the

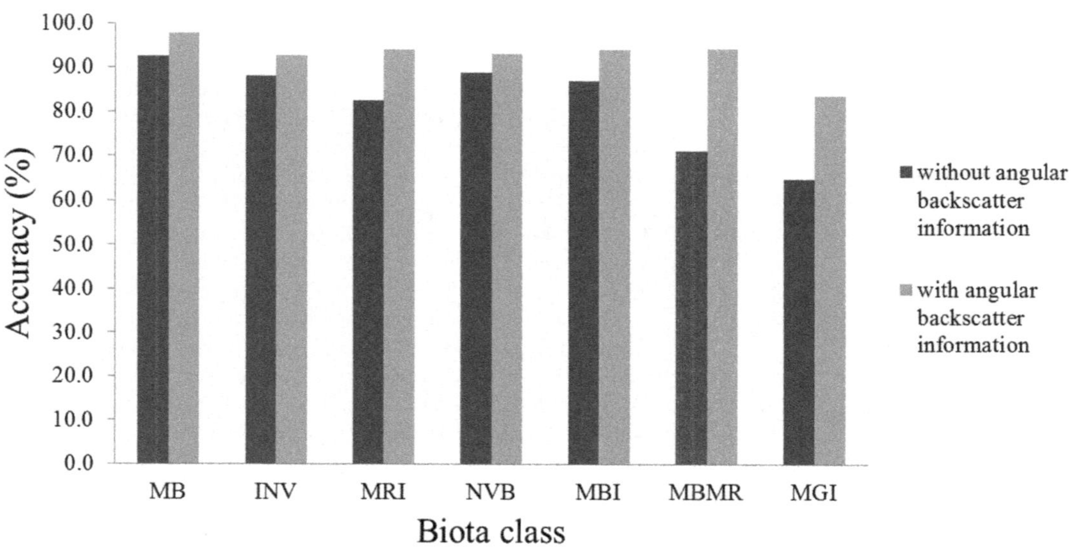


**Figure 5. Per class accuracy (mean of user and producer's accuracy) for models 1 and 2 (with and without the backscatter angular response features).** MB = Mixed Brown algae, INV = Invertebrates, MRI = Mixed Red algae and Invertebrates, NVB = No Visible Biota, MBI = Mixed Brown algae and Invertebrates, MBMR = Mixed Brown algae and Mixed Red algae, and MGI = Mixed Green algae and Invertebrates.

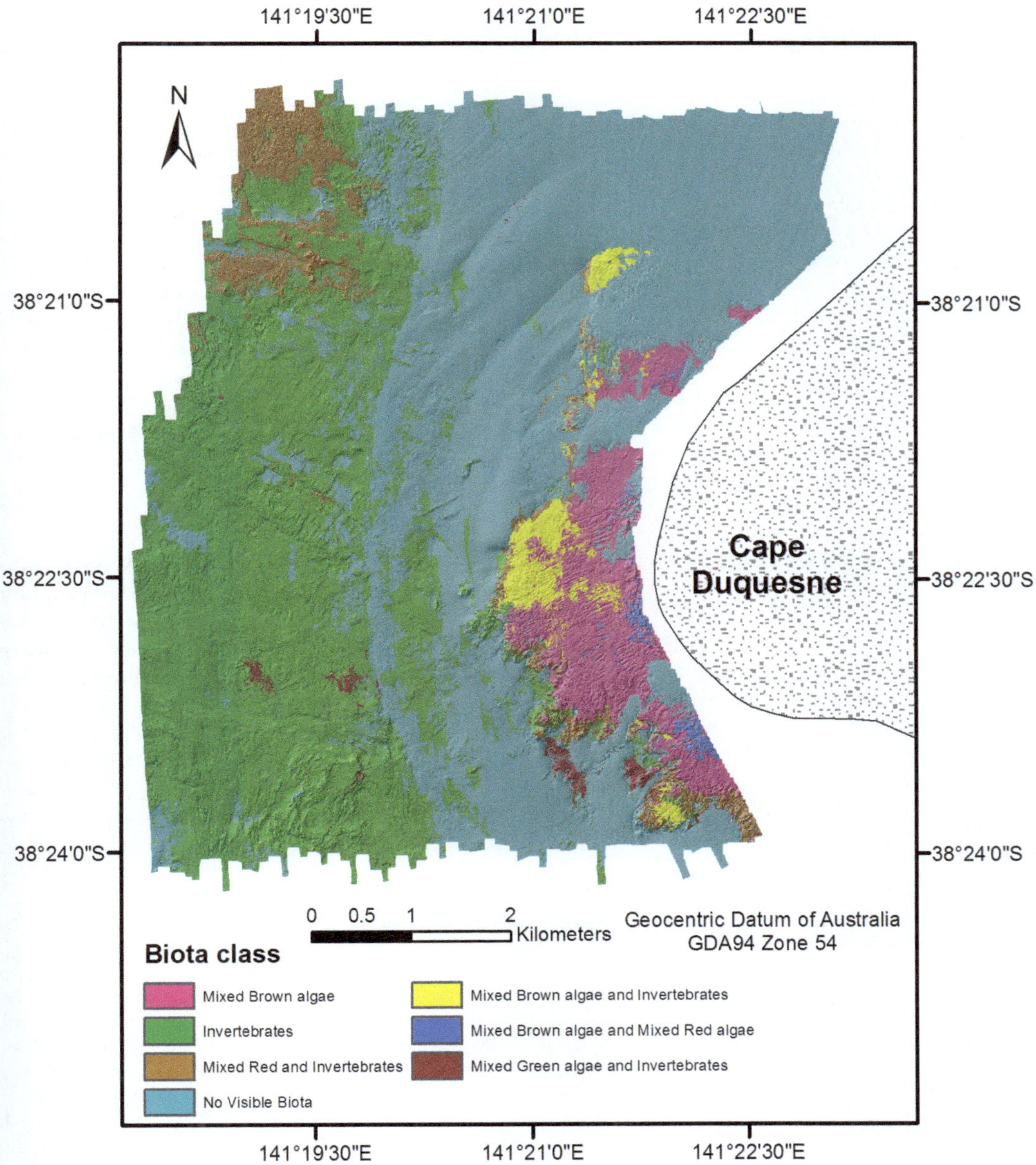


**Figure 6. Habitat map of biota classes produced from the simplest model of variable combinations (Model 6) overlaid with the hill shaded bathymetry.**

features derived from bathymetry and the mosaic, but were less relevant compared to the angular response mean. Interestingly, the addition of these three features to the model using only bathymetry and angular response mean decreased the classification accuracy (from Model 3; 90.2%, to Model 4; 89.7%). The inconsistency between the relevance of these features and the decrease in accuracy following their addition warrants future investigation. At present, it may be assumed that the unique areas that angular response skewness and angular response kurtosis appeared to single out in the south-western part of the study site (Figure 3) may be irrelevant in terms of habitat differences, and might have caused some inconsistencies in the final habitat map. The angular response slope probably does not contribute to this problem as it is more similar to the angular response mean and the original mosaic. Incorporating only the slope with bathymetry and the angular response mean might have produced a more successful model compared to Model 4.

Regardless of the success of using angular response features in this study, bathymetry was found to be the single-most important habitat predictor across all models. The high accuracy of the simplest model using only bathymetry and angular response mean in this study (Model 3), and the low correlation between these two features, indicates that they contain very different and complementary information to predict benthic habitats. These results reinforce the argument that benthic habitat mapping efforts should not be based on backscatter data information alone [57].

Finally, a number of observations may be made for the less relevant features in this study. Although bathymetry and the angular response mean were undeniably the most important features (Model 3), they did not produce the best accuracy alone. The highest accuracy was achieved with a mix of bathymetry and mosaic derivatives (HSI layers, rugosity, complexity and GLCM correlation; Model 6), indicating that these minor features described subtle variations in terrain complexity across biotic habitat types. The topographic features other than rugosity and complexity (i.e. slope, aspect, maximum curvature and bathymetric position index) did not make a significant contribution to this study, which contrasted with the findings of previous works on the distribution of habitats of shallow water mobile species [58,59].

The approach presented in this study and its accuracy results are dependent on two main factors: the types of biological benthic habitat present in the study site and the scheme originally used to classify the ground-truth data. Soft sediments are the dominant seabed types in the present study site; the shallower north eastern areas inside Discovery Bay were largely composed of fine, well sorted sand flats with some fine rippling, while sediments composing the deeper areas to the west of the site tended to be coarser and formed into broad (>40 cm), well defined sand waves. These sediment dominated areas seldom had visible epifauna present and were therefore clustered together under the "no visible biota (NVB)" biotic habitat class, according to the classification scheme chosen for the study. Yet, previous studies investigating infaunal communities of eastern Victoria suggested that these regions contain highly diverse temperate infaunal assemblages [60,61]. Although assigned to only one class in this study, researchers have found that unconsolidated, sandy sediments can be further allocated to sub-classes which are acoustically distinct based on sediment size, surface morphology, and compactness [7,62–65]. Furthermore, there is evidence to suggest that distinct assemblages of fish and invertebrates exist within and between biotopes defined by grain size and ripple characteristics [66,67]. The present study did not identify such potential diversity due to its classification scheme focusing on epifauna biota assemblages. This limitation emphasizes how model accuracy results need to be appreciated in the context of the ground-truth data being used and the classification scheme that was applied to them. However, the approach presented in this study is very flexible and could be similarly applied to soft-sediment classification schemes to advance the benthic characterisation of sediment types in benthic habitat mapping studies.

## Conclusions

The overall high accuracy of all models in this study indicated that the suggestion to integrate angular response features with bathymetry and backscatter mosaic and their derivatives is sound and effective. High model accuracy was obtained using just bathymetry and angular response mean, with this further increasing following addition of other angular response features and features considered relevant based on RF algorithm analyses. While the angular response mean over mid-range incidence angles proved an important contributor to improving our prediction of benthic classes, the other angular response features (slope, skewness and kurtosis) produced more mixed results. It is anticipated that this methodology will be applied to other datasets to assess whether these three features should be conserved or abandoned, and if other angular features should be developed.

The results also confirmed that bathymetry remains the most important predictor of marine biotic habitats, and highlighted that some bathymetry and mosaic derivatives were rather irrelevant to the classification process. These conclusions were reached by using a classification algorithm that allowed the model variables to be ranked by the importance of their contribution to the resulting model. Since the importance of any given feature might depend on the habitats present in the study area, we strongly recommend the use of RF algorithms, or other classifiers that have this additional capability.

## Supporting Information

**Supplementary S1 A zip file containing all Matlab codes used in this paper to construct mean angular response curves from homogeneous regions using the "proc" files (processed files from the CMST MB Process software) and user-set polygons (ArcMap shapefile format).** Users need to unzip this file and add all codes to their local Matlab path. The main function is 'inpoly_plosone_v1.m' (to see full instructions, type 'help inpoly_plosone_v1' in the Matlab Command Window). Running this code in Matlab will display a Matlab graphical user interface to run the analysis (successfully tested on Matlab R2013a 32-bit).

## Acknowledgments

We thank the crew from the Australian Maritime College research vessel *Bluefin*, which was used for the multibeam data collection. We also thank Iain Parnum (Curtin University) and Justy Siwabessy (Geoscience Australia) for the MB Process software, Abhishek Jaiantilal for the Matlab RF algorithm and Russel Congalton for the Kappa program. Thanks to Alex Rattray and Jacquomo Monk for their help with the ENVI software and the processing of bathymetry derivatives. R.C.H. was supported by the Malaysian Ministry of Higher Education and *Universiti Teknologi Malaysia* PhD scholarships.

## Author Contributions

Conceived and designed the experiments: DI RCH. Performed the experiments: DI RCH. Analyzed the data: RCH. Contributed reagents/materials/analysis tools: RCH DI. Wrote the paper: RCH DI LL AS.

## References

1. Jackson JBC, Kirby MX, Berger WH, Bjorndal KA, Botsford LW, et al. (2001) Historical overfishing and the recent collapse of coastal ecosystems. Science 293: 629–637.
2. Halpern BS, Walbridge S, Selkoe KA, Kappel CV, Micheli F, et al. (2008) A global map of human impact on marine ecosystems. Science 319: 948–952.
3. Kostylev VE, Todd BJ, Fader GBJ, Courtney RC, Cameron GDM, et al. (2001) Benthic habitat mapping on the Scotian Shelf based on multibeam bathymetry, surficial geology and sea floor photographs. Marine Ecology Progress Series 219: 121–137.
4. Brown CJ, Smith SJ, Lawton P, Anderson JT (2011) Benthic habitat mapping: A review of progress towards improved understanding of the spatial ecology of the seafloor using acoustic techniques. Estuarine, Coastal and Shelf Science 92: 502–520.
5. Le Bas TP, Huvenne VAI (2009) Acquisition and processing of backscatter data for habitat mapping – Comparison of multibeam and sidescan systems. Applied Acoustics 70: 1248–1257.
6. Hellequin L, Lunton X, Augustin JM (1997) Postprocessing and signal corrections for multibeam echosounder images. OCEANS '97 MTS/IEEE Conference Proceedings. pp. 23–26 vol. 21.
7. Hughes Clarke JE, Danforth BW, P.Valentine (1997) Areal seabed classification using backscatter angular response at 95 kHz. NATO SACLANTCEN Conference Proceedings Series CP-45, High Frequency Acoustics in Shallow Water: pp. 243–250.
8. Lockhart D, Pawlowski RJ, Saade EJ (2005) Advances in processing and collecting multibeam echosounder data for seabed habitat mapping. Benthic Habitats and the Effects of Fishing 41: 179–182.
9. Edwards BD, Dartnell P, Chezar H (2003) Characterizing benthic substrates of Santa Monica Bay with seafloor photography and multibeam sonar imagery. Marine Environmental Research 56: 47–66.
10. Dartnell P, Gardner JV (2004) Predicting seafloor facies from multibeam bathymetry and backscatter data. Photogrammetric Engineering & Remote Sensing 70: 1081–1091.

11. McGonigle C, Grabowski JH, Brown CJ, Weber TC, Quinn R (2011) Detection of deep water benthic macroalgae using image-based classification techniques on multibeam backscatter at Cashes Ledge, Gulf of Maine, USA. Estuarine, Coastal and Shelf Science 91: 87–101.
12. Preston JM, Christney AC, Beran LS, Collins WT (2004) Statistical seabed segmentation from images and echoes to objective clustering. Proceedings of the Seventh European Conference on Underwater Acoustics, ECUA 2004, 5–8 July 2004, Delft, The Netherlands: 813–818.
13. Brown CJ, Collier JS (2008) Mapping benthic habitat in regions of gradational substrata: An automated approach utilising geophysical, geological, and biological relationships. Estuarine, Coastal and Shelf Science 78: 203–214.
14. Blondel P, Gómez Sichi O (2009) Textural analyses of multibeam sonar imagery from Stanton Banks, Northern Ireland continental shelf. Applied Acoustics 70: 1288–1297.
15. Fonseca L, Mayer L (2007) Remote estimation of surficial seafloor properties through the application Angular Range Analysis to multibeam sonar data. Marine Geophysical Researches 28: 119–126.
16. Hamilton LJ, Parnum I (2011) Acoustic seabed segmentation from direct statistical clustering of entire multibeam sonar backscatter curves. Cont Shelf Res 31: 138–148.
17. Huang Z, Siwabessy J, Nichol S, Anderson T, Brooke B (2013) Predictive mapping of seabed cover types using angular response curves of multibeam backscatter data: Testing different feature analysis approaches. Continental Shelf Research 61–62: 12–22.
18. Lamarche G, Lurton X, Verdier A-L, Augustin J-M (2010) Quantitative characterisation of seafloor substrate and bedforms using advanced processing of multibeam backscatter—Application to Cook Strait, New Zealand. Cont Shelf Res 31: S93–S109.
19. Simons DG, Snellen M (2009) A Bayesian approach to seafloor classification using multi-beam echo-sounder backscatter data. Applied Acoustics 70: 1258–1268.
20. Wilson MFJ, O'Connell B, Brown C, Guinan JC, Grehan AJ (2007) Multiscale terrain analysis of multibeam bathymetry data for habitat mapping on the Continental Slope. Marine Geodesy 30: 3–35.
21. Cutter GR, Rzhanov Y, Mayer LA (2003) Automated segmentation of seafloor bathymetry from multibeam echosounder data using local Fourier histogram texture features. Journal of Experimental Marine Biology and Ecology 285–286: 355–370.
22. Holmes KW, Van Niel KP, Radford B, Kendrick GA, Grove SL (2008) Modelling distribution of marine benthos from hydroacoustics and underwater video. Cont Shelf Res 28: 1800–1810.
23. Beaman RJ, Daniell JJ, Harris PT (2005) Geology–benthos relationships on a temperate rocky bank, eastern Bass Strait, Australia. Marine and Freshwater Research 56: 943–958.
24. Lundblad ER, Wright DJ, Miller J, Larkin EM, Rinehart R, et al. (2006) A benthic terrain classification scheme for American Samoa. Marine Geodesy 29: 89–111.
25. Rattray A, Ierodiaconou D, Laurenson L, Burq S, Reston M (2009) Hydro-acoustic remote sensing of benthic biological communities on the shallow South East Australian continental shelf. Estuarine, Coastal and Shelf Science 84: 237–245.
26. Ierodiaconou D, Monk J, Rattray A, Laurenson L, Versace VL (2011) Comparison of automated classification techniques for predicting benthic biological communities using hydroacoustics and video observations. Cont Shelf Res 31: S28–S38.
27. Ierodiaconou D, Burq S, Reston M, Laurenson L (2007) Marine benthic habitat mapping using multibeam data, georeferenced video and image classification techniques in Victoria, Australia. Journal of Spatial Science 52: 93–104.
28. Lucieer V, Hill NA, Barrett NS, Nichol S (2013) Do marine substrates 'look' and 'sound' the same? Supervised classification of multibeam acoustic data using autonomous underwater vehicle images. Estuarine, Coastal and Shelf Science 117: 94–106.
29. Micallef A, Le Bas TP, Huvenne VAI, Blondel P, Hühnerbach V, et al. (2012) A multi-method approach for benthic habitat mapping of shallow coastal areas with high-resolution multibeam data. Cont Shelf Res 39–40: 14–26.
30. Fonseca L, Brown C, Calder B, Mayer L, Rzhanov Y (2009) Angular range analysis of acoustic themes from Stanton Banks Ireland: A link between visual interpretation and multibeam echosounder angular signatures. Applied Acoustics 70: 1298–1304.
31. Che Hasan R, Ierodiaconou D, Laurenson L (2012) Combining angular response classification and backscatter imagery segmentation for benthic biological habitat mapping. Estuarine, Coastal and Shelf Science 97: 1–9.
32. Rzhanov Y, Fonseca L, Mayer L (2012) Construction of seafloor thematic maps from multibeam acoustic backscatter angular response data. Computers & Geosciences 41: 181–187.
33. Marsh I, Brown C (2009) Neural network classification of multibeam backscatter and bathymetry data from Stanton Bank (Area IV). Applied Acoustics 70: 1269–1276.
34. Breiman L (2001) Random forests. Machine Learning 45: 5–32.
35. Liaw A, Wiener M (2002) Classification and Regression by randomForest. R News 2: 18–22.
36. Cheng L, Lek S, Lek-Ang S, Li Z (2011) Predicting fish assemblages and diversity in shallow lakes in the Yangtze River basin. Limnologica - Ecology and Management of Inland Waters.
37. Collin A, Archambault P, Long B (2011) Predicting Species Diversity of Benthic Communities within Turbid Nearshore Using Full-Waveform Bathymetric LiDAR and Machine Learners. PLoS ONE 6: e21265.
38. Seiler J, Friedman A, Steinberg D, Barrett N, Williams A, et al. (2012) Image-based continental shelf habitat mapping using novel automated data extraction techniques. Cont Shelf Res 45: 87–97.
39. Ierodiaconou D, Rattray A, Laurenson L, Monk J, Lind P (2007) Victorian marine habitat mapping project. Warrnambool, Victoria, Australia: Deakin University.
40. Schmidt J, Evans IS, Brinkmann J (2003) Comparison of polynomial models for land surface curvature calculation. International Journal of Geographical Information Science 17: 797–814.
41. Parnum IM (2007) Benthic habitat mapping using multibeam sonar systems. PhD thesis, Curtin University, Perth, Australia: 213pp.
42. Parnum IM, Gavrilov AN (2011) High-frequency multibeam echo-sounder measurements of seafloor backscatter in shallow water: Part 1 Data acquisition and processing. Underwater Technology: The International Journal of the Society for Underwater 30: 3–12.
43. Daily M (1983) Hue-saturation-intensity split-spectrum processing of Seasat radar imagery. Photogrammetric Engineering & Remote Sensing 49: 349–355.
44. Haralick RM, Shanmugam K, Dinstein IH (1973) Textural features for image classification. IEEE Transactions on Systems, Man and Cybernetics SMC-3: 610–621.
45. Lucieer V, Lamarche G (2011) Unsupervised fuzzy classification and object-based image analysis of multibeam data to map deep water substrates, Cook Strait, New Zealand. Cont Shelf Res 31: 1236–1247.
46. Meyer J, White S (2007) Lava morphology mapping by expert system classification of high-resolution side-scan sonar imagery from the East Pacific Rise, 9°–10°N. Marine Geophysical Researches 28: 81–93.
47. Masetti G, Calder B (2012) Remote identification of a shipwreck site from MBES backscatter. Journal of Environmental Management 111: 44–52.
48. Bins LS, Fonseca LMG, Erthal GJ, Ii FAM (1996) Satellite imagery segmentation: a region growing approach. VIII Brazilian Remote Sensing Symposium. Slavador, Brasil. pp. 677–680.
49. IMCRA (1998) Interim Marine and Coastal Regionalisation for Australia: An ecosystem-based classification for marine and coastal environments (Version 3.3). Environment Australia, Commonwealth Department of the Environment, Canberra.
50. Jaiantilal A (2009) Random Forest (Regression, Classification and Clustering) implementation for MATLAB (and Standalone). Available: http://codegooglecom/p/randomforest-matlab. Accessed 20 March 2014.
51. Congalton RG, Green K (2009) Assessing the Accuracy of Remotely Sensed Data: Principles and Practices. 2nd Edition. Boca Raton, Florida: CRC Press. 183pp p.
52. Congalton RG (1991) A review of assessing the accuracy of classifications of remotely sensed data. Remote Sensing of Environment 37: 35–46.
53. Li J, Heap AD, Potter A, Huang Z, Daniell JJ (2011) Can we improve the spatial predictions of seabed sediments? A case study of spatial interpolation of mud content across the southwest Australian margin. Cont Shelf Res 31: 1365–1376.
54. Le Gonidec Y, Lamarche G, Wright I (2003) Inhomogeneous substrate analysis using EM300 backscatter imagery. Marine Geophysical Researches 24: 311–327.
55. Kobryn HT, Wouters K, Beckley LE, Heege T (2013) Ningaloo Reef: Shallow Marine Habitats Mapped Using a Hyperspectral Sensor. PLoS ONE 8: e70105.
56. van der Meer FD, van der Werff HMA, van Ruitenbeek FJA, Hecker CA, Bakker WH, et al. (2012) Multi- and hyperspectral geologic remote sensing: A review. International Journal of Applied Earth Observation and Geoinformation 14: 112–128.
57. De Falco G, Tonielli R, Di Martino G, Innangi S, Simeone S, et al. (2010) Relationships between multibeam backscatter, sediment grain size and *Posidonia oceanica* seagrass distribution. Cont Shelf Res 30: 1941–1950.
58. Galparsoro I, Borja Á, Bald J, Liria P, Chust G (2009) Predicting suitable habitat for the European lobster (*Homarus gammarus*), on the Basque continental shelf (Bay of Biscay), using Ecological-Niche Factor Analysis. Ecological Modelling 220: 556–567.
59. Iampietro PJ, Kvitek RG, Morris E (2005) Recent advances in automated genus-specific marine habitat mapping enabled by high-resolution multibeam bathymetry. Marine Technology Society Journal 39: 83–93.
60. Coleman N, Gason ASH, Poore GCB (1997) High species richness in the shallow marine waters of south-east Australia. Marine Ecology Progress Series 154: 17–26.
61. Gray JS (1997) Marine biodiversity: patterns, threats and conservation needs. Biodiversity and Conservation 6: 153–175.
62. Lurton X, Dugelay S, Augustin JM (1994) Analysis of multibeam echo-sounder signals from the deep seafloor. OCEANS '94. pp. 213–218.
63. Parnum IM, Gavrilov AN (2011) High-frequency multibeam echo-sounder measurements of seafloor backscatter in shallow water: Part 2 Mosaic production, analysis and classification. Underwater Technology: The International Journal of the Society for Underwater 30: 13–26.
64. Freitas R, Rodrigues AM, Quintino V (2003) Benthic biotopes remote sensing using acoustics. Journal of Experimental Marine Biology and Ecology 285–286: 339–353.
65. Kendall MS, Jensen OP, Alexander C, Field D, McFall G, et al. (2005) Benthic Mapping Using Sonar, Video Transects, and an Innovative Approach to

Accuracy Assessment: A Characterization of Bottom Features in the Georgia Bight. Journal of Coastal Research 21: 1154–1165.
66. Kaiser MJ, Collie JS, Hall SJ, Jennings S, Poiner IR (2002) Modification of marine habitats by trawling activities: prognosis and solutions. Fish and Fisheries 3: 114–136.
67. Gratwicke B, Speight MR (2005) The relationship between fish species richness, abundance and habitat complexity in a range of shallow tropical marine habitats. Journal of Fish Biology 66: 650–667.
68. Weiss AD (2001) Topographic positions and landforms analysis (Conference Poster). ESRI International User Conference. San Diego, CA.

8

# Combining Natural History Collections with Fisher Knowledge for Community-Based Conservation in Fiji

**Abigail S. Golden[1], Waisea Naisilsisili[2], Isikele Ligairi[3], Joshua A. Drew[1]***

1 Dept. of Ecology, Evolution and Environmental Biology, Columbia University, New York, New York, United States of America, 2 Wildlife Conservation Society Fiji, Suva, Fiji, 3 University of the South Pacific, Suva, Fiji

## Abstract

Harnessing the traditional ecological knowledge (TEK) of local communities has the potential to enhance conservation planning in developing regions. Marine protected areas (MPAs) that incorporate traditional beliefs about reef tenure are generally more successful in reaching conservation goals and ensuring the participation of local fishermen on vulnerable tropical reef systems. Fiji possesses a unique system of traditional reef management in which local clans or villages, called *mataqali*, control individual units of a reef, known as *qoliqoli*, and make independent management decisions based on traditional beliefs and conservation concerns. This is an example of a system, known as customary marine tenure, which has attracted interest from conservation scientists hoping to set up MPAs in vulnerable regions. As one example of this grassroots participation, Nagigi village on the Fijian island of Vanua Levu has expressed interest in setting up an MPA in part of its *qoliqoli* because of concerns about overfishing. In response to this interest, we took a two-pronged approach to assessing Nagigi's fishery status and conservation needs, first conducting a fishery-independent species survey using destructive sampling and then focusing on fisheries targets identified through fisher interviews. These interviews allowed us to identify heavily targeted species, assess villagers' understanding of reef dynamics over 30 or 40 years of fisheries expansion, and evaluate village support and expectations for a proposed conservation program. Based on our findings we recommend a temporary closure to be in effect for at least three years, allowing one of the more important fishery targets, *Lethrinus harak* (Forsskål, 1775; Lethrinidae), to complete at least one generation within the reserve. The methodology of matching the proposed marine protected area with the life histories and ecologies of heavily targeted species identified through fisherman and -woman interviews can offer a template for future conservation projects that seek to synthesize indigenous peoples' needs and knowledge with ecological data.

**Editor:** Richard K.F. Unsworth, Seagrass Ecosystem Research Group, Swansea University, United Kingdom

**Funding:** Funding was provided by the Department of Ecology, Evolution and Environmental Biology of Columbia University and the Mindlin Foundation (http://mindlinfoundation.org). The funders had no role in study design, data collection and analysis, decision to publish, or preparation of the manuscript.

**Competing Interests:** The authors have declared that no competing interests exist.

* E-mail: jd2977@columbia.edu

## Introduction

The last thirty years have seen a surge in interest from ecologists and conservation scientists in trying to couple formal management strategies with local indigenous communities' strategies for exploiting and managing their natural resources. These combined strategies, collectively known as community-based management, include techniques such as the temporal restriction of wildlife harvests, protection of vulnerable life-history stages of an exploited species, and resource rotation [1]. Decisions concerning when or where resource exploitation is prohibited and what resources are to be utilized at any given time are traditionally made by local leaders recognized by their community for their high status and/or harvesting prowess [2,3]. These traditional management techniques vary across indigenous communities and between habitat types.

Many South Pacific peoples are highly dependent on marine products as a source of animal protein [4–6]. As a result, the health and food resources of these people are intertwined with the health of their coastal ecosystems, and communities have developed traditional management techniques to regulate their use. The most widespread form of community-based marine management in the region is the reef and lagoon tenure system, also known as customary marine tenure [5–8]. To manage their crucial marine resources, many cultures developed tenure systems in which a chief, clan, or family controlled a particular area of coast and regulated its exploitation for successive generations. Consequently, it was in the best interests of a particular tenure-holder to harvest from the reef in moderation, ensuring consistently high yields and avoiding social censure for overharvesting. In this way, Pacific Islander communities could prevent the situation described by Hardin [9] as the "tragedy of the commons," in which natural resources are progressively overutilized and degraded because no single individual bears as much of the cost of overharvesting as he or she reaps of its benefits. In Fiji, reef tenure takes the form of coastal tenure areas called *qoliqoli* that are legally controlled by individual patrilineal clans known as *mataqali* [8,10]. More recently, as development brought greater mobility, clans dispersed, and this control moved to individual villages, which could make the communal decision to temporarily ban fishing on portions of a *qoliqoli* in response to overfishing and other causes, thus helping to maintain a healthy coastal ecosystem over many generations.

One corollary of community-based management in indigenous cultures is the existence of bodies of traditional ecological knowledge (TEK) concerning the plants, animals, and environmental conditions these societies need to survive. TEK has been defined in various ways, but it is generally understood to consist of a body of knowledge passed down through generations about the relations of humans with other living beings and their environment [1,11]. Facets of a culture's TEK include the local names for common species, knowledge about timing and location of biologically significant events such as migrations, and the ways in which local people use, perceive, and manage their natural resources [12]. Since the 1970s, ecologists have increasingly begun to consider these aspects of indigenous cultures' TEK a significant resource for making informed decisions about the management and conservation of poorly studied ecosystems, such as Pacific coral reefs and sea grass meadows [5].

Despite the importance of TEK and customary marine tenure to indigenous communities, colonization by Western powers brought economic and cultural changes that led to degradation of traditional knowledge and management systems throughout the Pacific. According to Johannes [5], three main factors contributed to the decline of marine tenure systems in Pacific Island nations under colonial rule: (a) the introduction of a market economy; (b) the devaluation of traditional authorities; and (c) new management laws imposed by colonial powers. In Fiji, these processes began to put heavy fishing pressure on the country's marine resources following World War II, with concerted fisheries development coming to the fore in the 1950s [10,13]. International and Fijian national interests began to push for the increased commercialization of Fiji's fisheries and the shift to an export market, with the Fiji Development Bank encouraging the commercial development of even the country's remotest villages [8,10]. This transformation led to significant overexploitation of both traditional inshore fisheries and the international offshore tuna fishery, which according to many experts may be close to collapse [10,14]. In addition, cultural values have shifted such that some younger Fijians place less value on traditional management systems and see the ocean primarily through a commercial, rather than a traditional, lens [10]. Today, of Fiji's 410 *qoliqoli*, 70 are considered overexploited, with a further 250 fully developed at maximum sustainable yield capacity [15].

The increasing replacement of traditional fishing methods with new technologies, such as motorboats, nylon nets, spearguns, diving gear, and flashlights for night fishing, has also contributed to overharvesting in Pacific subsistence communities [16–18]. In places where traditional values are largely intact, such as the remote Fijian island of Ono-i-Lau, 400 kilometers from the country's capital, these new technologies have increased fishing efficiency but not overall catch [16]. In such places, subsistence still governs fishing effort and fishermen catch no more than they need for their own use [5,16]. However, in more central locations where a cash economy is in place, fishermen use these new technologies to overfish their local reefs and seagrass habitats, selling their surplus for cash, thereby eventually increasing the fishing effort necessary to make a living from the reef.

More recently, Fiji and other Pacific countries developed marine conservation and management programs, often in the form of marine protected areas (MPAs), in response to increased concerns about overexploitation and habitat degradation. In many Western countries, such fishing recommendations would be based on hard data detailing the ecological dynamics of the fishery, and regulation would be top-down and enforced by national legislation [19]. However, Pacific Island marine ecosystems offer unique challenges to such a Westernized approach to fisheries management, due to their greater biodiversity, remoteness, and interconnectivity [19]. For instance, such coastal areas are characterized by complex interactions between coral reefs and the mangrove forest and seagrass meadow habitats, which provide nursery and foraging areas for valuable reef fish [20]. To circumvent the challenges posed by this Western model of top-down management, researchers have begun to suggest the creation of networks of small, locally protected areas based on reef tenure divisions and incorporating traditional management systems [17,18]. In Fiji, these have taken the form of locally managed marine areas constructed around the country's existing *qoliqoli* system [3,8,10,15]. Such management programs attempt to blend modern scientific techniques with the input of local stakeholders and possessors of traditional ecological knowledge.

Nagigi Village, a small fishing community on the southern coast of Vanua Levu, Fiji's second-largest island, provides one such example of community-driven demand for marine protection (Figure 1). In recent years, village leaders have expressed interest in establishing a short-term marine protected area on part of Nagigi's *qoliqoli*. As one villager put it, "For the sake of future generations, if we want to have an abundance of resources again, we should encourage an MPA on the fishing grounds. Our main concern is that if we're not aware of what's done, future generations won't know what those species are or recognize the need to gain back what they've lost" (Interview 2). Our research combined destructive sampling of reef fish to create a survey of Nagigi's biodiversity with villager interviews in an attempt to answer the following questions related to Nagigi's subsistence and artisanal fishery:

- What methods characterize subsistence and artisanal fishing in a traditionally managed Pacific Islander society, and what are the benefits that people gain from these crafts?
- What species are heavily targeted in such a society, and therefore at risk for overfishing and population collapse?
- How do villagers perceive that their reef environment has changed throughout their lives, and what do they understand to have caused these changes?
- How do villagers conceptualize marine resource management, and what conservation measures might they support to halt and reverse any negative impacts they have observed?

To answer these questions we carried out a two-part investigation into the fisheries of Nagigi. The first component was a fishery independent survey of potential target species captured in and around the Nagigi *qoliqoli*. We then used this baseline data as a launching pad for a series of interviews with Nagigi fishers to understand the extent of the fishery, to explore any differences between genders in fishing and finally to assess fishers' attitudes towards the establishment of a no-take MPA in their waters.

## Methods

### Ethics Statement

All fish were collected under the auspices of the Columbia University Animal Care Board permit to Joshua Drew (AC-AAAF6300). Every effort was made to minimize the number of samples collected, the environmental damage caused by collection, and the suffering of the animals sacrificed. For our human subject data, prior to each interview we explained the potential, albeit minimal, risks to the interviewee as well as why and how the information would be collected and personal data secured. After explaining this in both English and Fijian we obtained a signed written consent form and the interview proceeded. These

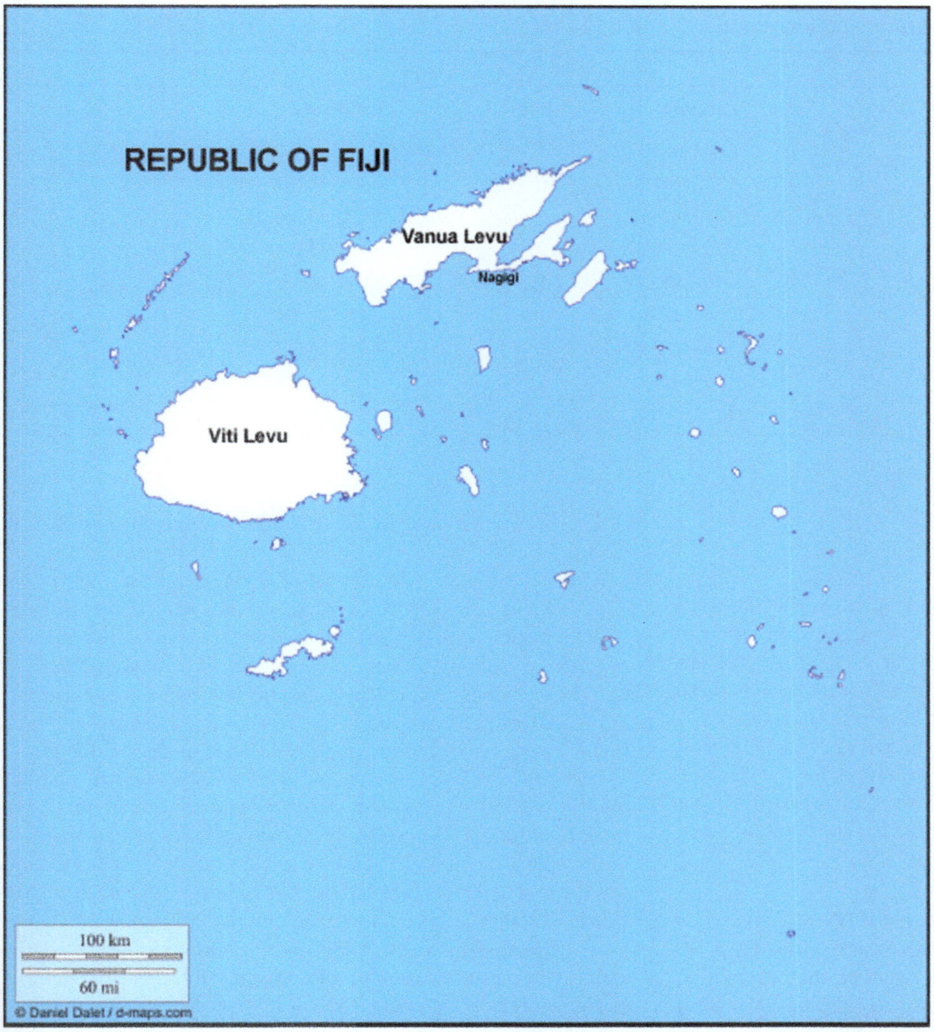


**Figure 1. Map of Fiji used with permission of Daniel Dalet, d-maps.com.**

procedures were reviewed and approved by the Columbia University Institutional Review Board (IRB-AAAL4860).

## Fish Collection

Data were collected over four days in June and July of 2013. Nine sites were selected within and outside the fringing reef adjacent to Nagigi village (Figure 1, Table 1). Fish specimens were collected using spearfishing and the application of the fish anesthetic MS-222, following Columbia University IRB protocols (Proposal IRB-AAAF6300). Additional species were observed on the reef but not sampled due to their conservation status or a lack of sampling opportunity. After collection fish were photographed and identified to species level using field guides [21]. Gill or muscle tissue was collected and stored in liquid nitrogen, while whole specimens were preserved in formalin and have since been accessioned to the collection of the American Museum of Natural History, New York.

## Fisher Interviews

Twenty-two individual fishers were interviewed singly and in groups across a total of 15 interviews. Interviews followed an IRB-approved questionnaire and participants' written consent was obtained in accordance with IRB protocol (Proposal IRB-AAAL4860). Most interviews took place in participants' homes, but some occurred while participants fished or gleaned on the reef at low tide and one interview took place during a kava-drinking session in front of the village hall. Interviews were generally about half an hour long, but some were considerably shorter or longer depending on context, language barriers, and participants' knowledge. For instance, one interview in which participants knew little to no English lasted only ten minutes, while the kava-drinking session lasted more than two hours. Three native Fijian speakers assisted the researchers as translators.

Of the participants, eight were male and 14 were female, and participants' average age was 50. We deliberately sought out older, expert fisherwomen because their knowledge has typically been discounted in studies of traditional ecological knowledge in the Pacific region, despite their important contribution to subsistence fisheries [22]. Participants had an average of 44 years' experience fishing on Nagigi's reef, ranging from one woman who had moved to the village a year ago to three villagers in their sixties who had lived in Nagigi their whole lives and presumably fished and gleaned on the reef from a young age. We deliberately chose participants who varied greatly in age and experience in order to gain impressions of fishers' knowledge across generations and experience levels.

**Table 1.** Sampling locations with latitude and longitude.

| Location # | Site Name | Latitude | Longitude |
|---|---|---|---|
| 1 | FJ_01 | s 16d 48.39″ | e 179 28.474″ |
| 2 | FJ_02 | s 16′ 48.608″ | e 179′ 28.545″ |
| 3 | FJ_03 | s 16° 48.451′ | e 179° 28.138′ |
| 4 | FJ_04 | s 16′48.377″ | e 179′ 28.757″ |
| 5 | FJ_05 | s 16′ 48.323″ | e 179. 28.554″ |
| 6 | FJ_06 | s 16′ 48.409″ | e 179′ 28.594″ |
| 7 | FJ_07 | s 16′ 48.340″ | e 179′ 28.598″ |
| 8 | FJ_08 | s 16′ 48.667″ | e 179′ 28.602″ |
| 9 | FJ_09 | 16° 48.180′S | 179° 28.738′E |

## Results

### Partial Species List

In total, 150 species were recorded on Nagigi's reef (Table S1). Of these, 96 (64%) were collected and accessioned to the American Museum of Natural History in New York and 15 (10%) were sighted by researchers but not collected. Sixty-three were mentioned in interviews by expert fishers. Seventeen species were both collected and mentioned in interviews (11.3% of total species collected). Fijian names were recorded for 82 species. Several additional generic Fijian names were recorded based on interviews (Table 2). The most species-rich families were Pomacentridae (16.7%) and Labridae (10.7%).

### Social Characteristics of the Nagigi Fishery

Across all 15 interviews, male villagers were more likely to describe spearfishing and night diving as their main activities, while female villagers were more likely to use fishing nets. Men and women both used fishing lines frequently. Men fished from boats more frequently, while women were more likely simply to walk or swim to the portion of the reef on which they planned to fish. Only women used *bilibili*, or traditional bamboo fishing rafts (Figure 2).

Most interviewees said they fished daily or at least once a week. The majority (11 out of 16) of participants said that they fished only for their own home consumption or for some combination of personal consumption and cash income. Three young men all mentioned selling fish, octopus, and bêche-de-mer (sea cucumber) at the market in Savusavu, the nearest town, as a source of "quick cash" (Interview 7). One couple have a small business selling fish at the Savusavu market to gain the income they need to maintain their house and educate their children. As the husband put it, "our bank is in the sea" (Interview 3). According to his wife, he is known in the village as an expert fisherman who can reliably bring home fish the length of a forearm (about 40–50 cm), unlike the rest of the village's fishermen, who only catch hand-length fish (about 20 cm). One 52-year-old woman, who is a community-acknowledged expert octopus fisherwoman, sells octopus and forearm-length fish at the market in Savusavu and keeps smaller fish for her own consumption (Interview 9). Another woman says that she fishes for her and her husband's consumption and to sell at market but also to share with her friends and fellow villagers: "anybody here wants

**Table 2.** Generic Fijian fish names.

| Fijian name | English definition | Families represented |
|---|---|---|
| *baludawa* | parrotfish | Scaridae |
| *boila* | moray eel | Muraenidae |
| *dridri* | several surgeonfishes | Acanthuridae |
| *kake* | several snappers | Lutjanidae |
| *kawakawa* | grouper | Serranidae |
| *labe* | several wrasses | Labridae |
| *lematua* | scorpionfish | Scorpaenidae |
| *nuru* | any small finger-length fish | Apogonidae, Pomacentridae, etc. |
| *rawarawa* | blue or green parrotfish | Scaridae |
| *rusarusa* | small rabbitfish | Siganidae |
| *sumusumu* | pufferfish | Tetraodontidae |
| *tivitivi* | butterflyfish | Chaetodontidae |
| *ulavi* | gray or white parrotfish longer than 30 cm | Scaridae |

Several generic Fijian names (*i.e.* those which represent multiple species) with approximate scientific definitions.

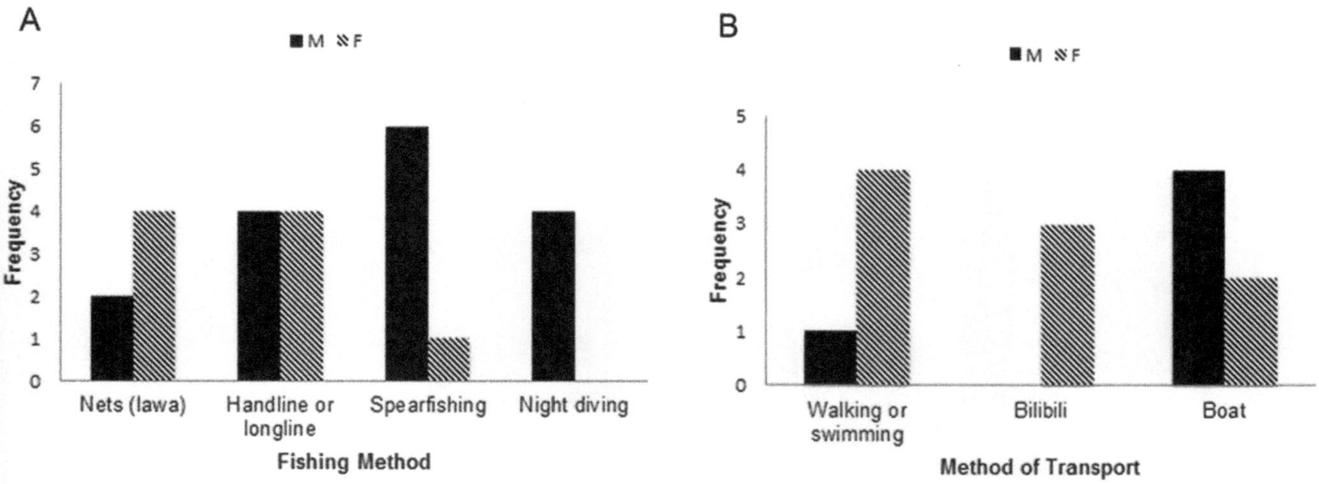


**Figure 2. Fishing and transport methods.** Frequency with which common fishing and transport methods were mentioned by men and women in 15 interviews.

to eat fish, they just come.I just keep a little bit for the two of us" (Interview 8).

Interviews indicate that fishing labor is primarily divided along gender lines, and that it is a frequent occupation for both men and women. There is no clear divide between those who fish for income (artisanal fishers) and those who fish for their own consumption (subsistence fishers); rather, villagers pursue both options as they need protein or cash. Fishing also appears to facilitate social relationships, as with the sharing of fish among friends and relatives, and upward socioeconomic mobility, as demonstrated by the couple who use their fishing income to pay their children's school fees.

## Identifying At-Risk Species

The most common species that participants reported targeting was *kuita* (octopus *Octopus spp.*), with *kabatia* (thumbprint emperor, *Lethrinus harak*), *kanace* (bluetail mullet, *Moolgarda engeli*) and *saqa* (giant trevally, *Caranx ignobilis*) only slightly less in demand (Table 3). Octopus fishing for market seems to be a primarily, though not exclusively, female skill, passed on from grandmothers to their children and grandchildren. Several non-finfish targets were mentioned repeatedly during my interviews, including octopus (*kuita*), sea turtle (*vonu*), shellfish, and bêche-de-mer (Figure 3). Only men described targeting turtle and bêche-de-mer, while the general category of reef gleaning, which includes the targeting of octopus, shellfish, and seaweed, was a more egalitarian pursuit. The bêche-de-mer fishery serves exclusively as a source for cash income in Fijian society, and no participants reported eating sea cucumbers. Holothurians are typically caught and dried in salt and then traded with foreign vessels for entry into Asian food markets. In contrast, turtle (*vonu*) meat is consumed in Fijian villages on celebratory occasions, and thus the fishery is largely driven by local demand and culture.

## Villagers' Perceptions of Reef Change

Nearly all participants said that fish size and abundance had decreased over the years they'd lived in Nagigi and that the required fishing effort had increased (Table 4). Participants described having to go farther or work longer to catch enough fish. In particular, villagers said that *kawakawa* (a generic Fijian name for small species of groupers, genera *Epinephelus* and *Cephalopholi*; Serranidae) and octopus had decreased in size and abundance. As one villager put it, "Before, they used to catch eight, nine [octopus] sometimes. But now—you can just catch two, three" (Interview 7). Several participants said that the boxfish (*Ostracion cubicus*) had disappeared from the reef entirely. *Varivoce* (humphead wrasse or *Cheilinus undulatus*) and *kalia* (bumphead parrotfish, *Bolbometopon muricatum*) have also become scarce. These findings echo range-wide population dynamics in these species: *C. undulatus* is listed as endangered on the IUCN Red List and *B. muricatum* is categorized as vulnerable [23,24] Two villagers reported that *nuru* (any fish shorter than a finger's length) have become scarce and that inshore coral heads, which are an important source of habitat for the *nuru*, are dead or degraded.

Three villagers cited the smashing of inshore coral heads while fishing, either purposefully or accidentally, as one cause of this habitat loss and fish population decrease. Interestingly, both members of one married couple mentioned this issue in separate interviews, but the husband described coral-smashing as a deliberate fishing technique while his wife believed that it occurs accidentally when fishers walk on the reef. (Interviews 8, 11).

Nagigi's *Turaga ni Koro* (elected village headman) proved to be an especially valuable source of information about environmental change. Although he no longer fishes, his role as headman requires him to act as a liaison between Nagigi Village and the Savusavu provincial office and to represent the village's interests. He also prosecutes poachers from outside Nagigi, people who fish without a license, and those who violate laws such as the bans on turtle fishing, use of the poisonous *duva* root, and targeting endangered species. Perhaps as a result, he proved highly knowledgeable about and observant of reef dynamics. He cited three sources of environmental stress that no other villagers mentioned: (1) development inland of Nagigi that flushes sediment and pesticides down Nagigi Creek to the reef, (2) an unusually hot season in 1998 that placed temperature stress on the coral, and (3) demand for prawns from nearby resorts that has spurred overharvesting of these organisms.

## Conservation Attitudes and Suggested Solutions

Interviewees unanimously supported the establishment of a marine protected area or *tabu* on Nagigi's reef. None of them, however, believed that an MPA should last permanently or should extend over the village's entire fishing grounds. Suggestions ranged from protections lasting one year to ten years. The *Turaga ni Koro*,

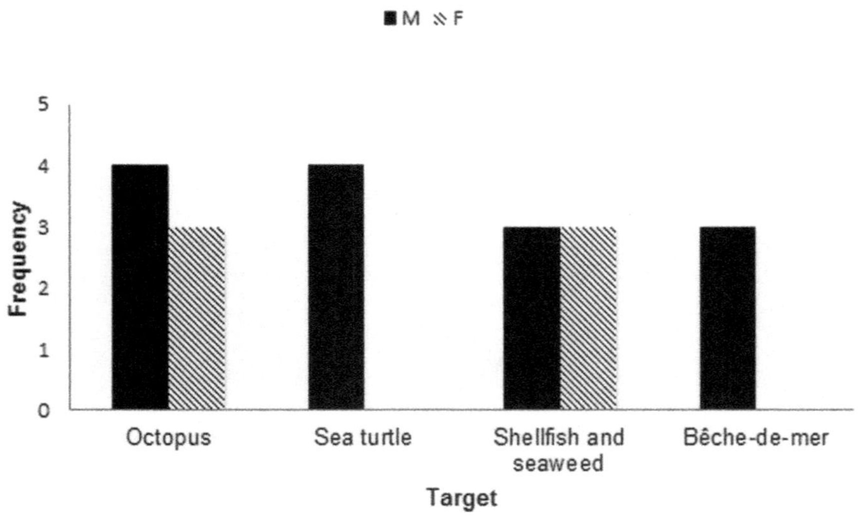


**Figure 3. Non-finfish targets of men and women's fishing.** Sea turtle and bêche-de-mer harvesting are primarily male pursuits while reef gleaning (for octopus, shellfish, seaweed, etc.) is more egalitarian.

who has been a vocal proponent of the MPA plan, believes that a closure should last for five years. Another participant thinks that only a one-year MPA is necessary, while his wife thinks it should be "the longer the better" (Interview 8). Three young men who fish to make "quick cash" believe that a three-year closure would be enough for most fish, including the IUCN red-listed *Bulbometopon muricatum*, to regain their former size. One interviewee had recently moved to Nagigi from a neighboring village, which had just decided to extend its five-year *tabu* for another five years after observing rebounding fish populations, and she was consequently very supportive of Nagigi's plan.

No interviewees expressed concern about losing their fishing income or subsistence catch while the MPA was in place. Several spoke about their apprehension for the future if conservation steps were not taken. One man expressed the expectation that the *tabu* area would provide a safe breeding area and would "overflow with fish," increasing the villagers' catch throughout their marine tenure area (Interview 11). One couple that moved to Nagigi two years ago had not heard of the plan to create an MPA, but when the project's purpose was explained to them, they agreed that it would be a good idea because they had noticed that Nagigi's fish populations were scarce. Only one interviewee expressed reservations about the MPA plan because she noticed that when a neighboring village set a *tabu* area on its reef, the villagers came to Nagigi's reef at night to fish clandestinely. She thinks that if an

**Table 3.** At-risk reef species.

| Fijian Name | Scientific Name | Number of Times Mentioned | Perceptions of Pop. Change (n = number of participants with this perception) |
|---|---|---|---|
| Kuita | *Octopus* sp. | 8 | Decreasing size (n = 1) and abundance (n = 3) |
| Kabatia | *Lethrinus harak* | 6 | Decreasing abundance (n = 1) |
| Saqa | *Caranx ignobilis* | 6 | Decreasing abundance (n = 1) |
| Kanace | *Moolgarda engeli* | 6 | Smaller, scarcer, and harder to catch (n = 1) |
| Ulavi | Gray or white parrotfish larger than 30 cm | 5 | Increasing abundance (n = 1) |
| Vonu | Sea turtles | 5 | Decreasing abundance (n = 2) |
| Labe | *Halichoeres trimaculatus* | 5 | N/A |
| Nuqa | *Siganus vermiculatus* | 5 | Decreasing abundance (n = 1) or increasing abundance (n = 1) |
| Kawakawa | *Epinephelus polyphekadion* | 4 | Decreasing size and abundance; increased fishing effort necessary (n = 5) |
| Ta | *Naso unicornis* | 4 | N/A |
| Tabace | *Acanthurus triostegus* | 4 | N/A |
| Dridri | 3 *Acanthurus* sp. | 4 | Increasing abundance (n = 1) |
| Vasua | *Tridacna gigas* (sea clams) | 4 | Decreasing abundance (n = 1) |
| Deou | *Upeneus vittatus* | 4 | N/A |

Most at-risk reef species based on the number of villagers who claimed to target them. Includes perceived changes in the population of these species and the number of interviewees who made these assessments.

**Table 4.** Villager perceptions of reef change across 12 interviews.

| Interview # | Average Years in Village | Perceptions of Reef Dynamics |
|---|---|---|
| 1 | 58 | Populations of *kabatia, kawakawa* and *ululoa* have decreased. Populations of *balagi, nuqa, dridri*, and *ulavi* have increased; interviewee perceived these as coming from outside the reef. Interviewee has seen no change in her octopus catch because she is an expert octopus fisherwoman and knows where the breeding ground is. |
| 2 | 49 | *Nuru* have disappeared and inshore corals are mostly dead. Fish abundance has decreased. In particular, *kawakawa* and *ose* have decreased in abundance and become smaller. |
| 3 | 50 | *Varivoce* and *kawakawa* have become harder to find. |
| 4 | 52 | Boxfish and sea prawns have become hard to find. |
| 5 | 62 | *Nuqa* is hard to find. Sea crabs and sea prawns aren't abundant. |
| 6 | 66 | Fish abundance has decreased. |
| 7 | 34 | Fish size has decreased. *Toto* have disappeared. There used to be lots of *kalia*, but now they are scarce. Octopus are smaller and scarcer: "Before, they used to catch eight, nine sometimes, but now—you can just catch two, three." *Balagi* and *ogo* are plentiful. It's still possible to catch large *kawakawa*, but fishers have to go outside the reef. |
| 8 | More than 30 | Fish abundance and size have decreased. Villagers have to go farther outside the reef to go fishing and put in more effort to catch the same amount of fish. |
| 9 | 52 | Octopus have become scarce. |
| 10 | About 2 | Fish have decreased in abundance. *Kanace, vonu*, and *kawakawaloa* used to be more abundant and easier to catch close to shore when the interviewee was growing up in Nagigi (about 30 years ago). |
| 11 | 60 | Fish don't come in close to the shore anymore, so they're harder to catch. *Nuru* are hard to find and the coral is mostly dead. Fish are smaller. *Ogo* and snapper in particular don't come in close to the shore. *Saqa, saku, yawa, vilu, vonu, vasua*, and octopus are scarce. |
| 12 | | *Varivoce* are scarce. In 1998, Fiji had a very hot season that noticeably raised the temperature of the ocean, causing the coral to die. Since then the coral has been recovering from that event. There is a quarry on Nagigi Creek upstream from the village that flushes sediment and waste down to the shore, killing the coral. A government research station along the creek uses weedkiller, which travels down to the sea when it rains and kills the coral. In 2001, the village held a traditional 3-month *tabu* (fishing moratorium) to honor the memory of a chief who had died. |

See Table S1 for translations of Fijian species names.

MPA is set up in Nagigi, the village's fishermen will do the same on neighboring fishing grounds.

## Discussion

### Fishing as a Way of Life

Our results highlight the way that fishing and reef gleaning possess a social significance in the political ecology of villagers of Nagigi beyond simply providing a source of protein and cash. Few of our interviewees reported that they did not fish at all, and those who fished infrequently usually qualified this by saying that they were "too old" or were pregnant or nursing, the latter of which groups is subject to strict food prohibitions [25]. This suggests that fishing is a basic way of life in Nagigi, but may wax and wane over villagers' life spans as they pass through various life stages such as motherhood and old age. Fishing also provides social and recreational benefits as well as economic ones, with some participants describing fishing as their hobby or saying that they loved to catch certain fish (Interview 8). Jones [6] notes a similar enjoyment of fishing among the Fijian fisherwomen in the Lau archipelago who were her ethnoarchaeological study subjects, and notes that Fijian fishing expeditions reflect cultural customs and social relations as well as the need for food and income.

Two married couples described fishing together for personal consumption or as part of a small business, and one man said that his wife had taught him how to fish with a handline, an unusual statement in a society in which teaching is usually vertical (i.e., parents to children) (Interview 11) [26]. There seems to be a significant inter- and intra-village trade in fish that is informal and based around relationships with family and friends. As one villager put it, "if I don't eat it, I'm happy to just give it out to others" (Interview 8). Another, a recent transplant to Nagigi from a neighboring village, told me that she does not fish at all, but that when she wants fish, she asks her brothers at home to bring her some (Interview 15). Because no cash exchanges hands, or does so only within the village, such trade necessarily is included in assessments of Fiji's artisanal fishery, complicating fisheries assessments and making management more difficult.

### Conserving At-Risk Species

The proposed marine protected area in Nagigi would consist of one square kilometer of intertidal habitat directly in front of the village (Figure 4). The proposed area includes some fringing reef, patch reef and seagrass beds. The inclusion of this diverse set of habitats promises to strengthen the proposed reserve, as it accounts for ontogenetic changes in habitat use by many species, including *L. harak* [27]. However, Jupiter and Egli [15] suggest that MPAs should be at least 1400 m on a side to make sure that they are larger than the home ranges of targeted species. *Lethrinus* spp., for instance, can move up to 700 m, and do so mostly at night, making them vulnerable to poaching and night fishing. Aswani and Hamilton [28] recommend that small reserves should be 4–6 km in diameter to be effective. And while small reserves ($<$ 1 $km^2$) can be effective, more recent research suggests that large

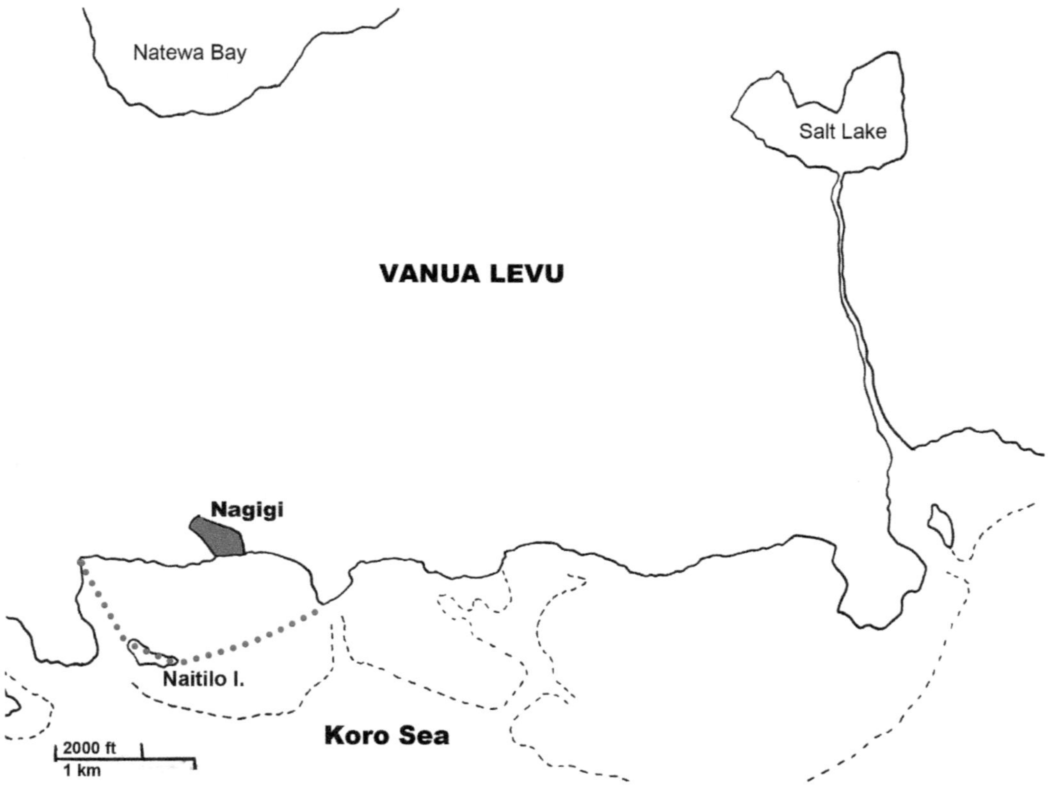


**Figure 4. Map of Nagigi showing proposed MPA site.** The proposed MPA includes 1 $km^2$ of reef flat and seagrass meadow directly in front of Nagigi village. It excludes the mangrove swamp to the east of the village and any area outside the fringing reef. Black dashed lines show fringing reefs, while red dotted line shows marine protected area site proposed by the *Turaga ni Koro*.

size is a strong indicator of reserve success [29,30]. Based on these recommendations, it seems unlikely that Nagigi's MPA will have a significant positive effect on reef fish populations without a twofold or threefold increase in size.

Even short-term MPAs are known to increase abundance and biomass of targeted species, leading to increased recruitment and migration of fish into neighboring reefs, but it is unclear how long an MPA of the size of Nagigi's would need to last in order to make these benefits tangible [15,17]. Several of the species most heavily targeted by Nagigi's villagers do not reach sexual maturity until the age of three or four years, suggesting that a three-year MPA would not be enough to ensure that an entire reproductive cohort could reach maturity without facing fishing pressure (Table 5). In *Lethrinus harak*, for instance, which is one of the most heavily targeted fish in Nagigi, females reach maturity within one or two years [31]. However, *L. harak* is a protogynous hermaphroditic species, and females begin to transition to males at age three or four [31]. In addition, *L. harak* utilizes both coral reef and seagrass habitats at different life stages, indicating that both habitat types are critical for *L. harak* conservation [27]. Based only on female maturation rate, and without taking this habitat use into account, one would anticipate that a three-year MPA would be enough to increase local populations of the species, but greater knowledge of *L. harak*'s life history and reproductive strategy reveals that a longer period and larger area of protection are needed.

Similarly the local population of *Upeneus vittatus*, another heavily targeted species, would benefit from a three-year MPA. *Upeneus vittatus* reaches maturity within a year and has a total life span of three years [32]. However, Nagigi's *U. vittatus* population is catadromous and highly geographically specific in its habitat use, traveling once a year from its habitat in Navatu Lake, northeast of Nagigi, to the mangrove swamp west of the village to spawn (Figure 4). This mangrove swamp is outside the proposed marine protected area and the fish face significant harvesting pressure from villagers during their spawning period, which occurs October through December. This harvesting event is invested with cultural significance in Nagigi, and the entire population of the village dresses up in fine clothes and garlands, known as *salusalu*, to harvest the spawning fish. Villagers believe that they must not use knives to gut *U. vittatus* during spawning season, or the fish will never return to their spawning site, and that they must not sell the fish but instead share them with those who are less fortunate (Interviews 7, 8). Although no villagers expressed concern about declines in the abundance or biomass of *U. vittatus*, these traditional beliefs and respect for the spawning aggregation could motivate village leaders to place restrictions on fishing this species if population levels begin to show signs of decline.

These recommendations about MPA size are predicated on the assumption that fishing effort would be dispersed on either side of the reserve. However, Cinner [33–35] points out that in some traditional fisheries, when alternative employment opportunities exist, some people opt out of the fishery altogether when reserves are established. For example, one could imagine a scenario in which fishers from Nagigi switch from fishing to farming should the costs of fishing in the non-reserve area become too high. Although these switching behaviors can be couched in terms of optimal foraging theory—namely, that the energetic and time costs of traveling farther to reach the fishery would outweigh the energetic or monetary benefits of fishing—we urge caution in interpreting people's activities strictly through the lens of optimal foraging. Although useful, this theory runs the risk of ignoring

**Table 5.** Age at maturity of five of the most frequently fished species in Nagigi, previously identified in Table 3.

| Fijian Name | Scientific Name | Age at first spawning |
|---|---|---|
| Kabatia | *Lethrinus harak* | Protogynous hermaphroditic; females reach sexual maturity at 1–2 years and transition to males beginning at age 3 or 4 [31] |
| Saqa | *Caranx ignobilis* | Reach maturity at about 3.5 years [46] |
| Vonu | *Chelonia mydas* | Estimated 30 years [47] |
| Nuqa | *Siganus vermiculatus* | Females: 1 year or younger [48] |
| Deou | *Upeneus vittatus* | Total life span 3 years; estimated age of sexual maturity 1 year [32] |

Includes four finfish and one sea turtle species, ranked by intensity of fishing pressure (highest to lowest).

cultural and social aspects of fishing that are not captured in such models.

## Perceptions and Realities of Reef Change

In general, villagers recognized an interlocking complex of economic causes as being primarily responsible for the overfishing of Nagigi's reefs. According to interviewees, the village's population has grown in recent decades as the cost of living in Fiji has increased, leading to increasing numbers of villagers fishing intensively on the reef for cash income (Table 6). The bêche-de-mer fishery and the practice of night fishing, in particular, draw village youth in search of "quick cash" because bêche-de-mer commands high prices on the foreign market and night diving yields higher catch per unit effort than lower-tech day diving (Interview 7). One interviewee reported that in a single hour of night fishing he could expect to catch 10 kg of fish because the fish are "not using as much energy, because they are sleeping the spear, you can put it closer to the head" (Interview 7). Johannes [18] observed a similar strategy among fishermen in Palau.

Two participants expressed particular worry about the increase in the bêche-de-mer harvest because holothurians release toxins when injured or killed that can harm reef fish; in fact, they have a well-documented history of use by indigenous societies in the South Pacific as a fish poison similar to *duva* (*Derris* spp., Fabaceae) root [36]. One participant said that he personally did not harvest bêche-de-mer because he "knows their role in the ocean," alluding to their importance in bioturbation [37]. Out of 15 interviews total, in nine of them the participants attributed overfishing to some part of this combination of the cost of living, the village's population increase, the desire for a "quick catch," the profitability of bêche-de-mer, and night fishing (Figure 5). This increasing reliance on the bêche-de-mer fishery for cash income is especially worrisome from an economic standpoint because holothurian populations are likely to decline as reefs degrade due to the impacts of ocean acidification and climate change [38].

It is difficult to correlate villagers' perceptions of reef change with quantitative measures of reef fish population dynamics because such data is scarce and patchy for the Pacific region [39]. However, regional- and national-level fisheries reports bear out villagers' perception that finfish abundance and biomass are in decline in Fiji, although data is not available for the island of Vanua Levu in particular. Richards *et al.* [39] note a 16.5% decrease in Fiji's artisanal finfish production by weight between 1988 and 1992, implying that finfish biomass and abundance decreased over this period. A general trend of population growth is well-documented in the South Pacific and demand for fisheries

**Table 6.** Perceived causes of reef dynamics over participants' lifetimes, by frequency with which they are cited and participants who cited them.

| Cause of Environmental Change | Number of Mentions | Mentioned by: |
|---|---|---|
| Increase in fishing pressure for market instead of subsistence, especially for "quick cash" | 9 | P#1, 2, 3, 4, 10, 14, 16, 19, 21 |
| Poaching by outsiders | 6 | P#4, 5, 6, 7, 19, 20 |
| Increasing population of Nagigi | 3 | P#1, 15, 17 |
| Coral smashing either as a fishing method to flush out *nuru* or by accident while walking on the reef | 3 | P#3, 15, 19 |
| Night fishing | 3 | P#13, 19, 20 |
| Profitability of bêche-de-mer fishery and toxicity of injured holothurians | 3 | P#4, 19, 20 |
| Increased cost of living | 2 | P#19, 20 |
| Changing climate patterns and sea level rise | 1 | P#1 |
| Demand for sea prawns from local resorts | 1 | P#20 |
| Unusually hot season in 1998 which placed stress on coral ecosystem | 1 | P#20 |
| Ongoing upstream development flushing sediment and weedkiller onto reef | 1 | P#20 |
| Use of *duva* root as fish poison | 2 | P#1, 19 |
| Use of nets with small openings that catch juveniles | 1 | P#21 |

Note that a few highly knowledgeable participants (#19, #20) provided the source for many observations.

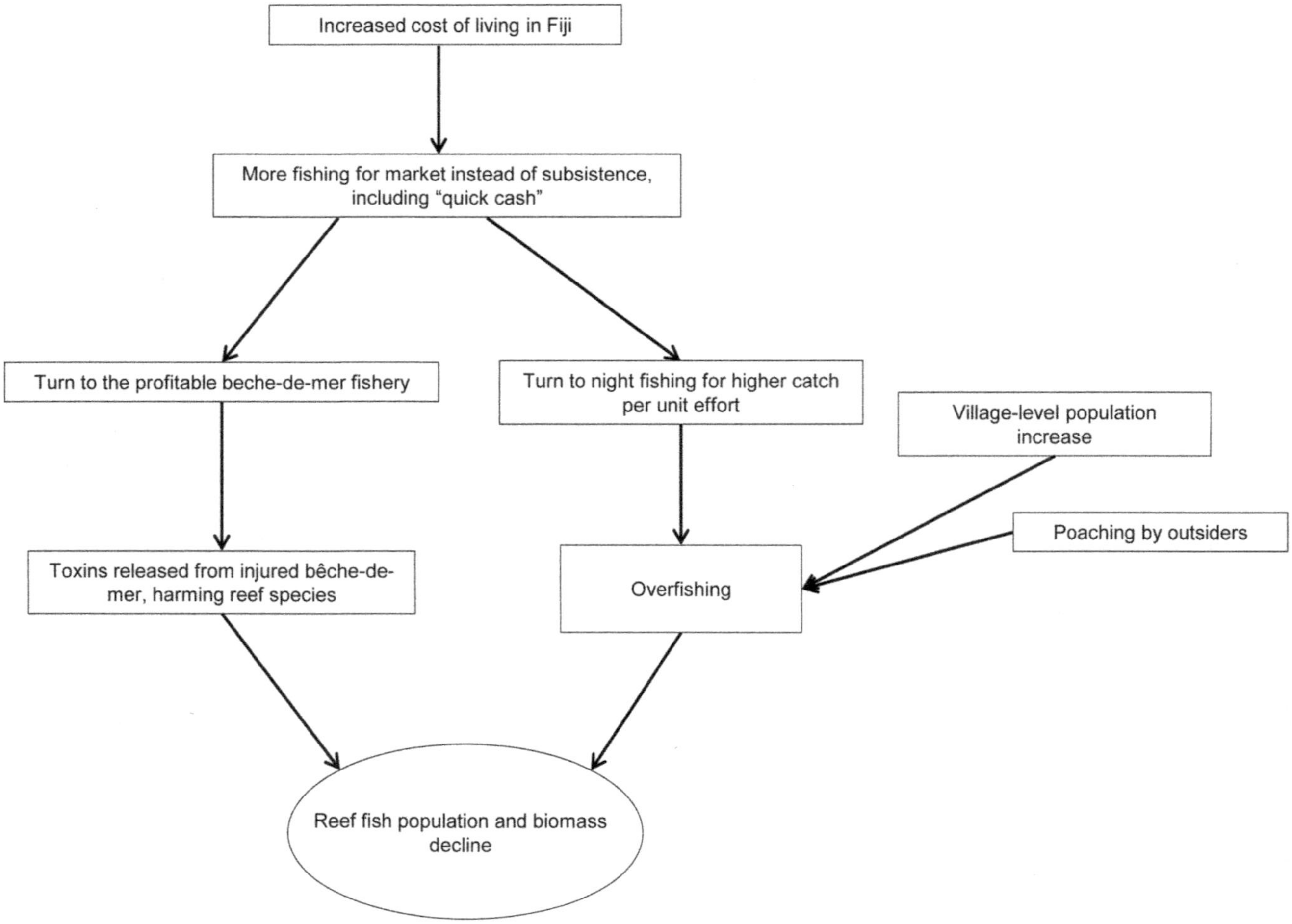


**Figure 5. Economic and social causes of fish population decline.** Flowchart representing the causal relationships between the economic and social factors to which villagers attribute fish population and biomass decline.

products is expected to remain high, exacerbating the problems of overfishing that villagers have perceived [40].

## Understanding Conservation Attitudes in Nagigi

A number of participants, when asked if Nagigi had ever had a *tabu* before during their lifetimes, mentioned the traditional reef closure for the death of a chief. According to this tradition, after a village chief dies, the fishing ground is closed for a hundred nights so that the fish population will increase. The start of the *tabu*, according to one account, is signified when members of the chief's funeral party come down to the reef after his burial and wash their hands in the sea. After the hundred nights, fishing begins again, and by this point the temporary closure has allowed local fish populations to increase enough that the villagers can fill two boats with fish in honor of the chief. These fish are divided into two parts and shared between the chief's *mataqali*, or clan, and the rest of the village (Interview 1). Such a save-and-spend model for marine closures is well established in Fijian culture and is known to have predated Western contact [41,42]. Nagigi last held a *tabu* for the death of a chief in 2001 (Table 4).

Any MPA project in Nagigi designed explicitly for biodiversity conservation will have to take into account the prevailing Fijian notion that closure areas are a kind of short-term "food bank" that can be opened and closed at will. Leaving aside the question of whether a three-to-five-year MPA will provide enough time for populations to rebound, as discussed above, it seems reasonably likely that Nagigi's villagers will decide to open the protected area for a short period at some point during the closure to pay for special projects or to feed visitors. Several interviewees stated explicitly that they think a future MPA should be opened if a chief dies or to if they have to prepare a feast for special visitors (Interviews 1, 9). Previous research in Fiji has shown that these project-specific openings, however brief, can have significant negative impacts on fish populations and can become a seductively convenient resource once villagers see how profitable they can be [15,41]. In a case documented by Jupiter *et al.* [42], inhabitants of Fiji's Kia Island collectively decided to suspend their local marine closure for a few days in order to raise money for community projects. Originally, the islanders aimed to raise FJD$12,000 (about US$7,500), but when they exceeded this goal on the first day of the fundraiser, they elected to continue the harvest. Residents fished in shifts for 24-hour periods six days a week during the five weeks, and netted an estimated FJD$200,000. Jupiter *et al.* [42] observed significant loss of fish biomass in large-bodied species such as acanthurids, carangids, and scarids during the harvest event, with negative effects lingering a year later. When presenting management options to village leaders in Nagigi, it is important to note these findings and warn them that marine protected areas may not have the desired effect of increasing fish populations if the MPA is opened to intense fishing pressure for even short periods of time.

One possible strategy to counter this spend-and-save effect is to extend some species-specific prohibitions beyond the term of MPA protection. For instance, long-term fishing protection for the endangered *Bolbometopon muricatum* and other similarly vulnerable species may offer lasting conservation benefits [25,43]. Precedent for such species-specific bans exists in Fijian culture in the form of traditional *tabus* that forbid certain groups from consuming specific species. For example, Nagigi contains a large demographic of Seventh Day Adventists, who do not consume or harvest any marine species that do not possess fins and scales (Interviews 8, 11). This means that they do not contribute to the harvesting pressure for sea turtles, clams, octopus, and other shellfish. The villagers of Nagigi also possess the cultural belief that the consumption of lizardfishes (Synodontidae), known as *dolo*, will render men impotent (Interview 7). As a result, the men of Nagigi will not eat lizardfish, though women will. And, throughout Fiji, *tabus* exist against women's consumption of various marine species during pregnancy and lactation, which help protect women and their infants from life-threatening ciguatera poisoning [25]. All of these existing models could lead to cultural acceptance for a species-specific ban designed to relieve overharvesting pressure.

While several MPAs have been created in Fiji with the tourism sector in mind, Nagigi's villagers have been silent on using ecotourism as a way to monetize their proposed MPA [44,45]. Several large dive resorts exist within 15 km of Nagigi and there is an airport approximately 20 km away in Savusavu, so the region has a well-developed tourism capability. However, creating Nagigi's MPA independent of any attempt to attract ecotourism could serve to insulate the reserve and Nagigi's villagers from dependence on a volatile global economy.

### Sources of Bias

The list of fish sampled for this paper should not be taken as a complete representation of the location's biodiversity due to the extremely short period of sampling and flaws in the sampling methods used. Fishes with an interstitial coral habitat, including the speciose groups Gobiidae and Blenniidae, are particularly underrepresented due to strong currents on sampling days meant collection of these groups with MS-222 was impractical. Although further sampling efforts will be necessary to determine the overall biodiversity of Nagigi's reef habitat, species targeted by local fishers are well-represented in the collection, making it a valuable if incomplete survey.

The interview-based portion of this study is also subject to limitations due to the brief survey period and small sample size. With only 22 participants, it is impossible to tell which beliefs about reef dynamics are common or unusual within the village, and the contributions of a few social demographics, especially the youth of the village, are lacking. The language barrier meant that some of questions may have been unclear to participants, and some of the nuances of their responses were undoubtedly lost in translation. The *Turaga ni Koro* facilitated introductions to 16 out of 22 of interview subjects. In most cases he conducted the researcher to participants' homes and either translated or waited outside while the researcher worked with another translator. In several instances, he interjected his own knowledge and opinions into the conversation when he sensed that a point needed clarification. He is well informed about village and regional issues, has lived in Nagigi for most of his life, and has been a vocal proponent of the plan to create a marine protected area on Nagigi's reef. Due to this advocacy, his choice of participants might have meant that we spoke mostly with villagers who were also supportive of the MPA idea. However, interviewees who were approached without the *Turaga ni Koro's* mediation also expressed strong support for the MPA project. Interviewees may also have been unwilling to mention that they targeted turtles, sharks, and other protected species such as *B. muricatum* for fear of prosecution for violating Fijian law.

## Conclusions and Implications

Based on its limited temporal and geographic scope, this study necessarily deals with only a small group of marine management stakeholders and makes conservation recommendations for only a small area of Fiji's coastline. However, we believe the synthesis of biodiversity sampling and fisherman interviews described here provides a template for future research projects that seek to integrate the traditional ecological knowledge of indigenous communities into conservation planning. Members of indigenous fishing communities can provide information about changes in reef health over time, the socioeconomic uses of marine resources, and the population dynamics of at-risk reef species—data that can be difficult and time-consuming to gather using conventional means, especially in developing countries. The necessity of gathering such information is highlighted by the fact that, in this study, only 11.3% of total species were captured by both methods of data collection. Without using both destructive sampling and interview techniques, a significant amount of data about the composition of Nagigi's reef ecosystem would have been lost for the purposes of this study. But, by combining these two methods, researchers can construct a clearer and more complete picture of the reef ecosystem and fishers' needs. Based on this information, researchers can then tailor their conservation recommendations to the life histories and habitat needs of targeted species, while taking into account the subsistence needs of local communities.

## Supporting Information

**Table S1** Reef Fishes of Nagigi. Partial list of the reef fishes of Nagigi by order and family, with common names and local Fijian names (if available). Species accessioned to the AMNH collection are in bold. C = collected, I = interview, S = sighted but not collected.

## Acknowledgments

We would like to thank Erin Eastwood, Amy McDermott, Madeline Thaler, and Helen Scales for their assistance with data collection. Metui Delaivoni, Nanesi Dirabe, and Metuisela Niu were gracious hosts to the Drew Lab in Nagigi Village. Elisa Bone of Columbia University and Michelle Kline of the University of California, Los Angeles assisted the authors as rigorous and insightful editors. Barbara Brown, Rad Arrindell and Lowell Flanders provided invaluable collection and curation assistance.

## Author Contributions

Conceived and designed the experiments: AG JD. Performed the experiments: AG WN IL JD. Analyzed the data: AG JD. Contributed reagents/materials/analysis tools: AG JD. Wrote the paper: AG JD. Translations: WN IL.

## References


1. Berkes F, Colding J, Folke C (2000) Rediscovery of Traditional Ecological Knowledge as Adaptive Management. Ecological Applications 10: 1251–1262.
2. Chapman MD (1987) Traditional Political Structure and Conservation in Oceania. Ambio 16: 201–205.
3. Thornton TF, Scheer AM (2012) Collaborative Engagement of Local and Traditional Knowledge and Science in Marine Environments: A Review. Ecology and Society 17.
4. Johannes RE, Yeeting B (2001) I-Kiribati knowledge and management of Tarawa's lagoon resources. Atoll Research Bulletin 489: 1–24.
5. Johannes RE (1978) Traditional marine conservation methods in Oceania and their demise. Annual Review of Ecology and Systematics 9: 349–364.
6. Jones S (2009) Food and gender in Fiji: ethnoarchaeological explorations. Lexington Books.
7. Johannes RE (2002) The Renaissance of Community-Based Marine Resource Management in Oceania. Annual Review of Ecology and Systematics 33: 317–340.
8. Zylich K, OMeara D, Jacquet J, Harper S, Zeller D (2002) Reconstruction of marine fisheries catches for the Republic of Fiji (1950–2009). Fisheries centre working paper series.
9. Hardin G (1968) The Tragedy of the Commons. Science 162: 1243–1248.
10. DeMers A, Kahui V (2012) An overview of Fiji's fisheries development. Marine Policy 36: 174–179.
11. Agrawal A (1995) Dismantling the divide between indigenous and scientific knowledge. Development and Change 26: 413–439.
12. Drew JA (2005) The use of traditional ecological knowledge in marine conservation. Conservation Biology 19: 1286–1293.
13. van Pel H (1955) Report on the Fisheries Industries of Fiji. Suva, Fiji: Government Press.
14. MacKenzie BR, Mosegaard H, Rosenberg AA (2009) Impending collapse of bluefin tuna in the northeast Atlantic and Mediterranean. Conservation Letters 2: 26–35.
15. Jupiter SD, Egli DP (2011) Ecosystem-Based Management in Fiji: Successes and Challenges after Five Years of Implementation. Journal of Marine Biology 2011: Article ID 940765.
16. Kuster C, Vuki VC, Zann LP (2005) Long-term trends in subsistence fishing patterns and coral reef fisheries yield from a remote Fijian island. Fisheries Research 76: 221–228.
17. Aswani S, Hamilton RJ (2004l) Integrating indigenous ecological knowledge and customary sea tenure with marine and social science for conservation of bumphead parrotfish (*Bolbometopon muricatum*) in the Roviana Lagoon, Solomon Islands. Environmental Conservation 31: 69–83.
18. Johannes RE (1981) Words of the Lagoon. Berkeley California: University of California Press.
19. Johannes RE (1998) The case for data-less marine resource management: examples from tropical nearshore finfisheries. Trends Ecology and Evolution 13: 243–246.
20. Unsworth RK, Cullen LC (2010) Recognising the necessity for Indo-Pacific seagrass conservation. Conservation Letters 3: 63–73.
21. Randall JE (2005) Reef and Shore Fishes of the South Pacific. Honolulu, HI: University of Hawaii Press.
22. Chapman MD (1987l) Women's Fishing in Oceania. Human Ecology 15: 267–288.
23. Sadovy Y, Kulbicki M, Labrosse P, Letourneur Y, Lokani P, Donaldson TJ (2003) The humphead wrasse, *Cheilinus undulatus*: synopsis of a threatened and poorly known giant coral reef fish. Reviews in Fish Biology and Fisheries 13: 327–364.
24. Olds AD, Connolly RM, Pitt KA, Maxwell PS, Aswani S, Albert S (2014) Incorporating Surrogate Species and Seascape Connectivity to Improve Marine Conservation Outcomes. Conservation Biology. 28: in press.
25. Henrich J, Henrich N (2010) The evolution of cultural adaptations: Fijian food taboos protect against dangerous marine toxins. Proceedings of the Royal Society of Londong B: Biological Sciences 277: 3715–3724.
26. Kline MA, Boyd R, Henrich J (2013) Teaching and the life history of cultural transmission in Fijian villages. Human Nature 24: 351–374.
27. Unsworth RK, De León PS, Garrard SL, Smith DJ, Bell JJ (2009) Habitat Usage of the Thumbprint Emporer *Lethrinus harak* (Forsskål, 1775) in an Indo-Pacific Coastal Seascape. Open Marine Biology Journal 3: 16–20.
28. Aswani S, Hamilton RJ (2004) The value of many small vs. few large marine protected areas in the Western Solomon Islands. SPC Traditional Marine Resource Management and Knowledge Information Bulletin 16: 3–16.
29. Edgar GJ, Stuart-Smith RD, Willis TJ, Kininmonth S, Baker SC, et al (2014) Global conservation outcomes depend on marine protected areas with five key features. Nature 506: 216–220.
30. Roberts CM, Hawkins JP (1997) How small can a marine reserve be and still be effective? Coral Reefs 16: 150.
31. Ebisawa A, Ozawa T (2009) Life-history traits of eight *Lethrinus* species from two local populations in waters off the Ryukyu Islands. Fisheries Science 75: 553–566.
32. Sabrah MM, El-Ganainy AA (2009) Observation on Biological Traits of Striped Goatfish (*Upeneus vittatus*) and Freckled Goatfish (*Upeneus tragula*) from the Gulf of Suez, Egypt. World Journal of Fish and Marine Sciences 1: 121–128.
33. Cinner JE (2007) Designing marine reserves to reflect local socioeconomic conditions: lessons from long-enduring customary management systems. Coral Reefs 26: 1035–1045.
34. Aswani S (1998) Patterns of marine harvest effort in southwestern New Georgia, Solomon Islands: resource management or optimal foraging? Ocean and Coastal Management 40: 207–235.
35. Lopes PFM, Begossi A (2011) Decision-making processes by small-scale fishermen on the southeast coast of Brazil. Fisheries Management and Ecology 18: 400–410.
36. Frey DG (1951) The use of sea cucumbers in poisoning fishes. Copeia: 175–176.
37. Uthicke S (1999) Sediment bioturbation and impact of feeding activity of *Holothuria (Halodeima) atra* and *Stichopus chloronotus*, two sediment feeding holothurians, at lizard island, great barrier reef. Bulletin of Marine Science 64: 129–141.
38. Dupont S, Ortega-Martínez O, Thorndyke M (2010) Impact of near-future ocean acidification on echinoderms. Ecotoxicology 19: 449–462.
39. Richards A, Lagibalavu M, Sharma S, Swamy K (1994) Fiji fisheries resources profiles. Fiji Fisheries Agency Forum proceedings xxxvii, 206pp.
40. Gillett R (2009) Fisheries in the economies of the Pacific island countries and territories.
41. Waterhouse J (1997) The King and People of Fiji. Honolulu, HI: University of Hawai'i Press.
42. Jupiter SD, Weeks R, Jenkins AP, Egli DP, Cakacaka A (2012) Effects of a single intensive harvest event on fish populations inside a customary marine closure. Coral Reefs 31: 321–334.
43. Dulvy NK, Polunin NV (2004) Using informal knowledge to infer human-induced rarity of a conspicuous reef fish. Animal Conservation 7: 365–374.
44. Brunnschweiler JM, Baensch H (2011) Seasonal and long-term changes in relative abundance of bull sharks from a tourist shark feeding site in Fiji. PLoS One 6: e16597.
45. Goetze JS, Fullwood LAF (2012) Fiji's largest marine reserve benefits reef sharks. Coral Reefs 32: 121–125.
46. Sudekum AE, Parrish JD, Radtke RL, Ralston S (1991) Life history and ecology of large jacks in undisturbed, shallow, oceanic communities. Fishery Bulletin 89: 493–513.
47. Chaloupka M, Limpus CJ, Miller JD (2002) Sea turtle growth dynamics in a spatially disjunct metapopulation. Coral Reefs 23: 325–335.
48. Gundermann N, Popper DM, Lichatowich T (1983) Biology and life cycle of *Siganus vermiculatus* (Siganidae, Pisces). Pacific Science 37: 165–180.

9

# Estimating Historical Eastern North Pacific Blue Whale Catches Using Spatial Calling Patterns


**Cole C. Monnahan[1]*, Trevor A. Branch[2], Kathleen M. Stafford[3], Yulia V. Ivashchenko[4,5], Erin M. Oleson[6]**

**1** Quantitative Ecology and Resource Management, University of Washington, Seattle, Washington, United States of America, **2** School of Aquatic and Fishery Sciences, University of Washington, Seattle, Washington, United States of America, **3** Applied Physics Lab, University of Washington, Seattle, Washington, United States of America, **4** National Marine Mammal Laboratory, Alaska Fisheries Science Center, Seattle, Washington, United States of America, **5** School of Environmental Science and Management, Southern Cross University, Lismore, New South Wales, Australia, **6** Protected Species Division, Pacific Islands Fisheries Science Center, Honolulu, Hawaii, United States of America



## Abstract

Blue whales (*Balaenoptera musculus*) were exploited extensively around the world and remain endangered. In the North Pacific their population structure is unclear and current status unknown, with the exception of a well-studied eastern North Pacific (ENP) population. Despite existing abundance estimates for the ENP population, it is difficult to estimate pre-exploitation abundance levels and gauge their recovery because historical catches of the ENP population are difficult to separate from catches of other populations in the North Pacific. We collated previously unreported Soviet catches and combined these with known catches to form the most current estimates of North Pacific blue whale catches. We split these conflated catches using recorded acoustic calls from throughout the North Pacific, the knowledge that the ENP population produces a different call than blue whales in the western North Pacific (WNP). The catches were split by estimating spatiotemporal occurrence of blue whales with generalized additive models fitted to acoustic call patterns, which predict the probability a catch belonged to the ENP population based on the proportion of calls of each population recorded by latitude, longitude, and month. When applied to the conflated historical catches, which totaled 9,773, we estimate that ENP blue whale catches totaled 3,411 (95% range 2,593 to 4,114) from 1905–1971, and amounted to 35% (95% range 27% to 42%) of all catches in the North Pacific. Thus most catches in the North Pacific were for WNP blue whales, totaling 6,362 (95% range 5,659 to 7,180). The uncertainty in the acoustic data influence the results substantially more than uncertainty in catch locations and dates, but the results are fairly insensitive to the ecological assumptions made in the analysis. The results of this study provide information for future studies investigating the recovery of these populations and the impact of continuing and future sources of anthropogenic mortality.



**Editor:** Isabelle Charrier, Universite Paris Sud, France

**Funding:** This publication is funded by the Joint Institute for the Study of the Atmosphere and Ocean (JISAO) under NOAA Cooperative Agreement NA10OAR4320148, Contribution No. 2151. The funders had no role in study design, data collection and analysis, decision to publish, or preparation of the manuscript.

**Competing Interests:** The authors have declared that no competing interests exist.



* E-mail: colemonnahan@gmail.com


## Introduction

The blue whale (*Balaenoptera musculus*) is an endangered species with three widely recognized subspecies found in the Southern Hemisphere (*B.m. intermedia*), Indian Ocean (*B.m. brevicauda*) and in the North Pacific and North Atlantic (*B.m. musculus*) [1,2]. Many populations are heavily depleted due to targeted whaling through their distribution. In the North Pacific, blue whales were hunted from 1905–1971 and despite decades without commercial catches the status of their recovery is unknown [3]. At present a large population known as the eastern North Pacific (ENP) population feeds each summer within 30 miles of the California coast and is believed to range from the equator to the Gulf of Alaska [4]. The ENP population is the one of the most accessible blue whale populations in the world and has been the focus of many studies, such as feeding mechanics and behavior [5], behavioral ecology [6] and responses to anthropogenic noise [7]. However, a formal assessment of their current abundance relative to historical levels has not been conducted, largely because of the difficulty in separating historical catches of the ENP population from catches from other populations in the western North Pacific (WNP).

A key step in separating catches is using vocalization occurrence data to estimate the migration pathways of the ENP and WNP populations over time and space. Analyses of these vocalizations observe two distinct NP blue whale 'song call' types (Figure 1) [4,8–10], which are assumed to be produced by the ENP and WNP populations. The ENP song call is comprised of the rhythmic repetition of a two part vocalization known as the 'AB call' type [11]. The AB call type components are produced exclusively by males in a variety of behavioral states [6]. The AB song calls are observed only in lone, traveling males and are produced year-round and thus likely have some kind of reproductive function, although their exact purpose is unknown [6]. The WNP song call is a single part call repeated as song in a similar way as the AB song of the ENP population, but has a clearly identifiable and distinct form (Figure 1) [8]. No studies have examined the behavioral context of WNP song, but it is assumed to be similar in function to the ENP song. Another common call type observed in the ENP population (and other populations of blue whales) is the downswept 'D' produced by foraging groups of both sexes [6]. A recent study found a temporal separation in the production of the AB and D call types at a summer feeding area, and argued that both were necessary for an accurate assessment of the timing of fine-scale seasonal movements into foraging regions [12]. Unfortunately there is no evidence of population differences in D-like calls as there is with song calls, and so we focused exclusively on the song call occurrence patterns.

Besides distinct song call types, the acoustic analyses also show that broad spatial migration patterns over time can be qualitatively distinguished from the data: the ENP population inhabits waters from the Costa Rica Dome to the Gulf of Alaska (GOA) and the WNP population is heard predominantly between Kamchatka and the GOA. We use the timing of occurrence of these song calls of the two populations to quantitatively model the seasonal movements by the two populations.

Additional data sources beyond vocalization differences support the hypothesis of two populations of blue whales in the North Pacific. Satellite tags, photographic identification, and sightings showed individuals moved between the Gulf of Alaska, the Californian coast, and the eastern tropical Pacific [13,14], although such data are not available for the WNP population. In addition, there are significant differences in length between whales in the ENP and WNP [15]. Finally, Gulf of California blue whales are genetically similar to those off California [16], although there are no studies investigating genetic differences between the ENP and WNP populations. Taken together there is substantial evidence that the ENP population is separate from blue whales in the WNP, which we used to validate the acoustics-based modeling approach taken in this study. While diverse sources of evidence point toward a single ENP population, the population structure of the central and western NP remains unclear, and there may be additional populations in these regions. For example, there is some speculation that a population off coastal Japan was extirpated and that another exists around Hawaii [17]. The acoustic and other data shed little light on these hypotheses, which do not affect our estimates of ENP catches, although we recognize that what we call the 'WNP population' may include additional population structure.

The recovery and resilience of the ENP population are unknown [3], but can be estimated using abundance estimates combined with historical catches [18]. Abundance estimates are available for the ENP population [19,20], but historical catches have not been split between ENP and WNP populations because blue whale population structure was unknown during commercial whaling and catches were reported only as blue whales. We estimated the probability of each catch belonging to the ENP population to separate the conflated catch series and produce an ENP and WNP catch series that can be used in future assessments. Qualitatively, it was clear from the acoustic data that whales caught off Japan and Kamchatka are highly unlikely to be from the ENP, and likewise blue whales caught off California were likely to be from the ENP population. However, splitting catches in the GOA is more complicated because the two populations overlap in space and time there during the feeding season [9].

In this study we collated the most up-to-date time series of historical catches of blue whales in the North Pacific, including previously unreported Soviet catches in the 1970s. Using song call occurrence data, we modeled the probability that a whale caught in a particular location and month belonged to the ENP population, and applied this model to the historical catches. The resulting time series of catches for ENP and WNP blue whales are the first to be obtained objectively for each population.

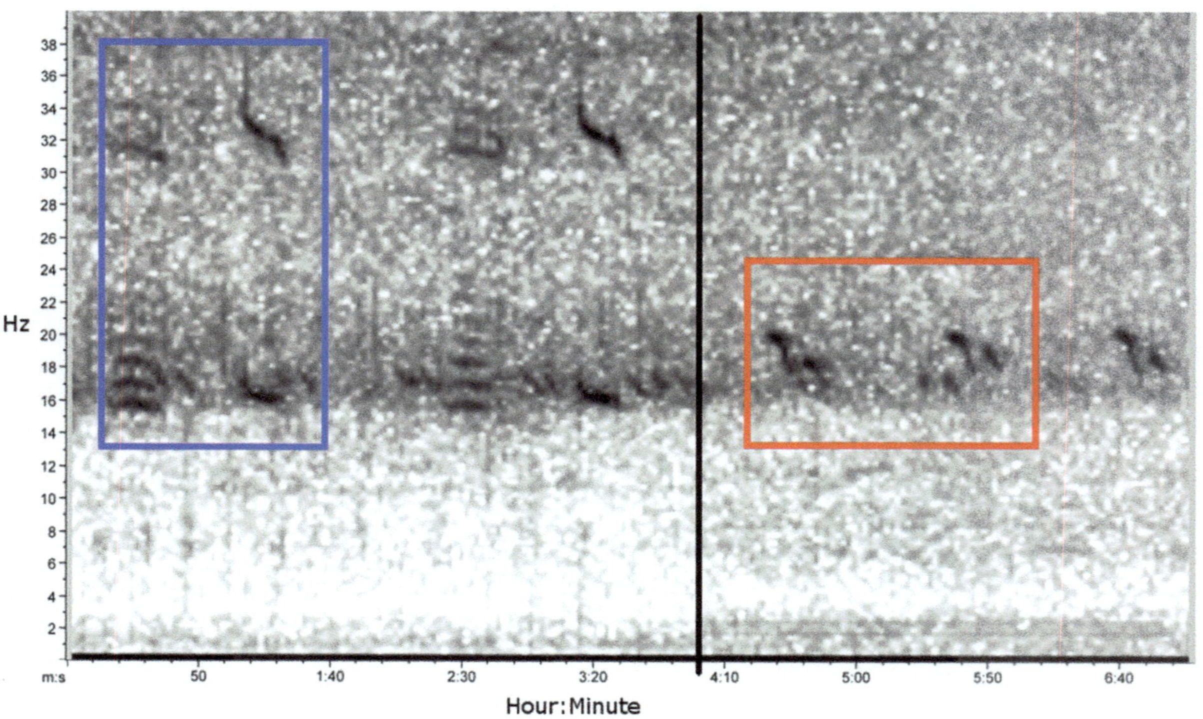


**Figure 1. Spectrogram showing the ENP and WNP blue whale song calls.** These calls were recorded on a hydrophone in the Gulf of Alaska at different times; the blue box shows the ENP song call and the red box the WNP song call. The clear distinction between the two is used to differentiate the presence and absence of the two populations.

## Materials and Methods

### Catch Data

Official catches for NP blue whales were obtained from two databases maintained by the International Whaling Commission (IWC): the 'summary database' [21] and the 'catch database' [22]. For both databases, the expedition code referred to either a pelagic fleet operating in a predefined IWC region, or catches taken back to a land station for shore-based processing; thus a pelagic vessel operating in multiple regions in the same year would be recorded as different expeditions. For most expeditions the start and end month of each operation were recorded annually.

The summary database contained annual catch totals by species (including blue whales) for each whaling expedition since 1900, but contained no detailed information about the exact location or date of individual catches. Catches with unspecified species accounted for 8,519 out of 570,146 (1.5%) total whales of all species in the North Pacific. We estimated how many unspecified catches were likely to be blue whales and included them in the analysis as follows: (1) For expeditions with no catches identified to species in a year, we used the proportion of blue whale catches relative to all other species in adjacent years and applied that to the number of unspecified catches. (2) For expeditions with unspecified catches and catches reported to species in the same year, we used the proportion of blue whales within that year. (3) For the substantial unspecified catches (4,415 whales) from coastal Japan after blue whaling started (1905–1909), we assumed blue whales were 16.4% of all whales. This was the average of the following two years (1910–1911), after which the proportion of blue whales caught off coastal Japan declined quickly. Using these methods we added 49 blue whale catches off British Columbia (1907 and 1914), 81 off the US west coast (1918–1938) and 782 off coastal Japan (1905–1909 and 1934–1936).

The catch database contained data for individual whales including species, location, date, expedition, length, length of fetus (if present), and sex (Figure 2). Detailed information was not available for all catches, particularly in earlier years and during the period of later Soviet whaling, thus the catch database contained only a subset of the summary database.

These databases contained all catches known to the IWC as of November 2011. However, it has been known since the mid-1990s that the USSR caught additional blue whales in the 1960s and 1970s that were misreported or not reported to the Bureau of International Whaling Statistics and are missing from the IWC databases [23]. These discrepancies were recently resolved for catches in the North Pacific with recovered original scientific reports, adding 738 (7.6% of total) blue whale catches to the previous version of the IWC annual database [24]. In many cases the Soviet reports gave clues about the likely date and locations of these catches, which were used to help infer likely catch locations. Not all of these reports were available and uncertainty remained surrounding the number, locations and dates of some Soviet catches during 1962–1971 [24]. We accounted for this uncertainty explicitly in the analysis as described in the next section. Outside of these years few, if any, blue whale catches were missing.

### Catch Uncertainty

In addition to missing data about Soviet catches, many other catches in the IWC catch databases are missing locations, dates or both (Table 1), and the original records are lost to history. This 'catch uncertainty' must be explicitly taken into account when separating catches to reliably quantify the uncertainty associated with the catch time series. We used a Monte Carlo method to integrate over all potential locations and months for each individual catch. Conceptually, possible sets of locations and months were determined for each catch, from which random samples could be generated. By generating and splitting many potential catch series, the catch uncertainty was propagated through into the uncertainty in the final results. The challenging part was to determine accurate sets of locations from which to draw randomly. We assigned catches into five categories ranging from the most uncertain location (IWC Region) to the most certain (exact position known), and treated each category as described below.

(1) 'IWC Region' ($n=3{,}978$ catches) was the most uncertain category and only occurred when no location information was available for an expedition in a year beyond the broad regions defined by the IWC. IWC regions are defined by longitude and latitude ranges (i.e. a rectangle, Figure 3), which cover both land as well as regions of ocean where no whaling ever occurred. Clearly, randomly drawing uniform locations from these large regions would lead to inaccurate and even nonsensical catch locations. Instead, we drew from locations within each IWC region where catches of blue whales and other species had actually been taken. There was too little variation in reported blue whale catch locations (Figure 2), while conversely some species such as sperm (*Physeter macrocephalus*), humpback (*Megaptera novaeangliae*) and bowhead (*Balaena mysticetus*) whales occupy very different habitations than blue whales. Therefore we randomly drew from the reported locations of catches of known locations of blue whales and other species most similar to the blue whales: blue ($n=4{,}640$), fin (*B. physalus*, $n=47{,}232$), sei (*B. borealis*, $n=50{,}875$), and common minke (*B. acutorostrata*, $n=4{,}742$) that were caught during 1905–1971 when blue whales were targeted. By drawing from actual catch locations from similar species, we accounted for spatial patterns of whaling effort, but implicitly assumed this effort was similar between species and that blue whales occupied the same spatial extent as other species. This assumption generally increased the spatial uncertainty of catches compared to the scenario where we randomly drew only from locations where blue whales were caught. Although catches in the IWC region category are the most uncertain, most of these catches came from Japanese and Korean waters ($n=2{,}355$) and Aleutian islands ($n=819$) which likely contain mostly WNP blue whales (Figure 3).

(2) The 'Partial Locations' category ($n=924$) was where an expedition reported locations for some species but not for some blue whales. Most of these catches ($n=829$) were from Soviet pelagic expeditions where the month is known and can be cross referenced with the catch database. However, there were a few cases ($n=95$) when an expedition reported some, but not all, of the catches in a year. We inferred locations of catches in this category by drawing, with replacement, from locations of all species in the same month.

(3) The 'Inferred' category ($n=447$) was where the location was not reported, but could be inferred with high precision. Most often this occurred when catches were reported as coming from a land station with a known location in a previous year, but no location reported in the current year. The unreported catches must have the same locations as previous years since the land stations were stationary. There were some catches ($n=19$) with unknown locations but known dates, in addition to known locations of catches of other species on the same dates. For these cases averages of locations of adjacent catches of other species were used without uncertainty.

(4) The 'Soviet' category ($n=224$) was specifically for Soviet catches from 1962-1971 that were updated in [24] and whose original reports allowed us to determine smaller rectangles than the broad IWC regions. We assumed the vessels could have been

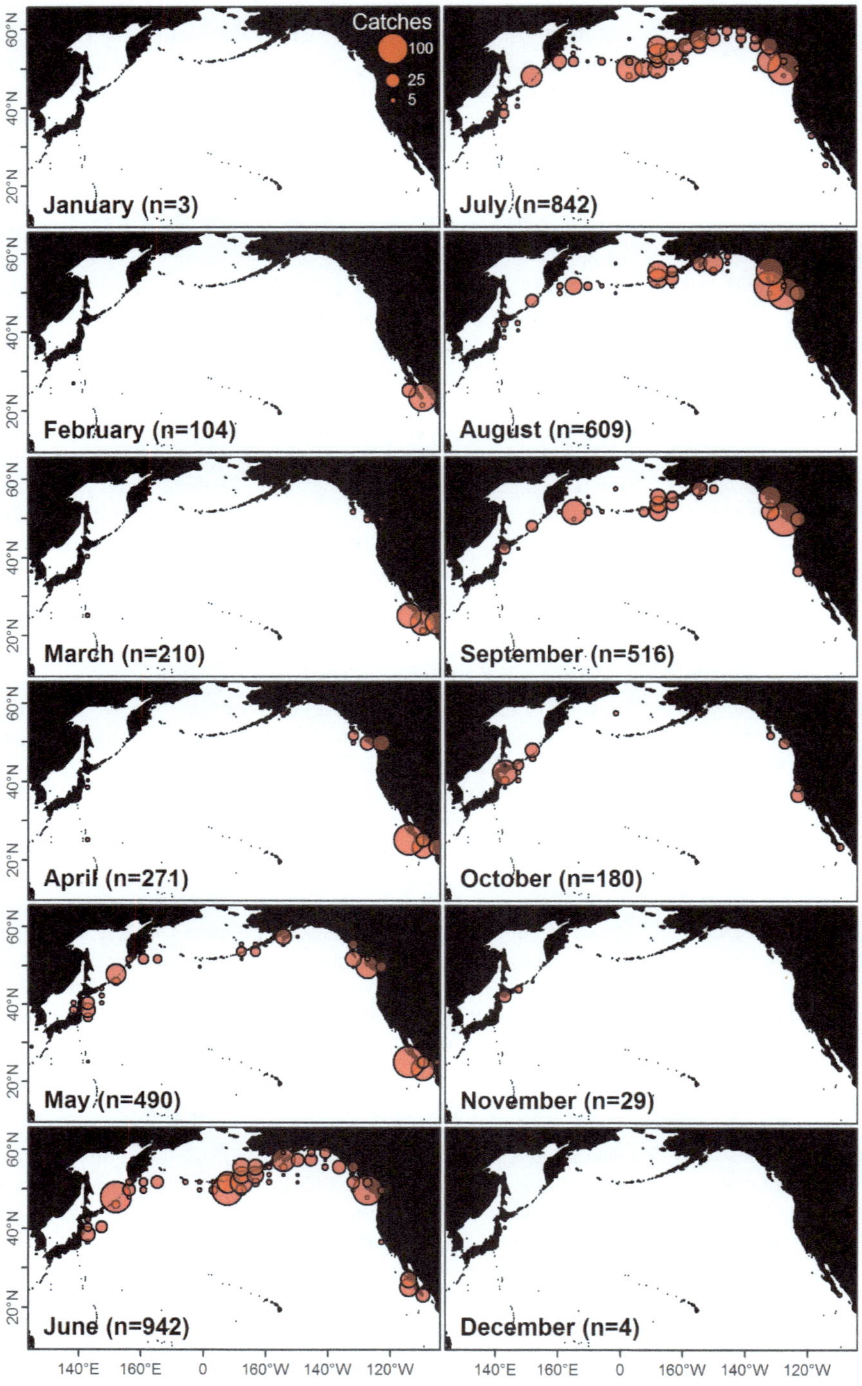


**Figure 2. Distribution of the blue whale catches with reported locations.** Monthly maps showing blue whale catch locations that were reported without uncertainty ($n = 4{,}200$; 43% total) to the International Whaling Commission. There are clear spatial and temporal patterns reflecting the migrations of the two populations and whaling seasons.

anywhere within these rectangles and therefore drew random uniform locations from each rectangle.

(5) The 'Certain' category ($n = 4{,}200$) was for catches where the exact location was reported in the catch database. These are, fortunately, the most common case (43% of the total), and we did not consider any uncertainty in these locations.

Blue whale catch dates were also often not reported ($n = 4{,}979$, 51%) in the IWC catch database. We applied the same Monte Carlo approach to account for uncertainty in date by sampling from the possible months. We used an expedition's operating period from the IWC summary database to determine the range of possible months. If the operating period for an expedition was missing, we assumed the catches could have been from any month. However, we assigned a probability to each potential month that reflects the fact that most blue whales were caught in summer because of defined whaling seasons, weather limitations, and seasonal availability of whales. This weighting was based on the proportions of whales caught in each month for the same set of

**Table 1.** North Pacific blue whale catches by categories of uncertainty.

| Location Uncertainty Category | Uncertain Month (%) | Certain Month (%) | Total (%) | Category Description |
|---|---|---|---|---|
| IWC Region | 3977 (41%) | 1 (0%) | 3978 (41%) | Known to be within a large rectangular region (Figure 3) |
| Partial Locations | 447 (5%) | 477 (5%) | 924 (9%) | Catches of other species by the same expedition and year are available. |
| Inferred | 447 (5%) | 0 (0%) | 447 (5%) | Able to infer accurate locations with high confidence. |
| Soviet | 108 (1%) | 116 (1%) | 224 (2%) | Original Soviet whaling logs give information about locations and months. |
| Certain | 0 (0%) | 4200 (43%) | 4200 (43%) | Reported to the IWC. |
| Total | 4979 (51%) | 4794 (49%) | 9773 (100%) | |

The percent of all catches is indicated in parentheses after the number of catches. Some exact catch positions and dates are unavailable and were inferred with varying levels of uncertainty for many cases.

species used for geographic uncertainty (blue, fin, sei and common minke whales).

Thus, there was a range of potential locations and dates for each catch depending on the level of detail reported (Table 1). A routine was written in the statistical framework R [25] which randomly samples from the sets of locations and months for all catches, and returns a dataset which represents a possible realization of the conflated catch series. Repeatedly running this routine to generate sets of potential catch series (i.e. a Monte Carlo approach) allowed us to include catch uncertainty in our estimates of ENP catches. This approach assumed the unknown catches were not systematically biased in space and time compared to the known catches. The generated sets of conflated catch series were then propagated through to the process of assigning catches to populations.

## Acoustic Data

Detections of ENP and WNP song calls came from six sources in the North Pacific (Table 2, Figures 4 and 5). For all hydrophones, data were converted into hourly presence or absence of ENP and WNP calls. Hydrophone sources were: (1) six hydrophones placed by the National Oceanic and Atmospheric Administration (NOAA) in the eastern tropical Pacific [26]; (2) sixteen US Navy Sound Surveillance System and other hydrophones spread across the North Pacific; note that their exact locations were classified and approximated based on previously published studies [8]; (3) six hydrophones in the Gulf of Alaska placed by NOAA - after 9 months, one of these hydrophones was discontinued and another was moved slightly [9]; (4) six hydrophones in the Channel Islands and Cortez and Tanner Banks deployed by Scripps Institution of Oceanography [12]; (5) one hydrophone moored off Oahu, Hawaii, described in [27] and used in [8]; and (6) a single hydrophone off Wake Island [8].

Presence was recorded when at least one call was detected in an hour of analyzed data, implying no differentiation in the number of calls within an hour. The presence data were then aggregated by month, and across years for the GOA and Channel Island

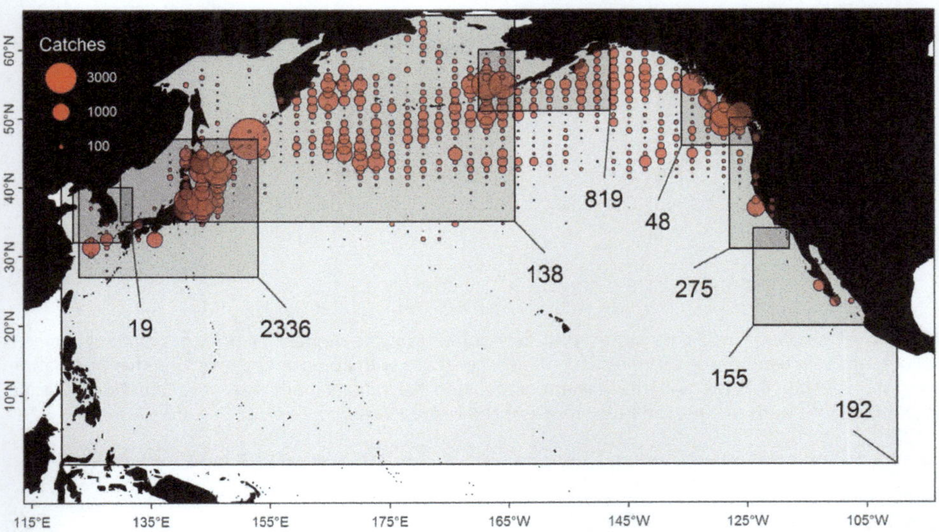


**Figure 3. Map of reported blue, fin, sei and common minke whale catch locations (1905–1971).** These catch locations were used to infer potential locations for blue whale catches without known locations. 41% of catches are known only to broad, management-defined regions (shaded boxes) which are shown with their corresponding blue whale catches.

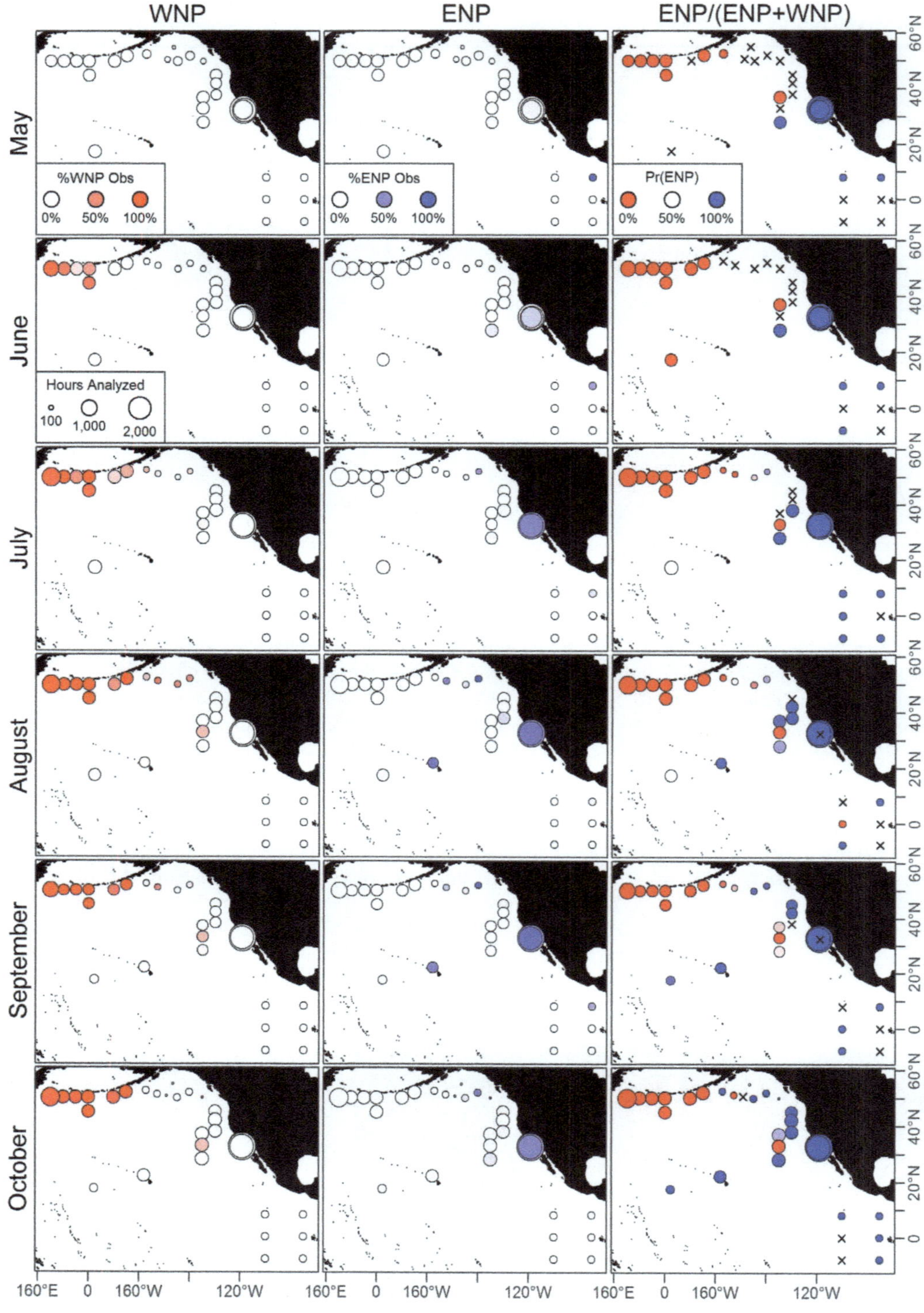


**Figure 4. Observed hydrophone data for May-October.** These data are used to fit separate acoustic models for both populations. Colored circles show hydrophones and the proportion of hours calls were observed while and "x" denotes those with no observed calls for either population. The western population migrates east along the Aleutian Islands and the eastern population migrations northwest into the North Pacific. Hydrophones with observed calls from both populations clearly show overlap for much of the North Pacific.

hydrophones. Specific details regarding hydrophone deployment and methods for processing the raw data can be found in the original references (Table 2). Taken together these 32 hydrophones covered a large expanse of the North Pacific (Figures 4–5), and provided a unique opportunity to model the large movement patterns undertaken by both populations of blue whales.

## Acoustic Presence Models

The acoustic data used here contained information about NP blue whales' distribution both in space and season, from which migration patterns could be inferred for the two populations. We used these patterns to infer to which population the conflated historical catches belonged. We developed a two-stage model that

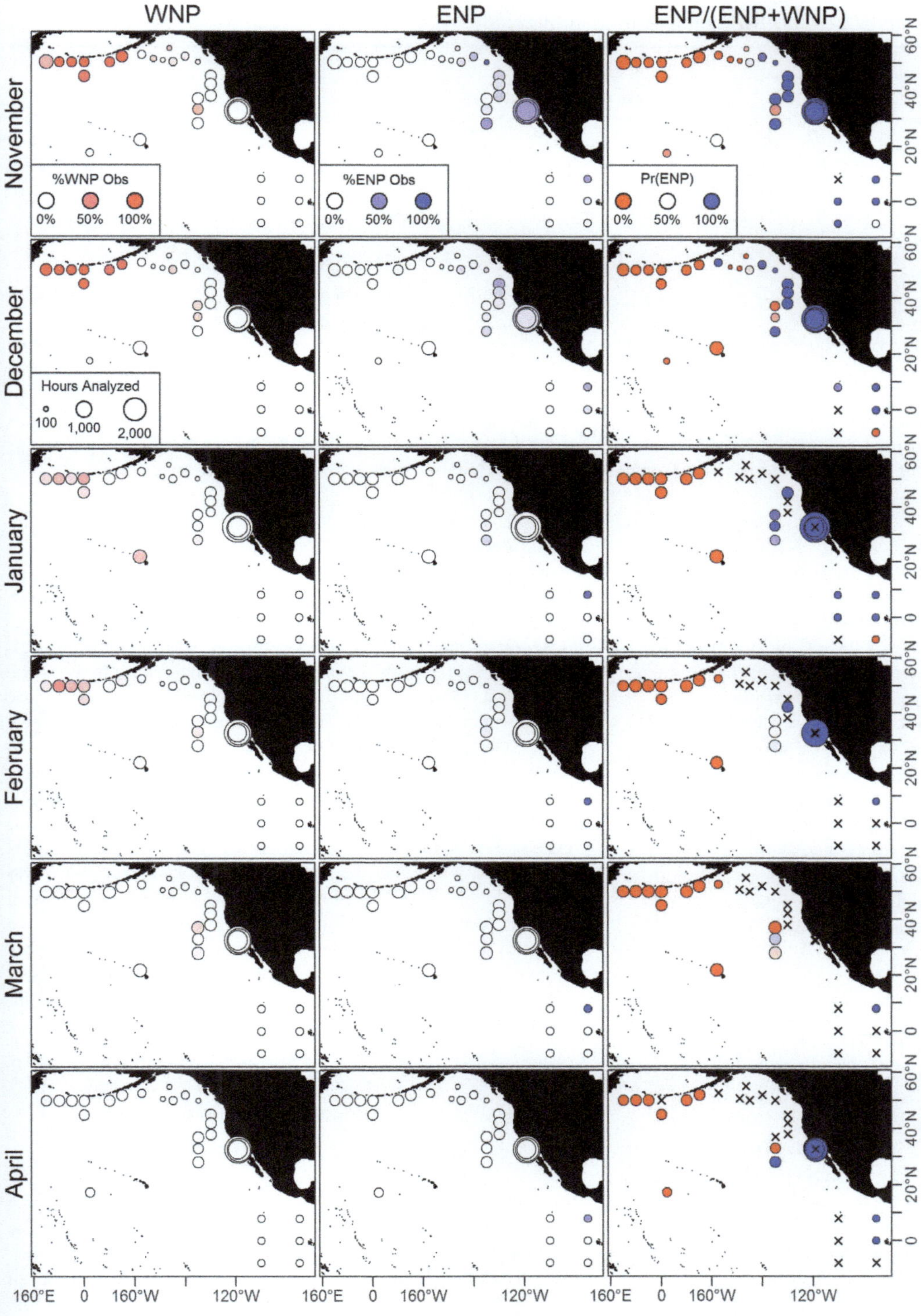


**Figure 5. Observed hydrophone data for November-April.** Caption as for Figure 4.

(1) estimated the presence or absence of each population in space and time and then (2) predicted the probability a catch belonged to the ENP population. The first stage fitted a spatial surface separately to the ENP and WNP populations based on the monthly call rates. These two independent 'acoustic models' predicted the probability of observing at least one song call in an hour at a given location and time. In the second stage the two acoustic models were combined together into a single 'prediction model' used specifically for predicting the probability that a conflated catch belongs to the ENP population, given its location and month. This section describes the motivation and development of the acoustic models and how they are combined into the prediction model.

To develop these acoustic models we assumed the spatiotemporal song call patterns observed in the data reflected an underlying occurrence and used the data to estimate this

**Table 2.** Summary of the groups of acoustic hydrophones used in this study.

| Group | # Hydrophones | Begin Date | End Date | Reference | Notes |
|---|---|---|---|---|---|
| ETP[1] | 6 | May-96 | May-97 | [4] | N/A |
| GOA[2] | 6 | Oct-99 | Jun-01 | [9] | One hydrophone removed after 5/2000, another moved in 5/2000 |
| SOSUS[3] | 13 | Nov-95 | Nov-96 | [8] | Hydrophone positions estimated as they are still 'protected' |
| Wake | 1 | Apr-92 | Dec-92 | [8] | Missing months |
| Kaneohe | 1 | Aug-92 | Apr-93 | [8,27] | Missing months |
| CH[4] | 6 | Aug-00 | Feb-04 | [12] | Down during 5 February to 15 April 2002 |

[1]Eastern Tropical Pacific;
[2]Gulf of Alaska;
[3]US Navy Sound Surveillance System;
[4]Channel Islands.
The GOA data are a random sample of 15% of total hours, while the rest are 100% of total hours. Several hydrophones were inoperable for periods of time due to hardware failures, leading to gaps in the data. There is little overlap in year between the groups. See specific references for further details of the data.

occurrence. The first choice in developing the acoustic models was deciding which likelihood function is most appropriate for the data. Since the data were presence-absence (a whale either was or was not detected in each hour), a binomial likelihood seemed like a natural choice. However, sequential hours of detection at a location may not have been independent, as is expected in a true binomial experiment, since a single traveling or resting whale will potentially be recorded calling for many consecutive hours. If a call was heard in one hour, it would likely be heard there in the following hour as well. As a result, the variance in the data was expected to be higher than a binomial process. In other words, the data were over-dispersed. We addressed this issue by adding an over-dispersion parameter to the binomial model to increase the expected variance, which had the effect of turning the binomial model into a beta-binomial model. The parameterization of the beta-binomial used in this study had a likelihood function of

$$L(p,\sigma|y,n,\text{month}) = \frac{\binom{n}{y}\Gamma(1/\sigma)\Gamma(y+p/\sigma)\Gamma(n+(1-p)/\sigma-y)}{\Gamma(n+1/\sigma)\Gamma(p/\sigma)\Gamma((1-p)/\sigma)}$$

where $y$ was the number of hours with at least one observed call, $n$the total hours analyzed, $0 \leq p \leq 1$ the probability of observing at least one song call in an hour, $\sigma > 0$ the over-dispersion term, and $\Gamma$ is the gamma function. For this distribution, $E[y] = np$ and $\text{Var}[y] = np(1-p)[1+\sigma(n-1)/(1+\sigma)]$. As $\sigma \to 0$, the variance becomes $np(1-p)$, which is the variance of a binomial model, demonstrating that the $\sigma$ parameter increases the variance as desired. The support in the data for a $\sigma$ term in the model was tested in the model selection phase.

In developing the acoustic models we used only covariates which were also available for the catches. Thus, while oceanographic conditions at the hydrophones likely could have explained the occurrence of blue whales [28], we could not obtain these oceanographic data for the historical catches. We therefore included only latitude, longitude, and date as covariates in the model, and the implications of this assumption were explored in the 'Ecological Uncertainty' section below. The positions were used as is, but it was natural to aggregate time into appropriately-sized units. We chose to bin time into months because anything longer in duration may have missed the fine-scale movements, and the weeks-long delay between presence and singing found in [12] precluded anything shorter.

The relationship between occurrence and spatial positions is complex, and unlikely to follow a smooth mathematical form (e.g. linear, quadratic, etc.). We therefore used nonparametric models to infer the relationship between explanatory and response variables. Typically the non-normal sampling and nonparametric nature proposed here would have implied a generalized additive model (GAM). However, GAMs are not flexible enough to accommodate beta-binomial sampling [29]. We therefore used a more flexible modeling platform called generalized additive models for location, shape and scale (GAMLSS) which was developed to extend GAM models to handle more complex distributions, including the beta-binomial [30]. The GAMLSS package in R [25,30] allowed additive predictors on both parameters of the beta-binomial distribution simultaneously, so that both varied spatially in a nonparametric way. We chose additive position terms and categorical month terms for both parameters as the most complex case (i.e. the full model) and fit, for example, the ENP model as:

$$\text{logit}(\hat{p}_{ENP}) = \text{cs(Longitude)} + \text{cs(Latitude)} + \text{Month}$$
$$\log(\hat{\sigma}_{ENP}) = \text{cs(Longitude)} + \text{cs(Latitude)} + \text{Month}$$

where logit and logwere the canonical link functions and *cs*is a cubic spline smoother. The same structure was also used to fit the full model for the WNP data. The resulting acoustic models predicted the probability of observing at least one song call in an hour for a given position and month.

Model selection proceeded by fitting several plausible models and comparing corrected Akaike information criterion (AICc) values [31], as well as examining model residuals using standard diagnostics. Model selection was not used to infer biologically meaningful predictors of blue whale habitat, as was done for example in [28], but rather to identify the most appropriate structure and complexity of the models of blue whale song call occurrence. To test the statistical support for over-dispersion we fitted both binomial and beta-binomial GAMLSS models.

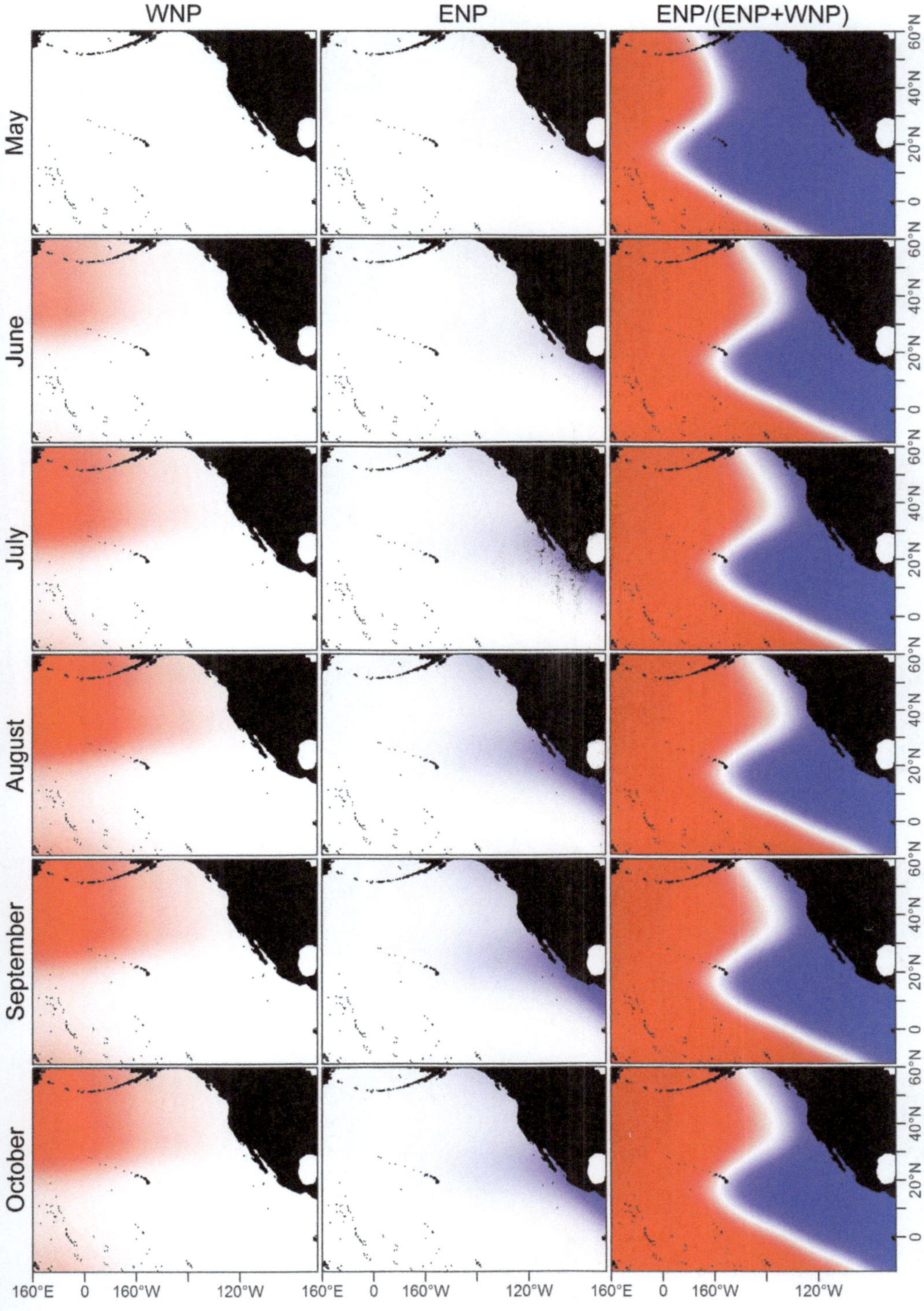


**Figure 6. Fits for ENP acoustic, WNP acoustic, and base case prediction models for May-October.** These models are fit to the acoustic data in Figures 4 and 5. The third column shows the base case prediction model which is the proportion of ENP to WNP calls, such that red areas correspond to predicted WNP occurrence and blue areas to ENP. The white line denotes where the model predicts an equal chance of observing an ENP or WNP blue whale.

Likewise, to test the additive nature of the models, a beta-binomial model with linear terms was also considered. Three variants of the beta-binomial GAMLSS model were also fit, having either constant, linear, or additive terms to investigate the spatial complexity of over-dispersion. The best supported acoustic models for the ENP and WNP represent the first of our two-stage model.

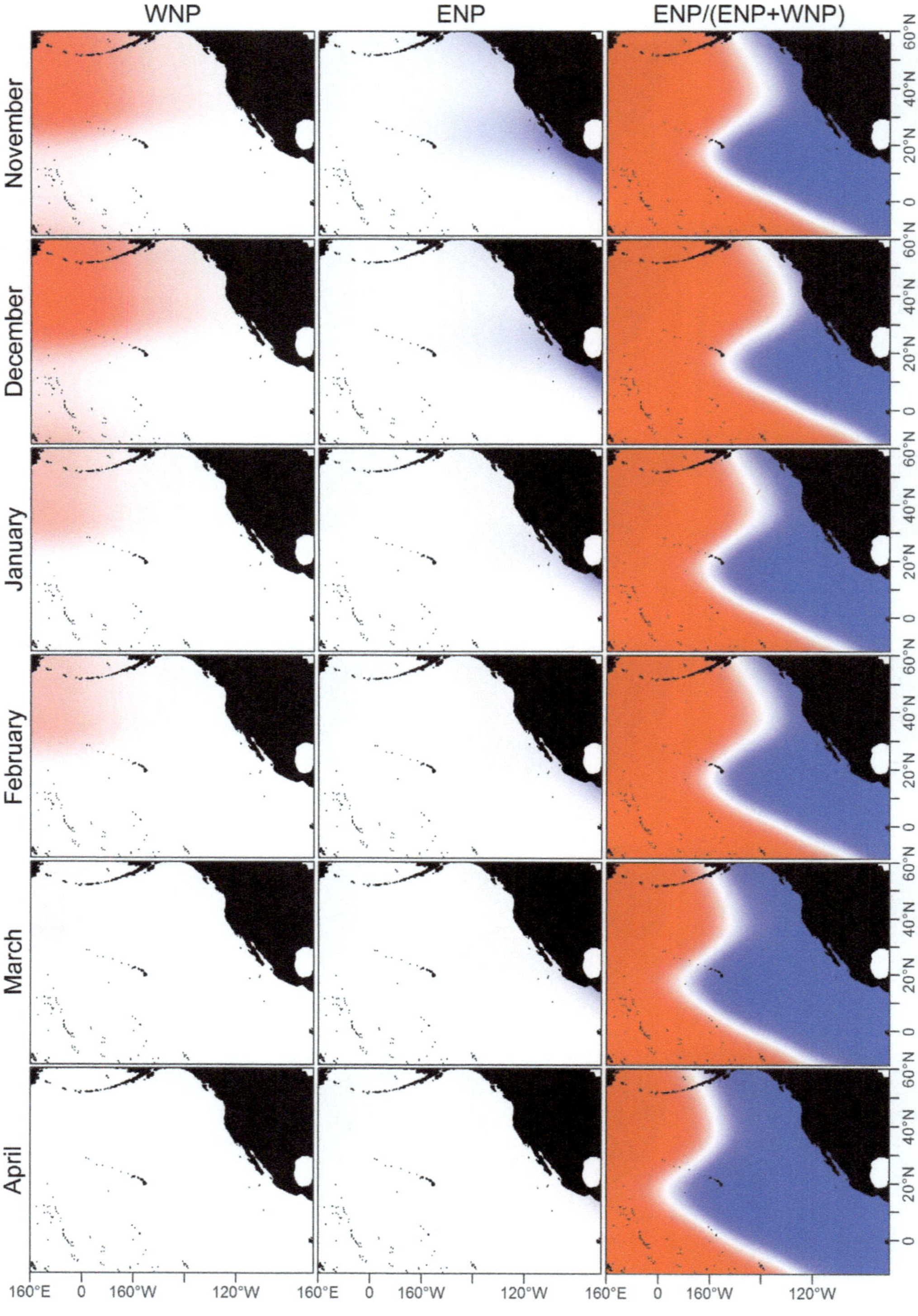


**Figure 7. Fits for ENP acoustic, WNP acoustic, and base case prediction models for November-April.** These models are fit to the acoustic data in Figures 4 and 5. The third column shows the base case prediction model which is the proportion of ENP to WNP calls, such that red areas correspond to predicted WNP occurrence and blue areas to ENP. The white line denotes where the model predicts an equal chance of observing an ENP or WNP blue whale.

## Prediction Model

The second stage of the modeling was to develop a model which predicts the probability that a catch belonged to the ENP given its location and month. To make these predictions we combined the ENP and WNP acoustic models together. Conceptually, if the ENP acoustic model predicts high occurrence in a region and the WNP model does not, we would assign catches in that region to the ENP population. Specifically if we let $d_{ENP}(t,z)$ and $d_{WNP}(t,z)$ be the density of whales at date $t$ and location $z$ then the probability a catch belonged to the ENP is:

**Table 3.** Model selection results.

| Distribution | Probability($p$) | Over-dispersion($\sigma$) | df[1] | ENP ΔAICc | WNP ΔAICc |
|---|---|---|---|---|---|
| Beta-Binomial | Additive | Additive | 40 | 0 | 0 |
| Beta-Binomial | Additive | Linear | 34 | 19 | 10 |
| Beta-Binomial | Additive | Constant | 21 | 33 | 36 |
| Beta-Binomial | Linear | Linear | 28 | 103 | 54 |
| Binomial | Additive | N/A | 20 | 24,063 | 12,675 |
| Binomial | Linear | N/A | 14 | 33,457 | 18,319 |

[1]degrees of freedom used in the model
Models were compared across likelihoods and parameter structures. The smallest ΔAICc indicated the model with the highest support was the most complex model. The binomial model in particular fits extremely poorly, confirming the need for an over-dispersed likelihood.

$$\Pr(\text{Catch is ENP}|t,z) = C(t,z) = \frac{d_{ENP}(t,z)}{d_{ENP}(t,z)+d_{WNP}(t,z)} = \frac{1}{1+\frac{d_{WNP}(t,z)}{d_{ENP}(t,z)}} = \frac{1}{1+D(t,z)}$$

Clearly the densities of the populations dropped during the period of historical whaling. However, the way the ratio of the densities, $D(t,z)$, changed over years is unclear because the abundance trends of both populations are poorly understood. For our base case we therefore assumed that both populations were depleted at an equal rate so $D(t,z)$ was independent of year and simplified to $D(m,z)$ for month $m$. We further assumed the ratio of densities was approximately proportional to the ratio of the probabilities of call occurrence:

$$D(m,z) \approx \alpha(m,z)\cdot\hat{P}(m,z),$$

Where $\hat{P}(m,z) = \hat{p}_{WNP}(m,z)/\hat{p}_{ENP}(m,z)$ is the ratio of estimated call probabilities and $\alpha(m,z)$ is an unknown scaling factor that may change over month or space in response to many external influences. Lacking any information on its true value, we assumed a default of $\alpha(m,z)=1$ for the base case and the sensitivity to this assumption was explored below. Thus the base case prediction model is

$$\hat{C}(m,z) = \frac{1}{1+\hat{P}(m,z)}.$$

The above formula combined the two acoustic models together into a prediction model that can be applied directly to the catches.

The GAMLSS models provide standard error estimates only at the positions of the hydrophones, but the predictions were to be interpolated and extrapolated, sometimes widely, to predict the catches. Since accurately quantifying the uncertainty in predictions is an important aspect of modeling these data, we used a common alternative approach known as nonparametric bootstrapping [32]. With this approach the acoustic and prediction models were refit to sets of acoustic data resampled with replacement. For each resampled set of data, the acoustic models were refit, and the predictions of ENP catches were recalculated. Since the Monte Carlo method for quantifying uncertainty in catches also relied on sampling repeatedly to account for uncertainty over potential positions and months, it was natural to incorporate the two resampling steps together. The algorithm used to accomplish this was:

1. Generate a random realization of a potential conflated catch series $\hat{C}_i$.
2. Bootstrap the acoustic data by sampling with replacement from rows of the acoustic data. Each row contained the proportion of hours with presence in a month for a hydrophone.
3. Refit the two GAMLSS acoustic models to the bootstrapped acoustic data.
4. Use the fitted bootstrapped models to predict an ENP catch series $\hat{C}_i^{ENP}$.
5. Aggregate the split catches by total and year to create time series.
6. Repeat steps 1-5 $n$ times.
7. Summarize the results using the median and percentiles over the set $\left\{\hat{C}_1^{ENP},\ldots,\hat{C}_n^{ENP}\right\}$.

This algorithm simultaneously incorporated both the catch and statistical uncertainty in allocating catches between the two populations and thus made it difficult to compare the relative sizes of these uncertainties. We therefore also ran the algorithm without step 3, in essence removing the effect of bootstrapping so that only catch uncertainty was estimated. Implementing this algorithm split the catches and quantified the relative size of the uncertainties.

## Ecological Uncertainty

There was one more important source of uncertainty that affected the results. In the base case results we made two big assumptions in setting $\alpha=1$: (1) the ratio of densities at a location was identical across years and equal to the ratio of call probabilities and (2) the behavioral context of song calls were identical over time and space, and between populations. By fixing $\alpha$ we implicitly ignored the potential uncertainty in these assumptions and their impact on the final results. Deviations from these ecological assumptions could arise from a variety of

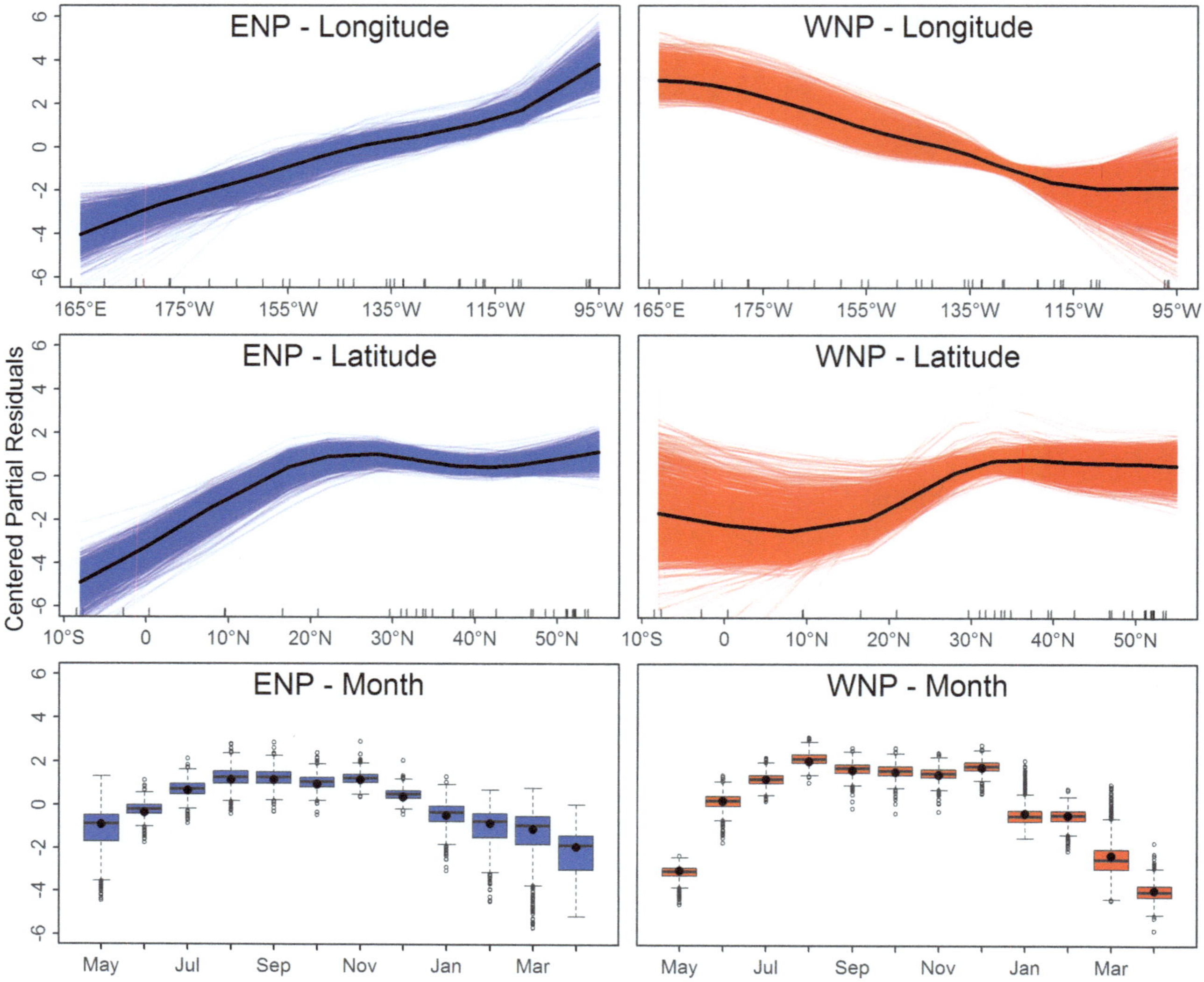


**Figure 8. Fits of the independent variables for the probability of observing a song call (*p*parameter).** The original model (black lines) and subsequent bootstrapped models (thin colored lines) are shown. Each panel shows the relationship between the dependent and independent variable after all other independent variables have been accounted for (i.e. the centered partial residuals). Higher relative values indicate a higher probability of observing a call. For longitude and latitude, *x*-axis tick marks show positions of the observed hydrophones with a small amount of noise added to prevent overplotting. See text for further discussion.

biological, environmental or anthropogenic factors. Examples of potential biological factors are different calling rates between populations, dynamic migration patterns (over year, space, or seasons) due to changing environmental conditions, and differences in the behavioral context of singing between populations. The largest potential anthropogenic factor was if whaling depleted the two populations at different rates so that the ratio of densities of the populations changed over time.

The impact of misspecifying $\alpha$ on the results depends on the size of $P$: in regions where only one population is found the impact will be small, and the impact is highest when the populations overlap the most. Thus the 'ecological uncertainty' caused by using an incorrect $\alpha$ value varies spatially with the estimate of the ratio of call probabilities. Consider the example where the truth is a prediction model with $\alpha_{\text{true}} = 2$, then following cases illustrate how assuming $\alpha = 1$ impacts predictions of catches. If $P = 0.001$ for some location, as would occur in predominantly ENP regions, then $\hat{C} \approx 0.999$ instead of the true value 0.998 and likewise if $P = 1000$, as in predominantly WNP regions, then $\hat{C} \approx 0.01$ instead of 0.005. Thus for extreme values of $P$ the effect of misspecifying $\alpha$ is mitigated by the nature of the structure of the prediction model. However, if $P = 1$ then $\hat{C} = 1/2$ instead of $1/3$, so the impact of ecological uncertainty is more pronounced in regions where both populations are singing at the same rates. Due to the complex relationship between $\alpha$ and the ratios of densities and call probabilities it is difficult to quantify and interpret $\alpha$ in terms of different biological and anthropogenic scenarios.

The veracity of the ecological assumptions about $\alpha$ could not be tested against data, but there was no reason to expect it would be constant across space for almost a century. We therefore conducted a sensitivity test where we assumed that $1/2 \leq \alpha \leq 2$ and that a randomly drawn $\alpha$ was constant across all catches within an iteration of our algorithm. Values were randomly chosen from a uniform distribution on the log scale so that $\alpha = 1$ was equally likely to be low or high.

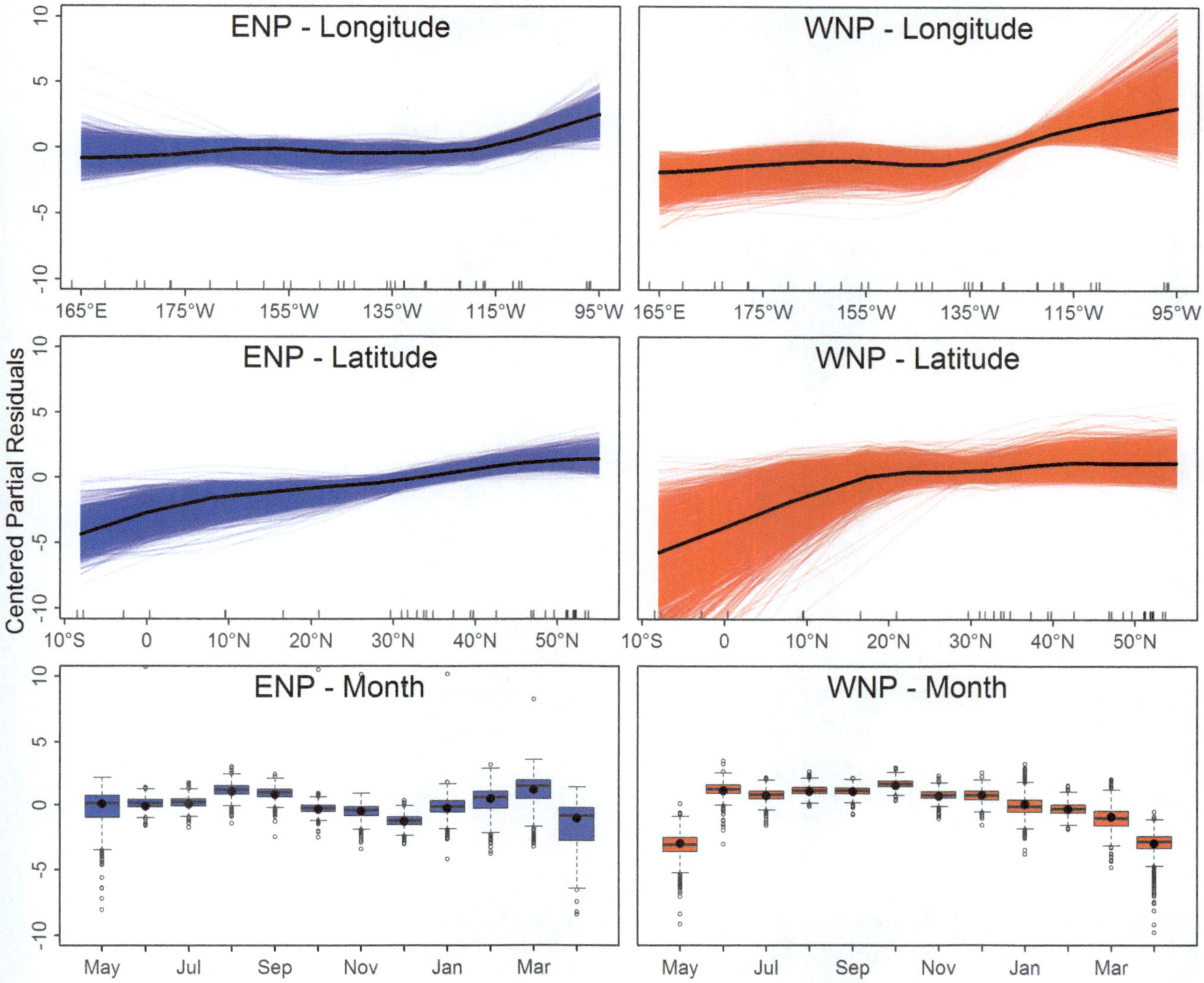


**Figure 9. As for Figure 8 except the results pertain to the over-dispersion parameter** $\sigma$. Higher relative values indicate a higher level of variance expected in repeated observations.

## Model Validation

We used information from previous studies to corroborate the catch splitting model developed here. The current distribution of ENP blue whales based on satellite tag and photographic re-sighting studies were compared to our model for ENP whale distribution as a qualitative check.

In addition, we used differences in mean total lengths to quantitatively validate the model. A previous study [15] found that WNP blue whales were longer than ENP blue whales, therefore whales with a low probability of being ENP should be longer than those with a high probability. We used the same selection criteria for mature females as in [15] for our length validation analysis, and performed two statistical analyses: a two-sample $t$-test comparing mean length differences by population, and a linear regression of model predictions against length. For the $t$-test we assigned catches to the ENP if their predicted probability was greater than 0.5, and WNP otherwise. Both statistical tests were performed for all 1000 bootstrapped sets of predictions, so that there was a distribution of mean length differences, regression lines, and $p$-values for both tests. If our model is correct, we should find that blue whales assigned to the ENP population are significantly shorter, and there should be a significant negative relationship between length and the predicted probability of being an ENP blue whale.

# Results

## Acoustic Presence Models

AICc provided clear justification for the more complex beta-binomial structure for acoustic models as well as additive terms on parameters $p$ and $\sigma$ in both the ENP and WNP acoustic models (Table 3). Therefore the full acoustic models were selected for both ENP and WNP populations and used throughout the rest of the analysis. These model predictions depended on the month, but generally classified parts of the Gulf of Alaska, the west coast of the US, and the eastern tropical Pacific as being predominantly ENP (Figures 6 and 7).

The algorithm of drawing random catch series and bootstrapping was run to produce 1000 ENP catch series. Some acoustic models failed to converged (1.73% ENP and 2.91% WNP), and since both acoustic models are needed for the prediction model, only cases where both converged for the same

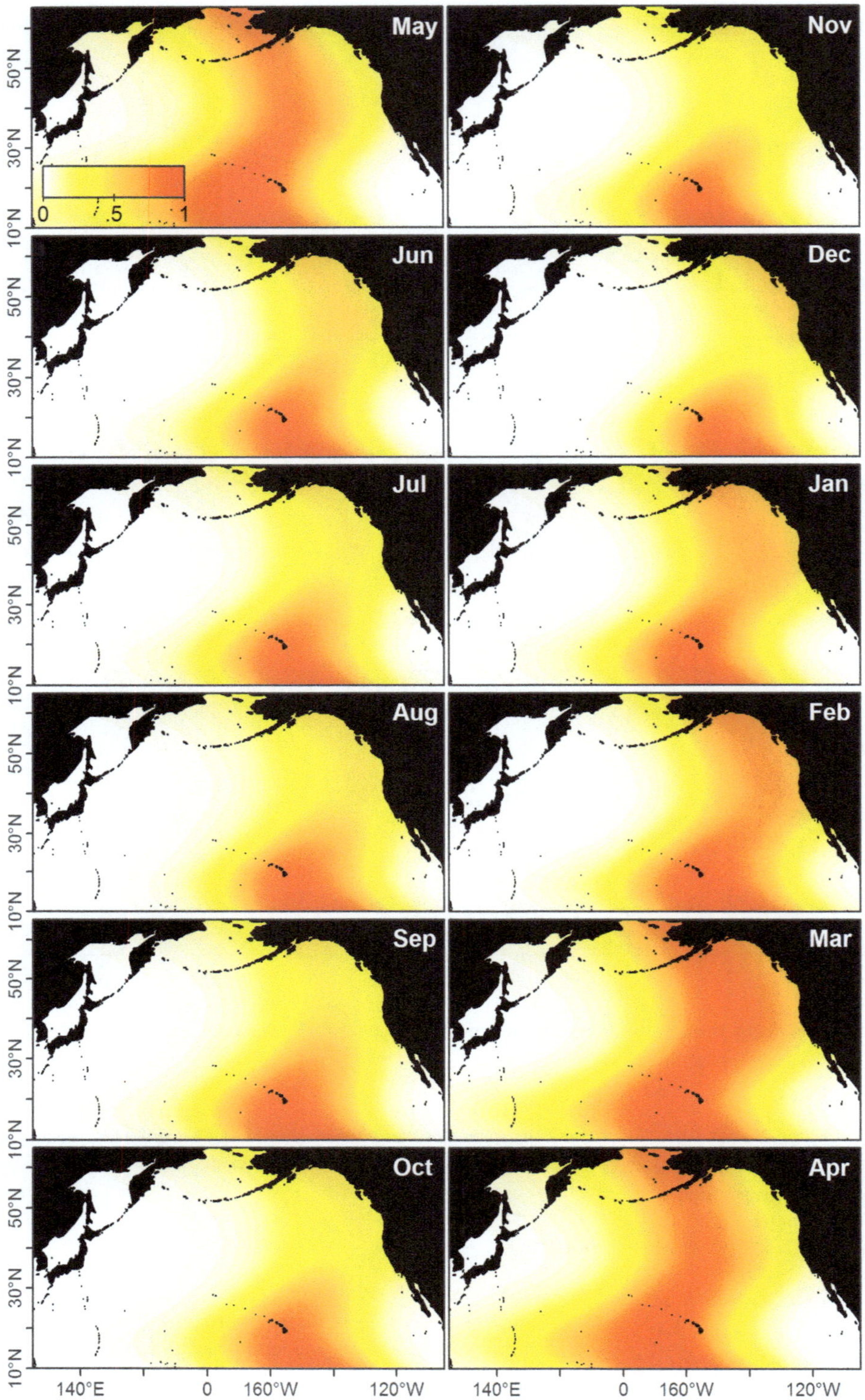


**Figure 10. Maps showing the statistical uncertainty arising from bootstrapping.** Uncertainty in the base case probability of a catch being from the ENP population quantified using bootstrapping (i.e. resampling the acoustic data with replacement and refitting the models). The $z$-axis is the range of the predictions (95th minus 5th percentile) across all 1000 bootstrapped models. A value of $z = 0$ indicates all bootstrap models predict the same value, while $z = 1$ indicates some bootstrap models predict 0 and others predict 1, so that there is high uncertainty at that location. There is higher uncertainty where overlap between the two populations exists (e.g. Gulf of Alaska) or there is limited data (e.g. Hawaii).

bootstrap sample could be used. A total of 1099 iterations of the algorithm were needed to arrive at 1000 successful ENP catch series.

GAMLSS model additive fits are interpreted by plotting the smoothers fit to "partial residuals" which have an arbitrary absolute scale and instead are judged via the relative change

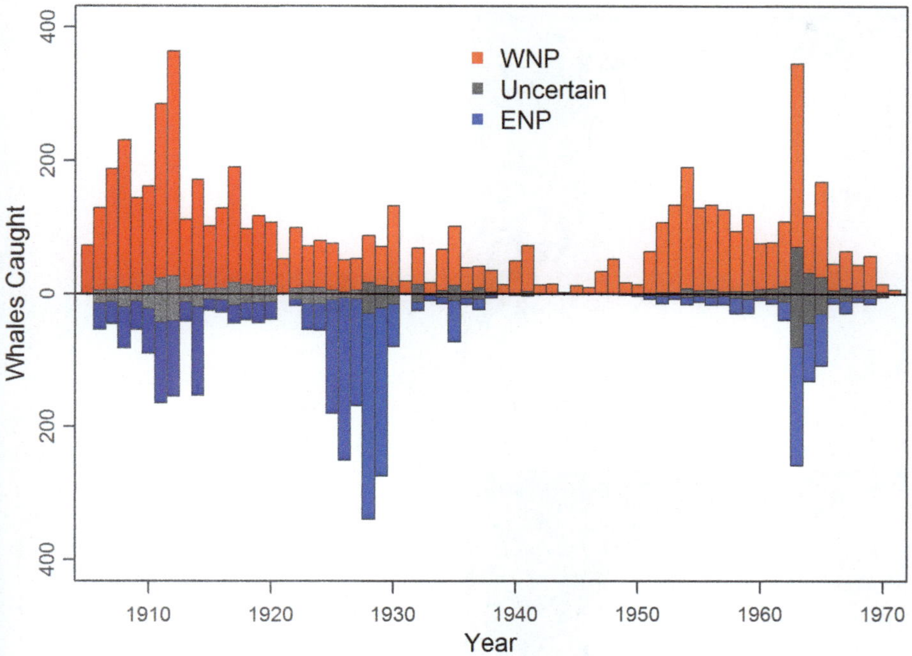


**Figure 11. Estimated annual catches of ENP and WNP blue whales for the base case.** Grey values contain 95% of the population uncertainty, and if above the line are more likely to be WNP, below the line more likely to be ENP. The uncertainty in the 1960s is caused by the unreported Soviet catches with high location uncertainty.

within a single independent variable (i.e. longitude, latitude, and month). In this case, increasing partial residual values indicated that the model predicted higher call rates for $p$, or higher variance for $\sigma$. For instance, both populations showed increased calling in the summer and fall months, ENP calls decreased from east to west, and WNP calls increased for east to west (Figures 8 and 9).

The variance in the parameters' additive fits across bootstrapped models provided a measure of statistical uncertainty. There was more uncertainty in the months with fewer observed calls and regions with ENP and WNP overlap and limited hours with observed calls (Figure 10). For this study, the uncertainty only impacts the results where it coincides with historical catches. Thus, while there is high uncertainty around Hawaii for all months, this has little influence on the results and more acoustic data here would likely have a minimal impact. The statistical uncertainty in the GOA in the summer months (the peak of catches) has the greatest influence on the uncertainty in the results.

## Catch Estimates

Each successful iteration of the algorithm provided distinct series of ENP catches and WNP catches, which were then aggregated by year (Tables 4–5). Most catches were taken in the early part of the century (1905–1930), with another spike in the 1950s and 1960s from Soviet pelagic whaling (Figure 11). Likewise, the uncertainty was highest during the early part of the 20$^{th}$ century when many catches were reported without location and/or month, and in the 1960s due to misreported Soviet catches which contained large catch uncertainty. Examples of specific ENP catch series demonstrate the general approach implemented here, including catch and statistical uncertainty predicted by the model (Figure 12). The catch uncertainty is reflected by the difference in locations of the catches between cases, and the statistical uncertainty by the differences in the region of uncertainty between cases.

Our methods allowed uncertainty to be estimated for the catch series, and incorporated into the results. The largest source of uncertainty for the base case came from bootstrapping the acoustic data. The catch uncertainty was relatively small compared to the statistical uncertainty, thus additional acoustic data has the potential to decrease the uncertainty in the results the most (Table 6). The ecological sensitivity run where $\alpha$ ranged uniformly between 0.5 and 2 contributed more uncertainty than catch uncertainty, but less than statistical uncertainty (Table 6). As expected a higher value of $\alpha$ for all algorithm iterations lead to smaller estimates for the total ENP catches (Figure 13).

We estimate that the total number of ENP whales caught from 1905–1971 in the North Pacific was 3,441 (95% range 2,855–3,920, out of 9,773 total) based on catch and statistical uncertainty. For the sensitivity case with ecological uncertainty included, the total number of ENP blue whales caught was 3,411 (95% range 2,593–4,114). Thus, the inclusion of ecological uncertainty increases the width of the 95% interval by 32% if $\alpha$ is assumed to range uniformly from 0.5 to 2. Catches of WNP blue whales totaled 6,332 (95% range 5,853–6,918) with catch and statistical uncertainty, while the case with ecological uncertainty was 6,362 (95% range 5,659–7,180).

## Model Validation

Movement of ENP whales between the ETP, California, and the Gulf of Alaska from satellite telemetry and photographic identification [13,14] corroborated our model's prediction that ENP whales occur throughout this range (Figures 6–7).

We applied the length analysis to 258 mature females, compared to 252 from [15], which were distributed sufficiently across time and space for the statistical tests (Figure 14a). For the two-sample $t$-tests, the median difference in mean lengths was 0.91 m (95% range 0.76–1.03 m, Figure 14b) and for all 1000 catch scenarios this difference was highly significant ($p < 0.00002$). The slopes of the linear regressions were all significantly negative,

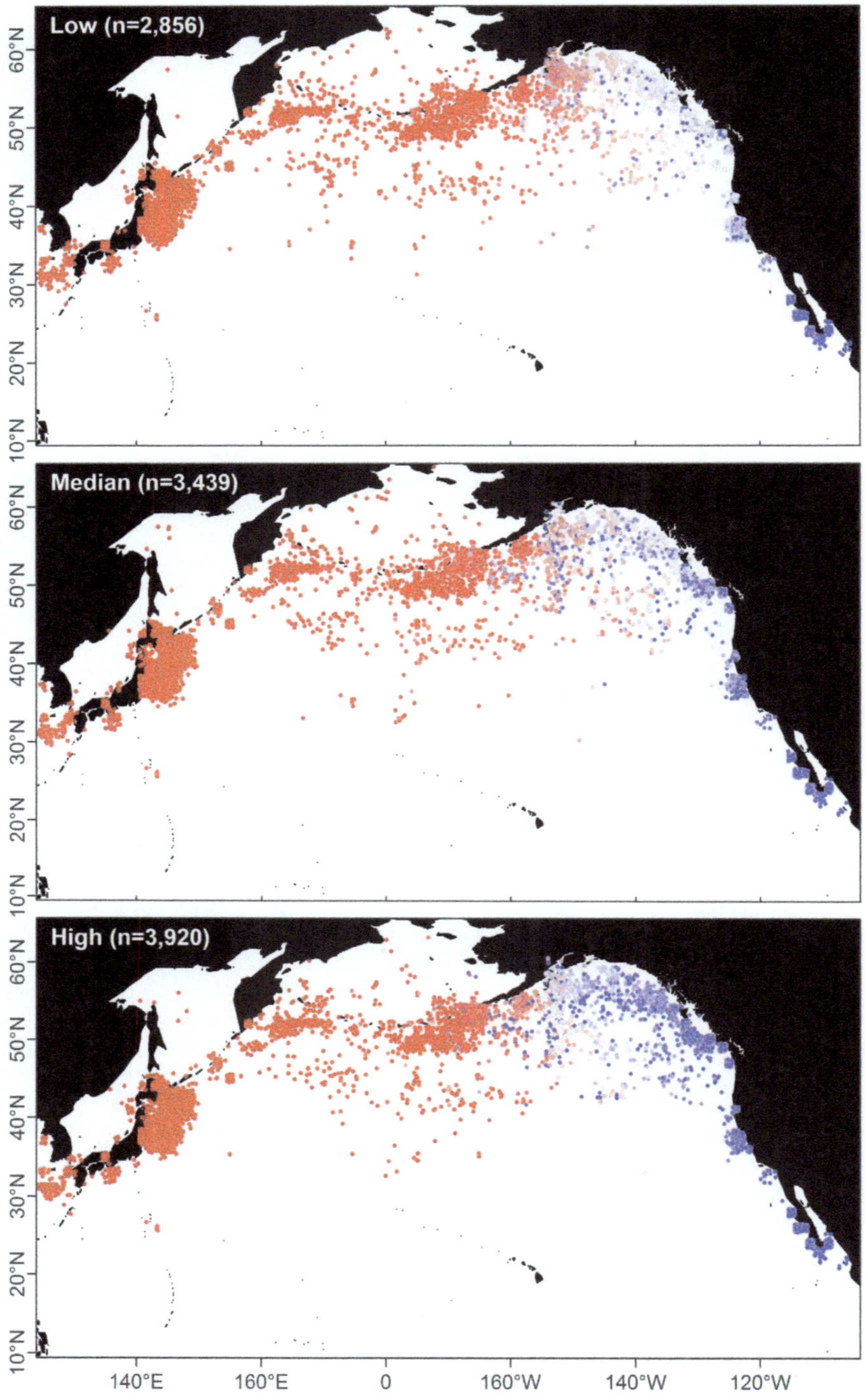


**Figure 12. Example model predictions.** Catches of ENP (blue) and WNP (red) blue whales for three realizations chosen to represent low, median, and high total ENP catches. Catch positions are plotted with a small amount of noise to prevent overplotting, and are different between the three cases because they are drawn randomly from potential locations. Nearby catches can have different predictions (colors) because months are aggregated.

demonstrating that the shorter a whale is, the more likely it from the ENP population (Figure 14c).

## Discussion

Our study is the first to split North Pacific blue whale catches by population, and was based on current day patterns in acoustic song types across the North Pacific. Our results based on acoustic calls were validated using the length data in the catches: those individuals we assigned to the ENP were shorter (on average by 0.91 m) and these differences were highly significant. We attributed roughly 35% of all blue whales catches to the ENP population, suggesting that the WNP population was likely substantially larger than the ENP population before whaling started. Current abundance estimates are not available for any

**Table 4.** Base case ENP catches by year showing the median as well as lower and upper 95% interval from the 1000 realizations.

| Year | 2.5% | Median | 97.5% | Year | 2.5% | Median | 97.5% |
|---|---|---|---|---|---|---|---|
| 1905 | 1 | 1 | 1 | 1939 | 0 | 0 | 1 |
| 1906 | 40 | 54 | 60 | 1940 | 1 | 2 | 3 |
| 1907 | 33 | 45 | 51 | 1941 | 1 | 3 | 6 |
| 1908 | 62 | 82 | 92 | 1942 | 1 | 1 | 1 |
| 1909 | 42 | 53 | 59 | 1943 | 0 | 0 | 0 |
| 1910 | 68 | 90 | 102 | 1944 | 0 | 0 | 0 |
| 1911 | 123 | 165 | 189 | 1945 | 0 | 0 | 0 |
| 1912 | 115 | 155 | 182 | 1946 | 0 | 0 | 1 |
| 1913 | 30 | 41 | 51 | 1947 | 0 | 0 | 1 |
| 1914 | 134 | 153 | 165 | 1948 | 0 | 0 | 1 |
| 1915 | 17 | 25 | 33 | 1949 | 1 | 2 | 2 |
| 1916 | 19 | 27 | 36 | 1950 | 3 | 3 | 4 |
| 1917 | 27 | 44 | 61 | 1951 | 5 | 8 | 9 |
| 1918 | 26 | 39 | 52 | 1952 | 11 | 14 | 16 |
| 1919 | 31 | 43 | 55 | 1953 | 6 | 7 | 10 |
| 1920 | 27 | 38 | 50 | 1954 | 10 | 15 | 24 |
| 1921 | 0 | 0 | 0 | 1955 | 9 | 12 | 17 |
| 1922 | 10 | 18 | 27 | 1956 | 12 | 17 | 23 |
| 1923 | 39 | 54 | 64 | 1957 | 13 | 16 | 20 |
| 1924 | 41 | 54 | 64 | 1958 | 24 | 30 | 34 |
| 1925 | 172 | 181 | 187 | 1959 | 22 | 29 | 33 |
| 1926 | 245 | 251 | 253 | 1960 | 4 | 9 | 17 |
| 1927 | 162 | 169 | 175 | 1961 | 8 | 15 | 22 |
| 1928 | 310 | 339 | 357 | 1962 | 28 | 40 | 51 |
| 1929 | 254 | 275 | 288 | 1963 | 179 | 259 | 330 |
| 1930 | 64 | 79 | 92 | 1964 | 89 | 132 | 165 |
| 1931 | 0 | 0 | 0 | 1965 | 79 | 108 | 134 |
| 1932 | 13 | 25 | 40 | 1966 | 9 | 15 | 21 |
| 1933 | 8 | 10 | 11 | 1967 | 19 | 29 | 38 |
| 1934 | 10 | 15 | 20 | 1968 | 6 | 12 | 17 |
| 1935 | 62 | 72 | 85 | 1969 | 10 | 15 | 21 |
| 1936 | 9 | 16 | 20 | 1970 | 1 | 4 | 7 |
| 1937 | 16 | 24 | 34 | 1971 | 0 | 0 | 0 |
| 1938 | 4 | 5 | 8 | Total | 2,765 | 3,439 | 3,993 |

These catches do not include uncertainty in $\alpha$.

**Table 5.** Base case WNP catches by year showing the median as well as lower and upper 95% interval from the 1000 realizations.

| Year | 2.5% | Median | 97.5% | Year | 2.5% | Median | 97.5% |
|---|---|---|---|---|---|---|---|
| 1905 | 73 | 73 | 73 | 1939 | 14 | 15 | 15 |
| 1906 | 123 | 129 | 143 | 1940 | 49 | 50 | 51 |
| 1907 | 182 | 188 | 200 | 1941 | 70 | 73 | 75 |
| 1908 | 221 | 231 | 251 | 1942 | 14 | 14 | 14 |
| 1909 | 139 | 145 | 156 | 1943 | 15 | 15 | 15 |
| 1910 | 150 | 162 | 184 | 1944 | 2 | 2 | 2 |
| 1911 | 261 | 285 | 327 | 1945 | 13 | 13 | 13 |
| 1912 | 337 | 364 | 404 | 1946 | 9 | 10 | 10 |
| 1913 | 103 | 113 | 124 | 1947 | 33 | 34 | 34 |
| 1914 | 160 | 172 | 191 | 1948 | 52 | 53 | 53 |
| 1915 | 95 | 103 | 111 | 1949 | 17 | 17 | 18 |
| 1916 | 121 | 130 | 138 | 1950 | 14 | 15 | 15 |
| 1917 | 174 | 191 | 208 | 1951 | 63 | 64 | 67 |
| 1918 | 85 | 98 | 111 | 1952 | 106 | 108 | 111 |
| 1919 | 106 | 118 | 130 | 1953 | 132 | 135 | 136 |
| 1920 | 96 | 108 | 119 | 1954 | 183 | 192 | 197 |
| 1921 | 53 | 53 | 53 | 1955 | 125 | 130 | 133 |
| 1922 | 91 | 100 | 108 | 1956 | 128 | 134 | 139 |
| 1923 | 62 | 72 | 87 | 1957 | 123 | 127 | 130 |
| 1924 | 71 | 81 | 94 | 1958 | 91 | 95 | 101 |
| 1925 | 70 | 76 | 85 | 1959 | 116 | 120 | 127 |
| 1926 | 49 | 51 | 57 | 1960 | 69 | 77 | 82 |
| 1927 | 48 | 54 | 61 | 1961 | 70 | 77 | 84 |
| 1928 | 71 | 89 | 118 | 1962 | 98 | 109 | 121 |
| 1929 | 59 | 72 | 93 | 1963 | 275 | 346 | 426 |
| 1930 | 121 | 134 | 149 | 1964 | 86 | 119 | 162 |
| 1931 | 20 | 20 | 20 | 1965 | 144 | 170 | 199 |
| 1932 | 55 | 70 | 82 | 1966 | 39 | 45 | 51 |
| 1933 | 16 | 17 | 19 | 1967 | 56 | 65 | 75 |
| 1934 | 62 | 67 | 72 | 1968 | 39 | 44 | 50 |
| 1935 | 89 | 102 | 112 | 1969 | 52 | 58 | 63 |
| 1936 | 36 | 40 | 47 | 1970 | 12 | 15 | 18 |
| 1937 | 32 | 42 | 50 | 1971 | 7 | 7 | 7 |
| 1938 | 33 | 36 | 37 | Total | 5,780 | 6,334 | 7,008 |

These catches do not include uncertainty in $\alpha$.

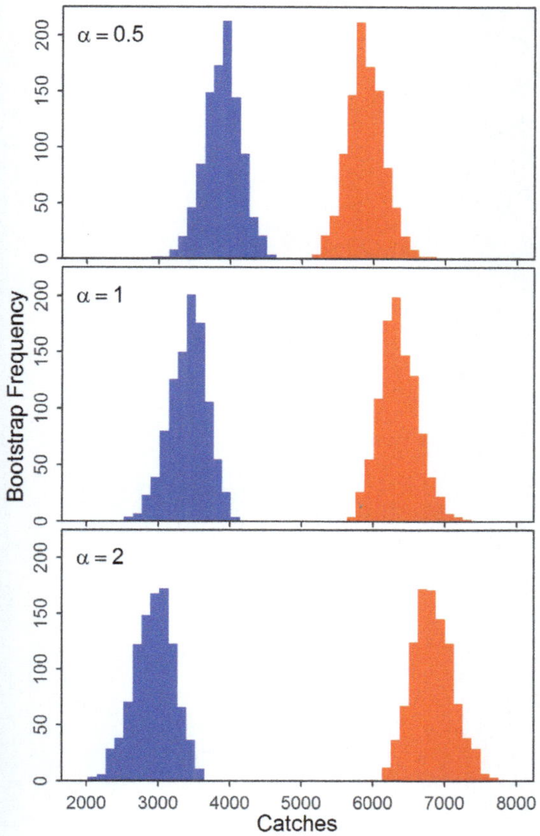


**Figure 13. Results of the sensitivity analysis to the conversion between song call frequency and density of whales.** The sensitivity to the base case assumption that $\alpha=1$ to the final results is explored with histograms of the total catches of the ENP (blue) and WNP (red) populations for three levels of $\alpha$. $\alpha$ is the ratio of the unknown WNP and ENP factors which convert density of whales to song call occurrence.

region within the WNP, so it is unclear whether the WNP blue whales are now more or less depleted than ENP blue whales.

Our results are based on the following ecological assumptions about movement patterns and acoustic behavior of the two populations. First, we assumed that the ENP and WNP populations had stable movement patterns over the entire span of the catches and acoustic recordings (1905–2004), despite intensive whaling and varying oceanographic and biological conditions. If ample acoustic data were available across many years, the effect of year could be quantified within the modeling framework provided here, but the acoustic data were collected in different time periods (Table 2), precluding such an analysis in this study. High maternal site fidelity in the Gulf of California has been shown for the ENP population, supporting the idea of spatial continuity in a breeding and calving region [16], but another study speculates a range shift into the Gulf of Alaska due to changes in abundance or oceanographic conditions [14]. It seems reasonable that slight changes in ranges for the ENP, particularly in the GOA are possible, but highly unlikely that they formerly occurred in the western North Pacific. For the WNP there is no information with which to judge how many populations there are, or their spatial fidelity.

Second, we assumed that the ratio of calling rates was proportional to the ratio of densities of the two populations across time and space, but cannot directly test the accuracy of this hypothesis. The relationship between singing and density is likely complex, especially if it is related to breeding. One study suggests that ENP blue whales arriving at feeding grounds may delay singing for weeks until adequate food has been obtained [12] and another in the same area found the proportion of whales singing increased during the feeding season [33]. Thus the relationship between density and ENP singing is weak for part of the feeding season, though this short-term effect is mitigated to some degree by binning the data into months. Little is known about the seasonal behavior of ENP blue whales, and none about WNP whales, so we could not explicitly include this in our analysis. The assumption may also be incorrect if there were a difference in the sex ratios or sex-specific mixing between populations. The historical catch records clearly indicated that both sexes were caught in similar ratios throughout the region modeled, so this is not believed to be a major issue in the analysis. In addition, the two populations could vocalize at different rates. Since they have similar morphologies, call types, seasonal pattern of call production, and life history strategies it is unlikely that the behavioral context of call production differs greatly between populations.

The veracity of the ecological assumptions discussed here was, and may remain, unknown. However, when we tested the influence of this factor in a sensitivity test allowing $\alpha$ to range from 0.5 to 2.0 (a fourfold range), the 95% confidence interval in the total ENP catches widened by only 32%. This suggests that the results are fairly robust to these assumptions. Similarly, the relatively small impact from catch uncertainty suggests the simplifying assumptions made, such as using reported locations without error and ignoring the selective preference of expeditions, likely had a minimal impact on the results.

Although the focus of this study was ENP blue whales, by examining all catch and acoustic data, new information about WNP blue whale exploitation emerged. We estimated 6,332 (95% range 5,852–6,918) or roughly 65% of blue whale catches in the North Pacific were likely from the WNP population. Unlike ENP blue whales the current abundance, status, and geographic range of these whales remains unknown. At present there is no dedicated effort to study WNP whales and acoustic data provide the only

**Table 6.** Comparison of the sources of uncertainty in total ENP catches.

| Uncertainty Source | 2.5% | Median | 97.5% | Cumulative % Uncertainty |
|---|---|---|---|---|
| Catch | 3,430 | 3,452 | 3,470 | 3% |
| Catch + Statistical | 2,855 | 3,441 | 3,920 | 70% |
| Catch + Statistical + Ecological | 2,593 | 3,411 | 4,114 | 100% |

Catch uncertainty arises from missing locations and dates, statistical uncertainty from song call occurrence data, and ecological from uncertainty in ecological assumptions over the range $1/2\leq\alpha\leq2$. Differences in totals compared to Table 4 are due to rounding and the order of percentile calculations.

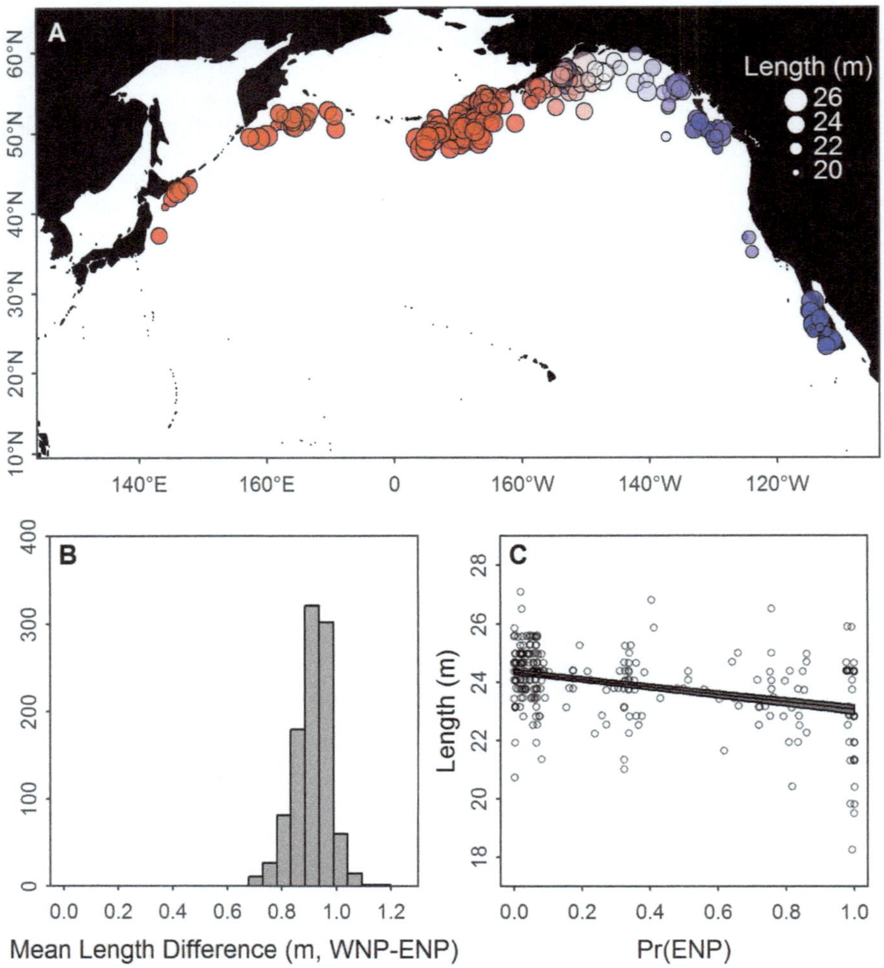


**Figure 14. Results of the length analysis of mature female blue whales.** (A) Locations, lengths, and median predicted population (color) for all years. (B) The distribution of mean length differences for all bootstrap iterations. (C) The 95% percentiles (gray polygon) and median line for all bootstrap linear models fit to the predicted population versus lengths. The circles represent the median probability of being ENP and observed lengths and are plotted with a small amount of noise to prevent overplotting.

information as to when and where these blue whales are found in the North Pacific.

We incorporated uncertainty into how the catches were split between populations, but there still may be missing catches, particularly off Japan in the early part of the century where many catches were not identified to species in 1905–1909. Future studies utilizing WNP catch data should visit those catches in further detail. Missing catches may also exist for the Soviet pelagic expeditions in the 1960s and 1970s, which would affect the results here, but is likely that most of these have been recovered and revised already and few more exist [24].

This study provides a new framework for differentiating spatial occurrence of populations of cetaceans using their distinct call types using data that are relatively easy to collect in comparison to broad scale sighting surveys, photo identification, and genetic biopsy sampling efforts. In doing so we provide the first estimate of catches for ENP blue whales that is based on data and includes uncertainty. Future work should combine these results with current abundance estimates to formally assess the recovery of the population and the potential risk of continued anthropogenic mortalities.

## Acknowledgments

We would like to thank André Punt and Brett McClintock for technical advice, Cherry Allison for sharing the IWC databases, and two anonymous reviewers for helpful suggestions. In particular, one reviewer suggested the length validation method that we used to strengthen our results.

## Author Contributions

Conceived and designed the experiments: CCM TAB. Analyzed the data: CCM. Wrote the paper: CCM TAB. Contributed data: KMS YVI EMO. Discussed results and contributed to manuscript: CCM TAB KMS EMO YVI.

## References

1. Rice DW (1998) Marine mammals of the world: systematics and distribution: Society of Marine Mammalogy.
2. Reilly SB, Bannister JL, Best PB, Brown M, RL BJ, et al. (2008) *Balaenoptera musculus*. In: IUCN 2012. IUCN Red List of Threatened Species. Version 2012.2.

3. Carretta JV, Oleson E, Weller DW, Lang AR, Forney KA, et al. (2012) U.S. Pacific marine mammal stock assessments. NOAA Technical Memorandum NOAA-TM-NMFS-SWFSC-504.
4. Stafford KM, Nieukirk SL, Fox CG (1999) An acoustic link between blue whales in the eastern tropical Pacific and the northeast Pacific. Marine Mammal Science 15: 1258–1268.
5. Goldbogen JA, Calambokidis J, Oleson E, Potvin J, Pyenson ND, et al. (2011) Mechanics, hydrodynamics and energetics of blue whale lunge feeding: efficiency dependence on krill density. Journal of Experimental Biology 214: 698–699.
6. Oleson EM, Calambokidis J, Burgess WC, McDonald MA, LeDuc CA, et al. (2007) Behavioral context of call production by eastern North Pacific blue whales. Marine Ecology-Progress Series 330: 269–284.
7. Melcon ML, Cummins AJ, Kerosky SM, Roche LK, Wiggins SM, et al. (2012) Blue whales respond to anthropogenic noise. PLoS One, 7: e32681.
8. Stafford KM, Nieukirk SL, Fox CG (2001) Geographic and seasonal variation of blue whale calls in the North Pacific. Journal of Cetacean Research and Management 3: 65–76.
9. Stafford KM (2003) Two types of blue whale calls recorded in the Gulf of Alaska. Marine Mammal Science 19: 682–693.
10. McDonald MA, Hildebrand JA, Mesnick SL (2006) Biogeographic characterization of blue whale song worldwide: using song to identify populations. Journal of Cetacean Research and Management 8: 55–65.
11. Rivers JA (1997) Blue whale, *Balaenoptera musculus*, vocalizations from the waters off central California. Marine Mammal Science 13: 186–195.
12. Oleson EM, Wiggins SM, Hildebrand JA (2007) Temporal separation of blue whale call types on a southern California feeding ground. Animal Behaviour 74: 881–894.
13. Bailey H, Mate BR, Palacios DM, Irvine L, Bograd SJ, et al. (2010) Behavioural estimation of blue whale movements in the Northeast Pacific from state-space model analysis of satellite tracks. Endangered Species Research 10: 93–106.
14. Calambokidis J, Barlow J, Ford JKB, Chandler TE, Douglas AB (2009) Insights into the population structure of blue whales in the Eastern North Pacific from recent sightings and photographic identification. Marine Mammal Science 25: 816–832.
15. Gilpatrick Jr JW, Perryman WL (2008) Geographic variation in external morphology of North Pacific and Southern Hemisphere blue whales (*Balaenoptera musculus*). Journal of Cetacean Research and Management 10: 9–21.
16. Costa-Urrutia P, Sanvito S, Victoria-Cota N, Enríquez-Paredes L, Gendron D (2013) Fine-scale population structure of blue whale wintering aggregations in the Gulf of California. PLoS One 8: e58315.
17. Reeves RR, Clapham PJ, Brownell Jr RL, Silber GK (1998) Recovery plan for the blue whale (*Balaenoptera musculus*). Publications, Agences and Staff of the U.S. Department of Commerce. Paper 118.
18. Baker CS, Clapham PJ (2004) Modelling the past and future of whales and whaling. Trends in Ecology & Evolution 19: 365–371.
19. Calambokidis J, Falcone E, Douglas A, Schlender L, Huggins J (2009) Photographic identification of humpback and blue whales off the U.S. West Coast: results and updated abundance estimates from 2008 field season. Final report for contract AB133F08SE2786. Southwest Fisheries Science Center.
20. Calambokidis J, Barlow J (2004) Abundance of blue and humpback whales in the eastern North Pacific estimated by capture-recapture and line-transect methods. Marine Mammal Science 20: 63–85.
21. Allison C (2011) IWC summary catch database Version 5.2.
22. Allison C (2011) IWC individual catch database Version 5.2.
23. Yablokov AV (1994) Validity of whaling data. Nature 367: 108–108.
24. Ivashchenko YV, Clapham PJ, Brownell Jr RL (2013) Soviet catches of whales in the North Pacific: revised totals. Journal of Cetacean Research and Management 13(1): 59–71.
25. R Core Team (2012) R: A language and environment for statistical computing. 1.7.0 ed. Vienna, Austria: R Foundation for Statistical Computing.
26. Stafford KM, Nieukirk SL, Fox CG (1999) Low-frequency whale sounds recorded on hydrophones moored in the eastern tropical Pacific. Journal of the Acoustical Society of America 106: 3687–3698.
27. McDonald MA, Fox CG (1999) Passive acoustic methods applied to fin whale population density estimation. Journal of the Acoustical Society of America 105: 2643–2651.
28. Gregr EJ, Trites AW (2001) Predictions of critical habitat for five whale species in the waters of coastal British Columbia. Canadian Journal of Fisheries and Aquatic Sciences 58: 1265–1285.
29. Wood SN (2006) Generalized additive models: an introduction with R: Chapman & Hall.
30. Rigby RA, Stasinopoulos DM (2005) Generalized additive models for location, scale and shape. Journal of the Royal Statistical Society Series C-Applied Statistics 54: 507–544.
31. Burnham KP, Anderson DR (2002) Model selection and multi-model inference. Berlin, Heidelberg, New York: Springer.
32. Efron B, Tibshirani R (1993) An introduction to the bootstrap: Chapman & Hall/CRC.
33. Oleson EM, Calambokidis J, Barlow J, Hildebrand JA (2007) Blue whale visual and acoustic encounter rates in the southern California bight. Marine Mammal Science 23: 574–597.

# The Diel Rhythms of Biosonar Behavior in the Yangtze Finless Porpoise (*Neophocaena asiaeorientalis asiaeorientalis*) in the Port of the Yangtze River: The Correlation between Prey Availability and Boat Traffic

**Zhitao Wang[1,2], Tomonari Akamatsu[3,4], Kexiong Wang[1]*, Ding Wang[1]***

1 The Key Laboratory of Aquatic Biodiversity and Conservation of the Chinese Academy of Sciences, Institute of Hydrobiology, Chinese Academy of Sciences, Wuhan, China, 2 University of Chinese Academy of Sciences, Beijing, China, 3 National Research Institute of Fisheries Engineering, Fisheries Research Agency, Ibaraki, Japan, 4 CREST, Japan Science and Technology Agency, Tokyo, Japan

**Abstract**

Information on the habitat use of the critically endangered Yangtze finless porpoise (*Neophocaena asiaeorientalis asiaeorientalis*) is critical for its conservation. The diel biosonar behavior of the porpoise in the port areas of the Yangtze River was examined along with simultaneous observations of fish density and boat traffic. Biosonar pulses from the porpoises were detected for 1233 min (5.77%) over a 21,380 min duration of effective observations. In total, 190 (5.63%) buzzes (an indication of prey capture attempts) were recorded among the 3372 identified click trains. Of the 168 echolocation encounters (bouts of click trains less than eight min apart), 150 (89.3%) involved single animals, indicating that solitary porpoises were frequently present and feeding in the port areas. Significant diel patterns were evident involving the biosonar behavior of the porpoises (including click trains and buzzes), fish density and boat traffic. The frequencies of the click trains and buzzes were significantly lower during the day than in the evening and at night, which suggests that porpoises in this region are primarily engaged in crepuscular and nocturnal foraging. The lack of a significant diel pattern in the echolocation encounters indicates the importance of the port in porpoise conservation. A forced feeding schedule may be associated with the lack of a significant correlation between porpoise acoustics and boat traffic. Overall, prey availability appears to be the primary factor that attracts porpoises. Additionally, porpoises tend to migrate or remain downstream in the morning and migrate or remain upstream in the evening, most likely to follow their prey. The findings of this study can be used to improve the conservation of the Yangtze finless porpoise.

**Editor:** Craig A. Radford, University of Auckland, New Zealand

**Funding:** This work was supported by grants from the Knowledge Innovation Program of the Chinese Academy of Sciences (KSCX2-EW-Z-4), the National Natural Science Foundation of China (31170501 and 31070347), the Special Fund for Agro-scientific Research in the Public Interest of the Ministry of Agriculture of China (201203086), Ministry of Science and Technology of China (2011BAG07B05-3), State Oceanic Administration of China (201105011-3), CREST, Japan Science and Technology Agency and the Ocean Park Conservation Foundation, Hong Kong (There is no grant code). The funders had no roles in designing the study, collecting and analyzing data or preparing and deciding where to publish the manuscript.

**Competing Interests:** The authors have declared that no competing interests exist.

* E-mail: wangk@ihb.ac.cn (KW); wangd@ihb.ac.cn (DW)

## Introduction

A genetically and morphologically distinct cetacean species, the Yangtze finless porpoise (*Neophocaena asiaeorientalis asiaeorientalis*) is listed as Critically Endangered on the International Union for the Conservation of Nature Red List of Threatened Species [1]. As an icon of the Yangtze ecosystem [2] and the only freshwater porpoise in the world, this species is currently found only in the main stem of the middle and lower reaches of the Yangtze River and the adjoining Poyang and Dongting Lakes [3,4]. Unfortunately, many of these porpoises live in areas that are undergoing rapid and increasing habitat destruction and suffering from sharply declining fish stocks. Consequently, these porpoise are at risk of extinction similar to the Yangtze River dolphin (*Lipotes vexillifer*), which was likely driven to extinction by human activity [1,5].

Observing free-ranging finless porpoises presents a logistical challenge, as these animals take swift surface breaths, lack a dorsal fin, are small in size and have barely visible natural markings. Indeed, these animals are among the most difficult cetacean species to observe visually [6,7].

Toothed whales (Odontoceti) rely primarily on their sophisticated acoustic sensing biosonar for communication, navigation and foraging [8]. The Yangtze finless porpoise uses sonar in an almost continuous manner (sound pressure level (SPL) of over 148 dB peak-to-peak re 1 μPa with an average inter-click-bout interval of 6.4 s) [9], which allows for acoustic monitoring of these animals.

Passive acoustic monitoring (PAM) is a rapidly evolving discipline that facilitates the investigation of the behavior of many aquatic mammals over a range of spatial and temporal scales (for reviews, see [10]), including at night and during adverse weather conditions. Additionally, this method enables round-the-clock monitoring for extended periods of time, which is logistically impossible for visual observations. Over the past decade, our understanding of porpoise behavior and distribution has improved greatly through the use of PAM [11–16].

Harbors and ports are common along the Yangtze River, and the structures of some of the platforms and installations extend down through the water column. These structures can attract large accumulations of plankton via illumination and hydrodynamic effects [17], provide shelter for other underwater organisms from currents or predators [18,19] and influence local biodiversity and ecosystems. The results from acoustic surveys from the Yangtze Freshwater Dolphin Expedition of 2012 suggest that finless porpoises typically occupy the port areas at night for prolonged periods and frequently emit buzz sounds (short-range sonar sounds with minimum inter-click intervals shorter than 10 ms) [14]. In acoustic studies of dolphins, buzz sounds have been used as indicators of attempts at prey capture [20], feeding activity [7,21] or foraging success [22,23], and Yangtze finless porpoises also emit buzzes during prey capture dives [24]. The detection of nocturnal buzz sounds suggests that finless porpoises feed near the port area at night [14]. Mounting evidence from studies of circadian rhythms has shown that a wide variety of taxa tend to switch from nocturnal to diurnal activity patterns to adapt to metabolic demands (for a review, see [25]) and/or their physical and social environment (see [26] for a review). However, with the exception of information collected from a buoy-based PAM system near the outlet of Poyang Lake into the Yangtze River [27], little is known about the diel patterns of porpoise distributions. These shortcomings should be addressed, particularly with regard to the diel rhythms of finless porpoises in the vicinity of the port area, because the area is frequently subjected to artificial activities. Knowledge of the time-specific habitat preferences of these animals can not only help us identify ways to mitigate possible human interference, especially during times of peak porpoise presence, but also inform present and future conservation needs for this species, which is facing significant extinction pressure.

Because regional studies can often provide useful insights into broader patterns, the primary goal of this study was to describe the potential diel patterns of porpoises in port areas. Moreover, earlier findings indicate that Yangtze finless porpoises tend to frequent habitats with high densities of fish [28]; the rate of porpoise detection was significantly higher in areas with fish than in areas without fish [12]. Additionally, the significant negative correlation between the presence of Yangtze finless porpoises and the presence of ships suggests that porpoises tend to avoid boat traffic [11], and indirect and direct evidence of the tendency to avoid passing vessels has also been documented [13,14]. We tested the hypothesis that prey availability triggers porpoise feeding and that boat traffic disrupts feeding. Porpoise movement patterns relative to the port were also analyzed.

## Methods

### Data Collection

**Porpoise acoustic data.** PAM of porpoise sonar sounds was conducted at the Pengze Maritime Safety Administration (MSA) Port (see Fig. 1) from 11:00 am on 8 June to 7:20 am on 23 June 2013. The MSA port consists of an anchored cruise vessel (27 m long and 5 m wide) and a rectangular wharf boat (39 m long and 9 m wide) connected to a waterway administration port of the same size. Each of these structures is water-based and arranged parallel to the south bank of the Yangtze River, approximately 5 m apart; no other water-based structures were located within 2 km of the area (Fig. 1). Three miniature stereo pulse event data loggers (A-tag; see details under "Porpoise sonar monitoring system" below) were used. Each data logger was tightly fastened to a stainless steel pole with plastic bands and deployed vertically at two sites on the cruise vessel (site A, 29°54′24″ N, 116°32′50″ E; site B, 29°54′25″ N, 116°32′51″ E) and at one site on the wharf boat (site C, 29°54′26″ N, 116°32′51″ E) (Fig. 1) with the hydrophone array of the A-tag deployed parallel to the current. The depth of the recording site ranged from 7 to 8 m, and each A-tag was deployed at a depth of 1.5 m depth (approximately 1 m deeper than the submerged part of the port) to minimize the possibility of obstructions to the sound.

### Fish Data

A maritime safety cruise boat-mounted single-beam Raymarine HD Digital Fishfinder (model C80, Raymarine UK Ltd., Fareham, UK) was used to log data while anchored at the port. The transducers of this echosounder can transmit 50 kHz as a narrow beam. However, to minimize the possible interference of the transmitted sound with the A-tags, the fish data were randomly sampled as echograms by photographing the monitoring screen of the Fishfinder at different diel phases (see below) from the 19$^{th}$ to the 23$^{rd}$ of June instead of being continuously monitored. The cruise boat was anchored along the wharf boat after each day's expedition to a random site (Fig. 1), which resulted in fish data from multiple sites rather than one fixed point to represent the density of fish at the port.

**Boat traffic data.** The diel pattern of boats upriver and downriver of the study site was either visually counted or photographed with a Pentax camera (model Optio W60, PENTAX Corporation, Tokyo, Japan) shooting at 1 min intervals from the 15$^{th}$ to the 23$^{rd}$ of June. The boat's safety lights made photography possible during the night. The camera was fixed at an advantageous location on the second deck of the cruise vessel (Fig. 1) and oriented to capture every vessel that passed during 1-min periods. As demonstrated by a preliminary study, the 1-min setting is sufficient to capture every passing vessel except for speed boats, which are rare in this region.

### Ethical Statement

The research was carried out under a research permit issued to the Institute of Hydrobiology of the Chinese Academy of Sciences by the Ministry of Agriculture of China. The approval number from the Fishery Bureau of the General Office of the Ministry of Agriculture was 2012–114.

### Porpoise Sonar Monitoring System

Two types of A-tags (including towed and fixed types) were used in this study. A brief description of these autonomous acoustic recorders is provided here, and a detailed description is provided by Akamatsu et al. (2005) [29]. Each A-tag (ML200-AS2, Marine Micro Technology, Saitama, Japan) consisted of two ultrasonic hydrophones, a band-pass filter (−3 dB with a 55–235 kHz range) and a high-gain amplifier (+60 dB). The hydrophones (model MHP-140ST; Marine Micro Technology, Saitama, Japan) had the following parameters: sensitivity at −201 dB re 1 V/μPa at 1 m distance and a frequency response range from 100 to 160 kHz within 5 dB with the primary and secondary hydrophones separated at a fixed range of either 195 mm (towed type) or 590 mm (fixed type) to form a stereo hydrophone array. The data logger had directional sensitivity with 0° on-axis and 90° off-axis detection thresholds of 129 dB and 140 dB peak-to-peak (reference pressure 1 μPa at 1 m), respectively [30,31]. This setup is considered suitable for monitoring the presence of porpoises, as

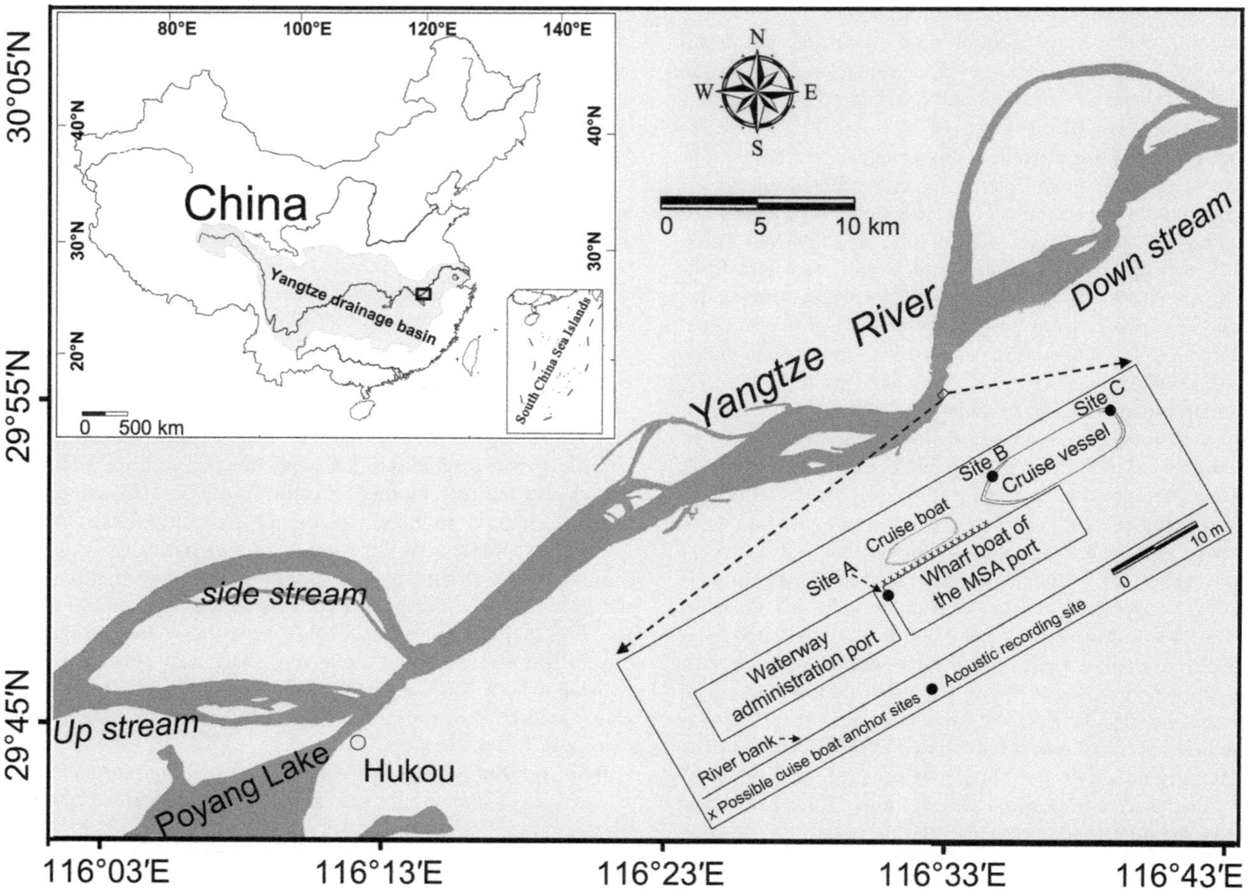


**Figure 1. Map of the static acoustic monitoring area.** The MSA port consisted of an anchored cruise vessel (27 m long and 5 m wide) and a rectangular wharf boat (39 m long and 9 m wide) connected to a waterway administration port of the same size. Three miniature stereo pulse event data loggers were deployed at two sites on the cruise vessel (sites A and B) and at one site on the wharf boat at the MSA port (site C). The possible cruise boat anchor sites represent possible sites for fish data monitoring. The inset map in the upper left corner shows the position of the study area in the middle region of the Yangtze drainage basin.

their apparent source levels are 163.7–185.6 dB peak-to-peak re 1 μPa [32]; the source levels of the 90° and 120° off-axis sonar signals were above the 160 dB peak-to-peak re 1 μPa [30,33]. Instead of recording the raw waveforms, the system registered ultrasonic pulse events. The peak sound pressure of each pulse, the absolute time of sound detection and the difference in arrival time (TD) of the same pulse at the two ultrasonic hydrophones were recorded up to 2000 (towed type) or 500 (fixed type) times per s. The dynamic range of the data logger was 129 dB to 160 dB peak-to-peak re 1 μPa with an internal thermal noise of approximately 134 dB, which was selected as the hardware detection threshold level. Once the primary hydrophone was triggered by a pulse sound that surpassed the preset hardware detection threshold level, the 10-bit (1024 count), high-speed counter would wait for the trigger at the other hydrophone and measure the TD between the stereo hydrophones at a resolution of 271 ns (towed type) or 1084 ns (fixed type). Each A-tag was equipped with a 128 MB flash memory card and a commercially available lithium battery (CR2) (towed type) or two UM-1 batteries (LR20) (fixed type). Depending on porpoise echolocation activity and background noise, this setup generally ensured autonomous monitoring operations for 40–60 h (towed type) or 1 month (fixed type) with a time drift of less than 1 s per day [29]. To ensure continuous recording, the three A-tags were never retrieved simultaneously for battery exchange. Theory and experimental evidence [15,16] suggest that the effective detection distance for porpoises is 300 m.

## Diel Classification

The 24 h of the day were divided into five diel phases (night1, morning, day, evening, night2) following the method proposed by Todd et al. (2009) [7]. The time points used to divide these diel phases were as follows: between night1 and morning = civil twilight start; between morning and day = 2*sunrise-civil twilight start; between day and evening = 2*sunset-civil twilight end; between evening and night2 = civil twilight end; between night1 and night2 = 12:00 pm. Sunrise and sunset referred to the times when the upper edge of the sun was on the horizon. Civil twilight started in the morning and ended in the evening when the center of the sun was geometrically 6° below the horizon. These data were obtained from the web site of the US Naval Observatory (http://aa.usno.navy.mil). The categories of night1 and night2 were combined to make up the rest of the diel phases of the night and reserved for further analysis.

## Data Analysis

During the off-line signal processing of the A-tag data, two-step procedures were applied using Igor Pro 5.01 software (Wave Metrics, Lake Oswego, OR, USA) to filter out the non-porpoise clicks and extract the porpoise sonar. A custom-made, multi-parameter filter program was used during the pilot step to classify the recorded pulses as originating from porpoises, boat traffic or other sources (for details, see [14]). Click trains were defined as sets of five clicks, and bouts consisting of fewer than five clicks were discarded [34,35]. During the second step, the number of false detections was further reduced by double-checking the detector output. We pooled all of the data from the three A-tags; if a signal was detected by more than one A-tag, only one was randomly selected to avoid over sampling. An echolocation encounter was defined as a series of click trains separated by eight min or less. Thus, if bouts of porpoise sonar were separated by more than eight min, they were considered to come from distinct echolocation encounters.

Because the A-tag hydrophone array was deployed parallel to the water current and the primary hydrophone was directed downstream, the recorded TD was assigned a value of zero if the sound came from a point midway between the two ultrasonic hydrophones (the passing point). In contrast, a positive or a negative value was recorded if the source was upstream or downstream (with the A-tag set as the reference point), respectively. The gradual changes in the TD of the sonar detected within each echolocation encounter indicated the underwater movement of the porpoise (acoustic trajectory). A trajectory lasting more than 20 s with at least three click trains was categorized into one of the following classes: upstream stay (staying upstream without a passing point, Fig. 2A), downstream stay (staying downstream without a passing point, Fig. 2B), straight up (moving from downstream to upstream with one passing point, Fig. 2C), straight down (moving from upstream to downstream with one passing point, Fig. 2D), milling and up (moving upstream after moving back and forth with at least two passing points, Fig. 2E) and milling and down (moving downstream after moving back and forth with at least two passing points, Fig. 2F). In some cases, more than one trajectory (multi-trajectories) was observed within an echolocation encounter, which facilitated porpoise counting [15]. Only the longest trajectory was counted if the multi-trajectories were indistinguishable at either the beginning and/or ending time during the analysis of trajectory duration.

For the analysis of fish resources, individual echo trace counting was performed. Individual echo traces with a target strength over −60 dB were counted, which corresponds to the approximate size of a fish with a swim bladder at 2 cm [36]. The finless porpoise in Hong Kong waters can feed on fish ranging from 5.5 cm to 58.6 cm [37], and the typical body length of the favorite prey of captive Yangtze finless porpoises is 7 cm (Wang CQ, personal communication), which is well above the selected target strength. Note that the beam pattern is quite narrow at 8° and transmits as a corn-shaped sound beam. We excluded echo traces shallower than 0.5 m because the beam width did not cover the entire 7 cm body length of the typical prey. Fish density was calculated as the number of individual echo traces in each echogram divided by the observed volume of the echosounder beam. During observation, the boat did not move to enable the clear echo traces of each fish to be obtained for a relatively longer duration than an observation from a moving vessel. We did not sort the echo traces by target strength because we did not measure the attack angle of the acoustic beam to the fish.

Porpoise acoustic detection data (click trains and buzzes), fish density in the port area and boat traffic (including upriver boats, downriver boats and the combination of both) for each day were first assigned to different diel phases (night1, morning, day, evening, night2 and night), and the data from each diel phase were further grouped into 10 min bins to account for the difference in sampling effort across diel phases. Echolocation encounter and trajectory were also assigned to different diel phases. During counting, echolocation encounters and trajectories that crossed boundaries between diel phases were assigned to the dominant phase that contained the greater portion. For the analysis of diel pattern, echolocation encounters and trajectories that spanned multiple days were segmented, and each segment was assigned an appropriate diel phase.

### Statistical Analysis

The Kolmogorov–Smirnov test and Levene's test for equality of variances were used to analyze the distribution of the data and the homogeneity of the variance, respectively. Nonparametric statistical analyses [38] were used because exploratory analysis showed that the data were not normally distributed (Kolmogorov-Smirnov test: $p<0.05$). The Kruskal–Wallis nonparametric test [38] was used to examine the overall difference across diel phases for all of the above-listed parameters. Duncan's multiple comparison test [38] was used for all post hoc comparisons to determine how these parameters varied with diel phase. Diel pattern analysis was conducted for 5 phases (night1, morning, day, evening and night2) and 4 phases (morning, day, evening and night). Spearman's rank order correlation [38] was employed to investigate the correlation among porpoise acoustics, boat traffic and fish density. For the correlation between porpoise acoustics and boat traffic, porpoise acoustics recorded simultaneously with boat traffic monitoring (from the $15^{th}$ to the $23^{rd}$ of June) were screened out and matched in the form of 10 min bins. As for the correlation between porpoise acoustics and fish density, the relevant data from the $19^{th}$ to the $23^{rd}$ of June were grouped and calculated for the 5 diel phase pattern. The Mann–Whitney *U*-test [38] was used to further analyze the presence or absence of clicks or buzzes with patterns of boat traffic. Statistical analyses were performed using Statistical Package for the Social Sciences (SPSS) 16.0 for Windows (SPSS Inc., Chicago, U.S.A.). Significance was accepted at $p<0.05$.

## Results

Porpoise sonar signals (both clicks and buzzes) were monitored every day (Fig. 3) for a total of 21,380 min of acoustic monitoring time; of that total, 1233 min (5.77% of the total recording time) contained porpoise sonar. A combined total of 3372 click trains were recorded, 190 (5.63%) of which were buzzes (Table 1). The average duration of the 168 echolocation encounters was 10.73±1.25 min (mean ± standard error, SE), with a maximum duration of 98.50 min (Table 1); 150 (89.3%) of these encounters involved one animal, 14 encounters involved two porpoises, and three encounters (1.8%) and one (0.6%) encounter involved three and four porpoises, respectively.

### Trajectory Types

Of the 18 multi-trajectory echolocation encounters, six involved two unambiguous multi-trajectories, resulting in a combined total of 174 measured trajectories (Table 2). One hundred and nine trajectories could be classified: nine (8.26%) and 14 (12.84%) were upstream stay and downstream stay, respectively; 30 (27.52%) and 35 (32.11%) were straight up and straight down, respectively; and 19 (17.43%) and two (1.83%) were milling and up and milling and down, respectively (Table 2).

Significant differences in the duration of the different trajectory types were observed (Kruskal-Wallis $\chi2=28.17$, df$=6$, $p<0.01$). Specifically, the duration of the milling and up trajectory was significantly longer than the straight down trajectory (Duncan's multiple-comparison test; $p<0.05$), and the upstream stay was

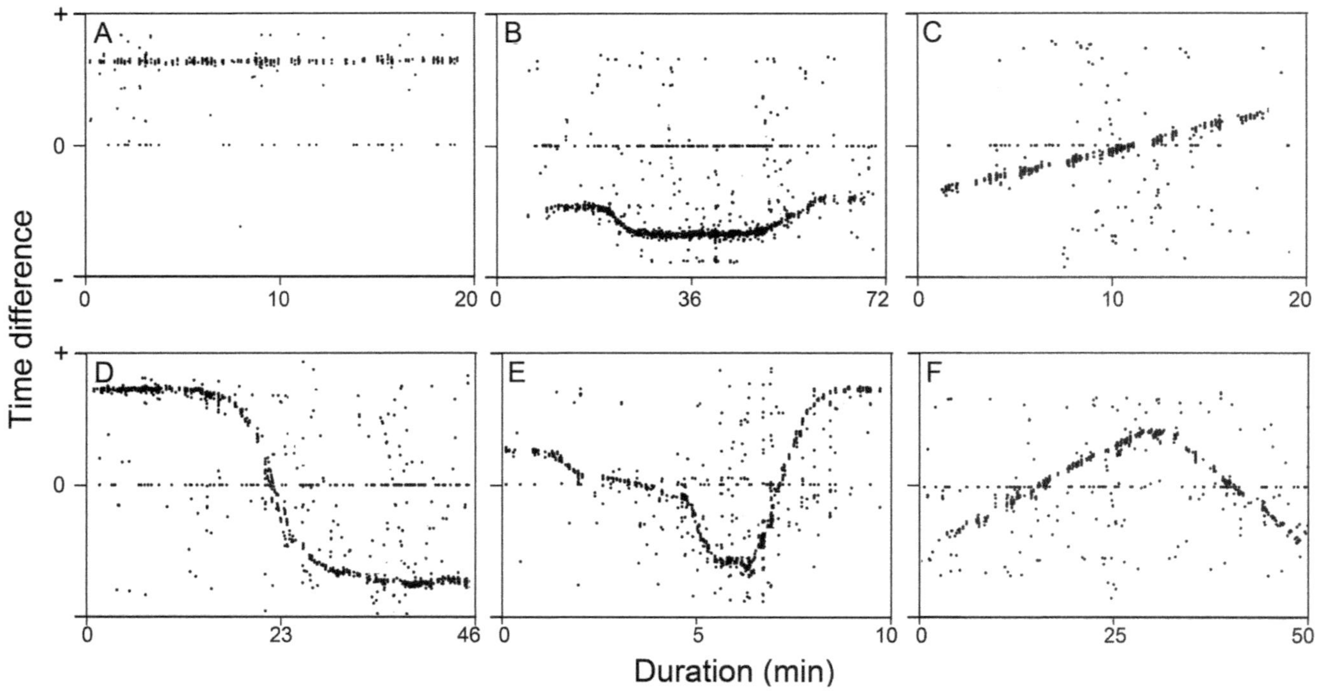


**Figure 2. Types of trajectory.** The positive and negative time differences indicate that the sound came from upstream or downstream, respectively. A value of zero denotes that the sound came from a point equidistant from the two ultrasonic hydrophones of the A-tag (the passing point). A = upstream stay, B = downstream stay, C = straight up, D = straight down, E = milling and up, F = milling and down.

significantly shorter than the milling and up (Duncan's multiple-comparison test; $p<0.05$) (Table 2).

## Diel Pattern of Biosonar Behavior and the Movement of Porpoises

Significant differences in the frequency of porpoise click trains (per 10 min) were observed among the diel patterns (5 phases: Kruskal-Wallis $\chi2 = 21.16$, df = 4, $p<0.01$; 4 phases: Kruskal-Wallis $\chi2 = 19.62$, df = 3, $p<0.01$). In particular, the frequency of click trains was significantly lower during the day (mean ± SE: 1.44±0.19 per 10 min) than during the evening (mean ± SE: 2.49±0.70 per 10 min), night2 (mean ± SE: 2.03±0.36 per 10 min) and night (mean ± SE: 1.69±0.21 per 10 min) (Duncan's multiple-comparison test; $p<0.05$) (Fig. 4A). The frequency of porpoise buzzes was significantly different among diel phases (5 phases: Kruskal-Wallis $\chi2 = 13.66$, df = 4, $p<0.01$; 4 phases: Kruskal-Wallis $\chi2 = 12.32$, df = 3, $p<0.01$). Buzz frequency was significantly lower during the day (mean ± SE: 0.06±0.02 per 10 min) than in the evening (mean ± SE: 0.23±0.10 per 10 min), night2 (mean ± SE: 0.17±0.06 per 10 min) and night (mean ± SE: 0.12±0.03 per 10 min) (Duncan's multiple-comparison test; $p<0.05$) (Fig. 4B).

No significant diel patterns were observed in either the frequency of echolocation encounters (5 phases: Kruskal-Wallis $\chi2 = 4.62$, df = 4, $p = 0.33$; 4 phases: Kruskal-Wallis $\chi2 = 5.39$, df = 3, $p = 0.15$) (Fig. 5) or the duration of the echolocation encounters (5 phases: Kruskal-Wallis $\chi2 = 3.03$, df = 4, $p = 0.55$; 4 phases: Kruskal-Wallis $\chi2 = 2.23$, df = 3, $p = 0.0.53$) (Table 3).

The stay trajectory types peaked in the morning and evening. The upstream stay, with a frequency of 0.24 counts per h (Fig. 6), accounted for 33% of the trajectories detected in the evening. The downstream stay, with a frequency of 0.16 counts per h (Fig. 6), accounted for 40% of the trajectories detected in the morning. The straight moving type also peaked in the morning and evening. The straight up type, with a frequency of 0.32 counts per h (Fig. 6), accounted for 44% of the trajectories identified in the evening, and the straight down type, with a frequency of 0.16 counts per h (Fig. 6), accounted for 40% of the identified trajectories in the morning. The milling trajectories were observed less often. The milling and up trajectory peaked in the evening and night2, with a frequency of 0.08 counts per h. The maximum occurrence rate of the milling and down trajectory was 0.01 counts per h, and no milling trajectories were observed in the morning (Fig. 6).

## Diel Pattern of Fish Density

Significant diel patterns in fish density were observed in the port area (5 phases: Kruskal-Wallis $\chi2 = 10.06$, df = 4, P = 0.04; 4 phases: Kruskal-Wallis $\chi2 = 10.16$, df = 3, $p = 0.02$). Specifically, fish densities in the morning (mean ± SE: 0.85±0.16 fish per $m^3$) and during the day (mean ± SE: 1.79±0.33) were significantly lower than fish densities in the evening (mean ± SE: 4.56±1.30), at night2 (mean ± SE: 4.27±0.18) and at night (mean ± SE: 4.58±0.55) (Duncan's multiple-comparison test; $p<0.05$) (Fig. 7).

## Diel Pattern of Boat Traffic

Significant diel patterns were observed in the frequency of boats passing upriver (5 phases: Kruskal-Wallis $\chi2 = 10.28$, df = 4, $p = 0.04$; 4 phases: Kruskal-Wallis $\chi2 = 10.28$, df = 3, $p = 0.02$) (Figs. 8 and 9A). In particular, the frequency was significantly lower in the evening (mean ± SE: 1.62±0.28 count per 10 min) than at night (mean ± SE: 2.51±0.09) (Duncan's multiple-comparison test; $p<0.05$) (Fig. 9A). The frequency of the passage of boats downriver was significantly different among diel phases (5 phases: Kruskal-Wallis $\chi2 = 133.59$, df = 4, $p<0.01$; 4 phases: Kruskal-Wallis $\chi2 = 122.56$, df = 3, $p<0.01$) (Figs. 8 and 9B). In particular, the frequency of boats passing downriver during the

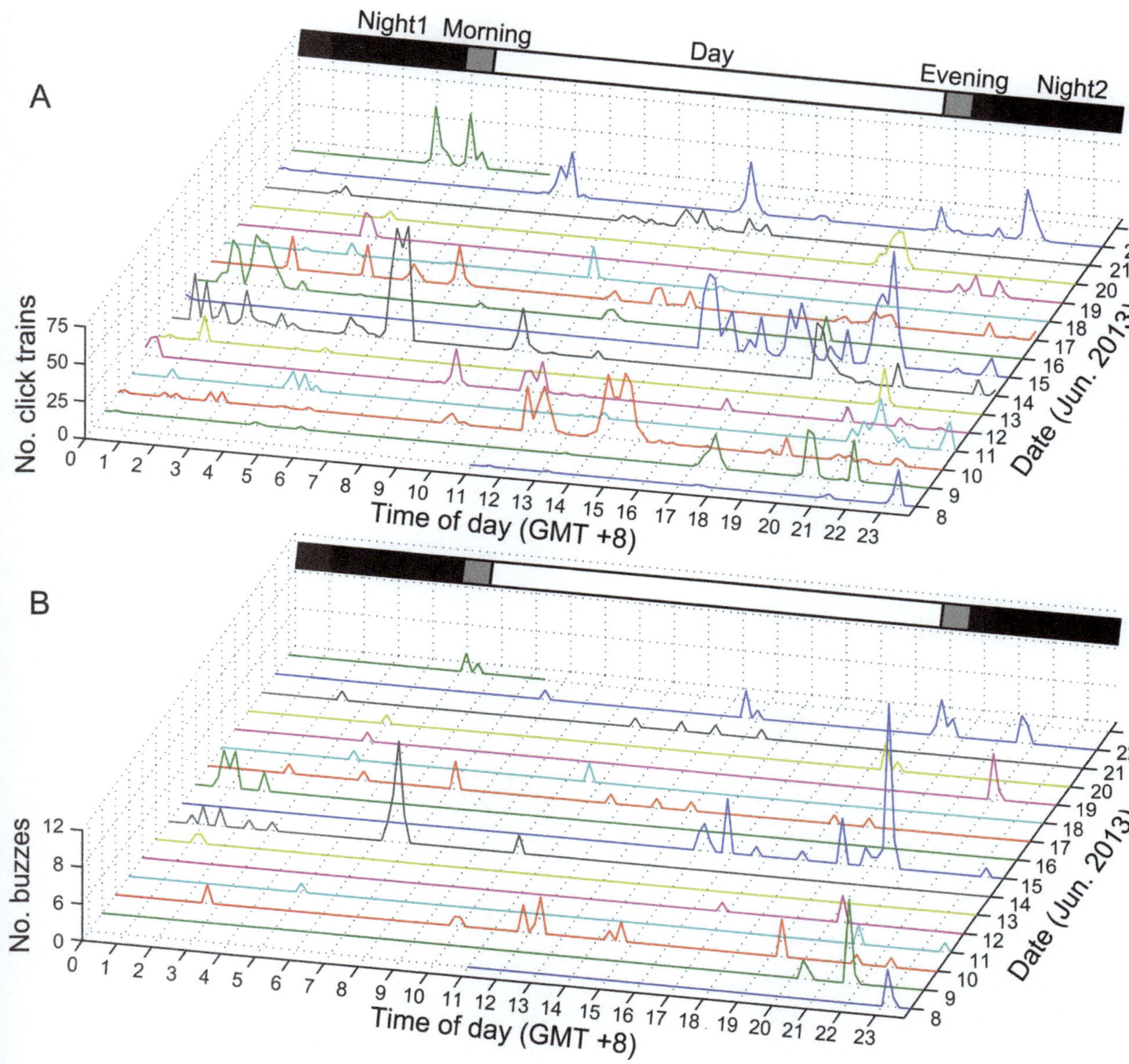


**Figure 3. The daily occurrence of porpoise click trains (A) and buzzes (B) by diel phase and time of day.** The data were grouped into 10 min bins. The horizontal bar at the top indicates the diel phase, whereas the bar at the bottom denotes the time of day (hour GMT +8).

day (mean ± SE: 1.70±0.09) was significantly lower than that of all of the other diel phases except the evening (mean ± SE: 2.31±0.26) (Duncan's multiple-comparison test; $p>0.05$) (Fig. 9B). Significant diel patterns were observed in the frequency of passage of all types of boats (5 phases: Kruskal-Wallis $\chi2=86.45$, df = 4, $p<0.01$; 4 phases: Kruskal-Wallis $\chi2=81.41$, df = 3, $p<0.01$) (Figs. 8 and 9C). In particular, the frequency of passage of all boats during the day (mean ± SE: 4.12±0.12) or in the evening (mean ± SE: 3.93±0.37) was significantly lower than that of the other diel phases (Duncan's multiple-comparison test; $p<0.05$); no significant differences in frequency were observed among these other diel phases or between the day and the evening (Duncan's multiple-comparison test; $p>0.05$) (Fig. 9C).

## Correlation among Porpoise Sonar, Fish Density and Boat Traffic

Twenty-two pairs of matched porpoise acoustic and fish density data were derived (including the 5 diel phases from the 19th to the 22nd plus night1 and the morning of the 23rd of June). Significant correlations were observed between fish density and the frequency of porpoise click trains (Spearman's rho = 0.771, $p=0.04$, n = 22) and between fish density and the frequency of porpoise buzzes (Spearman's rho = 0.829, $p=0.03$, n = 22) (Table 4).

A combined total of 682 simultaneously matched pairs of dolphin acoustic and boat traffic data (10 min bins) were obtained. The number of click trains was not significantly correlated with the passage of boats upriver, downriver or both (Spearman's rho = −0.016, $p=0.684$, n = 682, Spearman's rho = −0.045, $p=0.241$, n = 682 and Spearman's rho = 0.012, $p=0.755$, n = 682, respectively) (Table 4). Furthermore, no significant differences were observed in the presence or absence of click trains with the

**Table 1.** Descriptive statistics of porpoise acoustic detection (including minutes of clicks and number of click trains and buzzes) and echolocation encounters (including number, group size and duration) for each day.

| Date | Minutes of clicks | No. of click trains | No. of buzzes | Echolocation encounters | | | |
|---|---|---|---|---|---|---|---|
| | | | | N | Group size | Duration (min) | |
| | | | | | | mean±SE | Range |
| 6/8/2013 | 17 | 40 | 5 | 7 | 1 | 2.83±1.82 | 0.002–9.95 |
| 6/9/2013 | 58 | 155 | 13 | 9 | 1 | 10.00±4.79 | 0.001–38.51 |
| 6/10/2013 | 146 | 411 | 21 | 24 | 1–3 | 8.55±4.26 | 0.002–95.80 |
| 6/11/2013 | 96 | 187 | 4 | 10 | 1–2 | 13.02±7.22 | 0.002–76.67 |
| 6/12/2013 | 72 | 151 | 4 | 11 | 1 | 9.13±3.54 | 0.006–36.49 |
| 6/13/2013 | 24 | 55 | 2 | 5 | 1–3 | 7.55±4.34 | 0.007–23.38 |
| 6/14/2013 | 202 | 624 | 30 | 21 | 1–2 | 13.65±3.86 | 0.001–51.80 |
| 6/15/2013 | 178 | 654 | 47 | 10 | 1–4 | 25.78±7.75 | 1.437–61.23 |
| 6/16/2013 | 83 | 240 | 13 | 10 | 1–2 | 12.43±5.22 | 0.004–49.67 |
| 6/17/2013 | 86 | 201 | 10 | 17 | 1 | 6.65±1.60 | 0.001–23.56 |
| 6/18/2013 | 19 | 39 | 3 | 6 | 1 | 4.24±1.55 | 0.002–8.88 |
| 6/19/2013 | 30 | 69 | 7 | 4 | 1 | 14.13±3.39 | 8.083–20.65 |
| 6/20/2013 | 36 | 94 | 5 | 4 | 1–2 | 14.57±13.35 | 0.002–54.53 |
| 6/21/2013 | 56 | 94 | 5 | 13 | 1 | 6.67±2.32 | 0.008–24.69 |
| 6/22/2013 | 90 | 240 | 18 | 15 | 1 | 9.28±3.55 | 0.001–34.16 |
| 6/23/2013 | 40 | 118 | 3 | 2 | 1 | 35.30±1.95 | 33.35–37.26 |
| Mean±SE | 77.06±14.01 | 210.75±48.09 | 11.88±3.05 | 10.50±1.25 | | 10.73±1.25 | 0.001–95.80 |
| sum | 1233 | 3372 | 190 | 168 | | | |

The duration of echolocation encounters is given as the mean ± standard error (SE) and range.

**Table 2.** Descriptive statistics of the different trajectory types.

| | Upstream stay | Downstream stay | Straight up | Straight down | Milling and up | Milling and down | Unidentified | All |
|---|---|---|---|---|---|---|---|---|
| Mean±SE | 4.45±1.09 | 19.85±5.69 | 13.54±2.13 | 11.17±2.83 | 32.55±4.45 | 15.14±4.89 | 0.96±0.25 | 10.49±1.20 |
| Min–Max | 0.50–9.83 | 0.64–74.01 | 1.37–49.67 | 0.94–95.80 | 4.22–61.23 | 10.25–20.03 | 0.0001–8.68 | 0.0001–95.80 |
| N | 9 | 14 | 30 | 35 | 19 | 2 | 65 | 174 |

The duration of each type of trajectory (in min) is given as the mean ± standard error (SE) with minimum (Min) and maximum (Max) values.

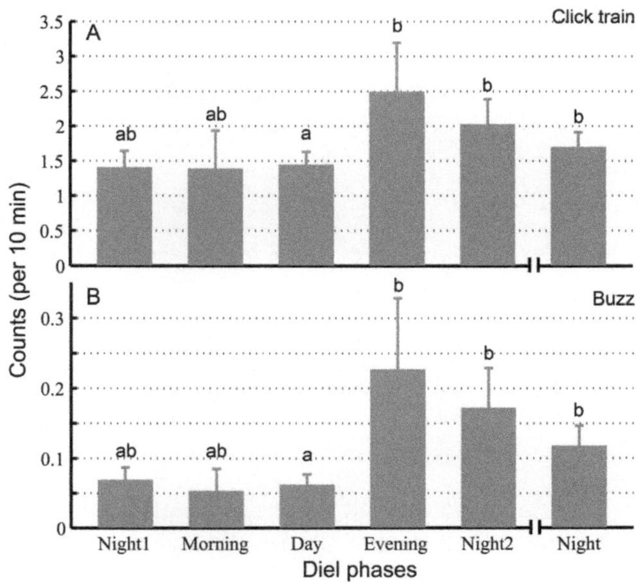


**Figure 4. Diel pattern of the frequency of click trains (A) and buzzes (B) per 10 min.** The data were combined and averaged across the entire acoustic monitoring period. Error bars (mean ± SE) with different lowercase letters refer to post hoc Duncan's multiple-comparison tests that yielded significant results ($p<0.05$).

frequency of the passage of boats upriver, downriver or both (Mann-Whitney *U*-test; $z=-1.455$, df = 682, $p=0.146$; Mann-Whitney *U*-test; $z=-0.188$, df = 682, $p=0.851$ and Mann-Whitney *U*-test; $z=-1.161$, df = 682, $p=0.245$, two-tailed, respectively). There was also no significant association between the number of porpoise buzzes and the passage of boats upriver (Spearman's rho = $-0.056$, $p=0.146$, n = 682), downriver (Spearman's rho = $-0.008$, $p=0.837$, n = 682) or both (Spearman's rho = $-0.044$, $p=0.251$, n = 682) (Table 4). No significant differences in the presence or absence of buzzes were observed with the frequency of the passage of boats upriver, downriver or both (Mann-Whitney *U*-test; $z=-0.342$, df = 682, $p=0.732$; Mann-Whitney *U*-test; $z=-1.145$, df = 682, $p=0.252$ and Mann-Whitney *U*-test; $z=-0.328$, df = 682, $p=0.743$, two-tailed, respectively).

## Discussion

Passive acoustic monitoring showed that Yangtze finless porpoises occupy port areas (Fig. 2) more routinely as described

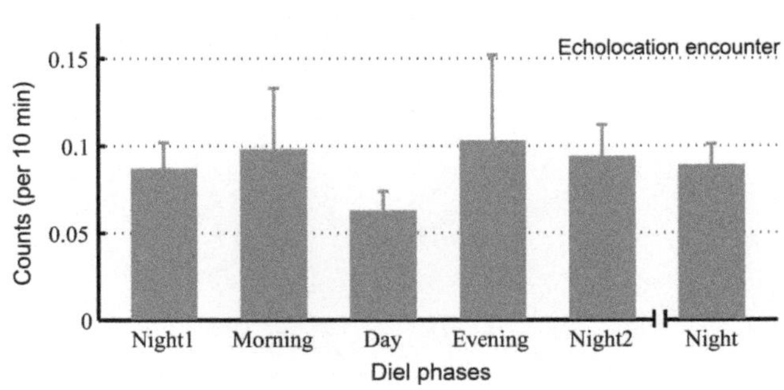


**Figure 5. Diel pattern of echolocation encounters.** The data were combined and averaged across the entire acoustic monitoring period.

**Table 3.** Descriptive statistics of the diel pattern of echolocation encounter duration.

| | Night1 | Morning | Day | Evening | Night2 | Night |
|---|---|---|---|---|---|---|
| Mean±SE | 9.25±2.08 | 12.53±5.36 | 10.87±2.03 | 10.53±3.67 | 10.77±2.62 | 10.02±1.67 |
| Min−Max | 0.001–49.67 | 0.01–39.64 | 0.01–95.80 | 0.001–41.80 | 0.001–76.67 | 0.001–76.67 |
| N | 37 | 8 | 76 | 11 | 38 | 75 |

The duration (in min) is given as the mean ± standard error (SE) with minimum (Min) and maximum (Max) values.

by Wang et al. (2014). The majority of the echolocation encounters involved single animals, indicating that porpoises tend to be solitary.

## Correlation between Porpoise Biosonar Behavior and Fish Density

Distinct diel patterns in porpoise acoustics were observed, with significantly more click trains and buzzes detected in the evening and at night than during the day. These findings suggest that porpoises are primarily engaged in crepuscular and nocturnal foraging in the port area. Considering that no significant diel patterns were observed in the frequency and duration of porpoise echolocation encounters, the diel pattern of porpoise acoustic behavior appears to be due to porpoises producing more click trains and buzzes at night [39] and possibly in the evening. These findings are consistent with findings from previous studies that show no reduced nocturnal activity among Yangtze finless porpoises in captivity [6]. The index of stomach contents of finless porpoises in Japan was also higher at night than during the day, with peak values observed at night [40].

Mounting evidence indicates that some marine mammals tend to feed at night. For example, Dall's porpoises (*Phocoenoides dalli*) are thought to forage mainly at night and/or at dawn, as greater amounts of stomach content were observed in animals captured in the morning [41]. Pantropical spotted dolphins (*Stenella attenuata*) have been shown to be primarily nocturnal and crepuscular feeders [42–44]. The feeding buzz rates and/or foraging-related behavior of wild harbor porpoises (*Phocoena phocoena*) [7,21,45], Cuvier's beaked whales (*Ziphius cavirostris*), Blainville's beaked whales (*Mesoplodon densirostris*) [46], Ganges River dolphins (*Platanista gangetica*) [34,47], Risso's dolphins (*Grampus griseus*) [48] and deep-diving echolocating odontocetes off of the Hawaiian islands [35] were higher at night than during the day. Free-ranging white-beaked dolphins (*Lagenorhynchus albirostris*) have been shown to forage only at night when monitored with bio-logging tags and passive acoustic methods [49].

In this study, porpoise biosonar behavior was significantly correlated with fish density, which supports previous findings that significantly more finless porpoises are encountered in areas with fish than in areas without fish [12]. Moreover, the higher fish densities observed in the evenings and at night in this study are consistent with more recent fixed echosounder detection findings that fish in the Yangtze River are more active (i.e., more fish are detected) and employ river bank-oriented movement at night [50].

The observation of no significant diel patterns in the frequency and duration of porpoise echolocation encounters indicates that the porpoises were always in the port area, even during the daytime, which highlights the importance of the port area.

## Diel Movement Patterns of Porpoises

Diel behavioral patterns have been correlated with spatio-temporal variations in food availability in common dolphins (*Delphinus capensis*) [51,52], pantropical spotted dolphins (*S. attenuata*) [42,43], spinner dolphins (*Stenella longirostris*) [53], dusky

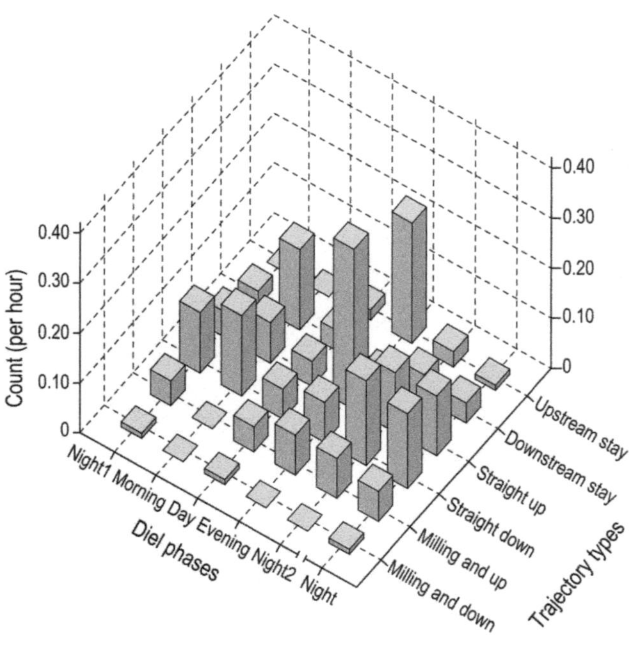


**Figure 6. Frequency of different types of trajectory by diel phase.** To calculate the frequency of occurrence of each trajectory type, all of the trajectories recorded in each diel phase were first classified by type and then divided by the total recording time for each diel phase.

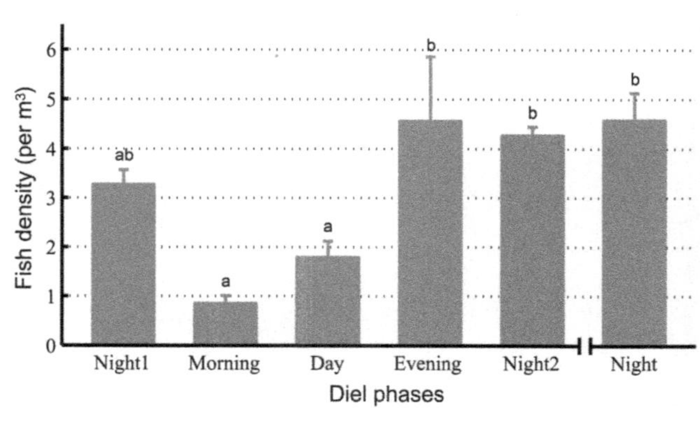


**Figure 7. Diel pattern of fish density in the port areas.** The data were combined and averaged across the entire echosounder monitoring period. Error bars (mean ± SE) with different lowercase letters refer to post hoc Duncan's multiple-comparison tests that yielded significant results ($p<0.05$).

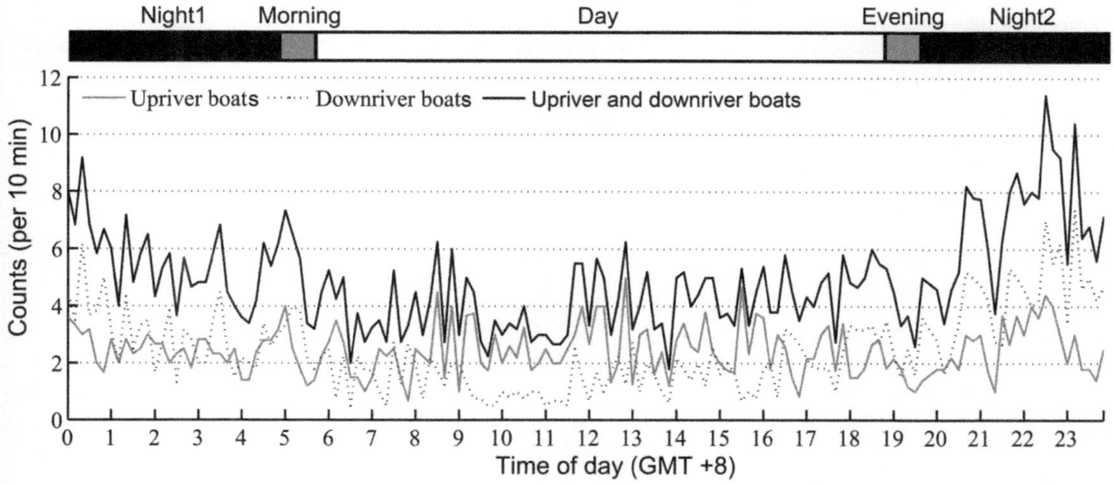


**Figure 8. Frequency of boat traffic (including upriver boats, downriver boats and the combination of upriver and downriver boats) by diel phase and time of day.** The data were pooled and averaged over the entire boat monitoring period. The horizontal bar at the top indicates the diel phase, and the bar at the bottom denotes the time of day (hour GMT +8).

dolphins (*Lagenorhynchus obscures*) [54], striped dolphins (*Stenella coeruleoalba*) [55], harbor porpoises (*P. phocoena*) [7], Risso's dolphins (*G. griseus*) [48], Heaviside's dolphins (*Cephalorhynchus heavisidii*) [56], Ganges River dolphins (*P. gangetica*) [34] and Hector's dolphins (*Cephalorhynchus hectori*) [57]. In this study, finless porpoises tended to migrate or stay downstream in the morning and then migrate or stay upstream in the evening. Milling and loop trajectories were possibly correlated with prey capture, and the lack of milling trajectories observed in the morning coincided with the lowest fish densities in the port. However, the milling trajectory peaked in the evening and night2, a phenomenon that was also observed in Ganges River dolphins [34], and the present study showed that the peak milling trajectory matched the peak fish density in the port. It is possible that these movements follow the local migration of prey, but further studies are needed to confirm this hypothesis.

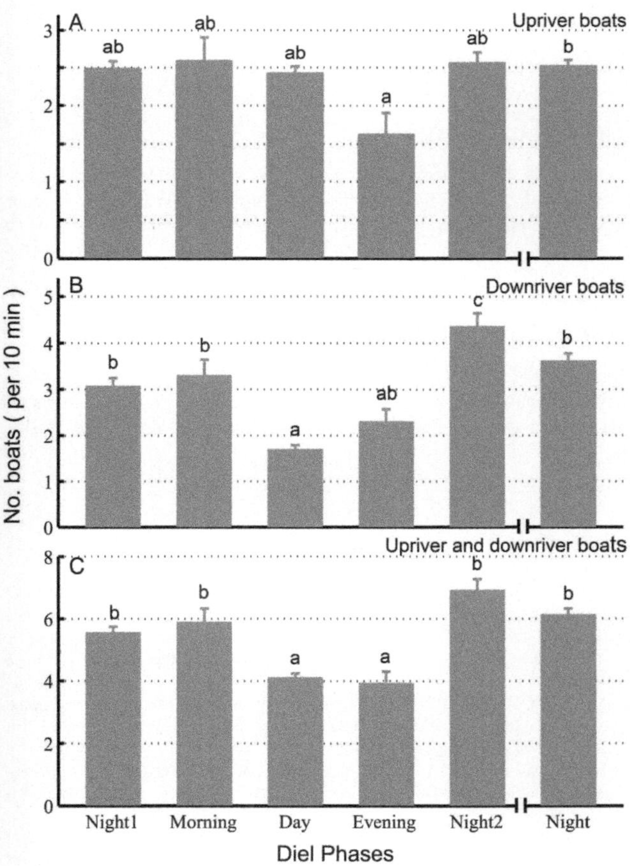


**Figure 9. Diel pattern of boat traffic (upriver boats in A, downriver boats in B, upriver and downriver boats in C).** The data were combined and averaged over the entire boat monitoring period. Error bars (mean ± SE) with different lowercase letters refer to post hoc Duncan's multiple-comparison tests that yielded significant results ($p<0.05$).

## Correlation between Porpoise Sonar and Boat Traffic

The boat traffic observed in this study primarily consisted of large sand carriers and very few fishing boats. This pattern may be due to the prohibition on fishing in the Yangtze River from the first of April through the 30th of June or to the nearby presence of the fishery law enforcement agency, which limits illegal fishing activities.

A broad range of organisms have been shown to exhibit diel foraging habits that are influenced by the trade-offs between foraging efficiency and predation risk [58]. In the Yangtze River system, however, finless porpoises are at the top of the food chain, which suggests that their diel foraging habits are not predicated on avoiding predators. The Yangtze River system has been severely affected by numerous types of anthropogenic activity, especially the high volume of boat traffic; it is likely that these activities are equivalent to predation risk. In this study, no significant correlation was observed between porpoise sonar and boat traffic, and no remarkable differences in boat traffic were observed in the presence or absence of porpoise acoustics (including click trains and buzzes). These findings suggest that prey availability in the port area are the primary factor that attracts porpoises. However, the lack of a significant correlation between porpoise acoustics and boat traffic does not rule out possible negative impacts of boat traffic on porpoise behavior. Indeed, indirect and direct evidence of finless porpoises avoiding passing vessels has been documented [13,14]. We should not exclude the effect of anthropological noise on the presence of porpoises. The lack of a significant correlation

**Table 4.** Correlation among dolphin acoustic, fish density and boat traffic.

| | | Fish density | Boat traffic | | |
|---|---|---|---|---|---|
| | | | Upriver boat | Downriver boat | All boats |
| Porpoise acoustic | Click train | * | n | n | n |
| | buzz | * | n | n | n |

*Represents significance at $p<0.05$; n represents not significant with $p>0.05$.

between porpoise acoustics and boat traffic may be due to a forced feeding schedule because the Yangtze River is significantly overfished and porpoises are most likely forced to tolerate disturbances from boats in the port areas to find prey.

Porpoise acoustics (click trains and buzzes) peak in the evening, suggesting that this is the preferred time for porpoise feeding. This finding is consistent with the plateau in fish density (Figs. 4,7 and 9).

### Limitations

The number of echolocation encounters is not necessarily equal to the number of individual porpoises. Interference resulting from the movement of porpoises in and out of the detection range of the instrument and periods of silence during which the porpoises were in the vicinity of the A-tags may have resulted in duplicate acoustic detections of the same animal. Additionally, non-vocal animals, even those within range of the equipment, were not detected [59], which resulted in a conservative estimate of the numbers of individual porpoises.

## Conclusions

Porpoise sonar was frequently detected in the port areas. Recordings tended to consist of feeding buzzes, and a majority of echolocation encounters involved single animals, indicating that solitary porpoises were frequently present and feeding in the port area. The distinct diel patterns of porpoise biosonar behavior showed that the click trains and buzzes occurred much more frequently in the evening and at night than during the day, which suggests that porpoises in the region were primarily engaged in crepuscular and nocturnal foraging. Porpoise sonar was significantly correlated with fish density but not with boat traffic; it is possible that overfishing in the Yangtze River has forced the porpoises to tolerate the disturbances caused by boats in the port areas to capture prey. Overall, prey availability appears to be the primary factor attracting porpoises, and the evening, when porpoise acoustics peak, seems to represent the preferred period for porpoise feeding. The plateau in fish density and the decrease in boat traffic in the evening further support this preference. Additionally, porpoises tend to stay or migrate downstream in the morning and stay or migrate upstream in the evening; it is possible that they are following prey.

No studies have examined possible seasonal (for example, changes due to the increased metabolic burden imposed by pregnancy), yearly or site-specific variations in patterns of porpoise activity. Longer-term and wider-ranging monitoring at more locations is needed to identify and mitigate potential threats to this species.

## Acknowledgments

We gratefully acknowledge the cooperation of the staff at the Pengze Maritime Safety Administration Bureau. Individual thanks are due to Wenjun Xu of the Central China Normal University for her assistance with the statistical analysis and Morisaka Tadamichi of Tokai University for his helpful critique of an earlier version of this manuscript.

## Author Contributions

Conceived and designed the experiments: ZW TA KW DW. Performed the experiments: ZW KW DW. Analyzed the data: ZW. Contributed reagents/materials/analysis tools: TA. Wrote the paper: ZW TA KW DW.

## References

1. Wang D, Turvey ST, Zhao X, Mei Z (2013) *Neophocaena asiaeorientalis ssp. asiaeorientalis*. In: IUCN 2013. IUCN Red List of Threatened Species. Version 2013.2. Available: http://www.iucnredlist.org. Accessed 2014 February 20.
2. Richard S (2010) Last stand on the Yangtze. Science 329: 378.
3. Zhao XJ, Barlow J, Taylor BL, Pitman RL, Wang KX, et al. (2008) Abundance and conservation status of the Yangtze finless porpoise in the Yangtze River, China. Biological Conservation 141: 3006–3018.
4. Wang D (2009) Population status, threats and conservation of the Yangtze finless porpoise. Chinese Science Bulletin 54: 3473–3484.
5. Turvey ST, Pitman RL, Taylor BL, Barlow J, Akamatsu T, et al. (2007) First human-caused extinction of a cetacean species? Biology Letters 3: 537–540.
6. Xiao JQ, Wang KX, Wang D (2005) Diurnal changes of behavior and respiration of Yangtze finless porpoises *(Neophocaena phocaenoides asiaeorientalis)* in captivity. Zoo Biology 24: 531–541.
7. Todd VLG, Pearse WD, Tregenza NC, Lepper PA, Todd IB (2009) Diel echolocation activity of harbour porpoises *(Phocoena phocoena)* around North Sea offshore gas installations. ICES Journal of Marine Science: Journal du Conseil 66: 734–745.
8. Au WWL, Hastings MC (2008) Principles of marine bioacoustics. New York: Springer Science.679 p.
9. Akamatsu T, Teilmann J, Miller LA, Tougaard J, Dietz R, et al. (2007) Comparison of echolocation behaviour between coastal and riverine porpoises. Deep Sea Research Part II: Topical studies in Oceanography 54: 290–297.
10. Zimmer WMX (2011) Passive Acoustic Monitoring of Cetaceans. United Kingdom: Cambridge University Press. 356 p.
11. Dong SY, Dong LJ, Li SH, Kimura S, Akamatsu T, et al. (2012) Effects of vessel traffic on the acoustic behavior of Yangtze finless porpoises *(Neophocaena phocaenoides asiaeorientalis)* in the confluence of Poyang lake and the Yangtze River using fixed passive acoustic observation methods. Acta Hydrobiologica Sinica 36: 246–254.
12. Kimura S, Akamatsu T, Li SH, Dong LJ, Wang KX, et al. (2012) Seasonal changes in the local distribution of Yangtze finless porpoises related to fish presence. Marine Mammal Science 28: 308–324.
13. Li SH, Akamatsu T, Wang D, Wang KX, Dong SY, et al. (2008) Indirect evidence of boat avoidance behavior of Yangtze finless porpoises. Bioacoustics 17: 174–176.
14. Wang ZT, Akamatsu T, Mei ZG, Dong LJ, Imaizumi T, et al. (2014) Frequent and prolonged nocturnal occupation of port areas by Yangtze finless porpoises *(Neophocaena asiaeorientalis)*: forced choice for feeding?. Integrative zoology 9: In press.
15. Kimura S, Akamatsu T, Wang KX, Wang D, Li SH, et al. (2009) Comparison of stationary acoustic monitoring and visual observation of finless porpoises. The Journal of the Acoustical Society of America 125: 547–553.
16. Akamatsu T, Wang D, Wang KX, Li SH, Dong SY, et al. (2008) Estimation of the detection probability for Yangtze finless porpoises *(Neophocaena phocaenoides asiaeorientalis)* with a passive acoustic method. The Journal of the Acoustical Society of America 123: 4403–4411.

17. Keenan SE, Benfield MC, Shaw RF (2003) Zooplanktivory by blue runner Caranx crysos: a potential energetic subsidy to Gulf of Mexico fish populations at petroleum platforms. In: Stanley DR, Scarboroughbull A, editors. Fisheries, Reefs, and Offshore Development. Bethesda, USA: American Fisheries Society. 167–180.
18. Fabi G, Grati F, Lucchetti A, Trovarelli L (2002) Evolution of the fish assemblage around a gas platform in the northern Adriatic Sea. ICES Journal of Marine Science 59: S309–S315.
19. Reubens JT, De Rijcke M, Degraer S, Vincx M (2014) Diel variation in feeding and movement patterns of juvenile Atlantic cod at offshore wind farms. Journal of Sea Research 85: 214–221.
20. Madsen PT, Johnson M, Aguilar de Soto N, Zimmer WMX, Tyack P (2005) Biosonar performance of foraging beaked whales (*Mesoplodon densirostris*). The Journal of Experimental Biology 208: 181–194.
21. Carlström J (2005) Diel variation in echolocation behavior of wild harbor porpoises. Marine Mammal Science 21: 1–12.
22. Miller PJ, Johnson MP, Tyack PL (2004) Sperm whale behaviour indicates the use of echolocation click buzzes 'creaks' in prey capture. Proceedings of the Royal Society of London Series B: Biological Sciences 271: 2239–2247.
23. Watwood SL, Miller PJ, Johnson M, Madsen PT, Tyack PL (2006) Deep-diving foraging behaviour of sperm whales (*Physeter macrocephalus*). Journal of Animal Ecology 75: 814–825.
24. Akamatsu T, Wang D, Wang KX, Li SH, Dong SY (2010) Scanning sonar of rolling porpoises during prey capture dives. The Journal of Experimental Biology 213: 146–152.
25. Green CB, Takahashi JS, Bass J (2008) The meter of metabolism. Cell 134: 728–742.
26. Yerushalmi S, Green RM (2009) Evidence for the adaptive significance of circadian rhythms. Ecology letters 12: 970–981.
27. Li SH, Dong SY, Kimura S, Akamatsu T, Wang KX, et al. (2010) Detection of Yangtze finless porpoise in the Poyang Lake mouth area via passive acoustic data-loggers. In: Ruiz-Garcia M, Shostell JM, editors. Biology, Evolution, and Conservation of River Dolphins within South America and Asia. Nova Science Pub Inc. 343–355.
28. Wei Z, Zhang X, Wang K, Zhao Q, Kuang X, et al. (2003) Habitat use and preliminary evaluation of the habitat status of the Yangtze finless porpoise (*Neophocaena phocaenoides asiaeorientalis*) in the Balijiang section of the Yangtze River, China. Acta Zoologica Sinica 49: 163–170.
29. Akamatsu T, Matsuda A, Suzuki S, Wang D, Wang KX, et al. (2005) New Stereo Acoustic Data Logger for Free-ranging Dolphins and Porpoises. Marine Technology Society Journal 39: 3–9.
30. Akamatsu T, Wang D, Wang KX (2005) Off-axis sonar beam pattern of free-ranging finless porpoises measured by a stereo pulse event data logger. The Journal of the Acoustical Society of America 117: 3325–3330.
31. Wang KX, Wang D, Akamatsu T, Li SH, Xiao JQ (2005) A passive acoustic monitoring method applied to observation and group size estimation of finless porpoises. the Journal of the Acoustical Society of America 118: 1180–1185.
32. Li SH, Wang D, Wang KX, Akamatsu T (2006) Sonar gain control in echolocating finless porpoises *(Neophocaena phocaenoides)* in an open water (L). The Journal of the Acoustical Society of America 120: 1803–1806.
33. Akamatsu T, Wang D, Wang KX, Naito Y (2000) A method for individual identification of echolocation signals in free-ranging finless porpoises carrying data loggers. the Journal of the Acoustical Society of America 108: 1353–1356.
34. Sasaki-Yamamoto Y, Akamatsu T, Ura T, Sugimatsu H, Kojima J, et al. (2013) Diel changes in the movement patterns of Ganges River dolphins monitored using stationed stereo acoustic data loggers. Marine Mammal Science 29: 589–605.
35. Au WWL, Giorli G, Chen J, Copeland A, Lammers M, et al. (2013) Nighttime foraging by deep diving echolocating odontocetes off the Hawaiian islands of Kauai and Ni'ihau as determined by passive acoustic monitors. The Journal of the Acoustical Society of America 133: 3119–3127.
36. Foote KG (1987) Fish target strengths for use in echo integrator surveys. The Journal of the Acoustical Society of America 82: 981–987.
37. Barros N, Jefferson TA, Parsons E (2002) Food habits of finless porpoises (*Neophocaena phocaenoides*) in Hong Kong waters. Raffles Bulletin of Zoology 50: 115–124.
38. Zar JH (1999) Biostatistical analysis. Upper Saddle River, NJ: Prentice-Hall. 929 p.
39. Kimura S, Akamatsu T, Wang D, Li SH, Wang KX, et al. (2013) Variation in the production rate of biosonar signals in freshwater porpoises. The Journal of the Acoustical Society of America 133: 3128–3134.
40. Shirakihara M, Seki K, Takemura A, Shirakihara K, Yoshida H, et al. (2008) Food habits of finless porpoises *neophocaena phocaenoides* in Western Kyushu, Japan. Journal of Mammalogy 89: 1248–1256.
41. Stroud R, Fiscus C, Kajimura H (1981) Food of the Pacific white-sided dolphin, Lagenorhynchus obliquidens, Dall's porpoise, *Phocoenoides dalli*, and northern fur seal, *Callorhinus ursinus*, off California and Washington. Fishery Bulletin US 78: 951–959.
42. Scott MD, Chivers SJ (2009) Movements and diving behavior of pelagic spotted dolphins. Marine Mammal Science 25: 137–160.
43. Baird RW, Ligon AD, Hooker SK, Gorgone AM (2001) Subsurface and nighttime behaviour of pantropical spotted dolphins in Hawai'i. Canadian Journal of Zoology 79: 988–996.
44. Robertson K, Chivers S (1997) Prey occurrence in pantropical spotted dolphins, *Stenella attenuata*, from the eastern tropical Pacific. Fishery Bulletin 95: 334–348.
45. Linnenschmidt M, Teilmann J, Akamatsu T, Dietz R, Miller LA (2013) Biosonar, dive, and foraging activity of satellite tracked harbor porpoises *(Phocoena phocoena)*. Marine Mammal Science 29: E77–E97.
46. Johnston D, McDonald M, Polovina J, Domokos R, Wiggins S, et al. (2008) Temporal patterns in the acoustic signals of beaked whales at Cross Seamount. Biology Letters 4: 208–211.
47. Pilleri G (1970) Observations on the behaviour of Platanista gangetica in the Indus and Brahmaputra rivers. Investigations on cetacea 2: 27–60.
48. Soldevilla MS, Wiggins SM, Hildebrand JA (2010) Spatial and temporal patterns of Risso's dolphin echolocation in the Southern California Bight. The Journal of the Acoustical Society of America 127: 124–132.
49. Rasmussen MH, Akamatsu T, Teilmann J, Vikingsson G, Miller LA (2013) Biosonar, diving and movements of two tagged white-beaked dolphin in Icelandic waters. Deep Sea Research Part II: Topical Studies in Oceanography 88–89: 97–105.
50. Lin PC, Gao X, Zhu QG, Wang T, Liu HZ (2013) Hydroacoustic survey on the spatial distribution pattern and day-night rhythmic behaviour of fishes in the Xiaonanhai reach of the upper Yangtze River. Journal of Applied Ichthyology 29: 1402–1407.
51. Evans WE (1974) Radio-telemetric studies of two species of small odontocete cetaceans In: Schevill W, editor. The whale problem: A status report. Cambridge, MA.: Harvard University Press. 385–394.
52. Henderson E, Hildebrand J, Smith M, Falcone E (2012) The behavioral context of common dolphin (*Delphinus sp.*) vocalizations. Marine Mammal Science 28: 439–460.
53. Benoit-Bird KJ, Au WWL (2003) Prey dynamics affect foraging by a pelagic predator (*Stenella longirostris*) over a range of spatial and temporal scales. Behavioral Ecology and Sociobiology 53: 364–373.
54. Benoit-Bird KJ, B.Wursig, Mcfadden CJ (2004) Dusky dolphin *(Lagenorhynchus obscurus)* foraging in two different habitats: Active acoustic detection of dolphins and their prey. Marine Mammal Science 20: 215–231.
55. Gannier A (1999) Diel variations of the striped dolphin distribution off the French Riviera (northwestern Mediterranean Sea). Aquatic Mammals 25: 123–134.
56. Leeney RH, Carslake D, Elwen SH (2011) Using static acoustic monitoring to describe echolocation behaviour of Heaviside's dolphins *(Cephalorhynchus heavisidii)* in Namibia. Aquatic Mammals 37: 151–160.
57. Dawson S, Fletcher D, Slooten E (2013) Habitat use and conservation of an Endangered dolphin. Endangered Species Research 21: 45–54.
58. Payne NL, van der Meulen DE, Gannon R, Semmens JM, Suthers IM, et al. (2013) Rain reverses diel activity rhythms in an estuarine teleost. Proceedings of the Royal Society B: Biological Sciences 280: 20122363.
59. Mackenzie DI, Nichols JD, Royle JA, Pollock KH, Bailey LL, et al. (2006) Occupancy estimation and modeling: inferring patterns and dynamics of species occurrence. San Diego, CA: Academic. 344 p.

# 11

# The Substantial First Impact of Bottom Fishing on Rare Biodiversity Hotspots: A Dilemma for Evidence-Based Conservation

**Robert Cook[1,5], Jose M. Fariñas-Franco[5], Fiona R. Gell[2], Rohan H. F. Holt[3], Terry Holt[4], Charles Lindenbaum[3], Joanne S. Porter[5], Ray Seed[1], Lucie R. Skates[3], Thomas B. Stringell[3], William G. Sanderson[3,5]***

**1** School of Ocean Sciences, Bangor University, Menai Bridge, Anglesey, United Kingdom, **2** Fisheries Directorate, Department of Environment Food and Agriculture, Isle of Man Government, St John's, Isle of Man, United Kingdom, **3** Natural Resources Wales, Bangor, Gwynedd, United Kingdom, **4** Center for Marine and Coastal Studies Ltd. (CMACS), Port Erin, Isle of Man, United Kingdom, **5** School of Life Sciences, Heriot-Watt University, Edinburgh, United Kingdom

## Abstract

This study describes the impact of the first passage of two types of bottom-towed fishing gear on rare protected shellfish-reefs formed by the horse mussel *Modiolus modiolus* (L.). One of the study sites was trawled and the other was scallop-dredged. Divers collected HD video imagery of epifauna from quadrats at the two study sites and directed infaunal samples from one site. The total number of epifaunal organisms was significantly reduced following a single pass of a trawl (90%) or scallop dredge (59%), as was the diversity of the associated community and the total number of *M. modiolus* at the trawled site. At both sites declines in anthozoans, hydrozoans, bivalves, echinoderms and ascidians accounted for most of the change. A year later, no recovery was evident at the trawled site and significantly fewer infaunal taxa (polychaetes, malacostracans, bivalves and ophuroids) were recorded in the trawl track. The severity of the two types of impact reflected the undisturbed status of the habitats compared to previous studies. As a 'priority habitat' the nature of the impacts described on *M. modiolus* communities are important to the development of conservation management policy and indicators of condition in Marine Protected Areas (EU Habitats Directive) as well as indicators of 'Good Environmental Status' under the European Union Marine Strategy Framework Directive. Conservation managers are under pressure to support decisions with good quality evidence. Elsewhere, indirect studies have shown declines of *M. modiolus* biogenic communities in fishing grounds. However, given the protected status of the rare habitat, premeditated demonstration of direct impact is unethical or illegal in Marine Protected Areas. This study therefore provides a unique opportunity to investigate the impact from fishing gear whilst at the same time reflecting on the dilemma of evidence-based conservation management.

**Editor:** Simon Thrush, National Institute of Water & Atmospheric Research, New Zealand

**Funding:** This project was resourced under a reciprocal agreement between the Countryside Council for Wales and the Isle of Man Government. Additional work was completed under the Joint Nature Conservation Committee (JNCC) contract C12-0233-0583 awarded to Heriot Watt University. The funders had no role in study design, data collection and analysis, decision to publish, or preparation of the manuscript.

**Competing Interests:** The authors have the following interests: Terry Holt is employed by CMACS Ltd. There are no patents, products in development or marketed products to declare. This does not alter the authors' adherence to all the PLOS ONE policies on sharing data and materials, as detailed online in the guide for authors.

* E-mail: w.g.sanderson@hw.ac.uk

## Introduction

Shellfish reefs are "one of, if not the most imperilled marine habitats on earth" [1]. The loss of 85% of the world's oyster reefs can be estimated from fisheries and other records [2], but declines in non-target shellfish reefs are harder to quantify. Horse mussels (*Modiolus modiolus* Linnaeus) are a non-target species in much of their range and widespread in the northern Atlantic and Pacific Oceans. Dense beds of *M. modiolus* are biogenic reefs [3] and have a more limited known distribution in the White Sea, Bay of Fundy, the Irish Sea, Scotland, Scandinavia and Iceland [4]–□□□[8]. In common with many other types of biogenic shellfish reefs, those formed by *M. modiolus* are known to be threatened and declining [9].

In open-coast locations with moderate to high tidal flow, *M. modiolus* reefs can form long-lived structures up to 3m above the surrounding seabed [4], [10]–□[12]. These habitats create high levels of physical complexity where clumps of dense *M. modiolus* provide substrata for an epifaunal community whilst the spaces between mussels accumulate sediment which supports a rich crevice and infauna of 200–300 species, at densities exceeding 22,000 individuals $m^{-2}$ [13], [14]. Tide-swept horse mussel reefs have therefore been identified as rare biodiversity hotspots compared to surrounding habitats, and networks of Marine Protected Areas (MPAs) are under development to support these and other habitats through international and national legislation (EC Habitats and Species Directive; Marine (Scotland) Act 2010; see also [12]). The maintenance of these so called 'Priority Habitats' [15] will also contribute to the achievement of 'Good Environmental Status' (GES) under the European Union (EU) Marine Strategy Framework Directive (MSFD; 2008/56/EC) such that indicators of their status are under consideration for development [15].

In the 21st century significant concerns have been raised that conservation practice is sometimes based upon anecdote and myth rather than systematic appraisal of evidence [16]. Systematic review has therefore been proposed to support "evidence-based conservation" [16], [17], and sources of evidence used by conservation managers have been increasingly scrutinised [18]. At the same time the impacts of fisheries on the seabed, and by implication the necessity to manage them, have been challenged by stakeholders on the basis of a lack of scientific evidence (see rebuttals by [19]–[][21]). Nevertheless, proponents of evidence-based conservation accept that management decisions still need to be made in the absence of good quality evidence [17] and in a marine context some have called for a "reversal of the burden of proof" in management decisions [22].

*M. modiolus* reefs and their associated communities have been found to decline in areas subjected to bottom-towed fishing gear [23]–[][25], as have oyster reefs [26]. Furthermore, where *M. modiolus* has been targeted as bait for cod, it has declined and not recovered [27]. The direct effects of bottom-towed fishing gear on sparse *M. modiolus* individuals has been shown [28] but where it occurs in high densities and forms reefs the direct impact of an individual pass of fishing gear has not been described. Indeed, the majority of direct impact studies of trawling and scallop dredging are from soft sediment and gravel communities [21], [29]–[][31] while those on complex, temperate biogenic habitats are rare [32].

Habitat rarity can prohibit elegant experimental approaches to support sensitive management [33], but providing the impact evidence - base may also be unethical or illegal if it is necessary to willingly damage a habitat or species in a protected area. In the present study, benthic marks attributed to the single passage of two types of bottom fishing gear were identified during routine monitoring operations on *M. modiolus* reefs. This provided a unique opportunity to investigate, directly, the scale of the epifaunal and infaunal impact under a null model. The study also provided an opportunity to reflect on the consequences of the absence of this kind of information to the conservation manager.

## Materials and Methods

No permits were required for the described study, which complied with all relevant regulations. Protected habitats were sampled in full consultation and collaboration with statutory conservation authorities (Countryside Council for Wales and Isle of Man Government).

### Site information

Previous survey data and side scan sonar outputs [34] were used to establish a study site in an extensive area of *Modiolus modiolus* bed 4.5 km off the Point of Ayre (Isle of Man; 54°26′.20N 004°18′.18W; Figure 1 & 2A). A steel marker with a hydroacoustic beacon (Sonardyne, Yateley, Hampshire, UK) was used for relocation and a corresponding hand-held hydroacoustic relocation device was used in conjunction with a compass to map the site to within 0.1 m accuracy (Figure 2A).

Side scan sonar imagery (Figure 2C) was used to identify an impact study site in June 2012 on another *M. modiolus* reef 5 km north of the Lleyn Peninsula (North Wales; 52 56′.99N 004°38′.56W; Figure 1 & 2C). Scallop dredging vessels had been recorded in the area during the preceding season (November 2011–April 2012) and the marks had not been recorded in all previous annual side scan sonar surveys.

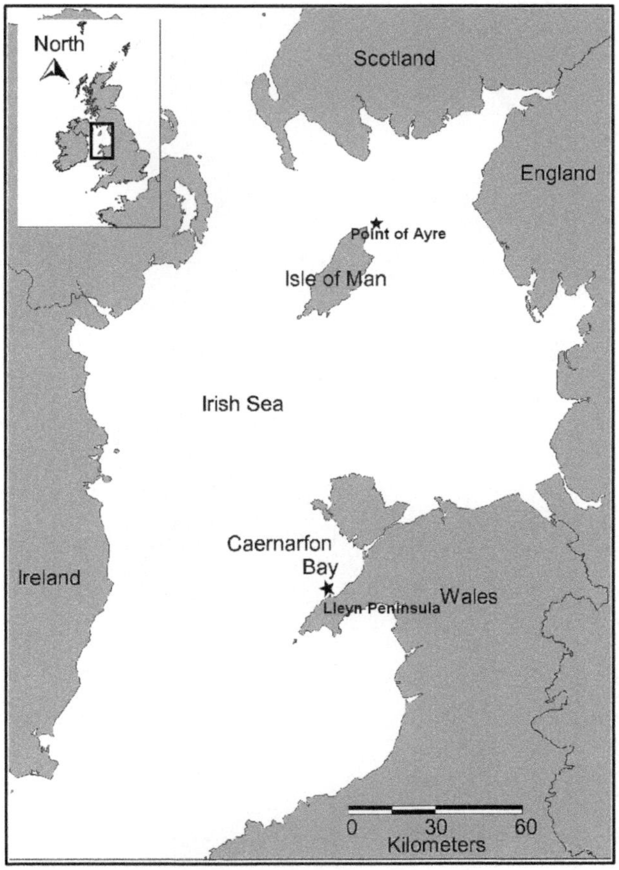


**Figure 1. Study sites.** Stars indicate *Modiolus modiolus* bed study sites north of the Point of Ayre (Isle of Man) and north of the Lleyn Peninsula in Caernarfon Bay (Wales).

The Point of Ayre (PoA) and north Lleyn Peninsula (nLP) sites both contained raised reef structures (1 m+) and high densities of *M. modiolus* (>350 $m^{-2}$ see (9), [13] and present study). PoA and nLP were 33 and 30 m below chart datum with peak tidal flows of 1 $ms^{-1}$ and 1.25 $ms^{-1}$ respectively and both were fully saline [35], [36].

### Records and samples

Divers systematically filmed the 25 cells that made up 0.25 $m^2$ quadrats at close range (<0.5 m) using high-definition handheld colour video-cameras (quadrats were then removed from the site).

At the PoA quadrat records were made between August and September in 2007, 2008 and 2009 at 12 positions relocated using fixed plastic pins on top of ridges of *M. modiolus*. In 2008 notification of the survey and a position was given to local shipping and fishing organisations in a Notice to Mariners a week before the survey. During the subsequent 2008 survey, 6 out of the 12 original quadrat positions were found to be impacted by a pair of clearly visible (to the diver) parallel furrows and a 'swept' area between that was tangential to the ridges of the natural bedform (Figure 2A & B) and consistent in size and orientation with the passage of an otter trawl. It is therefore likely that this occurred in response to the pre-survey information released.

Recording was conducted in a similar way at nLP in July 2012, except that quadrats were randomly placed in areas with conspicuous dredge marks and adjacent un-dredged areas.

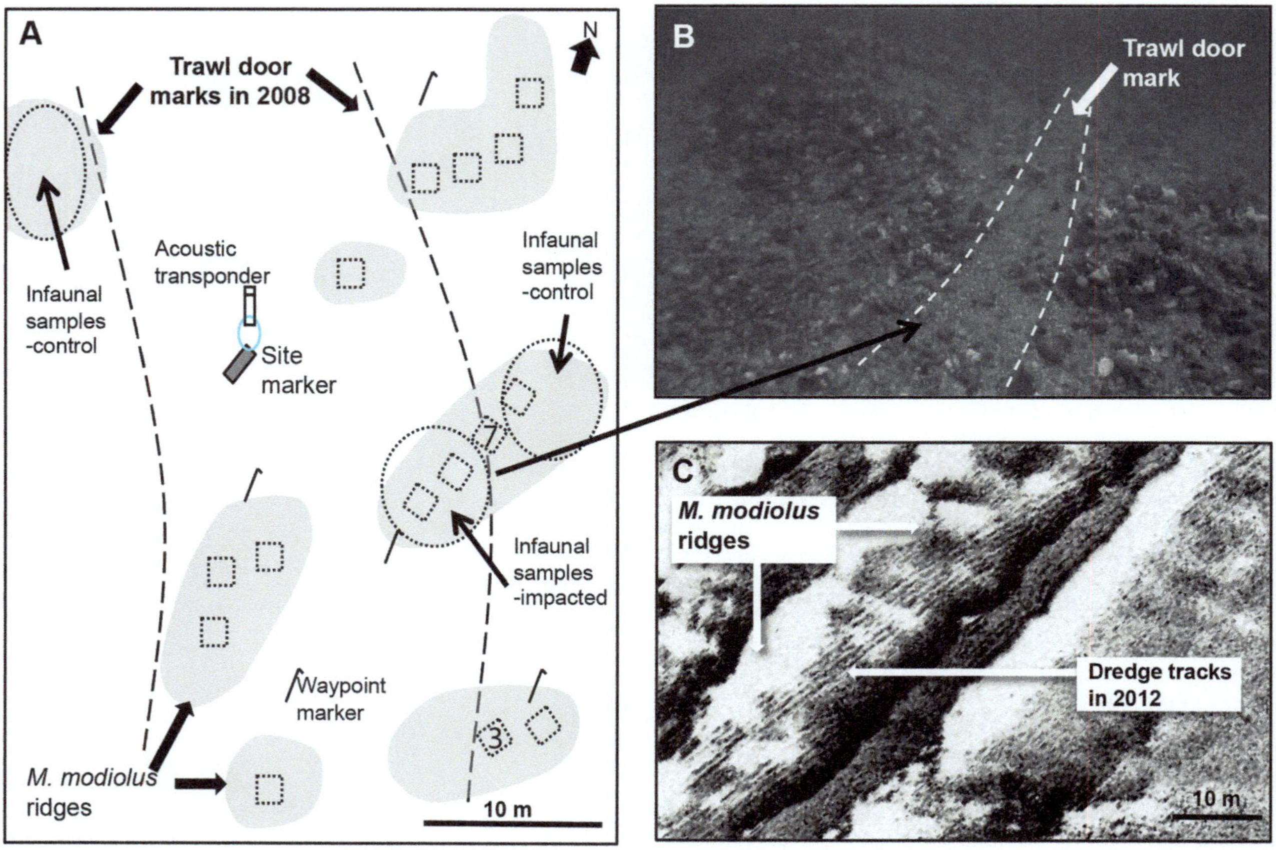


**Figure 2. Details of study sites.** (A) Map of fixed quadrat locations (dotted squares) on raised ridges (grey polygons) at Point of Ayre study site. Dotted ellipses indicate infaunal sample areas for impacted and control treatments. Two trawl door marks in 2008 are indicated by dashed lines. One trawl door mark in (A) is visible in the video-grab image (B) where the more extreme impact (compared to the net) in the path of the trawl door is also illustrated with dashed lines. The numbers "7" and "3" in (A) are quadrat numbers refered to in Figure 3 and Discussion (respectively). Metal waypoint pins enabled navigation around the site. (C) Side scan sonar image from 2012 at the study site off the north of the Lleyn Peninsula: marks from two gangs of scallop dredges are visible across the surface of the *Modiolus modiolus* ridges.

Frame grabs of each of the quadrat sub-cells were stitched together to create a high-resolution mosaic of the benthic community under each quadrat, from which conspicuous species were enumerated (Figure 3). Fauna were recorded to the highest taxonomic resolution possible and recording rules (file S1) were applied to reduce variability from cryptic organisms.

At PoA four random 0.0625 $m^2$ infaunal samples were taken in 2009 from each of three *M. modiolus* ridge locations: Two outside of the marks recorded in 2008 where there was no evidence of trawl damage (Figure 2A: "Control") and one where a ridge was found damaged in 2008 (Figure 2A: "Impacted"). Divers sampled to 20 cm depth and recovered material into 0.5 mm drawstring mesh bags. Samples were preserved in 5% buffered formaldehyde and sieved on a 0.5 mm mesh. Infaunal samples were sorted separately and all fauna identified to a coarse level of taxonomic resolution (Class level) sufficient to detect impacts [37]. Porifera, Hydrozoa, Anthozoa and Bryozoa were not used in subsequent analysis because they were better represented in the video analysis.

## Data treatment and statistical analysis

Multivariate analyses were conducted on Bray-Curtis similarity coefficients of square root transformed species abundance data, using PRIMER v6 with PERMANOVA+ software [38], [39]. Non-metric multidimensional scaling (MDS) was applied to Bray Curtis similarities using the Kruskal fit scheme [38] and, in the case of epifaunal data from PoA, a dummy variable was used to stabilise dispersion of sparse data [38]. Variation between impacted and unimpacted quadrats at the two sites were tested as fixed effects in one-way (nLP) and mixed two-way designs with year as random factor (PoA) using Permutational Multivariate Analysis of Variance (PERMANOVA) based on 9999 permutations and Type III sums of squares (SS). Type III SS is the most conservative SS method for PERMANOVA, fitting every term simultaneously and ensuring independence of all factors in unbalanced designs [39]. Within-site correlation differences through time in the PoA site were tested using PERMDISP (permutation of dispersion [39]). Taxa contributing to dissimilarities between treatments were investigated using a Similarity Permutation procedure (SIMPER; [38]).

Number of individuals (*N*), Shannon-Wiener's diversity (H′), Margalef's richness (d) and Pielou evenness (J) were imported into R (version 2.13.1, [41]) and tested for normality and heteroscedasticity. Effects of physical impact on diversity and evenness indices from quadrat records at both sites were tested ($\alpha$ of 0.05) by fitting linear mixed effects models (LMMs: lme4 package; [41]) with individual quadrats (both sites) and sampling year (PoA site) as random factors to account for spatial and temporal pseudoreplication. Impact (impacted vs non-impacted) was the categorical predictor (fixed factor) in the mixed model. Generalized LMMs with Poisson error distribution and logit link function were fitted to

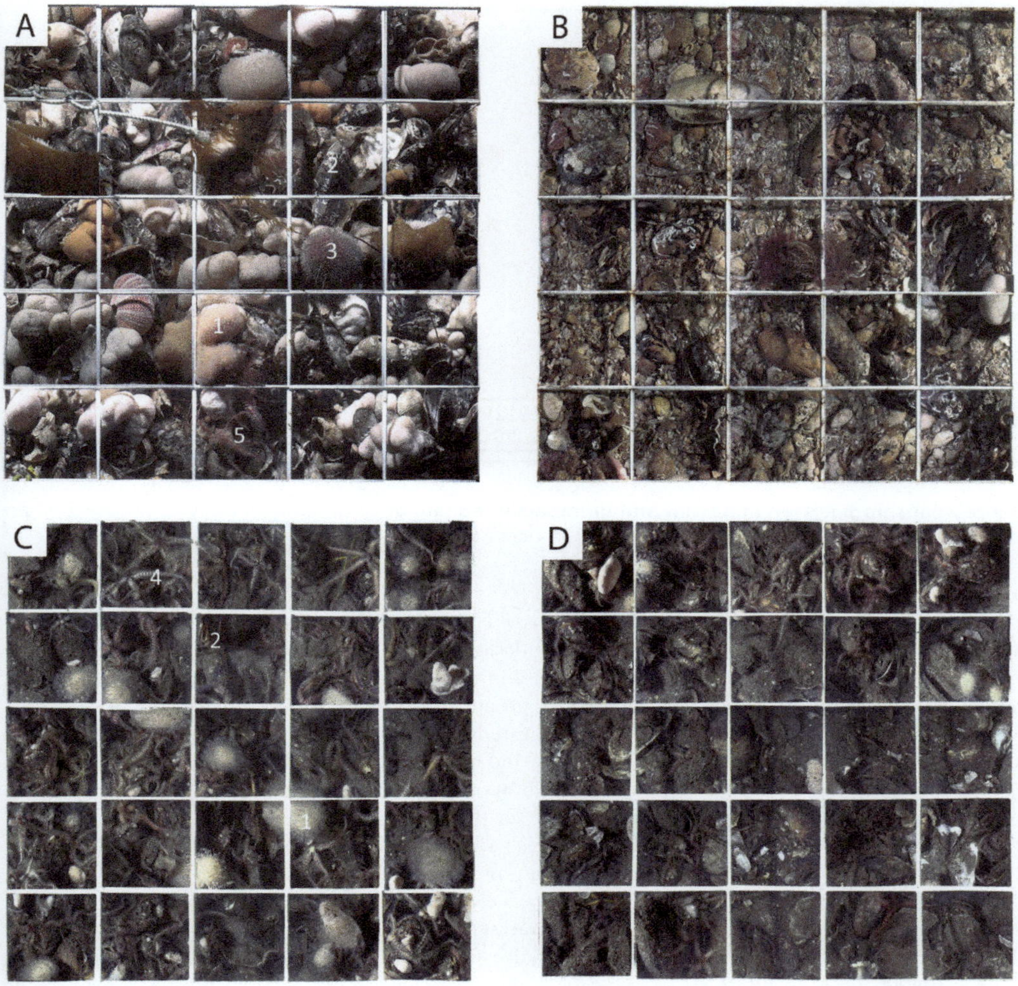


**Figure 3. Mosaic quadrat images of quadrats.** Quadrat 7 (indicated in Figure 2) from Point of Ayre in 2007 (A) and 2009 (B). (C) Unimpacted quadrat and (D) impacted quadrat from N. Lleyn Peninsula in 2012. Numbers indicate conspicuous epifauna: 1 *Alcyonium digitatum*, 2 *Modiolus modiolus*, 3 *Echinus esculentus*, 4 *Ophiothrix fragilis*, 5 *Antedon bifida*.

the abundance data (N; *M. modiolus* and epifauna) incorporating the same fixed and random factors as the LMMs [41] to cope with non-normal data in unbalanced, mixed-effect experiments [42]. Overdispersed Poisson models were refitted using Penalized Quasi Likelihood approximations (glmmPQL: MASS package [40]). The Akaike Information Criterion (AIC) was used to assess the effect of the physical impact on the null model for PoA and nLP while controlling for the random effects. Model selection was based on the lowest AIC score (Table S1). Infaunal count data from PoA cores conformed to the parametric assumptions and were therefore tested against impact treatments using standard one-way ANOVAs. All models were tested using residual plots to confirm that the assumptions of normality and sphericity of the residuals were met.

## Results

In total 29 different taxa were recorded in video quadrats at the two study sites. At both sites there were significant impact effects on community composition (pseudo $F = 24.37$, $p = 0.0001$; pseudo $F = 2.86$, $p = 0.03$ for PoA and nLP respectively). There was also significant variability among years in the structure of the community at PoA (pseudo $F = 2.52$, $p = 0.005$). PERMDISP analysis indicated significant larger dispersion across time in epifaunal community samples following impact (deviations from centroid: $F_{(1,36)} = 12.07$; $p<0.01$). However, individual pairwise tests at PoA showed significant difference in dispersion occurred only after the trawling event in 2008 (2007 and 2008: $t = 4.99$; 2007 and 2009: $t = 5.57$; $p<0.001$) with no significant within site differences between 2008 and 2009 ($t = 0.56$; $p = 0.69$).The average dissimilarity between impact treatments at PoA site was high (85%) in the SIMPER analysis and driven by reductions in all but one (Paguridae) of the taxa in the impacted quadrat records. More than 90% of the average differences between unimpacted and impacted quadrats were accounted for by reductions in *Alcyonium digitatum* (L.), Actinaria, *Antedon bifida* (Pennant), Hydrozoa and *Modiolus modiolus* (SIMPER). At nLP the impact was less pronounced with 31.3% average dissimilarity between impacted and unimpacted treatments and reductions in the abundance of *Modiolus modiolus*, *Alcyonium digitatum*, *Ophiothrix fragilis* (Abildgaard), *Ascidiella* sp., *Flustra foliacea* (L.), *Pyura* sp. and Anomiidae accounting for 57% of the dissimilarity between treatments (SIMPER). Some encrusting and low-lying taxa at nLP were more abundant in records from impacted quadrats because upright emergent epifauna had reduced and revealed them (e.g. increased *Crisia eburnea* (L.) contributed 5.5% to dissimilarity).

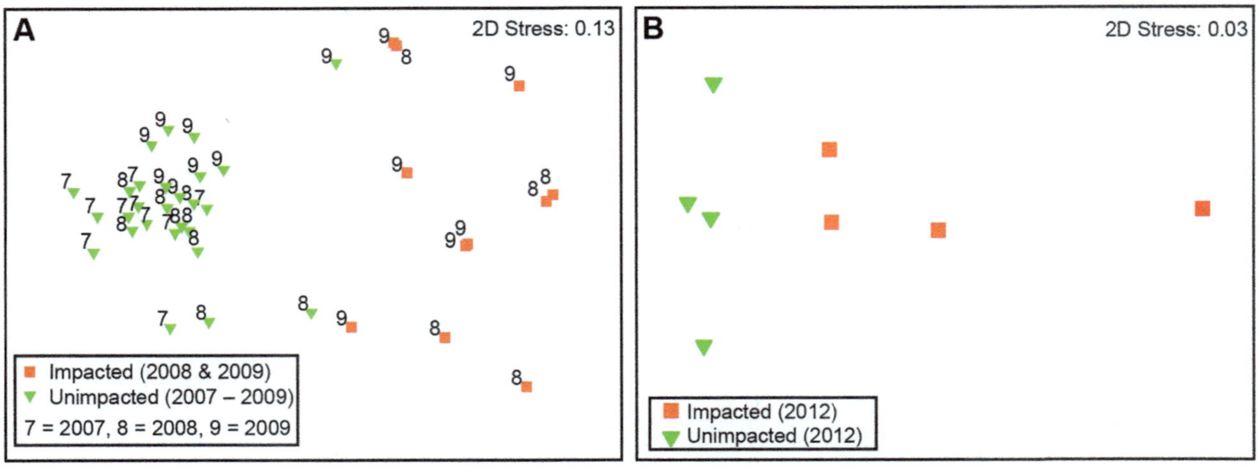


**Figure 4. MDS plot showing the relationship between impacted and unimpacted epifaunal communities.** (A) Point of Ayre. Dummy variable (present everywhere) used to create coherence in low abundance [impacted] data (see (36). (B) North of the Lleyn Peninsula.

Overall, for both *M. modiolus* reefs there was compelling evidence of physical impact on the epifaunal communities (Figure 3 and 4) and the significant differences in dispersion between 2007 and 2009 at PoA indicated no recovery.

In video quadrat data from PoA, significant reductions in number of individuals (N), numbers of upright emergent epifauna and total numbers of visible *M. modiolus* occurred at the impacted areas (Figure 5; Tables 1 and 2). Species richness (Margalef's d) and Shannon-Wiener's diversity (H′) and community evenness (J) were significantly lower in impacted quadrats (Figure 5; Tables 1 and 2). Overall, mean number of total individuals (N) was significantly reduced by 90.3% in trawled quadrats (2.63±1.96) compared to untrawled quadrats (27±12.23) (GLMM: $t = -11.41$; $p<0.001$). Most of the variation in N in impacted and unimpacted quadrats occurred between quadrat locations ($\sigma^2$site = 0.39), varying little between years ($\sigma^2$year = 0.09). At nLP there was a 59% lower mean abundance of total individuals (N) in video records from the scallop dredged areas (97.3±21.7 compared to 164±32.0; LMM) = 3.42; d.f. = 6; $p<0.05$). Lower abundances of *M. modiolus* and total upright emergent epifauna (mostly *A. digitatum* and *F. foliacea*) in dredged areas were significant only for the latter (LMM *M. modiolus* $t = 1.75$; $p = 0.13$; upright emergent epifauna $t = 3.06$; $p<0.05$; Figure 5). Shannon-Wiener's diversity (H′), Margalef's d richness and eveness (J) of the associated community were not significantly altered by impact (H′: $t = -1.74$, $p = 0.13$; d: $t = -1.55$, $p = 0.17$; J: $t = -1.14$, $p = 0.29$;).

Using low taxonomic resolution 19 broad groups were recorded from infaunal samples at PoA. The trawled infaunal community in 2009 varied significantly from the two control sites (PERMANOVA: pseudo $F = 9.02$, $p = 0.002$) a year after the impact was first observed. In the SIMPER analysis, reductions in the abundances of bivalves, malacostracans, ophuroids and polychaetes accounted for 60% of the average differences between impacted and unimpacted samples. Each of these reductions in abundance was significant (Figure 5D; ANOVA:Polychaeta, $F_{(2,9)} = 9.69$, $p<0.01$; Bivalvia, $F_{(2,9)} = 24.75$, $p<0.001$; Malacostraca, $F_{(2,9)} = 6.52$, $p<0.05$; Ophiuroidea, $F_{(2,9)} = 11.44$; $p<0.01$).

## Discussion

The present study investigated the effects of single passes of bottom-towed fishing gear on rare protected *Modiolus modiolus* reef communities. The null model was rejected because there were substantial declines in the abundance of epifauna in response to both trawl and scallop dredges as well as declines in all major taxonomic groups in the infaunal community at the trawled site. The present study provides the most direct evidence yet of physical impacts on the community associated with this type of complex habitat. Abrasion of epifauna is undoubtedly one mechanism responsible for the changes observed but loss of structure formed by *M. modiolus* and the role that the species plays in pelagic-benthic coupling also probably account for reductions in most taxonomic groups (especially at PoA). The post impact increase in Paguridae at PoA is consistent with increased scavenging in other fishing gear impact studies [43], [44]. The results are also consistent with indirect studies elsewhere in the world where *M. modiolus* and associated epifaunal declines have been documented in dredging and trawling grounds [24], [25], [45], [46] and where *M. modiolus* as a species (not forming biogenic structures) has been shown to decline in experimentally trawled areas [28]. Similarly, other biogenic reefs formed by oysters *Ostrea chilensis* and horse mussels *Modiolus areolatus* in the Faveux Strait (New Zealand) shown widespread reductions in the associated community and reef habitat following prolonged dredging [26]. Overall, it would seem that all complex temperate biogenic habitats such as shellfish reefs, maerl, sea grass beds and bryozoans reefs shown declines in response to dredge and trawls [2], [26], [32], [47], [48].

### First-pass impacts

The horse mussel reef off the north Lleyn Peninsula has existed for at least 150 years [10] probably because it has traditionally been distant from Irish Sea demersal (bottom-towed) fishing ports and since the late 1990s protected in an MPA and by a fisheries by-law [49], [50]. At the Point of Ayre, a combination of strong currents, a busy shipping lane and unsuitable habitat has left the study site largely un-fished and recent vessel monitoring data confirmed that the area was "non-impacted seabed" [51]. The present studies are therefore on relatively un-impacted, if not pristine sites and document the substantial impact of single first passes of bottom-towed fishing gear. The video method of recording, which clearly favours epifauna, undoubtedly underestimated the severity of the impact where multi-layered epifauna have been largely removed over the majority of the substratum (PoA). There were instances where one or two species of

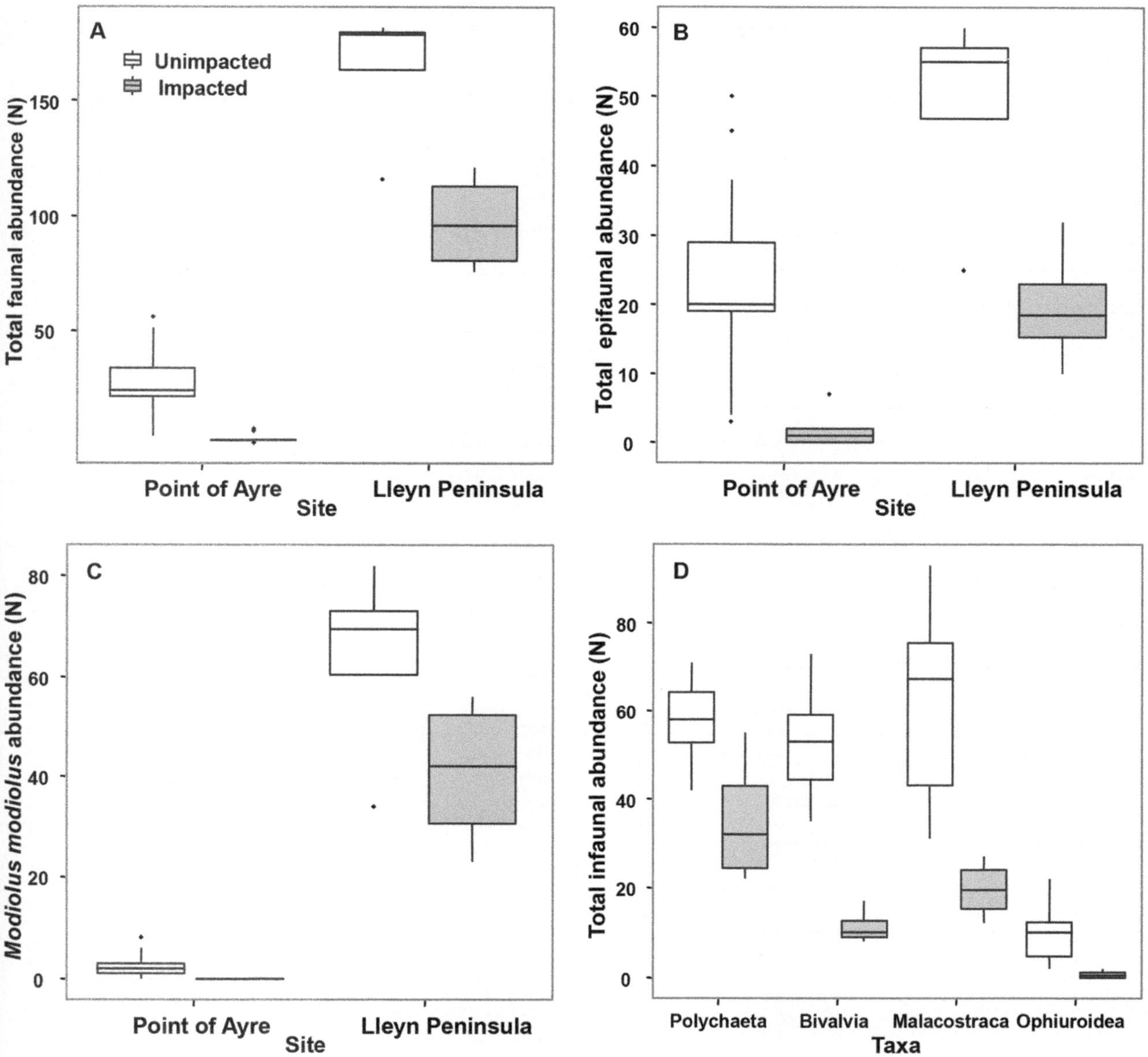


**Figure 5. Reductions in epifauna and infauna following bottom-towed fishing gear.** Total number of individuals (A), upright emergent epifauna (B) and numbers of *M. modiolus* (C) recorded on impacted and unimpacted 0.25×0.25 m video quadrats off Point of Ayre(PoA) and North Lleyn Peninsula (nLP). (D) Abundance of infaunal taxa contributing the most to the dissimilarities between impacted and unimpacted treatments at the PoA site (SIMPER). Box plots represent inter-quartile range, median, maximum and minimum values. The effect of physical impact was significant at $\alpha$ of 0.05 for all measures except *M. modiolus* abundance at nLP (Table 1).

understory turf were revealed and (erroneously) appeared to increase with impact (nLP). Didemnids, Bryozoa and Porifera all showed such increases at nLP, contributing 23% to dissimilarity between impacted and unimpacted quadrats in SIMPER analysis and probably contributed to a lack of significant change in diversity measures at nLP compared to PoA (Table 1). Without understanding the detailed structure of the understory this effect cannot be corrected or accounted for. Overall, the reductions of 90 and 59% of total epifaunal numbers supports the view that the majority of benthic impact occurs the first time an area is fished (e.g. [32], [52]) but also appears far greater in magnitude than modelled epifaunal biomass lost to fishing in surrounding Irish Sea habitats (8%: [53]). Given the sensitivity of the habitat, modern-day *M. modiolus* reefs are likely to be relics of their pre-fishing distribution. If so, then contemporary reference conditions for un-impacted benthic systems [53] may not account for these long-lived structure forming species in a classic 'shifting baselines' sense [54] and the bioengineering *M. modiolus* may well have occurred, albeit in lower densities, over more wide-spread areas as seen in other Atlantic studies [28], [46].

The magnitude of changes in the present study are similar to the differences in fauna between ridge and trough structures in naturally occurring beds (62%: [13]; [14]). In essence, the physical impact from bottom - towed gear removed ridge structure and appeared to reduce the community to a 'trough' habitat (*sensu* [14]) at PoA and, although declines in *M. modiolus* were not significant at nLP, clump structures were visibly flattened as well as showing significant epifaunal declines. The scale of change in epifaunal

**Table 1.** Abundance and diversity parameters measured for impacted and unimpacted *M. modiolus* communities.

| | Total abundance (N) | Margalef's richness (d) | Shannon-Wiener's diversity (H′) | Pielou's evenness (J) | Epifaunal abundance | *M. modiolus* abundance |
|---|---|---|---|---|---|---|
| ***nLP** | | | | | | |
| **Unimpacted** | 164.0±32.0 | 1.3±0.3 | 1.3±0.1 | 0.6±0.0 | 48.8±16.0 | 63.8±20.7 |
| **Impacted** | 97.2±21.7 | 1.8±0.6 | 1.5±0.2 | 0.7±0.1 | 19.7±9.2 | 40.7±15.4 |
| ***PoA** | | | | | | |
| *2007* | 33.6±13.6 | 1.8±0.2 | 1.5±0.2 | 0.8±0.1 | 29.0±12.0 | 2.1±1.3 |
| *2008* | | | | | | |
| **Unimpacted** | 23.8±10.6 | 1.6±0.3 | 1.4±0.2 | 0.8±0.1 | 21±13 | 1.6±1.3 |
| **Impacted** | 1.8±0.4 | 0.7±0.8 | 0.3±0.4 | 1.0±0.0 | 0.6±0.9 | 0.0±0.0 |
| *2009* | | | | | | |
| **Unimpacted** | 22±9.4 | 1.6±0.4 | 1.4±0.2 | 0.8±0.07 | 18±12 | 2.8±1.2 |
| **Impacted** | 3.3±2.5 | 1.0±0.7 | 0.5±0.5 | 0.9±0.2 | 2.2±0.9 | 0.0±0.0 |

*North Lleyn Peninsula (nLP) and Point of Ayre (PoA).
Mean +/− standard deviation values given.

**Table 2.** GLMM coefficients for diversity measures for impacted and unimpacted *M. modiolus* communities from PoA.

| Response | Random effects | Intercept | | Effect of disturbance (unimpacted vs impacted) |
|---|---|---|---|---|
| *Total abundance (N* | | | | |
| | $\sigma^2$year = 0.09 | Estimate | 3.24±0.1 | −2.24±0.20 |
| | $\sigma^2$site = 0.39 | t-value | 32.51 | −11.41 |
| | $\sigma^2$residual = 0.7 | p-value | <0.001 | <0.001 |
| **M. modiolus** *abundance* | | | | |
| | $\sigma^2$year = $1.23\times10^{-10}$ | Estimate | 0.38±0.16 | −28.68±0.3 |
| | $\sigma^2$site = 0.83 | t-value | 2.33 | −94.10 |
| | $\sigma^2$residual = $5.9\times10^{-10}$ | p-value | <0.05 | <0.001 |
| *Epifaunal abundance* | | | | |
| | $\sigma^2$year = $3.76\times10^{-6}$ | Estimate | 3.10±0.11 | −2.67±0.41 |
| | $\sigma^2$site = 0.85 | t-value | 27.22 | −6.59 |
| | $\sigma^2$residual = 1.86 | p-value | <0.001 | <0.001 |
| *Shannon-Wiener's diversity (H′)* | | | | |
| | $\sigma^2$year = 1.2 | Estimate | 1.48±0.06 | −1.06±0.11 |
| | $\sigma^2$site = 0.3 | t-value | 25.66 | −9.97 |
| | $\sigma^2$residual = 0.002 | p-value | <0.001 | <0.001 |
| *Margalef's richness (d)* | | | | |
| | $\sigma^2$year = $8.66\times10^{-6}$ | Estimate | 1.68±0.09 | −0.8±0.18 |
| | $\sigma^2$site = 0.46 | t-value | 18.92 | −4.49 |
| | $\sigma^2$residual = 0.002 | p-value | <0.001 | <0.001 |
| **Pielou's evenness (J)** | | | | |
| | $\sigma^2$year = 1.42 | Estimate | 0.82±0.02 | 0.11±0.043 |
| | $\sigma^2$site = 0.09 | t-value | 44.47 | 2.46 |
| | $\sigma^2$residual = 0.0004 | p-value | <0.001 | <0.05 |

Fixed factor = physical impact; random factors = time (year) and quadrat position. Estimate includes ± standard deviation (sd). Significance at $\alpha = 0.05$.
**See also Table S1**.

abundance, in particular, may be a useful indicator of condition in MPAs (EC Habitats and Species Directive; Marine (Scotland) Act 2010) but also Good Ecological Status (GES) under the EC Marine Strategy Framework Directive, where the biodiversity of 'special' habitats such as *Modiolus modiolus* reefs are currently being considered [15]. Large changes in the variance (see PERMDISP results) associated with epifaunal abundances might be expected across a reef experiencing low levels of physical impact that only cover part of it, whereas a substantial significant decline in mean epifaunal abundance (60% or more) would be expected across a reef experiencing fishing throughout; both scenarios would be incompatible with GES.

## Recovery and destabilisation

There was no evidence of recovery a year after impact was first recorded at Point of Ayre and since the long-lived structure forming species, *M. modiolus* (up to 48 years [55]), had significantly declined, this is unsurprising. Restorative experiments on *M. modiolus* beds have recently shown that semi-natural communities can recover within a year, but only when *M. modiolus* is translocated back into the habitat [56]. Irrespective of impact, there was also a change in the epifaunal community over time where *Alcyonium digitatum* and *Antedon bifida*, in particular, decreased in abundance. Although these observations probably reflect normal community dynamics they do highlight the need for investigations of the indirect effects of fishing gear such as re-suspended sediment [57] because large amounts of fine material are trapped in tidal areas under living *M. modiolus* reefs [4], [10]. It is possible that the mobilisation of this sediment could inhibit bivalve spat settlement [57] as well as the feeding of other species. Limited destabilisation of a ridge was observed beyond the initial impact in the present study where the tide appeared to have acted like wind in a sand-dune 'blow-out' (PoA, quadrat 3, 2009, Figure 2) and reduced the community to a crater next to a damaged section of reef.

## Scallop dredge vs trawl impact

On a heavily fished area of seabed in the Isle of Man it has been reported that otter trawls produced minimal bycatch and much less benthic damage compared to scallop dredges [58]: a contrast with the present study where the impact of the scallop dredge was not as great as the trawl. Although the present experimental design does not allow for direct comparison between fishing gears, it is nevertheless surprising that scallop dredges did not cause greater damage, given the metal dredge teeth at the leading edge of the gear. *M. modiolus* decline was not significant and evident physical structure remained in place post impact at nLP (Figure 3). It is plausible that dense *M. modiolus* reefs with mussel clumps [13], rapidly fill scallop dredges as bycatch and, affect the contact between the dredge teeth and the benthos (unlike a trawl door). Indeed, *M. modiolus* has been reported as a major component of an established scallop dredge fishery with 28% bycatch [59], but in a first-pass of fishing gear bycatch might be expected to be far greater when a dredge initially comes into contact with a reef. Overall, the levels of fishing disturbance to which the seabed has already been exposed may govern the impact of dredge gear.

## The dilemma of evidence-based conservation for rare habitats

To better understand the relative differences in gear impacts would require controlled, comparative fishing at similar sites. Such studies, however, are not ethically or legally possible when, as here, the protected biogenic habitat is rare and largely found within MPAs.

It is the authors' experience that the preclusion of premeditated impact studies on ethical and legal grounds can, paradoxically, be a significant obstruction to delivering the conservation objectives in MPAs because restrictions proposed by conservation managers can be challenged by stakeholders who perceive a lack of evidence of impact as evidence of absence of impact. Challenges to the adequacy of the evidence - base are apparent in the wider fisheries impact literature (eg [60] countered by [20] and [61] countered by [21]) and have also been used in political negotiation (e.g. [62]) but concerns about the evidence - base of conservation practice as a whole have also emerged (eg [17]). Calls for evidence-based conservation have acknowledged that conservation science lacks the resources to deliver meta-analyses in the same way as medical science [17], but limited scientific knowledge has long been used, for example, as an excuse to hinder the development of marine reserves [63]. In rare protected habitats where impact studies are unethical or illegal, the conservation manager is thus caught in an evidence trap when dealing with extant damaging commercial activities and no direct evidence of impact: a problem that is likely to be widespread and, ironically, unreported in the scientific literature.

## Acknowledgments

The Isle of Man government provided the Sir John Craster to support diving operations, skippered by Malcolm Daugherty and crewed by Jan Hesketh. The Countryside Council for Wales provided staff time and boat support on Pedryn skippered by Paul Turkentine. Josie Sanderson and Flora Kent helped with logistics and sample sorting whilst Anne-Marie Mahon, Paolo Pizzola and Lara Howe helped with diving work. Catherine Duigan and Lucy Kay of CCW and Liz Charter of DEFRA provided procedural and institutional support. Hayley Hinchen and a HBDSEG steering group supported the final production. Comments from two referees greatly improved the final paper.

## Author Contributions

Conceived and designed the experiments: RLC CL FRG RHFH JSP TH RS LRS TBS WGS. Performed the experiments: RLC RHFH CL JSP TBS WGS. Analyzed the data: RLC JMF LRS TBS WGS. Wrote the paper: RLC CL FRG WGS JMF TH RS TBS JSP. Logistics in the Isle of Man: FRG TH .

## References

1. Beck MW, Brumbaugh RD, Airoldi L, Carranza A, Coen LD, et al. (2009) Shellfish Reefs at Risk: A Global Analysis of Problems and Solutions, The Nature Conservancy, Arlington VA. 52 p.
2. Beck MW, Brumbaugh RD, Airoldi L, Carranza A, Coen LD, et al. (2011) Oyster Reefs at Risk and Recommendations for Conservation, Restoration, and Management. BioScience 61: 107–116.
3. European Commission DG Environment (2007) Interpretation Manual of European Union Habitats. EUR 27. 144 p.
4. Wildish DJ, Fader GBJ, Lawton P, MacDonald AJ (1998) The acoustic detection and characteristics of sublittoral bivalve reefs in the Bay of Fundy. Continental Shelf Research 18: 105–113.

5. Naumov AD (2006) Clams of the White Sea: Ecological and Faunistic Analysis, Proceedings of the Zoological Institute Russian Academy of Sciences. St. Petersburg: Zoological Institute RAS.
6. Rees EIS (2009) Background document for *Modiolus modiolus* beds. OSPAR Commission Biodiversity Series. London: OSPAR Commission.
7. Solyanko K, Spiridonov V, Naumov A (2011) Biomass, commonly occurring and dominant species of macrobenthos in Onega Bay (White Sea, Russia): data from three different decades. Marine Ecology 32: 36–48.
8. Ragnarsson SA, Burgos JM (2012) Separating the effects of a habitat modifier, *Modiolus modiolus* and substrate properties on the associated megafauna. Journal of Sea Research 72: 55–63.
9. OSPAR (2005) *Modiolus modiolus* horse mussel beds: Case Reports for the initial list of threatened and/or declining species and habitats in the OSPAR maritime area. OSPAR. 261 p.
10. Lindenbaum C, Bennell JD, Rees ELS, McClean D, Cook W, et al. (2008) Small-scale variation within a *Modiolus modiolus* (Mollusca : Bivalvia) reef in the Irish Sea: I. Seabed mapping and reef morphology. Journal of the Marine Biological Association of the United Kingdom 88: 133–141.
11. Robinson KA, Mackie ASY, Lindenbaum C, Darbyshire T, Van Landeghem KJJ, et al (2012) Seabed Habitats of the Southern Irish Sea, in: Harries, P.T., Baker, E.K. (Eds.), Seafloor Geomorphology as Benthic Habitat. Geohab Atlas of Seafloor Geomorphic Features and Benthic Habitats. London: Elsevier. pp. 523–527.
12. Hirst NE, Clark L, Sanderson WG (2012) The distribution of selected MPA search features and Priority Marine Features off the NE coast of Scotland. Scottish Natural Heritage Commissioned Report 500. Edinburgh: Scottish Natural Heritage. 132 p.
13. Rees EIS, Sanderson WG, Mackie ASY, Holt RHF (2008) Small-scale variation within a *Modiolus modiolus* (Mollusca: Bivalvia) reef in the Irish Sea. III. Crevice, sediment infauna and epifauna from targeted cores. Journal of the Marine Biological Association of the United Kingdom 88. 151–156.
14. Sanderson WG, Holt RHF, Kay L, Ramsay K, Perrins J, et al (2008) Small-scale variation within a Modiolus modiolus (Mollusca : Bivalvia) reef in the Irish Sea. II. Epifauna recorded by divers and cameras. Journal of the Marine Biological Association of the United Kingdom 88: 143–149.
15. Moffat C, Aish A, Hawkridge JM, Miles H, Mitchell PI, et al (2011) Advice on United Kingdom biodiversity indicators and targets for the Marine Strategy Framework Directive. Healthy and Biologically Diverse Sea Evidence Group Report to the Department for Environment, Food and Rural Affairs. 210 p.
16. Sutherland WJ, Pullin AS, Dolman PM, Knight TM (2004) The need for evidence-based conservation. Trends in Ecology and Evolution 19: 305–308.
17. Pullin AS, Knight TM, Stone DA, Charman K (2004) Do conservation managers use scientific evidence to support their decision-making? Biological Conservation 119: 245–252.
18. Cook CN, Carter RW, Fuller RA, Hockings M (2012) Managers consider multiple lines of evidence important for biodiversity management decisions. Journal of Environmental Management 113: 341–346.
19. DSCC (2005) Debunking claims of sustainability; High-seas bottom trawl red herrings. Marine Conservation Biology Institute. 16 p.
20. Rogers SI, Greenaway B (2005) A UK perspective on the development of marine ecosystem indicators. Marine Pollution Bulletin 50: 9–19.
21. Gray JS, Dayton P, Thrush S, Kaiser MJ (2006) On effects of trawling, benthos and sampling design. Marine Pollution Bulletin 52: 840–843.
22. The Royal Commission on Environmental Pollution (2004) Turning the tide: Addressing the impact of fisheries on the marine environment. Presented to parliament by command of Her Majesty. 480 p.
23. Service M, Magorrian BH (1997) The extent and temporal variation of disturbance to epibenthic communities in Strangford Lough, Northern Ireland. Journal of the Marine Biological Association of the United Kingdom 74: 1151–1164.
24. Roberts D, Allcock L, Fariñas Franco JM, Gorman E, Maggs CA, et al. (2011) Modiolus Restoration Research Project: Final Report and Recommendations, Report to Department of Agriculture and Rural Development and North Ireland Environment Agency by Queens University Belfast. 246 p.
25. Strain EMA, Allcock AL, Goodwin CE, Maggs CA, Picton BE, et al. (2012) The long-term impacts of fisheries on epifaunal assemblage function and structure, in a Special Area of Conservation. Journal of Sea Research 67: 58–68.
26. Cranfield HJ, Michael KP, Doonan IJ (1999) Changes in the distribution of epifaunal reefs and oysters during 130 years of dredging for oysters in Foveaux Strait, southern New Zealand. Aquatic Conservation: Marine and Freshwater Ecosystems 9: 461–483.
27. Wiborg K (1946) Undersøkelser overoskjellet (*Modiolus modiolus* L.)). I. Alrninnelig biologi, vekst og økonomisk betydning. Norw. Fishery mar. Invest VIII, 85.
28. Kenchington E, Gilkinson KD, MacIsaac KG, Bourbonnais-Boyce C, Kenchington T, et al. (2006) Effects of experimental otter trawling on benthic assemblages on Western Bank, Northwest Atlantic Ocean. Journal of Sea Research 56: 249–270.
29. Collie JS, Escanero GA, Valentine PC (1997) Effects of bottom fishing on the benthic megafauna of Georges Bank. Marine Ecology Progress Series 155: 159–172.
30. Bradshaw C, Veale LO, Hill AS, Brand AR (2001) The effect of scallop dredging on Irish Sea benthos: experiments using a closed area. Hydrobiologia 465: 129–138.
31. Kaiser MJ, Clarke KR, Hinz H, Austen MCV, Somerfield PJ, et al. (2006) Global analysis of response and recovery of benthic biota to fishing. Marine Ecology Progress Series 311: 1–14.
32. Hall-Spencer JM, Moore PG (2000) Scallop dredging has profound long-term impacts on maerl habitats. ICES Journal of Marine Science 57: 1407–1415.
33. Stringell TB, Bamber RN, Burton M, Lindenbaum C, Skates LR, et al. (2013) A tool for protected area management: multivariate control charts 'cope' with rare variable communities. Ecology & Evolution 3: 1667–1676.
34. Holt TJ, Shalla SHA (unpublished) Pre- and post-drilling survey of block IOM 112/19. A report to Elf Enterprise Caledonia Ltd. Port Erin Marine Laboratory, University of Liverpool, 1997.
35. Hydrographic Office (1992) Cardigan Bay: Northern Part. Admiralty Chart. United Kingdom Hydrographic Office, Taunton, no. 1971.
36. BERR (2008) Atlas of UK Marine Renewable Energy Resources. A strategicEnvironmental Assessment Report. Produced by ABPmer, The Met Office, Proudman Oceanographic Laboratory. 26 p.
37. Somerfield PJ, Clarke KR (1995) Taxonomic levels, in marine community studies, revisited. Marine Ecology Progress Series 127: 113–119.
38. Clarke KR, Gorley RN (2006) PRIMER v6: user manual/tutorial. Plymouth: Primer-E. 190 p.
39. Anderson MJ, Gorley RN, Clarke KR (2008) PERMANOVA + for PRIMER: guide to software and statistical methods. Plymouth: Primer-E. 214 p.
40. R Development Core Team (2012) R: A Language and Environment for Statistical Computing. Vienna: R Foundation for Statistical Computing. . Hothorn T, Bretz F, Westfall P (2008) Simultaneous inference in general parametric models. Biometrical Journal 50: 346–363.
41. Bates D, Maechler M, Bolker B (2011) lme4: Linear mixed-effects models using S4 classes. R package version 0.999375-42. Available: http://CRAN.R-project.org/package=lme4Bolker BM, Brooks ME, Clark CJ, Geanger SW, Poulsen JR, Stevens MHH, White JS (2008) Generalized linear mixed models: a practical guide for ecology and evolution. Trends in Ecology & Evolution 24: 127–135.
42. Venables WN, Ripley BD (2002) Modern Applied Statistics with S. Fourth Edition. Springer, New York. 497 p
43. Kaiser MJ, Spencer BE (1996) The effects of beam-trawl disturbance on infaunal communities in different habitats. Journal of Animal Ecology 65: 348–358.
44. Ramsay K, Kaiser MJ (1998) Demersal fishing disturbance increases predation risk for whelks (*Buccinum undatum* L.). Journal of Sea Research 39: 299–304.
45. Magorrian BH, Service M (1998) Analysis of underwater visual data to identify the impact of physical disturbance on horse mussel (*Modiolus modiolus*) beds. Marine Pollution Bulletin 36: 354–359.
46. Kenchington EL, Kenchington TJ, Henry LA, Fuller S, Gonzalez P (2007) Multi-decadal changes in the megabenthos of the Bay of Fundy: The effects of fishing. Journal of Sea Research 58: 220–240.
47. Cranfield HJ, Manighetti B, Michael KP, Hill ADA (2003). Effects of oyster dredging on the distribution of bryozoan biogenic reefs and associated sediments in Foveaux Strait, southern New Zealand. Continental Shelf Research 23: 1337–1357.
48. Martín MA, Sánchez-Lizaso JL, Ramos-Esplá AA (1997) Cuantificación del impacto de las artes de arrastre sobre la pradera de *Posidonia oceanica* (L.) Delile, 1813. Publicaciones especiales - Instituto Español de Oceanografía 23: 243–253.
49. CCW (2007) Press release on scallop prosecution. Available: http://www.ccgc.gov.uk/about-ccw/newsroom/press-releases/scallop-dredging.aspx. Accessed 2012 Dec 20.
50. WAG (2010) Welsh Assembly Government The Scallop Fishing (Wales) (No.2) Order 2010, Article 12. Available: http://www.legislation.gov.uk/wsi/2010/269/made. Accessed 2012 Dec 20.
51. Lambert GI, Hiddink JG, Hintzen NT, Hinz H, Kaiser MJ, et al. (2012) Implications of using alternative methods of vessel monitoring system (VMS) data analysis to describe fishing activities and impacts. ICES Journal of Marine Science 69: 682–693.
52. Jennings S, Kaiser MJ (1998) The effects of fishing on marine ecosystems. Advances in Marine Biology 34. 201–220.
53. Lambert GI, Jennings S, Kaiser MJ, Hinz H, Hiddink JG (2011) Quantification and prediction of the impact of fishing on epifaunal communities. Marine Ecology Progress Series 430: 71–86.
54. Pauly D (1995) Anecdotes and the shifting baseline syndrome of fisheries. Trends in Ecology and Evolution 10: 430.
55. Anwar NA, Richardson CA, Seed R (1990) Age determination, growth rate and population structure of the horse mussel (*Modiolus modiolus*) Journal of the Marine Biological Association of the UK 70: 441–457.
56. Fariñas-Franco JM, Allcock L, Smyth D, Roberts D (2013) Community convergence and recruitment of keystone species as performance indicators of artificial reefs. Journal of Sea Research 78: 59–74.
57. Jones JB (1992) Environmental-Impact of Trawling on The Seabed - A Review. New Zealand Journal of Marine and Freshwater Research 26: 59–67.
58. Hinz H, Murray LG, Malcolm FR, Kaiser MJ (2012) The environmental impacts of three different queen scallop (*Aequipecten opercularis*) fishing gears. Marine Environmental Research 73: 85–95.

59. Garcia EG, Ragnarsson SA, Eiriksson H (2006) Effects of scallop dredging on macrobenthic communities in west Iceland. ICES Journal of Marine Science 63: 434–443.
60. Løkkeberg S (2005) Impacts of trawling and scallop dredging on benthic communities. FAO Fisheries Technical Paper No. 472.
61. Hoydal K (2005) Oceans: fisheries not to blame for damage. Nature 433. 105.
62. Scottish Executive (2010) Scallop Fishing. The Scottish Government News Releases 2010. Available: http://www.scotland.gov.uk/News/Releases/2010/11/02153713. Accessed 2013 Jan 10.
63. Agardy T (2000) Information Needs For Marine Protected Areas: Scientific and Societal. Bulletin of Marine Science, 66(3): 875–888.

# 12

# Characteristics of Fishing Operations, Environment and Life History Contributing to Small Cetacean Bycatch in the Northeast Atlantic

**Susie Brown[1]*, David Reid[2], Emer Rogan[1]**

**1** School of Biological, Earth and Environmental Sciences, Cork Enterprise Centre, University College Cork, Distillery Fields, North Mall, Cork, Ireland, **2** Marine Institute Headquarters, Rinville, Oranmore, Galway, Ireland

## Abstract

Fisheries bycatch is a key threat to cetacean species globally. Managing the impact requires an understanding of the conditions under which animals are caught and the sections of the population affected. We used observer data collected on an albacore tuna gillnet fishery in the northeast Atlantic, to assess operational and environmental factors contributing to bycatch of common and striped dolphins, using generalised linear models and model averaging. Life history demographics of the captured animals were also investigated. In both species, young males dominated the catch. The age ratio of common dolphins was significantly different from that estimated for the population in the region, based on life tables ($G = 17.1$, d.f. $= 2$, $p = 0.002$). Skewed age and sex ratios may reflect varying vulnerability to capture, through differences in behaviour or segregation in populations. Adult females constituted the second largest portion of the bycatch for both species, with potential consequences for population sustainability. Depth was the most important parameter influencing bycatch of both species and reflected what is known about common and striped dolphin habitat use in the region as the probability of catching common dolphins decreased, and striped dolphins increased, with increasing depth. Striped dolphin capture was similarly influenced by the extent to which operations were conducted in daylight, with the probability of capture increasing with increased operations in the pre-sunset and post-sunrise period, potentially driven by increased ability of observers to record animals during daylight operations, or by diurnal movements increasing contact with the fishery. Effort, based on net length and soak time, had little influence on the probability of capturing either species. Our results illustrate the importance of assessing the demographic of the animals captured during observer programmes and, perhaps more importantly, suggest that effort restrictions alone may not be sufficient to eradicate bycatch in areas where driftnets and small cetaceans co-occur.

**Editor:** Richard Reina, Monash University, Australia

**Funding:** This analysis is funded under the Beaufort Ecosystem Approach to Fisheries Management award, as part of the Irish Government's National Development Plan (NDP), the Beaufort Marine Research Award is grant aided by the Department of Communications, Energy and Natural Resources (DCENR) and the Department of Agriculture, Fisheries and Food (DAFF) under the Strategy for Science Technology and Innovation (SSTI) and the Sea Change Strategy. This analysis was based on data collected under the European Commission funded EC-FAIR contract FAIR-CT95-0523: Assessment and Reduction of the By-catch of Small Cetaceans (BY-CARE). The funders had no role in study design, data collection and analysis, decision to publish, or preparation of the manuscript.

**Competing Interests:** The authors have declared that no competing interests exist.

* Email: s.brown@ucc.ie

## Introduction

Anthropogenic pressures on the marine environment are diverse and complex and include pollution, habitat disturbance, invasive species introductions, human induced climate change and fishing [1]. Whilst climate change and fishing pressure are acknowledged as being the main influences on marine ecosystems [2], overfishing has a significant impact on marine exploited communities [1], can act to increase the impact of climate change [3,4], and continues to be viewed as the key anthropogenic impact affecting marine ecosystems [5]. The pressures placed upon the marine environment continue to grow and as a consequence the need to improve marine management practices has become more urgent [6].

Many large top predators have been subjected to overfishing [7], contributing to the decline and collapse of target species around the world [1,7]. Alongside target species impacts, the capture of non-target or "bycatch" species [8] ensures commercial fishing is a key driver affecting the biodiversity of marine ecosystems, the loss of which is increasing on a global scale [9]. Bycatch of marine mammals [10,11], seabirds [12], elasmobranchs [13] and reptiles [14], has been documented in fisheries around the world [15], in extreme cases exceeding target catch [16,13] and resulting in declines of some species [11]. Management decisions must consider, amongst other factors, the risk posed by fishing to non-target species [2,12,17,18,19].

Impact on non-target species was a key reason behind restrictions on the use of driftnets by fleets in the European Union. The indiscriminate nature of driftnets and resulting high volume bycatch of non-target species [20] led the Council of the European Commission to restrict the length of driftnets targeting highly migratory species in 1991, culminating in a ban on use of the gear to target such species in 2002 [21]. Driftnets, for large pelagic species, operate legally in regions outside the EU, including the US Pacific, and are subject to strict management regimes which seek to monitor and mitigate marine mammal bycatch, through the use of season and area closures, observer programmes and acoustic devices such as "pingers" [22,23,24]. Driftnet operations in the southwest Atlantic are subject to much less regulation [25] and in the EU large scale driftnets reportedly operate illegally in the Mediterranean [26]. High numbers of marine mammals, seabirds, turtles, and non-target fish species such as sharks, occur as bycatch in driftnets [20,27]. Cetacean species are thought to be particularly vulnerable to the effects of bycatch due to the relatively low abundance of their populations, and because their long life spans, late maturation and low reproductive rates limit the capacity of their populations to recover from such pressures [28]. Successful management of the impacts of driftnets on cetaceans requires an understanding of the conditions under which animals are caught and the species and life history stages impacted.

Observer programmes document the cetacean species taken by fisheries and facilitate assessment of the number of species removed by that fishery. The resulting datasets can provide additional information regarding the operational and environmental characteristics of the capture events and, at times, the characteristics of the animals caught. In 1996, observers accompanied vessels of the Irish fleet deploying surface driftnets targeting albacore tuna (*Thunnus alalunga*) in the Bay of Biscay and Celtic Sea. Focusing on small cetaceans, we examined the resulting dataset with the aim of assessing the relative influence of a number of operational and environmental factors on cetacean bycatch. Additionally, the life history characteristics of the captured animals were assessed, with the objective of determining whether particular species, sexes or age groups were more prone to incidental capture. The objectives of this study were to identify the sections of the population potentially most at risk and the operational and environmental factors associated with small cetacean bycatch.

## Methods

### Observer programme

Observer coverage extended to 120 of an estimated 261 hauls (Figure 1). The number and species of cetacean, fish, seabirds and turtles caught, landed and discarded were recorded. This paper examines the small cetacean bycatch only. Full details of the observer programme procedures, methodologies and results are presented in [27].

Observers recorded a number of operational parameters relating to each set. The nets deployed were monofilament nylon twine with stretch mesh size of 17.8 cm. Additional operational parameters recorded included the length of net deployed; the depth of net deployed (based on the number of meshes); the time of setting and hauling; the length of time taken to set and haul the net; and soak time (calculated as the duration between half the time taken to set the net and half the time taken to haul it). Environmental variables recorded included latitude and longitude of net position; depth of water column; sea state (Beaufort scale) during setting and hauling of nets; and the date of the fishing operation.

Where possible, within the constraints of time and space, cetaceans were brought on board the fishing vessel for further examination. Life history parameters were collected including species; sex and total length (tip of beak to notch in tail fluke). Post-mortem examinations were carried out on a number of animals and biological samples, including teeth, were collected.

### Life history traits

Common dolphin (*Delphinus delphis*) and striped dolphin (*Stenella coeruleoalba*) dominated the cetacean bycatch and were subject to life history assessments. Age was estimated for 69 of the 150 common dolphins and 47 of the 66 striped dolphins using teeth (full details of the methodology used to estimate age via teeth are presented in [29]). For the remaining animals, age was estimated using the nose to tail length of the animals aged using teeth, informed by published length at age data [29,30,31,32]. Aged individuals were classified, by sex, into one of three broad age groups: juvenile; sub-adult; and adult (Table 1). Age groups were based broadly on growth phases and the approximate age when sexual maturity is reached [29,30,31,32].

To test the hypothesis that males and females were equally likely to be caught, sex ratios, for each species, were compared to a theoretical 1:1 ratio using a G test. To test the hypothesis that the age composition of the bycatch animals was similar to that of the population in the region, the proportion of each age group in the northeast Atlantic was estimated using published information on survivorship ($l_x$) in female common dolphins from the Bay of Biscay [28] under the assumption that survivorship would be similar between the two neighbouring areas. Survivorship was used to estimate the proportion of animals dying at each age ($d_x$) based on the following equation

$$d_x = l_{x+1} - l_x$$

Proportion dying ($d_x$) was used to estimate the proportion of the population in each of the three age groups which was expected to die. This was then compared to the proportions seen in the bycatch. The analysis was extended to examine the age ratios of the male common dolphins using the survivorship data for females, based on the assumption that there would be no difference in survivorship between the sexes. It was not possible to examine the age ratios of striped dolphins due to a lack of published life tables or similar data. G tests were used to make all comparisons and all analysis was conducted in software package R [33].

### Environmental and operational variables

The influence of operational and environmental variables on the occurrence of cetacean bycatch was assessed. Separate assessments were conducted for the occurrence of common dolphins and striped dolphins. Observers accompanied each vessel on a number of trips and between 3 and 10 sets were observed per trip, therefore the resulting data had a hierarchical structure. Sets took place on consecutive days, potentially introducing temporal correlation. The presence/absence of each species was initially assessed by fitting general estimating equations (GEE) [34], with an autoregressive correlation structure, under the geepack package [35,36,37]. A variable "VesselTrip" was created and included in the models to identify observations from the same fishing trip by the same vessel. The correlation of the bycatch between sequential hauls within the same vessel trip was low for both common (0.05) and striped dolphin (0.09); therefore final model fitting was with generalised linear mixed modelling (glmm), with VesselTrip fitted as a random effect, under the glmmML package [38].

Variables included in the models of common and striped dolphin occurrence were selected *a priori* from the parameters collected by the observers. The variables were checked for collinearity using pairwise Spearman's Rank correlation coefficients, with a correlation coefficient greater than 0.5 indicating variables were proxies for each other [39]. Five variables remained in the analysis when collinear variables were removed; depth, sea state, moon, daylight and effort. Eight collinear variables were removed from the assessment; latitude and longitude (correlated with depth); date of net set and haul (correlated with moon

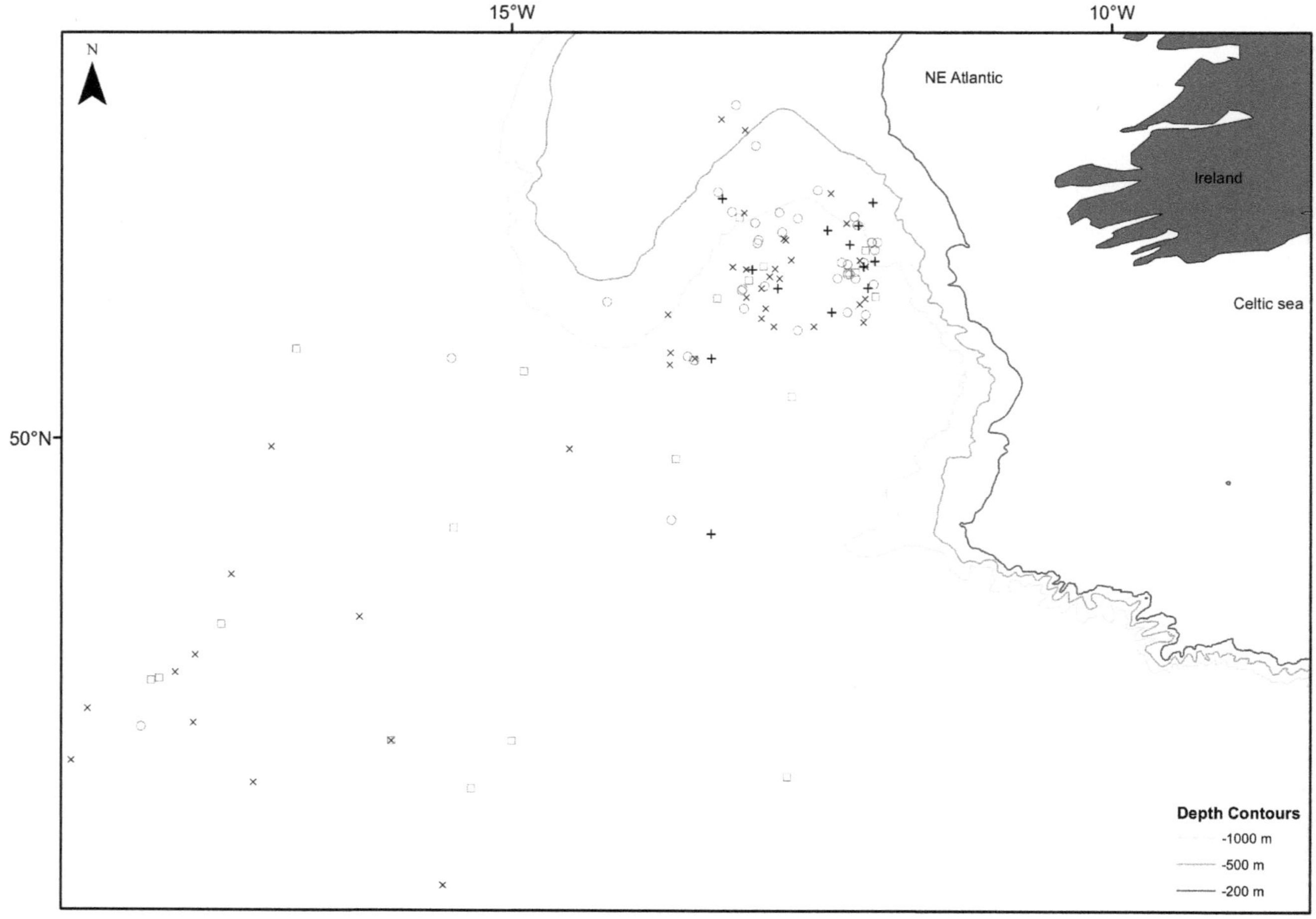


**Figure 1. Location of fishing operations and bycatch events.** Zero bycatch of common or striped dolphin (x symbol); common dolphin bycatch (empty circle); striped dolphin bycatch (empty square); common and striped dolphin bycatch (+ symbol).

illumination); time of net set and haul (correlated with daylight); and set and haul duration (correlated with soak time which was included in calculating effort).

Depth (environmental variable) was used to describe the depth of the water column where the fishing operation took place. Sea state (environmental variable) described the maximum sea state (Beaufort scale) recorded during net set and haul. Moon

**Table 1.** Age group classification of male and female common and striped dolphins determined through analysis of teeth and age at length data.

| Species | Sex | Group | Age | Length | Source of supporting information |
|---|---|---|---|---|---|
| **Common dolphin** | Male | Juvenile | ≤4 | ≤185 cm | [30] |
| | | Sub-adult | 5–9 | 186–199 cm | |
| | | Adult | ≥10 | ≥200 cm | |
| | Female | Juvenile | ≤4 | ≤172 cm | [29,31] |
| | | Sub-adult | 5–7 | 173–184 cm | |
| | | Adult | ≥8 | ≥180 cm | |
| **Striped dolphin** | Male | Juvenile | ≤4 | ≤185 cm | [32] |
| | | Sub-adult | 5–9 | 186–210 cm | |
| | | Adult | ≥10 | ≥210 cm | |
| | Female | Juvenile | ≤4 | ≤170 cm | [32] |
| | | Sub-adult | 5–10 | 171–195 cm | |
| | | Adult | ≥11 | ≥195 cm | |

(environmental variable) described the percentage illumination of the moon the night of the fishing operation. Daylight (operational variable) described the proportion of the set and haul occurring in daylight i.e. pre sunset and post sunrise. All nets were soaked at night but there was variation in the timing of net setting and hauling and therefore the extent to which these operations were conducted prior to sunset and post sunrise. The time taken to set nets ranged from 30 to 235 minutes and hauling time ranged from 150 to 660 minutes. Typically net setting began several minutes before sunset (mean 32 minutes before sunset, range 1–113 minutes) with only 23 net sets beginning after sunset. Net hauling continued after sunrise in every case (mean 270 minutes after sunrise, range 11–607 minutes). Effort (operational variable) described the length of the net deployed multiplied by soak time, where soak time was calculated as the time between the halfway point in the set and the halfway point in the haul. The explanatory variables were standardised to have a mean of zero and a standard deviation of one prior to analysis.

All possible model permutations were created and model selection was based on Akaike's Information Criterion AIC [40]. Models were ranked by AIC value and the model with the lowest AIC was considered the most parsimonious within the suite of models. Models within 2 ΔAIC of the most parsimonious model were considered to be similar in their empirical support and were included in the set of candidate models [41]. Akaike weights $w_i$ were calculated for each model and represented the proportional chance of that model being the best model within the set of candidate models [41]. The relative importance (RI) of each variable in determining the occurrence of the species in the bycatch was determined by summing Akaike weights over all candidate models containing the explanatory variable and ranged from 1 (most important) to 0 (least important) [41]. Since model comparisons resulted in a number of models similar in empirical support, we conducted model averaging across all models within 2ΔAIC of the most parsimonious model [41]. Model averaged parameter estimates were produced for each variable, with unconditional standard errors incorporating model uncertainty. Model averaging was conducted using the "MuMIn" package [42]. All analyses were conducted in software package R [33].

### Ethics statement

In Ireland, all cetacean species are protected under the Wildlife (Amendment) Act 1976–2005, therefore sampling was conducted under permit issued by the National Parks and Wildlife Service, Department of Arts, Heritage and the Gaeltacht. The study was entirely based on data collected post-mortem from cetacean carcasses bycaught in an albacore tuna fishery operating in the north east Atlantic. Sampling was conducted on-board fishing vessels and took place over a large area from 46° to 52°N and 11° to 18°W. Sampling did not involve observation or experimentation on live animals or captive animals, therefore ethical approval was not required.

## Results

### Life history characteristics

Cetacean bycatch was recorded in 79 of the 120 observed hauls with 242 individual cetaceans recorded across eight species. Common dolphin and striped dolphin were the most frequently occurring species with 150 common dolphins recorded across 51 sets and 66 striped dolphins recorded across 35 sets (as reported previously [27]). Other species captured were Atlantic white-sided dolphin (*Lagenorhynchus acutus*), Risso's dolphin (*Grampus griseus*), long-finned pilot whale (*Globicephala melas*), minke whale (*Balaenoptera acutorostrata*), bottlenose dolphin (*Tursiops truncatus*) and sperm whale (*Physeter macrocephalus*). Single individuals were caught in 23 sets with 56 sets having more than one individual. Multiple species were caught together in 18 sets. Common dolphin and striped dolphin were caught together in 13 sets.

Sex was determined for 116 common dolphins and the majority ($n = 72$) were male. The sex ratio was 1.64:1 and significantly different from the 1:1 ratio which would be expected if males and females were equally likely to be caught ($G = 9.04$, d.f. = 1, $p = 0.002$). Age was determined for 113 individuals and 66 were juveniles (43 male and 23 female), 10 were sub-adults (6 male and 4 female) and 37 were adults (20 male and 17 female) (Figure 2A). The age ratio of female common dolphins was significantly different from that estimated for the female population in the north east Atlantic, as determined from life tables ($G = 17.1$, d.f. = 2, $p = 0.002$) with more juveniles and fewer sub-adults caught than were estimated to be present in the population. This was also the case for male common dolphins ($G = 32.68$, d.f. = 2, $p < 0.0001$) although it must be stressed that this analysis was based on a life table for female common dolphins in the absence of the equivalent for males.

Common dolphins were most frequently captured in groups, with 33 sets catching multiple individuals and 18 sets capturing single individuals, including 6 sets which captured a single common dolphin alongside one or more individuals of another species. The average group size was 4 individuals, with the largest group catch being 13 individuals. The term "group" refers to an individual caught alongside at least one individual of the same species. It was not possible to tell whether individuals came from the same social groups. Time and space for processing animals was limited, therefore it was not possible to ascertain the sex of every individual within group catches. Sex was not determined for 33 of the 132 common dolphins which were caught in groups. The majority which were sexed were male ($n = 61$), with 38 female (Figure 2A). Mixed sex groups made up the bulk of group catches ($n = 18$) and mixed sex groups were most frequently made up of a mixture of juveniles and adults ($n = 8$). All male groups were the second most common ($n = 8$) (Table 2). The age composition of group catches was dominated by groups of adults and juveniles ($n = 13$) and these groups were predominantly mixed sex ($n = 10$).

Sex was determined for 60 of the 66 striped dolphins caught. Twice as many of the sexed individuals were male ($n = 40$) (Figure 2B) and the sex ratio was significantly different from a 1:1 ratio (2:1) ($G = 6.79$, d.f. = 1, $p = 0.009$) again suggesting that males were more likely to be caught than females. Age was determined for 59 individuals and 43 were juveniles (27 male and 16 female), 5 were sub-adult (4 male and 1 female) and 11 were adult (9 male and 2 female) (Figure 2B). It was not possible to test if these ratios were similar to that of the wider population of striped dolphins owing to the lack information on survivorship for the species.

Striped dolphins were caught as single individuals in 17 sets and as part of a group in 18 sets. The single individual catches included seven cases where a single striped dolphin was captured alongside at least one individual of another species. Groups were smaller than for common dolphins with an average group size of 2.7 and the largest group comprised five individuals. As with common dolphins the majority of animals occurring in group catches were male ($n = 28$) with close to half as many females caught ($n = 15$). Again, similar to common dolphins, mixed sex groups dominated the group catches ($n = 11$), with 6 of the mixed species groups composed entirely of juveniles. All male groups were the second most frequently captured group (Table 2). The age composition of

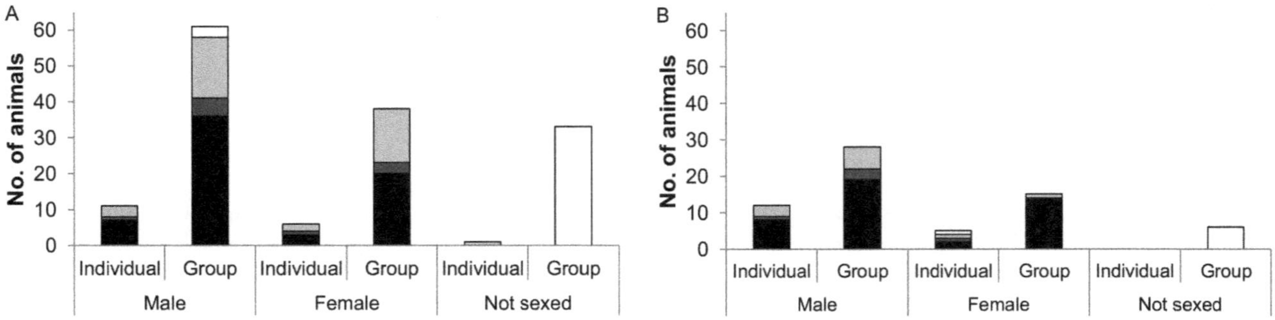


**Figure 2. The age group and sex of A) common and B) striped dolphins caught as individuals, or as part of groups.** Juvenile (black), sub-adult (dark grey), adult (light grey), not aged (white).

group catches differed from that of common dolphins with catches of adults and juveniles (n = 6) occurring almost as frequently as all juvenile groups (n = 7) (Table 2).

## Operational and environmental parameters

The most parsimonious model for common dolphin occurrence contained the variable depth. The variables sea state, light and moon also featured in the set of candidate models (models with ΔAIC ≤2 of the most parsimonious model) (Table 3). On the basis of model averaging, the relative importance of the variables in determining the occurrence of common dolphin bycatch was, in order of importance; depth; sea state; daylight; and moon (Table 4). The relative importance of depth and its occurrence in the most parsimonious model indicates that, of the variables examined, the depth of the water column where the fishing operation took place was the most important predictor of common dolphin bycatch. The probability of catching common dolphins decreased with increasing depth. Sea state and daylight had a similar level of importance. The probability of common dolphin bycatch increased with increasing sea state during setting and hauling of nets, and with increased time spent setting and hauling pre-sunset and post-sunrise. Moon illumination had very little importance in common dolphin bycatch with the probability of common bycatch occurring decreasing with increasing moon illumination. Effort did not feature in the set of candidate models for common dolphin bycatch.

The most parsimonious model for the occurrence of striped dolphins contained the variables depth and daylight (Table 5). The variables effort and moon also featured within the set of candidate models. The relative importance of the variables in determining the occurrence of striped dolphins was daylight; depth; moon; and effort (Table 6). The relative importance of depth and daylight and their occurrence in the most parsimonious model indicate that the depth of the water column where the fishing operation took place, and the extent to which fishing operations were conducted during daylight hours, were the most important predictors of striped dolphin bycatch. Daylight and depth had similar levels of importance. As with common dolphin bycatch, the probability of catching striped dolphins increased with increased time spent setting and hauling nets pre sunset and post sunrise. However, contrary to the pattern seen in common dolphin bycatch the probability of catching striped dolphins increased with increasing depth. Effort and moon had relatively low importance. Striped dolphin bycatch increased with effort but decreased with increasing moon illumination. The variable sea state did not feature in the candidate models for striped dolphin bycatch.

**Table 2.** The age class and sex composition of common dolphin and striped dolphin group catches.

| | **Common dolphin** | | **Striped dolphin** | |
|---|---|---|---|---|
| | **No. of groups** | **No. individuals within groups** | **No. of groups** | **No. individuals within groups** |
| **All juveniles** | 4 | 12 | 7 | 19 |
| **All sub-adult** | 0 | 0 | 0 | 0 |
| **All adult** | 3 | 6 | 0 | 0 |
| **Mix of juvenile & sub-adult** | 2 | 7 | 1 | 5 |
| **Mix of juvenile & adult** | 13 | 56 | 6 | 16 |
| **Mix of sub-adult & adult** | 1 | 3 | 0 | 0 |
| **Mix juvenile, sub-adult & adult** | 2 | 9 | 1 | 3 |
| **Not aged** | 8 | 39 | 3 | 6 |
| **All male** | 8 | 19 | 5 | 8 |
| **All female** | 1 | 2 | 0 | 0 |
| **Male & female** | 18 | 85 | 11 | 35 |
| **Not sexed** | 6 | 26 | 3 | 6 |

**Table 3.** Candidate models (ΔAIC≤2) of common dolphin occurrence in bycatch in the albacore tuna fishery.

| Rank | Model structure | AIC | ΔAIC | $w_i$ |
|---|---|---|---|---|
| 1 | Depth | 141.22 | 0.00 | 0.29 |
| 2 | Depth+Sea state | 141.65 | 0.43 | 0.24 |
| 3 | Depth+Light | 141.95 | 0.72 | 0.21 |
| 4 | Depth+Light+Sea state | 142.69 | 1.46 | 0.14 |
| 5 | Depth+Moon | 142.99 | 1.76 | 0.12 |

## Discussion

Common and striped dolphins dominated cetacean bycatch in the albacore tuna fishery. With the exception of harbour porpoise (*Phocoena phocoena*), which typically occur in shallower waters beyond the range of the tuna fishery, common dolphins are the most abundant small cetacean in the north east Atlantic in summer, and striped dolphins the second most abundant [43,44,45]. Common dolphins are present in both shelf and deep waters in the northeast Atlantic, whilst striped dolphins are restricted to deeper waters [44,46]. Of the variables examined, the depth of the water column where the fishing operation took place was the most important factor driving the occurrence of each species in bycatch in the albacore tuna fishery. The direction of influence reflects what is known about the distribution of the species across the depth gradient with striped dolphins more likely to be caught in deeper waters and common dolphins in shallower waters. Common dolphins have been recorded as bycatch in mobile [47,48,49] and static gears [45,50] in the northeast Atlantic. Although their abundance in the region and their distribution across the continental slope and deeper offshore waters is likely the key factor accounting for their prevalence in bycatch, as indicated by the importance of fishing depth, behavioural factors and fine scale movements within the region, may underpin the observed patterns.

The bycatch demographics were skewed towards males and young animals for both species, although a number of mature females were also caught. The prevalence of males and young animals in the bycatch suggests that some sections of the common and striped dolphin populations may be more vulnerable to capture than others. Several factors may drive the patterns observed in the bycatch. Firstly, the age and sex structure in the bycatch represents that of the wider population. Secondly, there is age and sex segregation in the wider population leading to different ratios in the area where the fishery operated. Thirdly, behavioural differences between age groups and sexes contributed to sections of the population being more vulnerable [51]. Finally, the size selective nature of gillnets contributed to the frequency of smaller animals in the bycatch. A single driver may be responsible for the observed pattern, but given the resultant complexities of multiple drivers, potentially acting in parallel, it is difficult to determine which scenario has greatest influence.

There is a lack of unbiased sources of demographic data for cetaceans [28] and little is known of the age and sex structure of the wider striped and common dolphin populations in the northeast Atlantic, as a whole. Delphinidae populations are, in general, segregated by age and sex [52] and skewed sex and age ratios in stranded and bycatch animals supports the hypothesis of age and sex segregation in the wider population of the northeast Atlantic, leading to regional variation in age and sex structure. The age distribution of common dolphins in this study is similar to the pattern in strandings along the French coast [28] but differs from that reported in pair trawls in northwest Spain, where animals in the sub-adult range were caught in greater numbers than juveniles and adults [47]. The prevalence of common dolphin males has been recorded in other gillnet fisheries [53] and in trawls [47]. Striped dolphin males dominated in Spanish driftnets in the western Mediterranean, with common dolphins exhibiting a more even sex ratio [54]. Whilst, the male bias in common dolphins recorded in trawlers operating off northwest Spain resulted from several large all male capture events and was thought to provide evidence of groups of bachelor males in the area [47]. When groups of mixed age where captured in the albacore tuna fishery they were most frequently composed of adult females with juveniles, indicating that the fishery may have overlapped with calving grounds or maternal feeding grounds [31]. The overlap between the albacore fishing season and the common dolphin calving season [29,30] provides further support to the theory that the fishery encountered these age groups more frequently than any other contributing to their abundance in bycatch.

Whilst the skewed age and sex distributions recorded in the bycatch are interesting, it is important to consider that not all individuals were aged and sexed. This was particularly true in the case of group catches where post mortem examination was constrained by the time and space available to process large numbers of animals on board. Furthermore, we based our

**Table 4.** Model averaging results of common dolphin occurrence in bycatch in the albacore tuna fishery.

| Variable | β | SE | RI |
|---|---|---|---|
| **Depth** | −1.08 | 0.27 | 1.00 |
| **Sea State** | 0.26 | 0.21 | 0.38 |
| **Daylight** | 0.24 | 0.21 | 0.35 |
| **Moon** | −0.14 | 0.22 | 0.12 |

β model averaged coefficients; SE standard error; RI relative importance.

**Table 5.** Candidate models (ΔAIC≤2) of striped dolphin occurrence in bycatch in the albacore tuna fishery.

| Rank | Model structure | AIC | ΔAIC | $w_i$ |
|---|---|---|---|---|
| 1 | Depth+Daylight | 141.53 | 0.00 | 0.36 |
| 2 | Depth+Effort+Daylight | 142.81 | 1.28 | 0.19 |
| 3 | Depth+Moon+Daylight | 143.12 | 1.59 | 0.16 |
| 4 | Daylight | 143.25 | 1.72 | 0.15 |
| 5 | Depth | 143.52 | 1.99 | 0.13 |

assessment of common dolphin population age structure on survivorship of females in the Bay of Biscay as determined through strandings [28]. Aside from the biases inherent in demographic data from stranded animals [28], the albacore fishery operated in waters north of the Bay of Biscay, off the western coast of Ireland. We assumed, for the purposes of the assessment, that survivorship was similar between these regions. Given the proximity of the regions and the apparent lack of genetic structure within the common dolphin population in the northeast Atlantic [55] this assumption seems reasonable. However, analysis of heavy metals [56,57] and stable isotopes [57] suggest the existence of a neritic stock on the continental shelf and an oceanic stock offshore. The albacore fishery operated in offshore waters and it is likely that bycaught common dolphins predominantly came from the oceanic stock. Conversely, strandings are potentially dominated by animals from the neritic stock [28]. Furthermore we assumed, for the purposes of the assessment, that survivorship was similar for males and females as data on the survivorship of males was not available. Given the prevalence of young male bycatch in the albacore tuna fishery and other fisheries operating in the northeast Atlantic [47,53] it is possible that survivorship in male common dolphins is lower than that of their female counterparts in the region.

Common and striped dolphins form social groups of different age classes so it is perhaps not surprising that group catches outnumbered incidental takes of single individuals throughout the range of the fishery. It is not possible to determine whether group catches were truly social groups, or whether other factors resulted in multiple individuals being caught in the same set. Nevertheless, the occurrence of both individual and group catches could suggest that some individuals within a social group, such as young animals, may be more vulnerable to capture than others, or may behave differently from other individuals within the group. These potential differences in behaviour between age groups and sexes may compound spatial segregation within populations. Behavioural responses to boats may contribute to bycatch as both species have been documented to respond to survey boats by approaching them [43,44,46]. A similar response elicited by fishing boats setting and hauling nets may increase the risk of being caught. Across all mammal populations, males and adolescents are less risk averse than females and other age groups [58,59] and may be more inclined to approach a boat during the setting or retrieval of gear, increasing the risk of capture. Sub-adults, or adolescents, were under represented in the albacore tuna bycatch, which in addition to the hypothesis of age and sex segregation amongst the population could indicate an influence of learned behaviour or experience. The younger animals, which dominated the bycatch, may have less precise echolocation abilities [31], may vary in their behavioural response to boats or may lack the physical skills or experience necessary to feed within the vicinity of the net without becoming entangled.

Gillnets are size selective and catch typically reflects stock structure and the mesh size deployed [60]. To our knowledge the influence of size selectivity on cetacean bycatch in gillnets has not been tested with regard to mesh size. Measures to reduce selectivity towards cetaceans typically focus on modifications to gear deployment and retrieval, or changes to fishing practices such as the height of static gear deployment in the water column, or the speed, depth and duration of trawls [17]. It is possible that mesh size also contributes to selectivity for cetaceans and may act to compound the influence of sex and age segregation and behaviour. The stretch mesh size of deployed gillnets was 17.8 cm and the majority of the common and striped dolphin bycatch were juvenile animals less than 185 cm in length. Mean length of juvenile bycatch was 136 cm for common dolphins and 146 cm for striped dolphins. In a size selectivity experiment for sandbar shark (*Carcharhinus plumbeus*), all mesh sizes captured all size classes but maximum selectivity of 17.8 cm mesh was between 95 and 124 cm size classes [60]. Sharks and dolphins have the same fusiform shape, and size selectivity may be similar. However, features of dolphin morphology, including long, narrow rostrums may increase the vulnerability of larger individuals to capture, and this may explain why adult dolphins were also captured in large numbers. Cetacean bycatch in the albacore fishery also included larger species which lack long, narrow rostrums, including minke whales, long-finned pilot whales and sperm whales and it is possible that the mechanism by which larger individuals, and

**Table 6.** Model averaging results of striped dolphin occurrence in bycatch in the albacore tuna fishery.

| Variable | β | SE | RI |
|---|---|---|---|
| **Daylight** | 0.47 | 0.27 | 0.87 |
| **Depth** | 0.42 | 0.23 | 0.85 |
| **Effort** | 0.23 | 0.24 | 0.19 |
| **Moon** | −0.17 | 0.22 | 0.16 |

β model averaged coefficients; SE standard error; RI relative importance.

larger species, are captured differs from that of smaller animals. It was hypothesised that in the case of sandbar sharks, larger animals got caught by wrapping themselves in the net, or breaking it and becoming caught in larger openings. A similar mechanism of entanglement may capture larger cetacean species and larger, older, common and striped dolphins.

The behaviour of dolphins around nets, and the mechanism by which they become entangled, is not fully understood but assessment of echolocation abilities and observations at sea suggest some species may be capable of detecting and avoiding nets [61,62]. A number of scenarios could explain how cetaceans become entangled in gear they are capable of detecting. It has been suggested that the animals may not echolocate while travelling, failing to detect the net while in transit; they may detect the presence of the net but not identify it as a barrier; they may become distracted while feeding in the vicinity of the net; or fish entangled in the net, and free swimming in its vicinity, may block detection of the net [61,62,63]. Common dolphin bycatch increased with increasing sea state during setting and hauling suggesting that increasing turbidity, resulting from increased sea state, may hamper net detection.

The timing of cetacean bycatch within a fishing operation is also poorly understood. Video technology may be useful for monitoring stages which cannot be observed from deck for mobile gears [64] but may not be appropriate for static nets. The operational variable effort, which described the length of net deployed and its soak time in the water, had little influence on the occurrence of striped dolphin and no influence on the occurrence of common dolphin bycatch. The variable moon illumination, which captured the brightness of the moon during the soak, also had very little influence. The importance of operational variables relating to the setting and hauling of the net, including the influence of sea state during setting and hauling on common dolphin bycatch and the importance of daylight during setting and hauling for both species, suggests that this may be the stage of the fishing operation when the species are most vulnerable.

A greater vulnerability towards fishing operations conducted at night has been documented in a number of fisheries operating in the northeast Atlantic [45,47,48,65] and our findings suggest that timing of operations could be important on a much finer scale. The increase in bycatch with increased daylight during setting and hauling may seem counter intuitive as visual acuity and therefore visual detection of the net should be greater in daylight than in darkness. Indeed, although illumination of the moon had comparatively little influence on the occurrence of either species, the probability of bycatch decreased with increasing lunar illumination. Two mechanisms may contribute to increased likelihood of cetacean bycatch during setting and hauling in daylight. Firstly, observers monitoring nets during daylight hours may be more likely to document animals which fall out of the net as it is being hauled on board. These "drop outs" could be missed during operations occurring in darkness if the animal fell out some distance from the boat. Likewise, carcases floating in the water are more likely to be spotted and documented during daylight than at night. Secondly, the species may be more likely to come into contact with the net during certain periods of the night as a result of diurnal behaviour or movement.

Little is known about the behaviour or movements of common and striped dolphins, worldwide. Observations from the Mediterranean [66,67,68] and Australia [69] may shed light on how diurnal movements of these species may contribute to their capture. Feeding in striped dolphins, in the French Riviera, was found to peak in the three hours before sunset and after sun rise [66]. The same study revealed that although the majority of animals remained in offshore areas a significant number of animals exhibited diurnal movements from inshore to offshore areas coinciding with these peaks in feeding activity. If similar patterns of feeding and movement are present in striped dolphins in the northeast Atlantic, operations conducted during the pre-sunset and post sunrise period could coincide with these activity peaks, thus encountering more animals. Although diurnal patterns in feeding behaviour have been detected in common dolphins [68,69], similar diurnal horizontal movements have not. Feeding in oceanic common dolphin peaks at dusk, as the animals feed on mesopelagic fish which migrate to the surface at night [70]. Common dolphins have been recorded appearing in the vicinity of prawn trawlers, in Australia, during net hauling, seemingly attracted by the sound of the engines [71]. The extent of setting and hauling occurring in daylight may be less important for common dolphins if the overall driver for their presence at the boat is their attraction to it, rather than any increase in horizontal movements during this period.

The complex nature of species distribution and the dynamic nature of the fishery in space and time mean it is an oversimplification to suggest that our model captures all variables driving bycatch of these species. Observer data have been used to examine the influence of operational and environmental variables in cetacean bycatch for a number of species and fisheries e.g. [64,72,73,74,75] and although observer data are important for examining the patterns in bycatch, several issues should be considered. There is the potential for boats carrying observers to vary their behaviour to reduce the likelihood of bycatch occurring, perhaps by avoiding areas where it has occurred in the past [28], however in the northeast Atlantic fishing for albacore occurs between June and October and the short duration of the albacore season, and of the fishing trips, makes avoidance less likely in this case. Secondly, whilst bycatch "hotspots" occur in many regions it is worth noting that lack of documented bycatch incidents in an area, does not mean that bycatch would not happen there under different circumstances [76] and this is particularly relevant when assessing the role of operational and environmental factors.

The lack of importance of effort in bycatch occurrence, considered alongside the importance of depth, indicates that the location of the net, rather than the length of the net or time in the water, influenced the probability of cetacean bycatch occurring in this fishery. If animals are more vulnerable during net setting or hauling, either as a result of diurnal movement, timing of feeding, or attraction to the boat, the length of net deployed would have little influence on numbers captured and restricting net length would have little impact as a mitigation measure. Prior to banning the use of pelagic driftnets to target highly migratory species, the European Commission prohibited the deployment of driftnets over 2.5 km in length. Our results support the decision that length restrictions alone are not sufficient to limit cetacean bycatch. It is also important, in the context of Ecosystem Based Fishery Management (EBFM), to emphasize that the EU restrictions on driftnets were introduced not only to reduce levels of cetacean bycatch but to address the unsustainable level of seabird, turtle and non-target fish species in this gear.

## Conclusion

This study illustrates that data collected during observer programmes can be utilised beyond estimating the number of animals taken by a fishery and that key patterns, in the factors influencing bycatch occurrence, can be elucidated. Analysis of bycatch should not only consider the operational and environmental factors which may drive it but, where possible, should consider the age and sex of the animals taken. Differing susceptibility across age classes and sexes could have important consequences for population recovery, as it may take longer to recover from depletion if animals of reproductive age [77], and particularly females [78,79] were impacted. Whilst driftnets are no longer deployed to target albacore tuna in the northeast Atlantic, striped and common dolphin bycatch persists in fisheries deploying other gears including pair trawls, otter trawls and set gillnets. Under EBFM, Ecological Risk Assessments for the Effects of Fishing (ERAEF) are increasingly used to examine the impact of commercial fisheries on non-target and bycatch species including cetaceans [19]. ERAEF includes assessment of the ability of a species to withstand fishing pressure and the likelihood that it will encounter fishing pressure based on spatial and temporal overlap with the fishery and susceptibility to the gear deployed. It is our view that observer programmes for all fisheries should, where possible, include detailed assessment of the sections of the populations affected as this, considered alongside the operational and environmental conditions under which bycatch occurs, could contribute to the ongoing refinement of ERAEF in the context of cetaceans.

## Acknowledgments

We are grateful to the observers who collected the data, Dave O'Leary, John Boyd, Mark Wilder, Graham Johnston; and the skippers and crews of the fishing boats in this study for their help and cooperation. We would like to thank Mick Mackey, Irene Gassner, Sarah-Jane Moore and Sinead Murphy for help with sample processing and ageing. Thanks to Sarah Kraak for helpful comments on a previous draft of this manuscript and to two anonymous reviewers who provided valuable and constructive comments and suggestions.

## Author Contributions

Analyzed the data: SB. Wrote the paper: SB ER DR. Conceived the analysis: ER.

## References


1. Jackson JBC, Kirby MX, Berger WH, Bjorndal KA, Botsford LW, et al. (2001) Historical overfishing and the recent collapse of coastal ecosystems. Science 293: 629–638.
2. Travers M, Watermeyer K, Shannon LJ, Shin Y-J (2010) Changes in food web structure under scenarios of overfishing in the southern Benguela: comparison of the Ecosim and OSMOSE modelling approaches. J Mar Syst 79: 101–111.
3. Cury PM, Shin Y-J, Planque B, Durant JM, Fromentin J-M, et al. (2008) Ecosystem oceanography for global change in fisheries. Trends Ecol Evol 23: 338–346.
4. Perry RI, Cury P, Brander K, Jennings S, Möllmann C, et al. (2010) Sensitivity of marine systems to climate and fishing: concepts, issues and management responses. J Mar Syst 79: 427–435.
5. Levin PS, Dufault A (2010) Eating up the food web. Fish Fish 11: 307–312.
6. Yates KL, Payo Payo A, Schoeman DS (2012) International, regional and national commitments meet local implementation: a case study of marine conservation in Northern Ireland. Mar Policy 38: 140–150.
7. Myers RA, Worm B (2003) Rapid worldwide depletion of predatory fish communities. Nature 423: 280–283.
8. Moore CH, Harvey ES, Van Niel KP (2009) Spatial prediction of demersal fish distributions: enhancing our understanding of species – environment relationships. ICES J Mar Sci 66: 2068–2075.
9. Worm B, Barbier EB, Beaumont N, Duffy JE, Folke C, et al. (2006) Impacts of biodiversity loss on ocean ecosystem services. Science 314: 787–790.
10. Read AJ, Drinker P, Northridge S (2006) Bycatch of marine mammals in U.S. and Global fisheries. Conserv Biol 20: 163–169.
11. Lewison RL, Crowder LB, Read AJ, Freeman SA (2004) Understanding impacts of fisheries bycatch on marine megafauna. Trends Ecol Evol 19: 598–604.
12. Anderson ORJ, Small CJ, Croxall JP, Dunn EK, Sullivan BJ, et al. (2011) Global seabird bycatch in longline fisheries. Endanger Species Res 14: 91–106.
13. Berkley SA, Compos WL (1988) Relative abundance and fishery potential of pelagic sharks along Florida's east coast. Mar Fish Rev 50: 9–16.
14. Lewison RL, Crowder LB (2007) Putting longline bycatch of sea turtles into perspective. Conserv Biol 21: 79–86.
15. Hamer DJ, Goldsworthy SD (2006) Seal-fishery operational interactions: identifying the environmental and operational aspects of a trawl fishery that contribute to by-catch and mortality of Australian fur seals (*Arctocephalus pusillus doriferus*). Biol Conserv 130: 517–529.
16. Carbonell A, Alemany F, Merella P, Quetglas A, Roman E (2003) The bycatch of sharks in the western Mediterranean (Balearic Islands) fishery. Fish Res 61: 7–18.
17. Hall MA, Alverson DL, Metuzals KI (2000) By-catch: problems and solutions. Mar Pollut Bull 41: 204–219.
18. Goldsworthy SD, Bulman C, He X, Larcombe J, Littnan C (2006) Trophic interactions between marine mammals and Australian fisheries: an ecosystem approach. In: Gales NJ, Hindell MA, Kirkwood R, editors. Marine Mammals: Fisheries Tourism and Management Issues. Melbourne, Australia: CSIRO Publishing. Pp. 62–99.
19. Brown SL, Reid D, Rogan E (2013) A risk-based approach to rapidly screen vulnerability of cetaceans to impacts from fisheries bycatch. Biol Conserv 168: 78–87.
20. Northridge S (1991) Driftnet fisheries and their impacts on non-target species: a worldwide review. FAO Fisheries Technical Paper No. 320. Rome, Italy: Food and Agriculture Organization of the United Nations.
21. Council of the European Union (1998) Council Regulation (EC) No. 1239/98 of 8 June 1998 amending Regulation (EC) No. 894/97 laying down certain technical measures for the conservation of fisheries resources.
22. Kastelein RA, Jennings N, Verboom WC, de Haan D, Schooneman NM (2006) Differences in the response of a striped dolphin (*Stenella coeruleoalba*) and a harbour porpoise (*Phocoena phocoena*) to an acoustic alarm. Mar Environ Res 61: 363–378.
23. Kastelein RA, van der Heul S, van der Veen J, Verboom WC, Jennings N, et al. (2007) Effects of acoustic alarms, designed to reduce small cetacean bycatch in gillnet fisheries, on the behaviour of North Sea fish species in a large tank. Mar Environ Res 64: 160–180.
24. Carretta JV, Forney KA, Oleson E, Martien K, Muto MM, et al. (2012) US Pacific Marine Stock Assessments: 2011. Technical Memorandum, NMFS-SWFSC-488. US Department of Commerce, National Oceanic and Atmospheric Administration, National Marine Fisheries Service.
25. Fiedler FN, Sales G, Giffoni BB, Monteiro-Filho EL, Secchi ER (2012) Driftnet fishery threats sea turtles in the Atlantic Ocean. Biol Conserv 21: 915–931.
26. Tudela S, Kai Kai A, Maynou F, El Andalossi M, Guglielmi P (2005) Driftnet fishing and biodiversity conservation: the case study of the large-scale Moroccan driftnet fleet operating in the Alboran Sea (SW Mediterranean). Biol Conserv 121: 65–78.
27. Rogan E, Mackey M (2007) Megafauna bycatch in drift nets for albacore tuna (Thunnus alalunga) in the NE Atlantic. Fish Res 86: 6–14.
28. Mannocci L, Dabin W, Augeraud-Véron E, Dupuy J-F, Barbraud C, et al. (2012) Assessing the Impact of Bycatch on Dolphin Populations: The Case of the Common Dolphin in the Eastern North Atlantic. PLoS ONE 7(2): e32615. doi:10.1371/journal.pone.0032615.
29. Murphy S, Winship A, Dabin W, Jepson PD, Deaville R, et al. (2009) Importance of biological parameters in assessing the status of *Delphinus delphis*. Mar Ecol Prog Ser 388: 273–291.
30. Murphy S, Collet A, Rogan E (2005) Mating strategy in the male common dolphin (*Delphinus delphis*): What gonadal analysis tells us. J Mammal 86: 1247–1258.
31. Murphy S, Rogan E (2006) External morphology of the short-beaked common dolphin, *Delphinus delphis:* growth, allometric relationships and sexual dimorphism. Acta Zool 87: 315–329.
32. Di-Méglio N, Romero-Alvarez R, Collet A (1996) Growth comparison in striped dolphins *Stenella coeruleoalba* from the Atlantic and Mediterranean coasts of France. Aquat Mamm 22: 11–21.
33. R Core Team (2013) R: A language and environment for statistical computing. R Foundation for Statistical Computing, Vienna, Austria. Available: http://www.R-project.org/. Accessed 2014 Feb 5.
34. Liang K, Zeger S (1986) Longitudinal data analysis using generalised linear models. Biometrika 73: 13–22.
35. Højsgaard S, Halekoh U, Yan J (2006) The R Package geepack for Generalized Estimating Equations. J Stat Softw 15: 1–11.

36. Yan J (2002) Geepack: Yet another Package for Generalised Estimating Equations. R-News 2/3: 12–14. Available: http://cran.r-project.org/doc/Rnews/Accessed 2014 Feb 5.
37. Yan J, Fine JP (2004) Estimating Equations for association structures. Stat Med 23: 859–880.
38. Broström G, Holmberg H (2011) glmmML: Generalized linear models with clustering. R package version 0.82–1. Available: http://CRAN.R-project.org/package = glmmML.
39. Booth GD, Niccolucci MJ, Schuster EG (1994) Identifying proxy sets in multiple linear regression: an aid to better coefficient interpretation. Research paper INT-470. United States Department of Agriculture, Forest Service, Ogden, USA.
40. Burnham KP, DR Anderson (2004) Multimodel inference understanding AIC and BIC in model selection. Sociol Methods Res 33: 261–304.
41. Burnham KP, Anderson DR (2002) Model selection and multimodel inference: a practical information-theoretic approach. Springer, New York, USA.
42. Bartón K (2013) MuMIn: Multi-model inference. R package version 1.9.5. http://CRAN.R-project.org/package = MuMIn.
43. Hammond PS, Macleod K, Berggren P, Borchers DL, Burt L, et al. (2013) Cetacean abundance and distribution in European Atlantic shelf waters to inform conservation and management. Biol Conserv 164: 107–122.
44. Hammond PS, McLeod K, Gillepsie D, Swift R, Winship A (2009) Cetacean Offshore Distribution and Abundance in the European Atlantic (CODA). Final Report. University of Saint Andrews, Scotland.
45. Lopéz A, Pierce GJ, Santos MB, Gracia J, Guerra A (2003) Fishery by-catches of marine mammals in Galician waters; results from on-board observations and an interview survey of fishermen. Biol Conserv 111: 25–40.
46. Ó Cadhla O, Mackey M, Aguilar de Soto N, Rogan E, Connolly N (2002) Cetaceans and seabirds of Ireland's Atlantic Margin. Volume II Cetacean Distribution and Abundance. Cork, Ireland: University College Cork.
47. Fernández-Contreras MM, Cardona L, Lockyer CH, Aguilar A (2010) Incidental bycatch of short-beaked common dolphins (*Delphinus delphis*) by pairtrawlers off northwestern Spain. ICES J Mar Sci 67: 1732–1738.
48. Morizur Y, Berrow SD, Tregenza NJC, Couperus AS, Pouvreau S (1999) Incidental catches of marine-mammals in pelagic trawl fisheries of the northeast Atlantic. Fish Res 41: 297–307.
49. Northridge S, Kingston A, Thomas L, Mackay A (2007) Second annual report on the UK cetacean bycatch monitoring scheme. Contract report to DEFRA on the work conducted 2005–2006. St. Andrews, Scotland: Sea Mammal Research Unit.
50. Tregenza NJC, Berrow SD, Hammond PS, Leaper R (1997) Common dolphin, *Delphinus delphis* L., bycatch in bottom set gillnets in the Celtic Sea. Rep Int Whal Commn 47: 835–839.
51. Atkins S, Cliff G, Pillay N (2013) Humpback dolphin bycatch in the shark nets in KwaZulu-Natal, South Africa. Biol Conserv 159: 442–449.
52. Perrin WF, Reilly SB (1984) Reproductive parameters of dolphins and small whales of the family Delphinidae. In Perrin WF, Brownell RL, DeMaster DP, editors. Reproduction in whales, dolphins and porpoises. Report of the International Whaling Commission Special Issue 6. 97–133.
53. Ferrero RC, Walker WA (1995) Growth and reproduction of the common dolphin, *Delphinus delphis* Linnaeus, in the offshore waters of the North Pacific Ocean. Fish Bull 93: 483–494.
54. Silvani L, Gazo M, Aguilar A (1999) Spanish driftnet fishing and incidental catches in the western Mediterranean. Biol Conserv 90: 79–85.
55. Mirimin L, Westgate A, Rogan E, Rosel P, Read A, et al. (2009) Population structure of short-beaked common dolphins (*Delphius delphis*) in the north Atlantic ocean as revealed by mitochondrial and nuclear genetic markers. Mar Biol 156: 821–834.
56. Lahaye V, Bustamante P, Spitz J, Dabin W, Pierce GJ, et al. (2005) Long-term dietary segregation of common dolphins Delphinus delphis in the Bay of Biscay, determined using cadmium as an ecological tracer. Mar Ecol Prog Ser 305: 275–285.
57. Caurant F, Chouvelon T, Lahaye V, Mendez-Fernandez P, Rogan E, et al. (2009) The use of ecological tracers for discriminating dolphin population structure: the case of the short-beaked common dolphin Delphinus delphis in European Atlantic waters. International Whaling Commission: Madeira.
58. Macrí S, Adriani W, Chiarotti F, Laviola G (2002) Risk taking during exploration of a plus-maze is greater in adolescent than in juvenile or adult mice. Anim Behav 64: 541–546.
59. Owens I (2002) Sex differences in mortality rate. Science 297: 2008–2009.
60. McAuley RB, Simpfendorfer CA, Wright IW (2007) Gillnet mesh selectivity of the sandbar shark (*Carcharhinus plumbeus*): implications for fisheries management. ICES J Mar Sci 64: 1702–1709.
61. Au WWL, Jones L (1991) Acoustic reflectivity of nets: implications concerning incidental take of dolphins. Mar Mamm Sci 7: 258–273.
62. Cox TM, Read AJ, Swanner D, Urian K, Waples D (2003) Behavioural responses of bottlenose dolphins, *Tursiops truncatus*, to gillnets and acoustic alarms. Biol Conserv 115: 203–212.
63. Mooney TA, Au WWL, Nachtigall PE, Trippel EA (2007) Acoustic and stiffness properties of gillnets as they relate to small cetacean bycatch. ICES J Mar Sci 64: 1324–1332.
64. Rossman MC (2010) Estimated bycatch of small cetaceans in northeast US bottom trawl fishing gear during 2000–2005. J Northw Atl Fish Sci 42: 77–101.
65. de Boer MN, Saulino JT, Leopold MF, Reijnders PJH, Simmonds MP (2012) Interactions between short-beaked common dolphin (*Delphinus delphis*) and the winter pelagic pair-trawl fishery off southwest England (UK). Int J Biodivers Conserv 4: 481–499.
66. Gannier A (1999) Diel variations of the striped dolphin distribution off the French Riviera (Northwestern Mediterranean Sea). Aquat Mamm 25: 123–134.
67. Neumann DR (2001) The activity budget of free-ranging common dolphins (*Delphinus delphis*) in the northwestern Bay of Plenty, New Zealand. Aquat Mamm 27: 121–136.
68. Stockin KA, Binedell V, Wiseman N, Brunton DH, Orams MB (2009) Behaviour of free-ranging common dolphins (*Delphinus* sp.) in the Hauraki Gulf, New Zealand. Mar Mamm Sci 25: 283–301.
69. Filby NE, Bossley M, Stockin KA (2013) Behaviour of free-ranging short-beaked common dolphins (*Delphinus delphis*) in Gulf St. Vincent, South Australia. Aust J Zool 61(4): 291–300 http://dx.doi.org/10.1071/ZO12033.
70. Pusineri C, Magnin V, Meynier L, Spitz J, Hassani S, et al. (2007) Food and feeding ecology of the common dolphin (*Delphinus delphis*) in the oceanic northeast Atlantic and comparison with its diet in neritic areas. Mar Mamm Sci 23: 30–47.
71. Svane I, Hammett Z, Lauer P (2005) Impacts of trawling on benthic macrofauna and flora of the Spencer Gulf prawn fishing grounds. Estuar Coast Shelf Sci 82: 621–631.
72. Bjorge A, Skern-Mauritzen M, Rossman MC (2013) Estimated bycatch of harbour porpoise (*Phocoena phocoena*) in two coastal gillnet fisheries in Norway, 2006–2008. Mitigation and implications for conservation. Biol Conserv 161: 164–173.
73. Thompson FN, Abraham ER, Berkenbusch K (2013) Common Dolphin (Delphinus delphis) Bycatch in New Zealand Commercial Trawl Fisheries. PLoS ONE 8(5): e64438. doi:10.1371/journal.pone.0064438.
74. Orphanides CD (2009) Protected species bycatch estimating approaches: estimating harbor porpoise bycatch in U. S. northwestern Atlantic gillnet fisheries. J Northw Atl Fish Sci 42: 55–76.
75. Palka D, Orphanides CD, Warden ML (2009) Summary of harbour porpoise (*Phocoena phocoena*) bycatch and levels of compliance in the northeast and mid-Atlantic gillnet fisheries after the implementation of the Take Reduction Plan: 1 January 1999–31 May 2007. NOAA Technical Memorandum NMFS-NE-212. Woods Hole, Massachusetts: Northeast Fisheries Science Centre.
76. Northridge S, Morizur Y, Souami Y, Van Canneyt O (2006) "Petracet" Pelagic Fisheries and Cetaceans. Final report to the European Commission. Project EC/FISH/2003/09. Lymington, UK: MacAlister Elliott and Partners.
77. Wade PR, Reeves RR, Mesnick SL (2012) Social and behavioural factors in cetacean responses to overexploitation: are Odontocetes less "resilient" than Mysticetes? J Mar Biol doi:10.1155/2012/567276.
78. Van der Hoop JM, Moore MJ, Barco SG, Cole TVN, Daoust P-Y, et al. (2013) Assessment of management to mitigate anthropogenic effects on large whales. Conserv Biol 27: 121–133.
79. Leung ES, Chilvers BL, Nakagawa S, Moore AB, Robertson BC (2012) Sexual segregation in juvenile New Zealand Sea Lion foraging ranges: Implications for intraspecific competition, population dynamics and conservation. PLoS ONE 7(9): e45389. doi:10.1371/journal.pone.0045389.

# 13

# Seafood Substitutions Obscure Patterns of Mercury Contamination in Patagonian Toothfish (*Dissostichus eleginoides*) or "Chilean Sea Bass"

**Peter B. Marko[1]*, Holly A. Nance[2], Peter van den Hurk[3]**

**1** Department of Biology, University of Hawai'i at Mānoa, Honolulu, Hawai'i, United States of America, **2** Indian River State College, Fort Pierce, Florida, United States of America, **3** Department of Biological Sciences, Clemson University, Clemson, South Carolina, United States of America

## Abstract

Seafood mislabeling distorts the true abundance of fish in the sea, defrauds consumers, and can also cause unwanted exposure to harmful pollutants. By combining genetic data with analyses of total mercury content, we have investigated how species substitutions and fishery-stock substitutions obscure mercury contamination in Patagonian toothfish (*Dissostichus eleginoides*), also known as "Chilean sea bass". Patagonian toothfish show wide variation in mercury concentrations such that consumers may be exposed to either acceptable or unacceptable levels of mercury depending on the geographic origins of the fish and the allowable limits of different countries. Most notably, stocks of Patagonian toothfish in Chile accumulate significantly more mercury than stocks closer to the South Pole, including the South Georgia/Shag Rocks stock, a fishery certified by the Marine Stewardship Council (MSC) as sustainably fished. Consistent with the documented geography of mercury contamination, our analysis showed that, on average, retail fish labeled as MSC-certified Patagonian toothfish had only half the mercury of uncertified fish. However, consideration of genetic data that were informative about seafood substitutions revealed a complex pattern of contamination hidden from consumers: *species* substitutions artificially inflated the expected difference in mercury levels between MSC-certified and uncertified fish whereas *fishery stock* substitutions artificially reduced the expected difference in mercury content between MSC-certified and uncertified fish that were actually *D. eleginoides*. Among MSC-certified fish that were actually *D. eleginoides*, several with exogenous mtDNA haplotypes (i.e., not known from the certified fishery) had mercury concentrations on par with uncertified fish from Chile. Overall, our analysis of mercury was consistent with inferences from the genetic data about the geographic origins of the fish, demonstrated the potential negative impact of seafood mislabeling on unwanted mercury exposure for consumers, and showed that fishery-stock substitutions may expose consumers to significantly greater mercury concentrations in retail-acquired fish than species substitutions.

**Editor:** Jaymie Meliker, Stony Brook University, Graduate Program in Public Health, United States of America

**Funding:** This work was supported by the National Science Foundation (OCE-0961996). The funder had no role in study design, data collection and analysis, decision to publish, or preparation of the manuscript.

**Competing Interests:** The authors have declared that no competing interests exist.

* Email: pmarko@hawaii.edu

## Introduction

Despite the many health benefits of eating fish, most commercially harvested fish are contaminated with mercury [1]. The most common form of mercury in fish is methylmercury, a neurotoxin that is especially dangerous to the developing nervous system [2]. Although present in only small quantities in the environment, mercury accumulates in living organisms. Among fish, accumulation of mercury is prevalent but variable (Fig. 1), primarily due to differences in trophic level and body size, such that mercury concentrations tend to be high in larger, longer-lived predatory fish [3]. Therefore, the amount of fish and the particular species of fish consumed are considered the most important factors determining the health risk associated with eating seafood contaminated with mercury. Consequently, the US Food and Drug Administration (FDA) advises pregnant and nursing women, women who may become pregnant, and young children not to eat species that have mean mercury concentrations near 1.0 ppm (shark, swordfish, king mackerel, and tilefish), and not to consume more than 12 ounces per week of other species that have lower mercury levels.

Less widely appreciated is the fact that variation in mercury concentration within species can also be large due to the same causal factors mentioned above [4–6]. As a consequence, even though the mean mercury level for many species commonly consumed is only ~0.3 ppm, the range of mercury concentrations reported from several of these moderately contaminated species often exceeds import limits (Fig. 1). For example, stocks of the southern hemisphere Patagonian toothfish (*Dissostichus eleginoides*), or "Chilean sea bass", have been shown to have anywhere from ~0.15 ppm Hg to >1.0 ppm Hg, often well beyond the allowable level of 0.5 ppm for imports into New Zealand, Canada, and Australia, and sometimes greater than the 1.0 ppm limit set by the European Union and the United States [4,6–10].

As with many other species of fish that absorb toxins at a greater rate than they are eliminated, mercury levels in Patagonian toothfish are explained in part by body size: the larger, and presumably the older a fish is, the greater the mercury concentration [7–10]. However, two recent studies have shown that mercury accumulation in Patagonian toothfish also has a strong geographic component. Specifically, fish from higher latitudes in the Southern Ocean have lower length-normalized mercury concentrations compared to fish from lower latitudes [9–10]. In particular, fish from the South Georgia/Shag Rocks (SGSR) stock have mean mercury levels far below allowable import levels worldwide, with a mean THg of 0.23 ppm (SE = 0.01 ppm) compared to fish from nearby Chile, which have more than three times as much mercury in their tissues (0.73 ppm, SE = 0.10 ppm) [9]. Mercury levels in fish taken from the SGSR stock are of particular interest given that several consumer-education (e.g., Seafood Watch) and seafood sustainability certification programs (e.g., Marine Stewardship Council or MSC) consider fish from SGSR populations as eco-friendly seafood choices, from the perspectives of sustainability and relatively low by-catch. This combination makes fish from certified-sustainable southern ocean stocks a potentially attractive choice for consumers interested in purchasing sustainably harvested seafood with consistently low levels of mercury.

Accurate labeling of seafood is essential to allow consumers to reliably choose sustainable fisheries [11–15], but labels may also serve to protect consumers from unhealthy mercury exposure. For example, several common *species* substitutions (i.e., the substitution of less desirable species for more expensive ones) can result in unintended high mercury consumption, such as the substitution of tilefish or king mackerel for grouper [12,17–18]. However, some *fishery-stock* substitutions also have the potential to significantly increase mercury exposure to consumers. For example, Gulf of Mexico tilefish averages 1.45 ppm Hg, but THg in Atlantic tilefish is an order of magnitude lower (Fig. 1). Tuna species from the Atlantic, Pacific and Mediterranean oceans also have significantly different length-normalized mercury concentrations [4].

Here, we report measured mercury concentrations in Patagonian toothfish or "Chilean sea bass" acquired from retail merchants in the USA. As an indicator for species and fishery-stock substitutions, we used genetic data from the same samples that showed 20% of fish labeled as "Chilean sea bass" were not *D. eleginoides* (8% of fish sold as MSC-certified Chilean sea bass were other species whereas 46% of uncertified fish were substitutes), and among those certified fish positively identified with DNA techniques as *D. eleginoides*, 15% had mtDNA haplotypes that were not present in the previously described MSC-certified stock [16]. Our analysis of mercury contamination in the genetically-identified samples shows that the pattern of mercury contamination is consistent with the interpretation of mislabeling based on mtDNA haplotype identities, and demonstrates that in addition to species substitutions, fishery-stock substitutions have the potential to significantly obscure mercury exposure for consumers.

## Methods

### Samples

The tissue samples used in this study were a subset of those used in an analysis of the genetic composition of retail-acquired "Chilean Sea Bass" [16]. All of the samples were purchased "fresh" (i.e., from the thawed seafood counter) from retailers in 10 US states as filets (body wall muscle), which were immediately stored in 90% ethanol to preserve mtDNA. Samples were acquired from retailers selling either MSC-certified Chilean sea bass (and were clearly labeled as such) or from retailers selling fish as Chilean sea bass, but not as MSC-certified. We measured the total mercury (THg) concentration in 25 of the MSC-certified fish and 13 of the non-MSC fish.

### Total mercury concentration

Samples stored in ethanol were freeze-dried in a VirTis lyophilizer for 72 h. Of the dried fish tissue, 70–100 mg weighed samples were digested in sealed Teflon containers with 6 ml nitric acid in a 1 ppm Au background according to EPA method 3052 [19]. Samples were microwaved at 400 W for 25 min, allowed to cool and taken up with water. Samples were then analyzed for total mercury content by ICP-MS according to EPA method 200.8 [20]. For every 5 samples a blank was run, and two reference samples of oyster tissue were included in the sample series. All quality control samples fell within acceptable ranges, and the limit of detection was 0.096 μg/g. To make our dry-weight measurements of THg comparable to wet-weight concentrations in the literature, we converted our THg measurements into wet-weight concentrations by dividing by the widely accepted conversion factor of 3.6 [21].

### Statistical Analysis

Because some comparisons involved unequal sample sizes, we first tested for unequal variances with an F-test and then used t-tests (either for equal or unequal variances) to examine the data for differences in THg among groups of samples. P-values were calculated as one-tailed given the expectation that MSC-certified fish (which should have originated from SGSR fishery) should have lower levels of THg than fish that originated from non-certified fisheries at lower latitudes in the southern hemisphere [9]. Because we performed three t-tests on overlapping subsets of the data, we considered P-values significant if $<0.05/3$ or 0.0167.

## Results

Measurements of THg for the retail-acquired Chilean sea bass in our study varied widely among samples, from a low of 0.07 ppm to a high of 1.9 ppm. Overall, fish labeled as MSC-certified Chilean sea bass had less than half the THg of uncertified fish (Table 1, Fig. 2), a highly significant result (t-test for unequal variances: $t = -2.97$, one-tailed $p = 0.0050$). However, among only those fish that were genetically verified as *D. eleginoides*, the difference in THg between MSC-certified and uncertified fish was only marginally significant at the 0.05 level and not significantly different at the 0.0167 level (t-test for unequal variances: $t = -2.15$, one-tailed $p = 0.0320$).

For MSC-certified fish that were actually *D. eleginoides*, we found an uneven distribution of THg contamination: those with mtDNA haplotypes not known from the MSC-certified fishery (haplotypes E, F, I, & J) had twice as much mercury (0.63 ppm) as *D. eleginoides* with haplotypes known from the South Georgia stock (0.31 ppm) (Fig. 3; t-test for equal variances: $t = -2.87$, one-tailed $p = 0.0047$). The single fish with haplotype E was unusual among fish with haplotypes unknown from the certified fishery in that it had relatively low THg (Fig. 3).

## Discussion

Our study of retail acquired Patagonian toothfish or "Chilean sea bass" demonstrated that both species substitutions and fishery-stock substitutions obscure a complex pattern of mercury contamination. First, although fish labeled as MSC-certified "Chilean sea bass" appeared to have significantly lower total

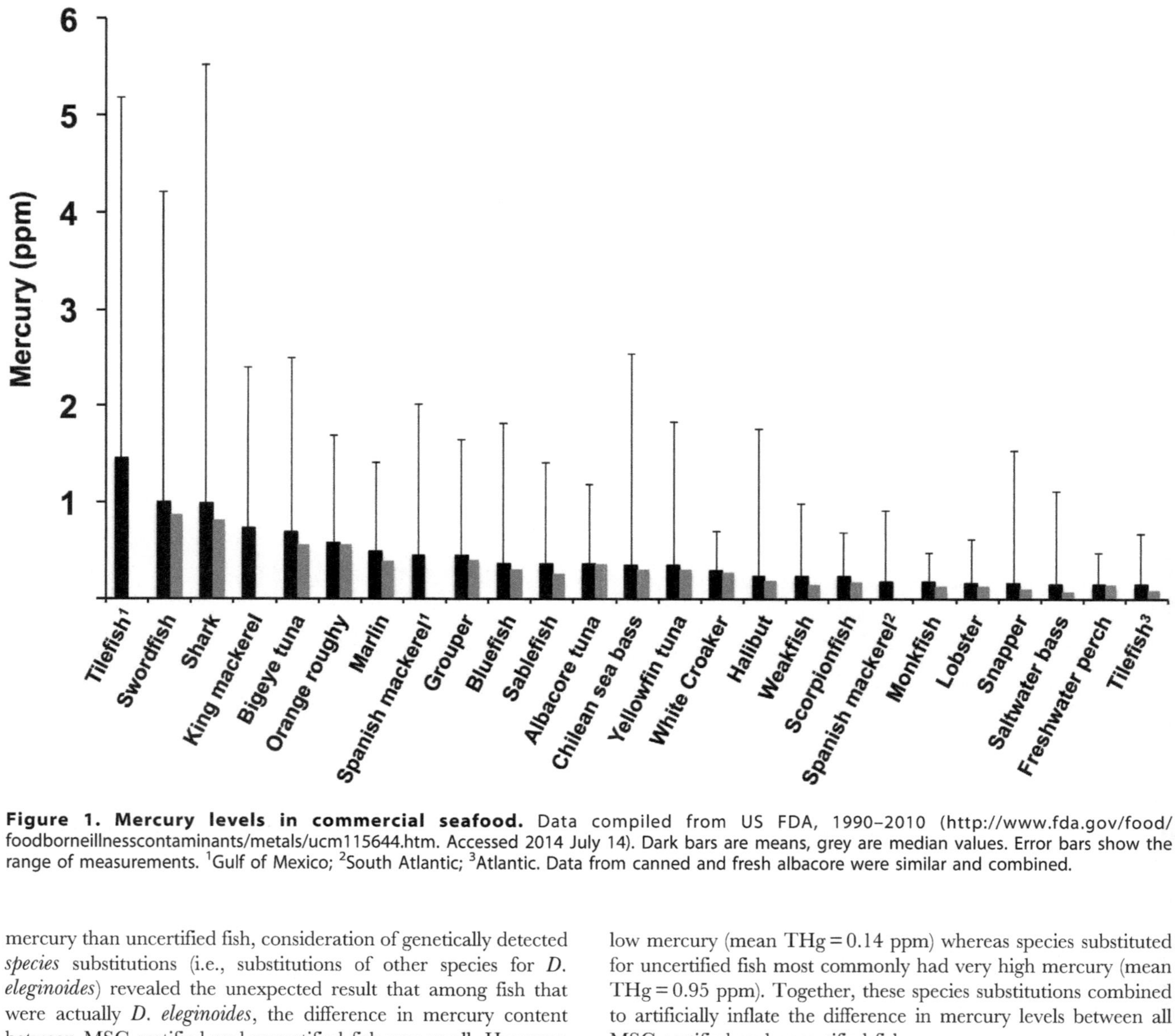


**Figure 1. Mercury levels in commercial seafood.** Data compiled from US FDA, 1990–2010 (http://www.fda.gov/food/foodborneillnesscontaminants/metals/ucm115644.htm. Accessed 2014 July 14). Dark bars are means, grey are median values. Error bars show the range of measurements. [1]Gulf of Mexico; [2]South Atlantic; [3]Atlantic. Data from canned and fresh albacore were similar and combined.

mercury than uncertified fish, consideration of genetically detected *species* substitutions (i.e., substitutions of other species for *D. eleginoides*) revealed the unexpected result that among fish that were actually *D. eleginoides*, the difference in mercury content between MSC-certified and uncertified fish was small. However, this unexpected result is explained in part by the finding that the species that were substituted for MSC-certified fish had relatively low mercury (mean THg = 0.14 ppm) whereas species substituted for uncertified fish most commonly had very high mercury (mean THg = 0.95 ppm). Together, these species substitutions combined to artificially inflate the difference in mercury levels between all MSC-certified and uncertified fish.

In contrast, *fishery-stock* substitutions (substitutions of *D. eleginoides* from uncertified stocks for *D. eleginoides* from certified

**Table 1.** Total mercury (THg) concentration of fish sold as Chilean sea bass (CSB) with and without MSC labels, indicating their origin from the certified South Georgia (SG) fishery based on genetic analyses.

| Samples | N | THg (ppm) | SD |
|---|---|---|---|
| *MSC-labeled fish* | | | |
| i) All fish | 25 | 0.35 | 0.24 |
| ii) CSB only | 22 | 0.42 | 0.24 |
| Known SG haplotypes | 17 | 0.31 | 0.20 |
| Unknown SG haplotypes | 5 | 0.63 | 0.25 |
| Excluding Heard Island haplotype | 4 | 0.72 | 0.16 |
| *Non-MSC fish* | | | |
| i) All fish | 13 | 0.89 | 0.60 |
| ii) CSB only | 8 | 0.80 | 0.56 |

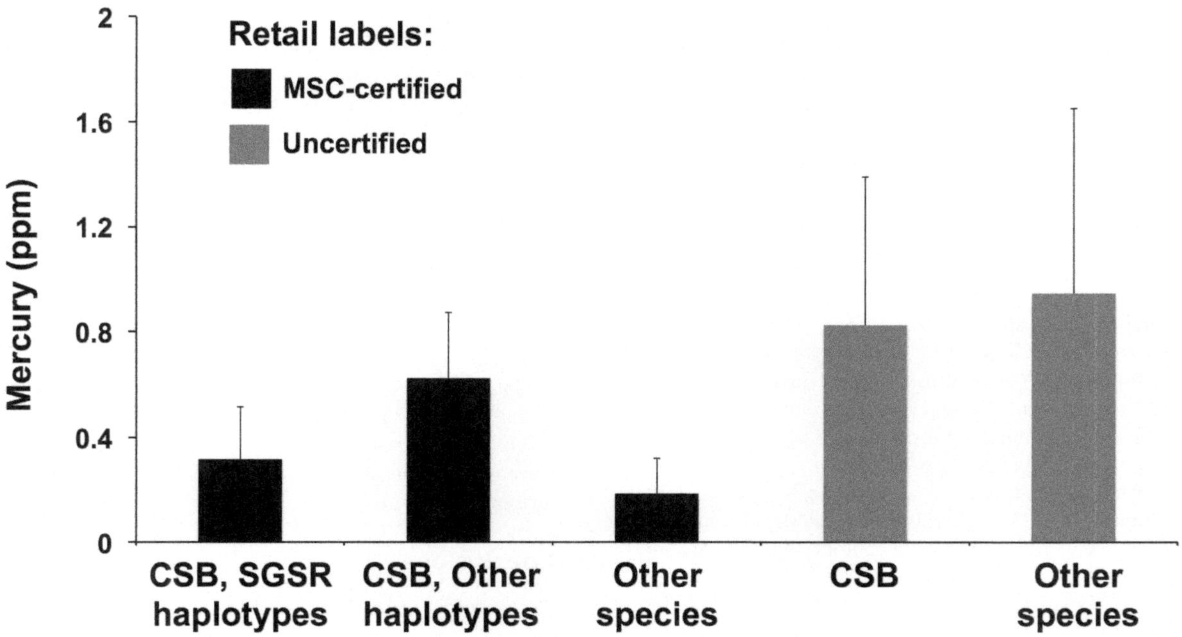


**Figure 2. Mercury concentrations among samples of retailed acquired Patagonian toothfish (*Dissostichus eleginoides*) or Chilean sea bass (CSB) in this study.** Error bars are standard deviations.

stocks) also likely contributed to the apparently small difference in mercury content between MSC-certified and uncertified fish that were verified genetically as *D. eleginoides*: several MSC-certified fish that were genetically verified as *D. eleginoides* had relatively high mercury, twice the concentration of most certified fish in our study (Fig. 3). Given that all of these toothfish with unexpectedly high mercury also had mtDNA haplotypes unknown from the certified SGSR fishery, their high mercury content is consistent with the idea that they did not originate from the certified fishery [16]. Two of these four fish with exogenous haplotypes had haplotype F, the dominant haplotype on the Patagonian shelf [22], where fish are known to accumulate higher mercury levels than compared to SGSR fish [9]. One fish labeled as MSC-certified (but which did not possess a haplotype known from the certified

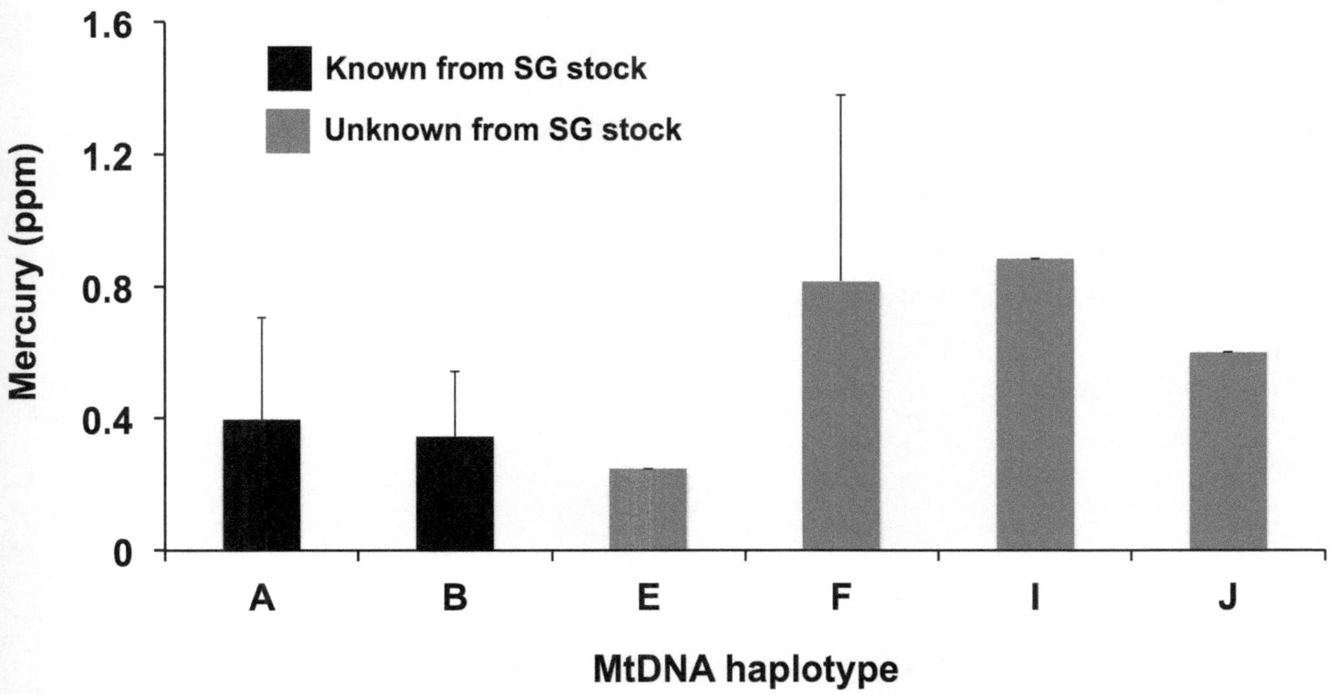


**Figure 3. Mercury concentrations among mitochondrial DNA haplotypes from MSC-certified retail samples of Patagonian toothfish (*Dissostichus eleginoides*) or Chilean sea bass.** Error bars are standard deviations.

SGSR fishery) had a haplotype known only from Heard Island [16]. As with SGSR, Heard Island lies south of the Antarctic Polar Front, an oceanographic barrier that likely restricts gene flow in *D. eleginoides* [22] and might protect the Southern Ocean from mercury contamination [9]. Consistent with a high-latitude origin, this putative Heard Island fish had correspondingly low mercury (0.24 ppm, Fig. 1), outside the 99% confidence interval (0.26–1.18 ppm) for the four other fish with haplotypes not known from the certified fishery. Although the Heard Island longline toothfish fishery is now MSC-certified, it was not certified at the time the samples in our study were purchased [16].

The potential for species substitutions to cause unintended mercury exposure for consumers has been described before, but our study of Patagonian toothfish is novel in that it shows that fishery-stock substitutions, which are more difficult to detect than species substitutions, can expose consumers to increased mercury concentrations in retail-acquired fish. For example, although species substitutions result in a decrease in mean mercury concentrations of ~7% (for MSC-certified fish) and an increase of ~11% (for uncertified fish), fishery-stock substitutions (among MSC-labeled fish) resulted in an increase in mean mercury of ~100% (Table 1). Because Patagonian toothfish from Chile often exceed the 1.0 ppm FDA mercury import threshold (and vastly exceed limits for Canada, New Zealand, and Australia) [9], our results suggest that in addition to species' identities, the geographic origins of fish may be worthy of greater consideration in seafood consumption studies.

A potential confounding factor in our results may be that we were not able to correct mercury concentrations with body size of the sampled fish, because the samples were obtained from fish filets that were sold at seafood counters in retail stores. Therefore, an alternative explanation for the unusually high THg of some fish labeled as having originated from the SGSR stock is that they were simply very large fish from SGSR. However, given that the MSC-labeled fish with exogenous mtDNA haplotypes (excluding the fish with haplotype E, known only from Heard Island) had an average mercury concentration of 0.72 ppm (Table 1), those fish must have had average lengths of approximately 160 cm if actually from SGSR (see Fig. 2 from 9), more than 45 cm longer than any fish obtained in earlier studies of mercury in SGSR fish [9], a fishery with a modal catch length of only 75 cm [23]. Furthermore, these MSC-labeled fish with unusually high mercury levels also had mtDNA haplotypes not known from the SGSR stock (two of which are the most common haplotype in Chilean waters) and mercury levels expected for Chilean fish ranging in size from 75 to 100 cm, suggesting the mislabeling hypothesis is a better explanation for MSC-certified fish with unexpectedly high mercury levels.

Although all of our samples were stored in ethanol, our mercury measurements were consistent with previous measurements from Patagonian toothfish. For example, we found a mean mercury level of 0.31 ppm for toothfish with haplotypes known from the SGSR stock whereas a mean mercury level of 0.23 ppm was previously reported for frozen fish (i.e., not preserved in ethanol) sampled from the same fishery [9]. The relative amounts of mercury between fish from north and south of the Antarctic Polar Front in the south Atlantic/Southern Ocean were also similar between our study and earlier work: frozen samples from Chile had 3.1 times the mercury content of fish from SGSR [9], similar to a factor of 2.7 in our study in a comparison of uncertified but genetically identified *D. eleginoides* (most of which were labeled as having originated from Chile) to MSC-certified Chilean sea bass that had mtDNA haplotypes known from the certified fishery. Therefore, our results are consistent with substantially increased mercury exposure (twice as much) from fish that did not originate further south [9–10].

## Conclusions

Our study found considerable retail variation in mercury levels in "Chilean sea bass" that is related to the likely geographic origins of the fish, but also indicated that species substitutions and fishery-stock substitutions may obscure apparent levels of mercury among fish labeled as both MSC-certified and uncertified Patagonian toothfish. Although on average, MSC-certified fish is a healthier option with respect to mercury contamination than compared to uncertified fish, our study showed that fishery-stock substitutions, can result in a larger proportional increase in mercury consumption than species substitutions for consumers, and that variation in mercury contamination among fishery stocks may be considered in future seafood consumption guidelines.

## Acknowledgments

K. Barr, B. Wilbur, J. McGuire, B. Sowers, C. Edwards and S. McAlister collected some of the samples. We thank J. Meliker and two anonymous reviewers for reading and commenting on the manuscript.

## Author Contributions

Conceived and designed the experiments: PM PH HN. Performed the experiments: PH. Analyzed the data: PH PM. Contributed reagents/materials/analysis tools: PM PH HN. Contributed to the writing of the manuscript: PM PH HN. Sample acquisition: HN.

## References

1. Hightower JM, Moore D (2002) Mercury Levels in High-End Consumers of Fish. Environmental Health Perspectives 111: 604–608.
2. Costa LG (1988) Interactions of neurotoxicants with neurotransmitter systems. Toxicology 49: 359–366.
3. Sackett DK, Cope WG, Rice JA, Aday DD (2013) The influence of fish length on tissue mercury dynamics: implications for natural resource management and human health risk. International Journal of Environmental Research on Public Health. 10(2): 638–659.
4. Sunderland EM (2007). Mercury exposure from domestic and imported estuarine and marine fish in the U.S. seafood market. Environmental Health Perspectives, 115(2): 235–42.
5. Piraino MN, Taylor DL (2009) Bioaccumulation and trophic transfer of mercury in striped bass (*Morone saxatilis*) and tautog (*Tautoga onitis*) from the Narragansett Bay (Rhode Island, USA). Marine Environmental Research, 67: 117–128.
6. Karimi R, Fitzgerald TP, Fisher NS (2012) A Quantitative Synthesis of Mercury in Commercial Seafood and Implications for Exposure in the United States. Environmental Health Perspectives 120: 1512–1519.
7. Méndez E, Giudice H, Pereira A, Inocente G, Medina D (2001) Preliminary Report on the Total Mercury Content of Patagonian Toothfish (*Dissostichus eleginoides*). Journal of Food Composition and Analysis 14: 547–549.
8. McArthur T, Butler EC V, Jackson GD (2003) Mercury in the marine food chain in the Southern Ocean at Macquarie Island: an analysis of a top predator, Patagonian toothfish (*Dissostichus eleginoides*) and a mid-trophic species, the warty squid (*Moroteuthis ingens*). Polar Biology 27: 1–5.
9. Dawson Guynn KD, Peterson MS (2007) Mercury concentrations in the Patagonian toothfish, *Dissostichus eleginoides* Smitt 1898, among three distinct stocks. Polar Biology 31: 269–274.
10. Hanchet SM, Tracey D, Dunn A, Horn P, Smith N (2011) Mercury concentrations of two toothfish and three of its prey species from the Pacific sector of the Antarctic. Antarctic Science 24: 34–42.
11. Marko PB, Lee SC, Rice AM, Gramling JM, Fitzhenry TM, et al. (2004) Mislabelling of a depleted reef fish. Nature 430: 309–310.
12. Lowenstein JH, Amato G, Kolokotronis SO (2009) The real maccoyii: identifying tuna sushi with DNA barcodes – contrasting characteristic attributes and genetic distances. PLoS ONE 4: e7866.

13. von der Heyden S (2009) Why do we need to integrate population genetics into South African marine protected area planning? African Journal of Marine Science 31: 263–269.
14. Miller DD, Mariani S (2014) Smoke, poor transparency and mislabeled in the European cod. Seafood 8: 517–521.
15. Hanner RH, Becker S, Ivanova NV, Steinke D (2011) Fish Barcode of Life and seafood identification: geographically dispersed case studies reveal systemic market substitution across Canada Mitochondrial DNA 22: 106–122.
16. Marko PB, Nance HA, Guynn KD (2011) Genetic detection of mislabeled fish from a certified sustainable fishery. Current Biology 21: R621–2.
17. Mystery fish: the label said red snapper, the said baloney. Consumer Reports Magazine. Published December 2011. Available: http://www.consumerreports.org/cro/magazine-archive/2011/december/food/fake-fish/overview/index.htm. Accessed 2014 July 14.
18. What you need to know about mercury in fish and shellfish. U.S. Food and Drug Administration. Published March 2004. Available: http://www.fda.gov/food/foodborneillnesscontaminants/buystoreservesafefood/ucm110591.htm. Accessed 2014 July 14.
19. Microwave assisted acid digestion of siliceous and organically based matrices. Method 3052. US-Environmental Protection Agency. Published December 1996. Available: http://www.epa.gov/wastes/hazard/testmethods/sw846/pdfs/3052.pdf. Accessed 2014 July 14.
20. Determination of trace elements in waters and wastes by Inductive Coupled Plasma - Mass Spectrometry. Method 200.8. US-Environmental Protection Agency. Published 1994. Available: http://water.epa.gov/scitech/methods/cwa/bioindicators/upload/2007_07_10_methods_method_200_8.pdf. Accessed 2014 July 14.
21. Streit B (1998) Bioaccumulation of contaminants in fish. In: Braunbeck T, Hinton DE, Streit B, editors. Fish Ecotoxicology. Basel: Birkhauser. 353–399.
22. Shaw PW, Arkhipkin AI, Al-Khairulla H (2004) Genetic structuring of Patagonian toothfish populations in the Southwest Atlantic Ocean: the effect of the Antarctic Polar Front and deep-water troughs as barriers to genetic exchange. Molecular Ecology 13: 3293–3303.
23. Hillary RM, Kirkwood GP, Agnew DJ (2006) An Assessment of toothfish in subarea 48.3 using Casal. CCAMLR Science 13: 65–95.

# Spatial Access Priority Mapping (SAPM) with Fishers: A Quantitative GIS Method for Participatory Planning

**Katherine L. Yates[1]*, David S. Schoeman[2]**

1 Environmental Science Research Institute, University of Ulster, Coleraine, United Kingdom, 2 Faculty of Science, Health, Education and Engineering, University of the Sunshine Coast, Maroochydore DC, Queensland, Australia

## Abstract

Spatial management tools, such as marine spatial planning and marine protected areas, are playing an increasingly important role in attempts to improve marine management and accommodate conflicting needs. Robust data are needed to inform decisions among different planning options, and early inclusion of stakeholder involvement is widely regarded as vital for success. One of the biggest stakeholder groups, and the most likely to be adversely impacted by spatial restrictions, is the fishing community. In order to take their priorities into account, planners need to understand spatial variation in their perceived value of the sea. Here a readily accessible, novel method for quantitatively mapping fishers' spatial access priorities is presented. Spatial access priority mapping, or SAPM, uses only basic functions of standard spreadsheet and GIS software. Unlike the use of remote-sensing data, SAPM actively engages fishers in participatory mapping, documenting rather than inferring their priorities. By so doing, SAPM also facilitates the gathering of other useful data, such as local ecological knowledge. The method was tested and validated in Northern Ireland, where over 100 fishers participated in a semi-structured questionnaire and mapping exercise. The response rate was excellent, 97%, demonstrating fishers' willingness to be involved. The resultant maps are easily accessible and instantly informative, providing a very clear visual indication of which areas are most important for the fishers. The maps also provide quantitative data, which can be used to analyse the relative impact of different management options on the fishing industry and can be incorporated into planning software, such as MARXAN, to ensure that conservation goals can be met at minimum negative impact to the industry. This research shows how spatial access priority mapping can facilitate the early engagement of fishers and the ready incorporation of their priorities into the decision-making process in a transparent, quantitative way.

**Editor:** Jan Geert Hiddink, Bangor University, United Kingdom

**Funding:** Research was carried out as part of a PhD, funded by the Department of Education and Learning, Northern Ireland. Additional field work resources and equipment were provided by the Agri-Foods and Biosciences Institute, Northern Ireland, as part of a service agreement with the University. The funders had no role in study design, data collection and analysis, decision to publish, or preparation of the manuscript.

**Competing Interests:** The authors have declared that no competing interests exist.

* E-mail: yates-k1@email.ulster.ac.uk

## Introduction

As the human population grows, pressure on the marine environment escalates and impacts of our unsustainable use can be seen in biodiversity loss, reduced water quality and declining fish stocks [1–6]. Efforts to improve management of the marine environment, and the resources within it, are becoming increasingly spatial, specifically through the development of marine spatial planning (MSP) and the designation of marine protected areas (MPAs). Whilst these management tools offer many potential benefits [7–15], too often these fail to be delivered [16,17], or the planning process itself fails [18,19].

One reason for the failure of MSP and MPAs is lack of adequate stakeholder involvement, even though this is often stated as vital for a successful planning process [20]. Incorporating stakeholders is important because it allows the planner to better appreciate the context of the plan and the potential impacts of different planning options. This greater appreciation provides the opportunity to develop more inclusive plans, which take into account the needs of all parties. By so doing, a planner should be able to both reduce conflict and help to develop a sense of stakeholder ownership over the project. Compliance with a chosen plan can be enhanced by this perception of ownership and inclusiveness [21]. Engaging with stakeholders provides the opportunity to explain the need for management changes, increasing understanding amongst stakeholders, which can again improve compliance [21]. Involving stakeholders can also facilitate generation of information that may not have otherwise been available, such as data on the distribution of species. This information can be incorporated into the decision-making process, which enhances planners' abilities to find the most efficient solutions, thereby maximising benefits whilst minimising conflict and negative impacts. Stakeholder involvement, therefore, is vital not only to increase a project's chances of success, but also to enhance the quality of decisions and reduce implementation cost [20,22–25].

In the marine environment, one of the largest commercial stakeholder groups is the fishing community. They are also arguably the group that are most likely to suffer direct adverse economic and social impacts of spatial restrictions. For planners to be able to assess the impact of different options on the fishing community, they need spatial information on the relative importance of different parts of the sea. However reliable data for even the distribution of target species is all too often unavailable [26], with arguably even less known about the distribution of fishing effort, and practically nothing about priorities of the fishers themselves. To fill some of the gaps, researchers and managers have begun to involve fishers in participatory mapping studies, and the importance of such initiatives for effective management is increasingly accepted [27].

Participatory mapping projects have documented a diverse range of features, including: the locations of fishing grounds; the distribution of habitats and species; key management areas such as spawning grounds; and benthic and oceanographic features [26,28–36]. These projects provide invaluable information for marine management. However, the maps are of limited use to a planner trying to take into account stakeholder priorities. First, these maps are frequently qualitative, making transparent quantitative analysis of the different options difficult. Second, even when quantitative data on stocks are available, they do not explicitly provide information on fishers' access priorities: the distribution of resources does not necessarily relate directly or linearly to the importance of different areas to fishers or to the impact of restricting access to them. Nor should a planner assume to understand the ways in which other variables interact with the spatial distribution of the resource to determine the importance of an area to a fisher.

The development of vessel monitoring systems (VMS) attached to fishing boats has begun to provide very accurate, high-resolution quantitative data on the distribution of fishing effort [37]. VMS is very useful, but alone it still only provides information on where most fishing occurs, which is not necessarily the same as which parts are the most valuable to fishers. A more complete picture of the gross economic value can be obtained by combining VMS data with daily logbook entries, then mapping the catch value of different parts of the sea to the fishery [38].

In many situations, however, VMS data are still not available: at present, VMS are generally deployed only on larger vessels and in developed countries. Detailed landings data are not always collected, nor are daily logbook systems always in place. Indeed, even in developed countries such as the UK, the extent and spatial distribution of inshore fishing effort can be poorly known (pers. obs). When data are collected, they might still be inaccessible [39]. Thus, VMS-based methods are only possible sometimes, for some fisheries. Even then, resultant data infer the value of different areas to fishers, instead of documenting them. Alone, VMS data are not able to account for all the factors that might affect how important an area is for a fisher. Grounds closer to home, for example, could be more important than their gross value implies because of reduced costs required to access them in both economic (e.g. less fuel) and/or social (e.g. coming home each night) terms. Other areas might provide relatively low revenue, but still be very important because they are the only areas available at a given time of year, due to weather conditions; in such cases, maintaining access is necessary for maintaining a regular flow of income. Furthermore, VMS documents current spatial fishing patterns, which in many cases are already restricted by area-based quotas, by-laws, and existing protected areas. Therefore VMS may not actually represent fishers preferred fishing grounds. Finally, VMS methods do not facilitate the active involvement of the fishers; they thus neither meet the recognised and increasingly legislated need to involve stakeholders, nor provide the opportunity for stakeholders to feel part of the decision-making process.

Here a novel method, spatial access priority mapping (SAPM), is described and demonstrated for the Northern Irish fleet. SAPM requires no VMS, logbook or landings data, is suitable for single or multiple fisheries of any type, and can be performed using only basic features of standard spreadsheet and GIS software. It actively engages fishers, documenting rather than inferring their priorities, and produces quantitative maps that can be combined easily with other data if available and desired.

## Study Area – the Northern Ireland Context

Marine management in Northern Ireland faces many challenges. Recent European legislation, such as the Marine Strategy Framework Directive, is setting new standards for the environmental status of coastal waters. A Marine Bill is currently pending in the Northern Ireland Assembly that will provide the Department of the Environment (DOE) with new powers to develop marine spatial plans and to designate marine conservation zones (MCZs), a new type of marine protected area. The European Common Fisheries Policy, which has significant impact on the Northern Irish Fleet, is under review. The Northern Ireland Inshore Fisheries Strategy is also under review by the Fisheries Division of the Department for Agriculture and Rural Development (DARD). In addition, there is significant focus on the development of marine renewable energy in Northern Irish waters [40–42].

Whilst these developments bring many potential benefits for the Northern Ireland community, they will likely also bring spatial restrictions for fishers. In order to properly take fishers into account, and to make the most efficient plans possible, planners need to understand which parts of the sea are most important to fishers. However data are patchy at best. VMS is currently only in place for vessels 15 m and longer. Associated data have been processed to produce effort maps for some of their range (Irish Sea, but not the Clyde or other areas) [43]. Information on the inshore fishing fleet, which relies on smaller vessels, is limited, with no current monitoring of effort either in terms of magnitude or distribution. The distribution and stock levels of *Nephrops* are regularly monitored [44], and there are annual surveys for scallops [45] and ground fish [46]. There is, however, currently no information on the status of lobster or crab stocks, despite these resources supporting almost a third of the active vessels in the fleet. Access to the data that do exist is also restricted, sometimes even between Government Departments (pers. obs.).

Issues between the Northern Ireland Government Departments, particularly poor communication and co-operation between DARD and DOE, have caused problems for marine management in Northern Ireland in the past [47]. With DOE responsible for developing the spatial management plans, and DARD responsible for managing the fisheries that these plans are likely to impact, there is potential for further conflict. Thus, there is a real need for universally available data and a transparent way of incorporating the priorities of the fishers into the planning process.

The majority of the Northern Irish fleet is based at three main ports: Kilkeel is the largest, followed by Portavogie and Ardglass. In addition, there are over twenty minor ports at which small numbers of pot boats are based (Figure 1). A few Northern Ireland registered boats also fish out of ports in neighbouring countries, mainly Greencastle in the Republic of Ireland. There are 367 vessels registered in the Northern Ireland fleet, 224 of which are officially recorded as active (DARD, pers.comm). The majority of those 224 vessels are under 15 m (127, 57%) and would thus not be covered by current VMS data.

The most important fishery in terms of revenue and employment is *Nephrops*, with 99 of the 224 vessels focusing virtually exclusively on *Nephrops*, and a further 11 mixed-fishery vessels having some *Nephrops* interest (DARD, pers.comm). Pot-fishing accounts for the second highest number of vessels, with 69 of the 224 registered as exclusively potting, and 8 mixed-fishery vessels having potting interest (DARD, pers.comm). There are two main Fish Producer Organisations (FPOs): the Anglo-Northern Irish

**Figure 1. Map of the home ports from which commercially active, Northern Ireland registered vessels fish.** Map of the home ports from which commercially active, Northern Ireland registered vessels fish Data obtained from the Department of Agriculture and Rural Development, Northern Ireland, 2012.

Fish Producer Organisation (ANIFPO); and the Northern Ireland Fish Producer organisation (NIFPO). There are two pot fishers associations, the North Coast Lobster Fishermen's Association and the North-East Lobster Fishermen's Co-Operative, and a Scallop Fishermen's Association. All of these representative bodies reported attempting to engage with the Government Departments to improve management.

## Methods

### Interviews

Research was approved by the School of Environmental Sciences Ethics Committee, University of Ulster, Coleraine. All interviews were conducted in accordance with ethics procedures, and research plans, including the consent procedure, were submitted and approved prior to the initiation of the project, in accordance with School and University policy. Prior to each interview, information sheets detailing the research were provided to fishers and discussed, and then verbal consent was obtained from participants. Written consent was not sought, partly to maintain the informal setting that assisted with the response rate, but mainly because the respondents were not required to give their names during the interview, thus requiring a signature (a form of identification) was considered inappropriate. Verbal consent was recorded in the response form. All interviews were conducted by the lead author, ensuring that the procedure was uniform for all respondents, and all individual responses were recorded in a confidential database. The pre-interview information sheet and questionnaire are available in Files S1 & S2.

Interviews consisted of two parts: a semi-structured questionnaire; and a mapping exercise. Each interview lasted between 30 minutes and two hours. The vast majority of fishers interviewed (>90%) were approached directly at ports. A number of other fishers responded to flyers that had been distributed by mail. All respondents were skippers or vessel owners. With the exception of three interviews conducted in fisher's homes (2.9%), all interviews were conducted in ports (97.1%): on vessels, at the quayside, in a cafe, or in the ANIFPO or NIFPO offices.

During the mapping exercise, fishers were provided with both paper admiralty charts and digitised admiralty charts within a GIS. Fishers indicated their priority areas (as polygons) directly onto the digitised charts. They were encouraged to identify as many or as few areas as they liked, and to cover as much area as they wanted. However, each respondent was advised that the greater the total area, the lower the priority that could be assigned to each unit of area; thus they were encouraged to be specific. Those fishers active in multiple fisheries were asked to indicate the target species of each area. They were given the option of assigning areas different levels of importance and/or indicating how much of the year they spent in each area, which was then converted into a percentage. This then became the importance value, such that an area that was fished for a fifth of the year received an importance value of 20. On completion of the mapping exercise, respondents were encouraged to review their results and to make changes as desired.

### Generating the Maps

The priority (SAP $km^{-2}$) of each of the areas indicated by a particular respondent was calculated by multiplying the number of full-time crew on the associated boat by the importance value, and then dividing by the number of square kilometres the area covered (Figures 2 & 3 show illustrative examples). Part-time or seasonal crew were assigned a *pro rata* proportion of a full-time member's weighting, so that a boat that reported four full-time staff throughout the year and one extra full-time staff for six months of the year was recorded as having a total of 4.5 crew. Each individual (full time) fisher has a total SAP of 100, such that a boat with a crew of 4.5 has a total SAP of 450 associated with it, and an area that has an SAP of 10 $km^{-2}$ represents a tenth of a fisher per $km^2$.

In order to create maps that represented the whole fleet, the results obtained from the individual respondents need to be scaled up (Figure 4 shows an illustrative example). Based on initial examination of data and discussions with the fishers, vessels were divided into six size categories. All vessels under 9 m were pot boats. With the exception of two specialist offshore crab boats, no vessels over 10 m fished pots. In the $\geq$9 m <10 m category there was a mix of pot boats and small dredge or trawl boats. Dredge and trawl boats of similar size generally have very similar numbers of crew, but pot boats of the same size tend to have fewer crew and are often single handed. A two-sided t-test demonstrated that there were significantly fewer crew on pot boats between 9 and 10 m than other gear types ($t = 3.05$, $df = 20$, $p = 0.007$). Consequently, this size category was further divided by gear: pots or not pots. Data was obtained for only one pelagic vessel, of which there are three in Northern Ireland; however, an additional category was assigned for pelagic vessels as they have much higher crew numbers (8–10) than any other type. Thus vessels were divided into a total of eight categories (Figure 5) and the average number of crew per vessel for each of these categories was calculated using the sample data.

A list of vessels at each port was provided by the fisheries division of DARD, along with the length and registered gear type of each vessel. The list showed 224 active registered vessels, but during the research a number of inaccuracies were found with this list and it was updated to include additional boats: 14 that were originally listed as 'no recorded fish effort this year', and 8 that were not listed. The final vessel list contained 246 vessels and was broken down by port, gear type and length. Using this and the calculated average crew per vessel, an estimate was made of the total number of fishers on all active vessels by each port, gear and length combination.

In each length/gear/port combination, the estimated number of crew was divided by the actual number of crew sampled in order to create a final weighting scheme, which was used to

**Port A: Illustrative example of Basic SAPM, single areas, no weighting and no scaling**

Port A has three small pot-fishing boats, all single-handed (1 fisher) and all in the same size category. All pot fishers at the port were interviewed so no scaling up is required. They all chose only one area.

Fisher 1 (yellow) chose a total area of $3km^2$. Fisher 2 (Blue) chose an area of $4km^2$ and Fisher 3 (red) chose an area of $10km^2$.

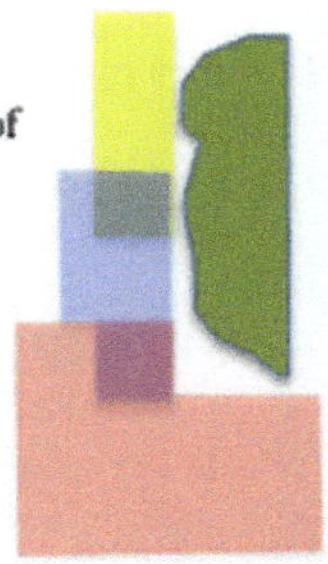

$$SAP = \frac{Crew \times Importance}{Area\ (km^2)}$$

As each fisher chose just one area, that area is inherently 100% of their importance, so their SAPs are as follows:

Fisher 1:
$(1\ crew \times 100\ importance) \div 3km^2 = 33.3\ SAP\ km^{-2}$
Fisher 2:
$(1\ crew \times 100\ importance) \div 4km^2 = 25\ SAP km^{-2}$
Fisher 3:
$(1\ crew \times 100\ importance) \div 10km^2 = 10\ SAPkm^{-2}$

Each fisher has the same total priority, but these priorities are spread over different areas: the larger the area the lower the resultant SAP $km^{-2}$. This is justified because the value of a given area to a fisher depends on the total area they fish: Fisher 1 losing access to $1km^2$ of their fishing area would suffer a significantly larger impact than Fisher 2 losing access to $1km^2$ of their area.

To calculate the total SAP for pot fishers at Port A the individual SAPs are summed. Thus the $1km^2$ overlap of priority areas identified by Fishers 1 and 2 will have a SAP of 58.3 $km^{-2}$ (33.3$km^{-2}$ + 25$km^{-2}$) and the overlap in priority areas identified by Fishers 2 and 3 will have a total of SAP 35 $km^{-2}$ (25 $km^{-2}$ + 10 $km^{-2}$).

**Figure 2. Illustrative example of basic SAPM.** Illustrative example of basic SAPM, no weighting or scaling involved.

**Port B: Illustrative example of SAPM with multiple areas and varying importance**

Port B has two small pot-fishing boats, both are single-handed and both skippers have been interviewed. In this example, however, they each chose multiple areas and stated that those different areas have different levels of importance.

Fisher 1 chose two areas: $2km^2$ and $3km^2$. They stated that they spend 3 months (25% of the year) in the smaller area and 9 months (75% of the year) in the larger.
Smaller area:
$(1\ crew \times 25\ importance) \div 2km^2 = 12.5\ SAP\ km^{-2}$
Larger area:
$(1\ crew \times 75\ importance) \div 3km^2 = 25\ SAP\ km^{-2}$

Fisher 2 chose three areas: $4km^2$, $2km^2$ and $1km^2$ and stated they get 80% of their profit from the largest area and 10% from each of the other two.
Largest area:
$(1\ crew \times 80\ importance) \div 4km^2 = 20\ SAP\ km^{-2}$
Middle size area:
$(1\ crew \times 10\ importance) \div 2km^2 = 5\ SAP\ km^{-2}$
Smallest area:
$(1\ crew \times 10\ importance) \div 1km^2 = 10\ SAP\ km^{-2}$

Note that the total SAP for each fisher still adds up to 100.

**Figure 3. Illustrative example of SAPM with multiple areas and varying importance.** Illustrative example of SAPM with multiple areas and varying importance.

upscale the maps generated by respondents to represent the whole active fleet (Figure 4). Where no skipper of a vessel with a particular gear-length-port combination had been interviewed, responses from skippers at that port, using that gear, whose vessels most closely matched the required length, were upweighted appropriately (Figure 6 shows an illustrative example). On the very few occasions (3) where no skipper of vessels of that gear type and port had been interviewed, appropriate response(s) from the nearest port were used.

Priorities for each individual area were then multiplied by their corresponding weights to obtain a final weighted priority score. Resulting shape files were rasterised, at a resolution of 1 $km^2$, and all raster layers were combined using the raster calculator to construct the overall access priority map for the Northern Ireland fleet. Access priority maps for different sectors of the fleet were generated by selecting only the layers that related to that sector. To demonstrate the simplicity of the approach, all calculations were performed using tables and basic cell formulae in an Excel spreadsheet. Other freely available software such as R could be used as an alternative, making the analysis much quicker without changing the method. All spatial analysis was conducted in ArcMap 9.3.

## Validation

In order to assess whether the sample of respondents was representative of the whole fleet, the three main characteristics of vessels were compared. A two-sided t-test was conducted to ascertain whether there was any significant difference in the vessel length distribution between the sample and the population. Chi-squared tests were used to test for any significant difference between the sample and the population in terms of the frequencies of gear types and registered ports. The smallest ports were grouped geographically to make this possible.

SAP maps from individual fishers within a given length-gear-port combination were overlaid to inspect for consistency of responses and verify the validity of up-scaling to represent all vessels. Variation is to be expected between different combinations, and even some minimal variation within combinations. However if the responses within a combination were very different up-scaling would not have been appropriate.

VMS-based effort maps (2009 and 2010) for boats 15 m and longer were provided by AFBI and were used to visually validate the areas indicated by those boats. It was not assumed that VMS is always an accurate representation of spatial access priorities. Indeed, as previously mentioned, an area may be fished for only a short time but may actually have a high value because of the quality of the catch or because it is the only site available at that time of year. Nevertheless, because it would be expected that all areas fished would have some priority associated with them, a visual inspection of the two sets of data provided a validity check

**Port C: Illustrative example of SAPM using weighting to scale up sample result to represent all fishers**

Port C has 11 *Nephrops* trawl boats in two different size classes. Interviews were obtained for six of the 11 boats. Each fisher chose one area.

NB. Average (mean) crew numbers are calculated based on results for the whole fleet (not just this port).

| Size class (m) | Mean no. crew | No. registered active boats | No. with interview |
|---|---|---|---|
| 10 - 14.9 | 3.0 | 6 | 3 |
| 15 - 19.9 | 3.5 | 5 | 3 |

Fisher 1:
14.5m boat, 3.5 crew, one area of 175 $km^2$
$(3.5\ crew \times 100\ importance) \div 175km^2 = 2\ SAP\ km^{-2}$

Fisher 2:
13.8 m boat, 3 crew, one area of 80$km^2$
$(3\ crew \times 100\ importance) \div 80km^2 = 3.75\ SAP\ km^{-2}$

Fisher 3:
14.9m boat, 3 crew, one area of 120$km^2$
$(3\ crew \times 100\ importance) \div 120km^2 = 2.5\ SAP\ km^{-2}$

In previous examples, all fishers were interviewed. Here three boats in the 10 - 14.9m class were included in interviews, but there are six registered at the port, so those three need to be scaled up to represent all six. This is done using a weighting, which is calculated by diving estimated number of crew by sampled number of crew.

$$Weighting\ (W) = \frac{estimated\ crew}{sample\ crew}$$

*Estimated crew* are those expected to be there: the mean number of crew for that gear-size combination multiplied by the no of registered, active vessels of that gear-size combination at the port. *Sample crew* is the total number of crew on all the vessels of that gear/size/port combination for which interview data has been collected.

$$W = \frac{3\ mean\ crew \times 6\ registered\ boats}{3.5 + 3 + 3\ sample\ crew} = \frac{18}{9.5} = 1.89$$

This weighting is then used to scale up the SAP's to account for the other vessels.

$$Weighted\ SAP = orginal\ SAP\ x\ W$$

Area 1: $2\ SAP\ km^{-2} \times 1.89 = 3.78\ SAP\ km^{-2}$
Area 2: $3.75\ SAP\ km^{-2} \times 1.89 = 7.09\ SAP\ km^{-2}$
Area 3: $2.5\ SAP\ km^{-2} \times 1.89 = 4.73\ SAP\ km^{-2}$

Weighting for 15 - 19.9m size class:
In this class interviews have been obtained for three boats, which have the following crew numbers: 3,4,3
$Estimated\ crew = 3.5\ mean\ crew \times 5\ boats = 17.5$
$Sample\ crew = 3 + 4 + 3 = 10$

$$W = \frac{17.5\ estimated\ crew}{10\ sample\ crew} = 1.75$$

**Figure 4. Illustrative example of SAPM using weighting to scale up the sample to represent the population.** Illustrative example of SAPM using weighting to scale up the sample (interviewed fishers) to represent the population (whole fleet).

for overall coverage. It would also be expected that a similar overall pattern would be evident.

Copies of the amalgamated maps for the whole fleet and for the four main fisheries, with the associated report [48], were made available to all members of the Northern Ireland fishing fleet to review. Digital copies were provided to the POs, fishers' associations and co-operatives, for distribution to their members. Hard copies were provided to the harbour offices of the POs and to harbour cafes, where they were made available for fishers to review. In addition, individuals were able to download digital copies of maps and reports from the University of Ulster website, the availability of which was advertised in the national fishing press. Feedback was provided in a number of ways: POs and associations solicited and passed on responses from members, some fishers responded using forms and prepaid envelopes that were provided along with the hard copies sent to POs and cafes, a number of individuals submitted responses by email, and others made direct contact with the lead author to discuss the results.

## Results

Interviews were conducted in 20 different ports. The response rate (97%) was excellent. Fishers were very willing, often keen, to participate once the project was explained. Indeed, many expressed concern that those making the decisions understood neither the industry nor the fishers' needs, and were glad to have an avenue to express their opinions. In total, 103 individual skippers and owners of active vessels were interviewed. Only three skippers declined, two saying they were too busy and the other refusing to engage with the process. The respondents skippered and/or owned 118 vessels, representing 48% of the active fleet (updated list), and those vessels fish out of 25 different ports (Figure 7).

Overall, the sample is considered to be an accurate representation of the fleet (Tables 1& 2, Figure 7). The two-sided t-test on vessel length showed no significant difference between the sample and the population ($t = 0.06$, $df = 334$, $p = 0.995$). Chi-squared tests also showed no significant dependence between sample and fleet coverage either in terms of the registered port ($X^2 = 18.232$, $df = 19$, $p = 0.507$), or the type of gear used ($X^2 = 3.095$, $df = 9$, $p = 0.96$).

The map for the whole fleet produced by SAPM (Figure 8) shows the spatial heterogeneity in access priority and clearly highlights areas whose closure would likely cause greatest contention with the fleet. Unsurprisingly, results show that spatial priorities vary greatly between different fisheries (Figure 9) and, as closures are frequently only for certain types of gear, this can provide valuable input for planners and managers deciding what activities to restrict and where.

Comparisons among results from individual respondents showed a high degree of spatial overlap within a given length-gear-port combination: Figure 10 shows an example from a combination within the *Nephrops* fishery. Overlap was high across all of the three main ports for the *Nephrops* fishery, within each size category. Between size categories, overlap reduced, with smaller vessels having a corresponding reduced range from port. Overlap was also high across the three main ports for the scallop fishery (dredge boats). However, overlap across ports for pot-fishing was very low. This is unsurprising, as the small vessels generally have a

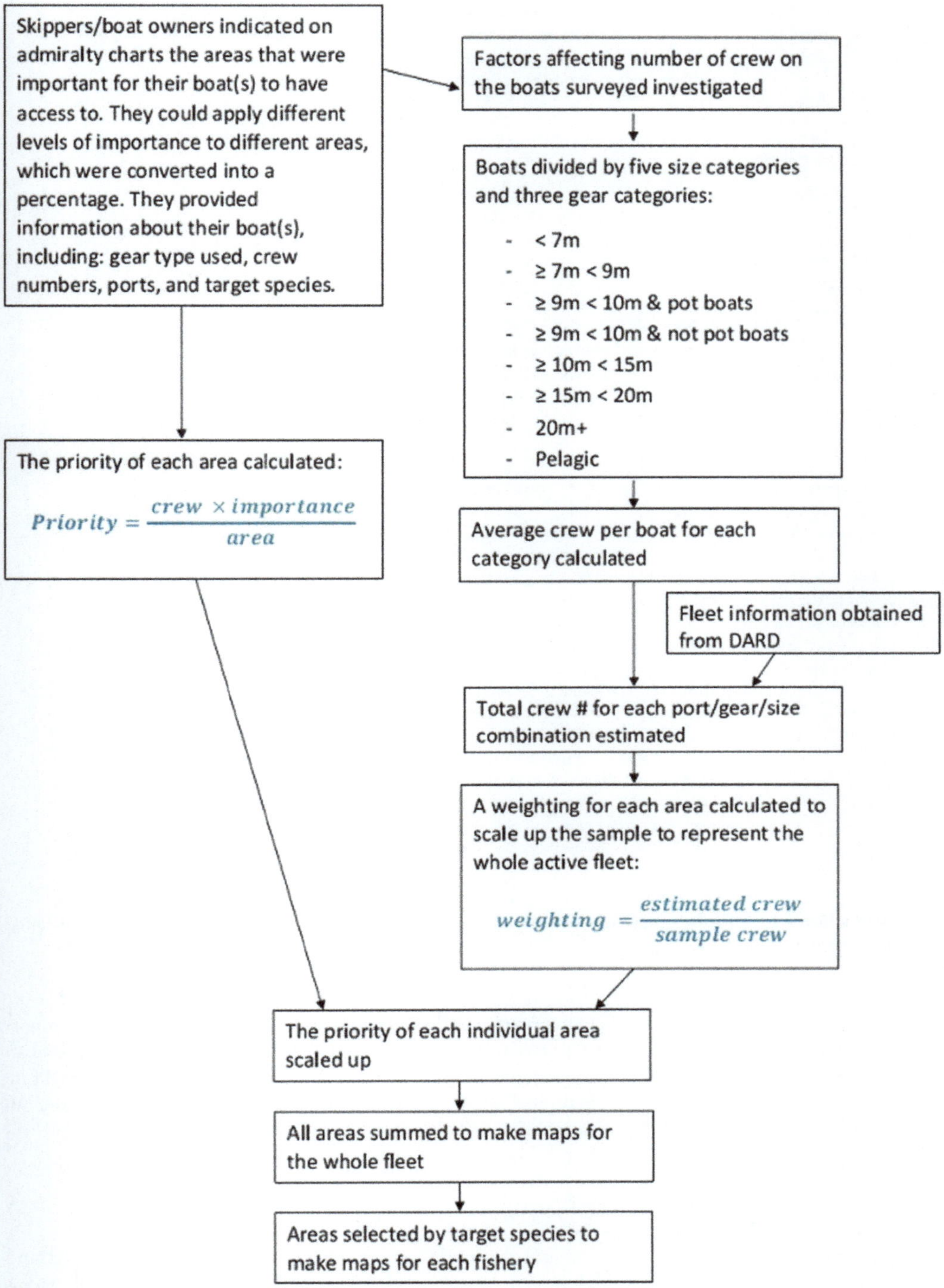


**Figure 5. Flow diagram of the Spatial Access Priority Mapping Method.** Flow diagram of the Spatial Access Priority Mapping Method use to map the access priorities of the Northern Ireland fleet in 2012. DARD stands for the Department of Agriculture and Rural Development, Northern Ireland.

very restricted range and fishers can be territorial between and sometimes even within ports. This stresses the importance of obtaining good spatial representation of the pot-fishing community, which can be difficult when small ports have just one or two fishers operating from them. In this study, the involvement of the Lobster Associations was invaluable in ensuring good coverage.

Overall, visual congruency between the access priority map for the over 15 m fleet and the VMS effort map is high, with all areas fished by the fleet being covered on the access priority map (Figure 11). The access priority map covers a larger total area than the VMS, which is unsurprising, and is most likely a combination of the inherently higher accuracy of the VMS and the wider scope of values that the access priority map captures.

Feedback on the initial report and maps was provided by 21 active skippers/owners, a 9% response rate, as well as by four retired fishers. Active skippers/owners represented all of the main

**Port D: Illustrative example of SAPM when not all port/size/gear combinations have been sampled**
Port D has 25 *Nephrops* trawl boats in five size categories.

NB. Average (mean) crew numbers are calculated based on results for the whole fleet (not just this port).

| Size class (m) | Mean no. crew | No. registered active boats | No. with interview |
|---|---|---|---|
| 10 - 14.9 | 3.5 | 6 | 3 |
| 15 - 19.9 | 4 | 2 | 0 |
| 20 - 24.9 | 4.2 | 12 | 5 |
| 25 - 29.9 | 4.5 | 5 | 3 |
| 30 - 34.9 | 4.6 | 1 | 0 |

As it has not been possible to sample all combinations, vessel(s) within the closest length class will have their weighting increased to represent the missing class. This is achieved by increasing their estimated crew.

*Estimated crew* is average number of crew for that gear-size combination multiplied by the number of vessels of that gear-size combination registered active at that port.

For the 15 - 19.9 m size category there are two 'closest' lengths, so both are used. Estimated crew for the class: mean crew × registered vessels = 4 × 2 = 8. Those estimated crew are divided evenly between the two closet classes, i.e. four extra crew are added to each.

Estimated crew for 10 - 14.9m class is therefore:

$$(mean\ crew \times registered\ boats) + additional\ crew = (3.5 \times 6) + 4 = 34\ Crew$$

Estimated crew for 20 - 24.9m class:

$$(mean\ crew \times registered\ boats) + additional\ crew = (4.2 \times 5) + 4 = 54.4\ Crew$$

For the 30 - 34.9 m class there is only one closest size class, 25 - 29.9 m, so the estimated crew, 4.6, is added entirely to that:

$$(mean\ crew \times registered\ boats) + additional\ crew = (4.5 \times 5) + 4.6 = 27.1\ Crew$$

This is in effect reallocating the estimated crew to classes for which sample data have been obtained. As estimated crew is used to calculate the weighting (see Box 3, example Port C), by proportionally increasing the estimated crew the SAP values are increased to take into account the missing vessels.

Interviewing every fisher would be the ideal, but rarely will that be possible. Collecting data for multiple vessels for each combination is the next best thing, but even that will not always be possible. There be just one or two vessels in a given combination and it might not be possible to find the skippers and/or they might not wish to partake.

Reallocating the estimated crew from unsampled combinations to the closest length class(es) of the same fishery (gear) and port allows the best possible estimation of their SAPs. Omission of those combinations altogether would lead to underrepresentation of those fishers and their fishery.

There will be occasions when reallocation to the closet length class may not be appropriate because of size based spatial restrictions. If there was, for example, a restriction in place that meant only vessels under 15m could fish within three miles of the shore, then it would be inappropriate to divide the estimated crew of the 15 - 19.9m class between both the 10 - 14.9m and 20 - 24.9m classes, as they are far less likely to prioritize areas they can not currently access. Thus, in this case, all the estimated crew should be allocated to the 20 - 24.9m class.

**Figure 6. Illustrative example of SAPM when not all gear-port-size combinations have been sampled.** Illustrative example of SAPM when not all gear-port-size combinations have been sampled.

fisheries and comprised: 8 trawl fishers; 9 pot fishers; 1 dredge/scallop fisher; and 3 fishers who targeted multiple fisheries. The response was very positive, with the result perceived by all respondents to be a good reflection of the fleet's priorities. However, many fishers expressed concern that Government Departments would not use fishers' inputs in forthcoming plans.

### Data Sharing

An important aspect of this highly applied research was to provide data to feed into impending management changes. Once the maps, and associated report, had been reviewed by the fishing community and their comments had been added, it was formally presented to a group of representatives from the relevant Northern Ireland Government Departments: DARD, AFBI, DOE, the Northern Ireland Environment Agency and the Department of Enterprise, Trade and Investment (Marine Renewable Energy Division). In addition to the digital and hard copies of the report, they were also provided with the raster data layers for the five maps (the whole fleet and the four main component fisheries) for use in their GISs. All agencies stated intentions to use these data in forthcoming marine spatial planning and MCZ-designation activities.

The report and data layers were also provided to the UK Marine Management Organisation, Natural England, the UK Joint Nature and Conservation Committee, and the Council for Conservation and the Countryside (Northern Ireland). The report is freely available to download [47], and interested parties can request the data layers from the lead author. Data layers will be deposited in the UK data archive in 2014. Raw data and all information relating to individual fishers are confidential and will not be made available.

## Discussion

More than 100 skippers and owners of Northern Ireland registered vessels were interviewed about their spatial access priorities: their boats represent almost half the active Northern Irish fleet. Data from these interviews was used to generate spatial access priority maps for both the whole fleet and for each of the main fisheries. These maps provide quantitative data on which parts of the sea are most important to the fishers, and they can be used to maximise the efficiency of protected area site selection and to minimise potential conflicts between conservation needs and those of the fishing industry. The maps can also be used to inform marine spatial plans, highlight areas of tension between fisheries and contribute to fisheries management plans.

This research demonstrates the willingness of Northern Ireland fishers to be involved in both research and management planning. The response rate was excellent, despite early concerns by many involved in local marine management that it would be difficult to

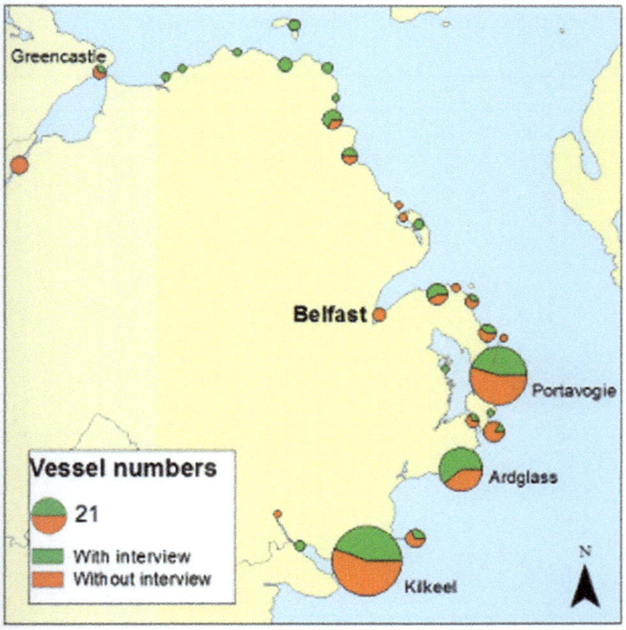


**Figure 7. The proportion of vessels for which interviews were obtained or not obtained.** Map of the Northern Irish ports, plus Greencastle, showing the proportion of vessels for which interviews were obtain for against those for which they were not. The size of the symbol for each port represents the number of commercially active, Northern Ireland fishing vessels registered there.

get the fishers to participate. In fact, fishers were extremely forthcoming with a range of information and many were pleased to have the opportunity to "have their say". The nature of the project may have been a factor in this. First, it was independent research, and thus it bypassed any historical animosity between the Government Departments and the fishers. Second, the research sought explicitly to map fishers' access priorities so that these could be taken into account in future planning processes; it was therefore in fishers' interests to take part. Third, the method was easy to explain to respondents: it was simple, intuitive and representative. This latter characteristic not only increases the chance of participation but also, increases the reliability of fishers' responses because they are far more likely to genuinely understand what is expected of them. Probably the main factor, however, was that the research went to them: interviews took place on fishers' boats, at the quayside, or in the cafes they frequent. This allowed the fishers to participate at a time and in a space that suited them, and it greatly reduced both the actual and opportunity costs of their participation. Indeed, many fishers complained that in order to participate they would normally have had to travel to meetings held in Belfast, at their own expense, and lose a day's fishing. That situation is not unique to Northern Ireland [31] and can lead to poor and/or skewed representation, with only certain fishers able or willing to contribute.

The response rate for the review of the results was much lower than for the face-to-face interviews. This is not surprising: response rates to distance surveys are normally much lower than face to face interviews [49], because in face-to-face interviews the researcher has the opportunity to explain the research and the value of the respondents' involvement. A survey discussing fishing revenue and expense data, distributed to 284 vessel owner/operators in Seattle, for example, had just 34 surveys returned, a 12% response rate [50]. It is also important to bear in mind that this response rate is artificially low because it is calculated by dividing the 21 respondents by all active skippers (246), when in fact it is unknown how many of those skippers were actually aware of the research, the review period or their opportunity to respond. The responses received also suggest that the lower response rate may in part reflect a general agreement with the results, combined with cynicism over the incorporation of the results into policy, making individuals less likely to take the time to respond.

The use of a both paper and digitised maps during the mapping process proved very effective, demonstrating the utility of this dual approach. Other mapping studies have annotated paper maps during interviews and then transformed them into digital form [36,51], but direct annotation into the GIS is more resource efficient and removes the possibility of introducing error or bias during a later digitising process. Use of digitised maps also provides much greater flexibility, allowing the interviewer to carry multiple maps and work at different scales, depending on the type of fisher being interviewed. A pot fisher, for example, may work an area smaller than 5 $km^2$ whereas a *Nephrops* fisher can be utilising the entire Irish Sea. Carrying around physical copies of sufficient maps, each with different resolutions to be appropriate for every fisher in each area is impractical, especially in the context of conducting interviews in small wheelhouses on boats. Digitised maps do not have this issue and can also be enhanced by adding or

**Table 1.** Comparision of the number of registered active Northern Ireland fishing vessels and the number with interviews, by size class.

| Size, in meters | Number active | Number with interviews | % with interview |
|---|---|---|---|
| under 7 | 48 | 21 | 43.8 |
| 7 to <9 | 17 | 10 | 58.8 |
| 9 to <10 | 49 | 22 | 44.9 |
| 10 to <15 | 32 | 19 | 59.4 |
| 15 to <20 | 67 | 35 | 52.2 |
| 20 to <30 | 29 | 9 | 31.0 |
| 30+ | 4 | 2 | 50.0 |
| **Total** | **246** | **118** | **47.9** |

The total number of registered active Northern Ireland fishing vessels and the number with interviews, by size class. Fleet information supplied by the Department of Agriculture and Rural Development, Northern Ireland, and update during the research.

**Table 2.** Comparision of the number of registered active Northern Ireland fishing vessels and the number with interviews, by gear type.

| Gear | Number active | Number with interview | % with interview |
|---|---|---|---|
| Dredge | 5 | 4 | 80.0 |
| Dredge & Nephrops trawl | 23 | 13 | 56.5 |
| Gill Net | 13 | 3 | 23.1 |
| Hand Fishing | 1 | 0 | 0.0 |
| Nephrops & Whitefish trawl | 5 | 2 | 40.0 |
| Nephrops trawl | 104 | 49 | 47.1 |
| Pelagic trawl | 3 | 1 | 33.3 |
| Nephrops trawl & Gill net | 1 | 0 | 0.0 |
| Pots | 80 | 42 | 52.5 |
| Pots & Nephrops trawl | 4 | 1 | 25.0 |
| Pots, Dredge & Nephrops trawl | 2 | 2 | 100.0 |
| Scallop Dredge | 2 | 1 | 50.0 |
| Seine netting | 1 | 0 | 0.0 |
| Pots & Dredge | 2 | 0 | 0.0 |
| **Total** | **246** | **118** | **47.9** |

The total number of registered active Northern Ireland fishing vessels and the number with interviews, by each gear type. Fleet information supplied by the Department of Agriculture and Rural Development, Northern Ireland, and update during the research. Please note that many scallop dredgers are registered as dredge or dredge and nephrops.

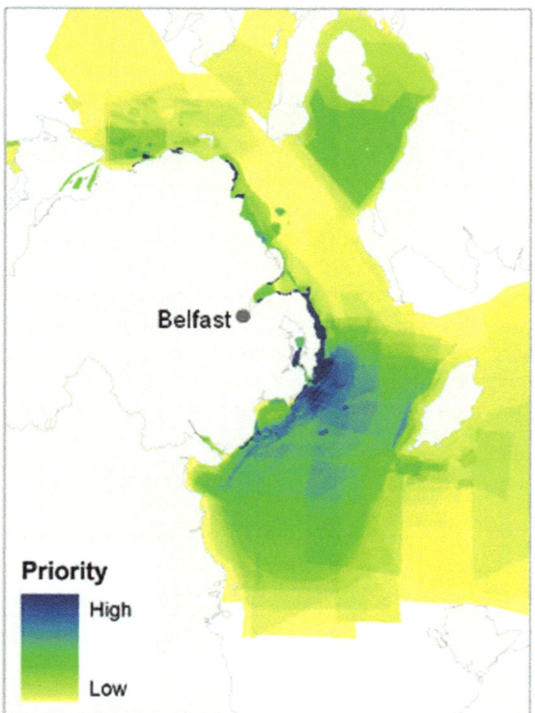


**Figure 8. Map of the spatial access priorities of the Northern Irish fishing fleet.** Map of the spatial access priorities of the Northern Irish fishing fleet generated from data obtained from 103 interviews with skippers and boat owners in 2012. It shows the main areas used by the fleet. There are a number vessels that have a much greater range than this. All areas were included in mapping process; however, for the purpose of maintaining an appropriate scale for visual display, they have not been shown.

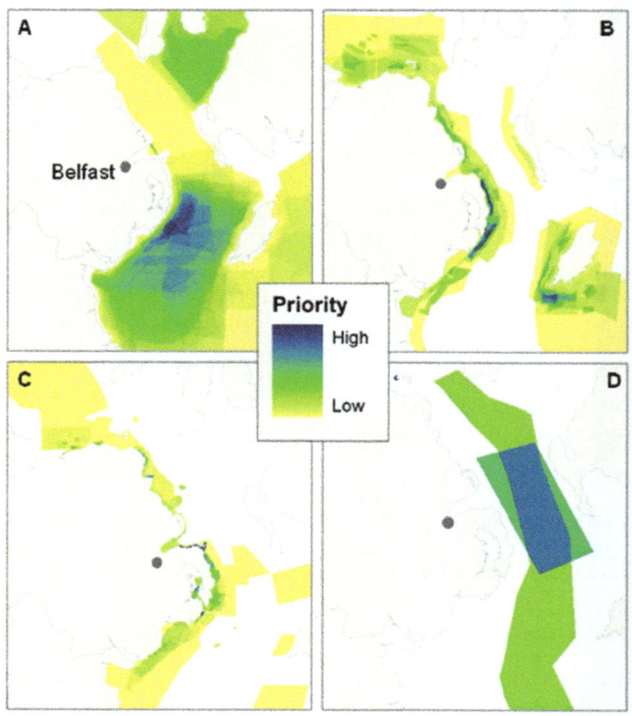


**Figure 9. Maps of the spatial access priorities of the four main fisheries within the Northern Irish fishing fleet.** Maps of the spatial access priorities of the four main fisheries within the Northern Irish fishing fleet: A) Nephrops, B) Pot fishing, C) Scallops and D) White fish. Maps were generated using data obtained from interviews with a total of 103 Skippers and boat owners in 2012. They show the main areas used by the fleet. There are a small minority of Nephrops vessels that have a greater range than this. All areas were included in mapping process; however, for the purpose of maintaining an appropriate scale for visual display, they have not been shown.

**Figure 10. Spatial overlap of fishers' access priority areas within a given gear-length-port combination.** Map of the access priority areas chosen by nine *Nephrops* fishers, from Northern Ireland. The nine fishers' vessels were within a given port-length combination and the map shows the spatial overlap in their responses.

highlighting reference points, such as lighthouses. Furthermore, when using digital maps, built-in tools such as "zoom" and "pan" have been shown to facilitate the mapping process by revealing more scale-dependent detail [26]. Fishers also appreciated the ability to quickly and accurately measure distances on the digital map when they were drawing on their areas. Having some broad-scale paper maps available did prove useful, mainly as a comfortable starting point for those less technologically inclined, and the use of a combination seems to provide the best of both worlds.

SAPM facilitated the documentation of fishers' actual access priorities, as opposed to interpreting their priorities from VMS-based mapping methods, and thus provides a fuller picture of which parts of the sea are most important to them. VMS alone maps only effort. Even when it is possible to incorporate log-book data with VMS data to estimate gross catch value of the different parts of the sea, this value is certainly not the only feature that makes an area important to fishers. Moreover, there are many boats that simply do not have VMS because they are too small. In Northern Ireland, over half the fleet would have their access priorities unaccounted for if planning were to use VMS-based maps alone.

Notwithstanding the limitations of VMS data, they remain vital for understanding the distribution of effort and are very important both for management and for informing the planning process. For example, they can be used to help predict the displacement impact of spatial restrictions. SAPM can be an important compliment to VMS based methods, when they are available, helping to compensate for their limitations. When they are not, it can be readily used as a standalone method.

This research also demonstrated how the SAPM method can be used to combine priorities across different types of fishery and to generate maps for the overall fleet, as well as for the separate fisheries. That SAPM can be used in any situation makes it suitable for enabling the direct comparison, and even combination, of the access priorities of multiple fleets, which otherwise might not have been possible due to differences in the associated data. Examples include: fleets that have VMS and fleets that do not, or fleets from different jurisdictions that access the same area but whose regulators have different fisheries data-capture and analysis methodologies.

Possibly the greatest advantage of SAPM over VMS is that it directly engages fishers in the process and allows them to contribute quantitative pre-planning data. The importance of stakeholder involvement is well recognised [20,23–25,33,52] and is increasingly becoming a legislative requirement, incorporated into both national and international policy [23,53]. Recent European Directives, such as the EU Public Participation Directive (2003/35/EC), have set new legal standards for stakeholder involvement, and almost every sustainability conference closes with a commitment to enhancing public participation [25].

Stakeholder involvement is sought because it has been shown to increase the quality and durability of environmental decisions, with the resultant plans better adapted to local socio-cultural and environmental conditions [23–25]. It can also reduce conflict and increase compliance [21,33]. Indeed stakeholder participation and involvement have been widely identified as essential components of effective ecosystem management [54,55], with differences in the extent and quality of participation being one of the key factors for variation in the quality of management approaches [20].

Stakeholder participation can also lead to the generation of additional data. Whilst undertaking SAPM, it would be relatively simple to gather other information, such as fishers' opinions of management options, or local ecological knowledge, which can enhance understanding of the ecosystems being managed. Both can provide highly valuable input to the planning processes. Inclusion of stakeholder knowledge such as this can also help to integrate stakeholders into the planning process and to promote a sense of ownership [56]. Certainly here in Northern Ireland, fishers were more than happy to contribute, and discussions indicated that the inclusion of their knowledge and opinions into the planning process would enhance buy-in.

Stakeholder involvement is not without its opponents, with counterarguments suggesting that the participatory process may not in reality lead to better-quality plans, and that the high levels of participation may actually increase conflict by having disputing parties at the negotiating table [20]. There are concerns that it will frustrate the planner by slowing down the process and that the strength of the final agreement will be diluted to balance competing interests [20]. Protected area designs that accommodate social needs may also result in outcomes that do not adequately conserve ecosystems [57]. Yet evidence suggests that this need not be the case. The incorporation of stakeholders during the planning process can increase the likelihood of plan implementation and the presence of certain stakeholders, particularly industry, has been shown to significantly increase the quality of local ecosystem plans [20]. Likewise, research shows that the use of fisheries data in MPA selection can reduce the negative impacts on the industry whilst still meeting conservation goals [58,59], and that incorporation of socio-economic considerations into systematic conservation planning can improve overall cost effectiveness [60]. It has also been frequently demonstrated that when fishers are involved in the research and decision-making process, management guidelines are more likely to be effective [26,61,62].

Another concern might be the credibility of stakeholder involvement, questioned on the basis that many may not have sufficient expertise to meaningfully engage in what are often

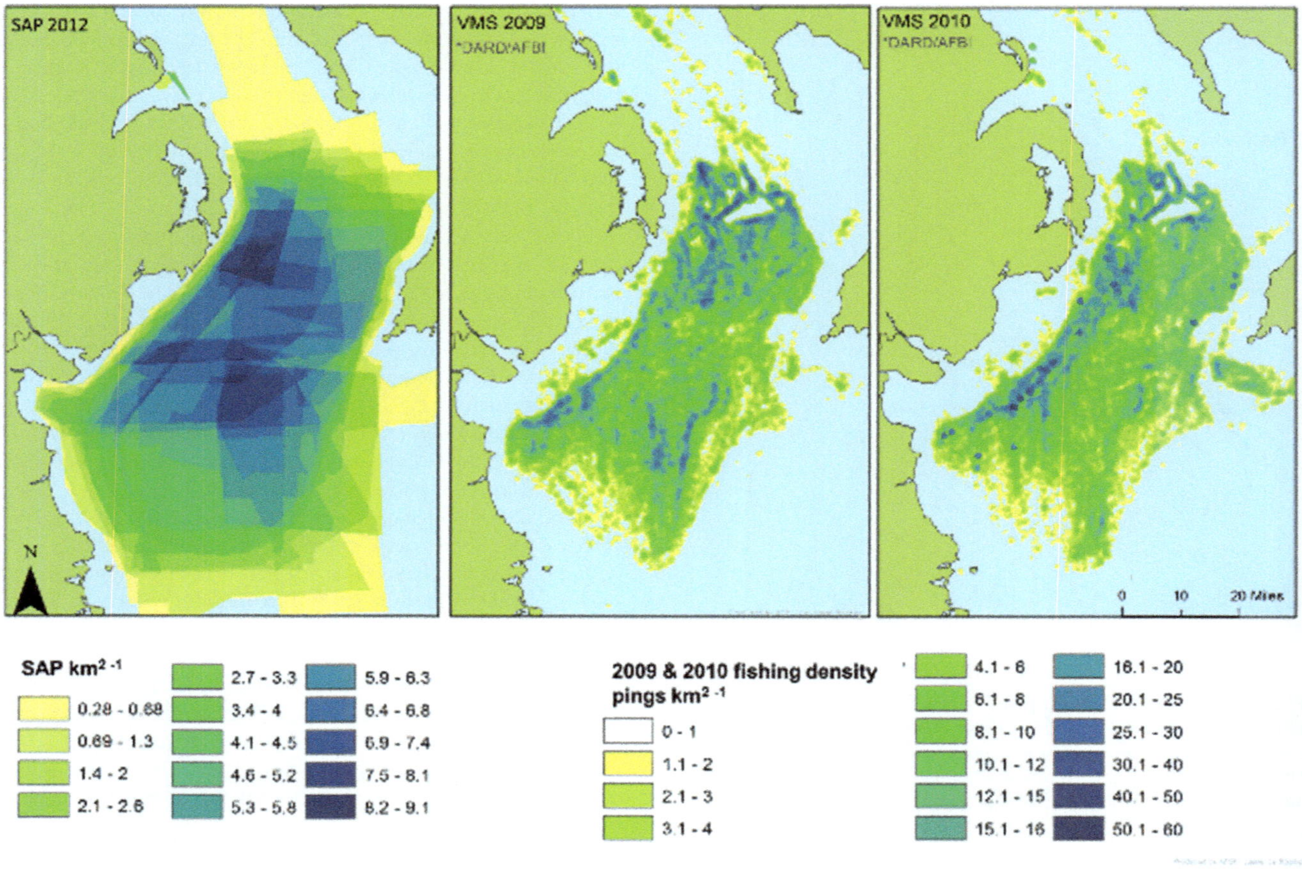


**Figure 11. Spatial access priority map and VMS effort maps for the over 15 m *Nephrops* fleet.** The spatial access priority map (2012) and the VMS effort maps (2009 and 2010) for over 15 m *Nephrops* fleet. VMS maps supplied by Laurence Rooney at AFBI. Processed VMS data were not available for all areas delineated by the fishers so this figure compares only main fishing areas where it was.

technical debates [63]. However, in this case it is actually the fishers that are the experts, often with decades of fishing experience. This acumen of fishers is increasingly recognised, with more and more studies documenting their extensive ecological knowledge [52,64,65] and surely who better to indicate which parts of the sea are most important to fishers than the fishers themselves?

With stakeholder consultation becoming more and more integrated into governance, the issue of consultation fatigue can also arise: when stakeholders are asked to take part in multiple or multi-stage participatory processes in which they perceive their involvement gains them little reward or capacity to influence the decisions being made [66]. Use of SAPM can help prevent that. It collects data in a systematic way, one of the recommendations for overcoming barriers to stakeholder participation [67]. Participation in SAPM is also of low cost to fishers and the method is readily explained. Resultant data are baseline preplanning data, so they can be used for multiple different planning projects, helping to avoid excessive or repeated demands on fishers' time. The maps generated can easily be used to demonstrate the results of the research and how that can feed into the planning process. They can also be readily incorporated into conservation planning software, such as MARXAN, where their influence on selecting potential MPA site can be illustrated.

As mentioned earlier, the cost of participation at centrally based stakeholder events can be a barrier to participation to fishers. Participation tends to be higher amongst those with more extreme positions and amongst those who are more dominant, influential and financially powerful within a community [68], so those fishers who do attend central events are unlikely to be truly representative of the fleet as a whole. By actually going to the fishers, SAPM can help to ensure a more diverse range of fishers is involved in the planning process. It also allows much larger numbers to partake: whilst it would be impractical for DOE to have 100 fishers around a planning table, SAPM allowed those fishers to contribute in a way that made them feel part of that process.

Other studies have looked at incorporating fishers' priorities into marine planning. In particular, a series of studies incorporating Californian fishers' knowledge and priorities into MPA selection has been conducted by Klein [58,69] and Scholz [31,32,70]. Fishers were asked to map their fishing grounds and to indicate relative importance [70], in a similar way to SAPM. The study area was then divided into 1381 planning units and the fishers data was used to calculate the relative importance of those planning units, across all fisheries and within a fishery [58]. Surrogate layers for relative fishing effort for each fishery were produced, and by summing the layers for the different fisheries they were able to develop an index of relative effort across all fisheries. In this method, each fishery receives equal weighting, such that a fishery that supported 10 fishers would have the same input into the final index as one that supported 1000. No weighting is applied to the individual fishers' areas either, so that results from a skipper whose vessel supported 10 crew would have the same weight as result from a skippers who vessel supported two.

This approach does increase the likelihood that all fisheries would be affected equally, which in this instance was a specific goal of the stakeholders (Klein, pers.comm). However, it is also often very desirable to minimise total negative impact on the fishing community and protect the greatest number of livelihoods. In later studies, Klein *et al* [69] and Scholz *et al* [32] incorporated a weighting system. They weighted the area(s) each fisher indicated by their vessel's revenue, which means that those fishers with vessels that caught more and made more profit had proportionally more input. This is similar to SAPM, where responses are weighted by the number of crew, so that those vessels with more crew have more total priority. Weighting by vessel revenue enabled Klein *et al* [69] and Scholz *et al* [32] to produce layers that showed the monetary value, in US dollars, of a given planning unit to an individual fisher. They summed those individual layers to create cost layers, which they used in Marxan to develop MPA planning options that minimised the negative impacts of spatial restrictions on the fishing industry.

The main difference between previous approaches and SAPM is that SAPM can allow an assessment of both relative and total access priority, and the production of surrogate cost layers, without the need for additional fisheries data. This is because SAPM uses information derived from interview (crew numbers), rather than revenue or landings, to weight the responses. Like the cost layer(s) created in both Klein *et al* [69] and Scholz *et al* [32], the SAPM generated surrogate cost layer(s) can be used in Marxan to minimise total negative impact of MPA designation on the fishing community. If the goal was to increase the likelihood that all fisheries are affected equally, like Klein *et al* [58], SAPM can also be used to map and assess relative priority, by dividing the SAP $km^{-2}$ by the total SAP of a fishery and/or a fleet. Moreover, SAPM can also be used to assess the equality of spatial restrictions by analysing the SAPs of different port-gear-length combinations and comparing how much a given restriction or set of restrictions affects each group. When it is possible to access other data, such as revenue data, SAPM can be easily modified to incorporate it, see below for some possible adaptations to the method.

Another major difference is that SAPM specifically takes account of the different characteristics of the vessels (length/port/gear) when weighting and scaling responses. This is important because size of a vessel and the port from which it operates are both likely to have an impact on its preferred grounds, even when fishing for the same target species. Furthermore, the previous studies did not ascertain whether their samples (the fishers they interviewed) were representative of the fleet. Indeed, Scholz *et al* [32] state that to reach their sampling goal as efficiently and cheaply as possible they targeted fishers with the largest catches [32]. Whilst understandable, this suggests that, within a given fishery, the fishers with the smallest vessels had a lower chance of being interviewed. Smaller vessels tend to be more restricted in terms of fishing grounds and may have totally different spatial access priorities than larger vessels. By requiring a representative sample of the fleet SAPM helps to ensure all fishers are included, essential when aspiring to equitable stakeholder involvement.

Use of SAPM does not remove the need for subsequent engagement with the fishing community, nor will it remove all the potential conflict associated with MSP and MPA designation. It is a beginning, not an end: providing data to feed into a planning process and allowing a planner to assess the impact of different potential plans on fishers. Its use should improve the transparency of the planning process, enhance buy-in and increase trust.

### Possible Adaptations to the Method

The weighting system used in SAPM is highly flexible, allowing for the incorporation of livelihoods or economic values beyond the vessels' crews. In Northern Ireland, for example, fish processors employ an estimated 500 full-time staff [43]. To incorporate these into the analysis, those 500 would be first separated into their associated fisheries. Then, based on the percentage of fishers within each length-gear-port combination for that fishery, they would be divided between each combination and added to the estimated number of crew. Doing this would proportionally increase the priority of fisheries with associated local processing, such as *Nephrops*, and cause a relative decrease in those without, reflecting the higher numbers of livelihoods that depend on those sectors. Similarly, if other data, such as landings, are available, the weighting system could also be modified accordingly. The total economic value of landings for a given fishery could be divided by the number of fishers in that fishery to obtain a value per fisher, which would then be multiplied by the weighting obtained from the standard method for each length/gear/port combination.

### Future Research

This research did not aim to investigate the fishers' motivations for selecting different areas or for assigning priorities. Nor were they restricted to assigning priorities for a specific time span. It would be informative to expand future SAPM projects to include a debriefing session, where researchers explore with fishers their rationale and motivation for selecting areas and assigning priorities. An improved understanding of what makes areas important to fishers may help with predicting how they respond to spatial and temporal restrictions. Future SAPM projects could also incorporate the gathering of local ecological and historical knowledge, including: mapping where the fishers historically accessed resources and how their SAPs have changed over time, discussing when and why they changed, and comparing answers to legislative, ecological and climatic trajectories.

### Conclusions

Spatial management can have many benefits for conservation and fisheries alike, but in order to realise them stakeholders need to be engaged in a strategic manner from an early stage. SAPM offers a tool that not only gathers quantitative data on fishers' access priorities, but also facilitates stakeholder engagement and increases the opportunity for dialogue between fishers and managers. The resultant maps provide a transparent, easy to demonstrate way of incorporating fishers' priorities, especially when combined with spatial planning software such as MARXAN. The use of SAPM should facilitate strategic conservation planning, maximising benefits whilst both minimising negative impacts on stakeholders and helping to ensure those impacts are spread equitably. We hope that the inclusive planning that SAPM encourages will lead to more successful solutions to the inevitable conflicts of interests that conservation planning encounters.

## Supporting Information

**File S1** Pre-interview Information sheet. Pre-interview information sheet provided to fishers prior to participation in a survey targeting skippers and owners of commercial fishing vessels registered in Northern Ireland. After the potential participant had read the information sheet, the content was discussed to ensure understanding and obtain verbal consent, before beginning the interview. Research was completed during 2012.

**File S2** Questionnaire. Questionnaire used to guide semi-structured interviews of skippers and owners of Northern Ireland registered commercial fishing vessels. Interviews were preceded with pre-interview information sheets. 103 interviews were conducted in total, during 2012. The responses from some of these questions were used to develop Spatial Access Priority Maps, this study, whilst others were used to answer research questions explored in subsequent studies.

## Acknowledgments

First and foremost the authors would like to thank all the fishers that generously gave up their time to make this research possible. Many thanks to everyone in ANIFPO, NIFPO and the North Coast Lobster Association; in particular we thank Mr Alan McCulla and Mr David Galbraith. Thanks also to members of DOE, NIEA, DARD and AFBI for their assistance, with special thanks to Laurence Rooney, at AFBI, for the provision of the VMS images. We would like to thank our colleagues Dr Thomas Smyth, Dr Sara Benettii and Dr Ruth Thurston for reviewing an earlier draft of the manuscript. We would also like to thank Dr N. Clare Eno and two other anonymous referees for their comments and suggestions. Input from all the reviewers greatly added to the final manuscript.

## Author Contributions

Conceived and designed the experiments: KLY. Performed the experiments: KLY. Analyzed the data: KLY. Contributed reagents/materials/analysis tools: KLY. Wrote the paper: KLY. Introduced the first author to initial key contacts: DS. Provided assistance with the funding proposal that made this research possible: DS. Provided significant critical feedback on multiple drafts, improving the argument and the robustness of the discussion: DS. Corrected grammatical errors and made suggested improvements to prose: DS. Suggested the development of a flow chart to make the methods simpler to follow: DS.

## References

1. Airoldi L, Beck MW (2007) Loss, status and trends for coastal marine habitats of Europe. Oceanography and Marine Biology: An Annual Review 45: 345–405.
2. Casey RA, Myers JM (1998) Near Extinction of a Large, Widely Distributed Fish. Science 281: 690–692. Available: http://www.sciencemag.org/cgi/doi/10.1126/science.281.5377.690.
3. Pauly D, Watson R, Alder J (2005) Global trends in world fisheries: impacts on marine ecosystems and food security. Philosophical transactions of the Royal Society of London Series B, Biological sciences 360: 5–12. Available: http://www.pubmedcentral.nih.gov/articlerender.fcgi?artid=1636108&tool=pmcentrez&rendertype=abstract.
4. Jackson JB, Kirby MX, Berger WH, Bjorndal KA, Botsford LW, et al. (2001) Historical overfishing and the recent collapse of coastal ecosystems. Science (New York, NY) 293: 629–637. Available: http://www.ncbi.nlm.nih.gov/pubmed/11474098.
5. Halpern BS, Walbridge S, Selkoe KA, Kappel CV, Micheli F, et al. (2008) A global map of human impact on marine ecosystems. Science (New York, NY) 319: 948–952. Available: http://www.ncbi.nlm.nih.gov/pubmed/18276889.
6. Worm B, Sandow M, Oschlies A, Lotze HK, Myers RA (2005) Global patterns of predator diversity in the open oceans. Science (New York, NY) 309: 1365–1369. Available: http://www.ncbi.nlm.nih.gov/pubmed/16051749.
7. Beukers-Stewart B, Vause B, Mosley M, Rossetti H, Brand A (2005) Benefits of closed area protection for a population of scallops. Marine Ecology Progress Series 298: 189–204. Available: http://www.int-res.com/abstracts/meps/v298/p189-204/.
8. Halpern BS, Warner RR (2002) Marine reserves have rapid and lasting effects. Ecology Letters 5: 361–366. Available: http://doi.wiley.com/10.1046/j.1461-0248.2002.00326.x.
9. Roberts CM, Hawkins JP (2000) Fully-protected marine reserves: a guide. WWF Endangered Seas Campaign, Washington, USA and Environment Department, University of York, UK. Avalible: http://billhutten.s3.amazonaws.com/fw/docs/345.pdf Accessed 2012 Jan 20.
10. Roberts CM, Bohnsack JA, Gell F, Hawkins JP, Goodridge R (2001) Effects of marine reserves on adjacent fisheries. Science (New York, NY) 294: 1920–1923. Available: http://www.ncbi.nlm.nih.gov/pubmed/11729316.
11. Worm B, Hilborn R, Baum JK, Branch TA, Collie JS, et al. (2009) Rebuilding global fisheries. Science (New York, NY) 325: 578–585. Available: http://www.ncbi.nlm.nih.gov/pubmed/19644114.
12. Hyrenbach KD, Forney KA, Dayton PK (2000) Marine Protected Areas and Ocean Basin Management. Aquatic Conservation: Marine and Freshwater Ecosystems 10: 437.
13. Heupel MR, Simpfendorfer CA (2005) Using Acoustic Monitoring to Evaluate MPAs for Shark Nursery Areas: The Importance of Long-Term Data. Marine Technology Society Journal 39: 10.
14. Rudd MA, Tupper MH (2002) The Impact of Nassau Grouper Size and Abundance on Scuba Diver Site Selection and MPA Economics. Coastal Management 30: 133.
15. Kelleher G, editor (1999) Guidelines for Marine Protected Areas. Gland, Switzerland: World Commision on Protected Areas. IUCN - The World Conservation Union. IUCN website. Avaliable: http://data.iucn.org/dbtw-wpd/edocs/PAG-003.pdf. Accessed 2012 Oct 10.
16. Jameson SC, Tupper MH, Ridley JM (2002) The three screen doors: Can marine "protected" areas be effective? Marine pollution bulletin 44: 1177–1183.
17. Hargreaves-Allen V, Mourato S, Milner-Gulland EJ (2011) A global evaluation of coral reef management performance: are MPAs producing conservation and socio-economic improvements? Environmental management 47: 684–700. Available: http://www.ncbi.nlm.nih.gov/pubmed/21365274.
18. Helvey M (2004) Seeking Consensus on Designing Marine Protected Areas: Keeping the Fishing Community Engaged. Coastal Management 32: 173–190. Available: http://www.tandfonline.com/doi/abs/10.1080/08920750490276236.
19. Weible CM (2008) Caught in a Maelstrom: Implementing California Marine Protected Areas. Coastal Management 36: 350–373. Available: http://www.tandfonline.com/doi/abs/10.1080/08920750802266387.
20. Brody SD (2003) Measuring the Effects of Stakeholder Participation on the Quality of Local Plans Based on the Principles of Collaborative Ecosystem Management. Journal of Planning Education and Research 22: 407–419. doi:10.1177/0739456X03253022.
21. Innes JE (1996) Planning through consensus building: A new view of the comprehensive planning ideal. Journal of the American Planning Association 62: 460–472.
22. Reed MS, Dougill AJ, Baker TR (2008) Participatory indicator development: what can ecologists and local communities learn from each other? Ecological applications: a publication of the Ecological Society of America 18: 1253–1269. Available: http://www.ncbi.nlm.nih.gov/pubmed/18686585.
23. Beierle TC (2002) The quality of stakeholder-based decisions. Risk analysis: an official publication of the Society for Risk Analysis 22: 739–749. Available: http://www.ncbi.nlm.nih.gov/pubmed/12224747.
24. Koontz TM, Thomas CW (2006) What Do We Know and Need to Know about the Environmental Outcomes of Collaborative Management? Public Administration Review 66: 111–121. Available: http://doi.wiley.com/10.1111/j.1540-6210.2006.00671.x.
25. Newig J (2007) Does public participation in environmental decisions lead to improved environmental qulaity? Towards an analytical framework. Research and Practice for a Sustainable Future 1: 51–71.
26. Moreno-báez M, Orr BJ, Cudney-bueno R, Shaw WW (2010) Using fishers' local knowledge to aid managment at regional scales: Spatial Distribution of small-scale Fisheries in the Northern Gulf of California Mexico. Bulletin of Marine Science 86: 1–16.
27. Maurstad A (2002) Fishing in murky waters: ethics and politics of research on fisher knowledge. Marine Policy 26: 159–166.
28. Aswani S, Hamilton RJ (2004) Integrating indigenous ecological knowledge and customary sea tenure with marine and social science for conservation of bumphead parrotfish (Bolbometopon muricatum) in the Roviana Lagoon, Solomon Islands. Environmental Conservation 31: 69–83. Available: http://www.journals.cambridge.org/abstract_S037689290400116X.
29. Aswani S, Lauer M (2006) Incorporating Fishermen's Local Knowledge and Behavior into Geographical Information Systems GIS for Designing Marine Protected Areas in Oceania. Human Organization 65: 81–102.
30. Hall GB, Close CH (2007) Local knowledge assessment for a small-scale fishery using geographic information systems. Fisheries Research 83: 11–22. Available: http://linkinghub.elsevier.com/retrieve/pii/S016578360600289X.
31. Scholz A, Bonzon K, Fujita R, Benjamin N, Woodling N, et al. (2004) Participatory socioeconomic analysis: drawing on fishermen's knowledge for marine protected area planning in California. Marine Policy 28: 335–349. Available: http://linkinghub.elsevier.com/retrieve/pii/S0308597X03001118.
32. Scholz AJ, Steinback C, Kruse S A, Mertens M, Silverman H (2011) Incorporation of spatial and economic analyses of human-use data in the design of marine protected areas. Conservation biology: the journal of the Society for Conservation Biology 25: 485–492. Available: http://www.ncbi.nlm.nih.gov/pubmed/21175829.
33. Reed MS (2008) Stakeholder participation for environmental management: A literature review. Biological Conservation 141: 2417–2431. Available: http://linkinghub.elsevier.com/retrieve/pii/S0006320708002693.
34. Gerhardinger LC, Hostim-Silva M, Medeiros RP, Matarezi J, Bertoncini ÁA, et al. (2009) Fishers' resource mapping and goliath grouper Epinephelus itajara (Serranidae) conservation in Brazil. Neotropical Ichthyology 7: 93–102.

Available: http://www.scielo.br/scielo.php?script = sci_arttext&pid = S1679-62252009000100012&lng = en&nrm = iso&tlng = en.
35. Des Clers S, Lewin S, Edwards D, Searle S, Lieberknecht L & Murphy D (2008) FisherMap Mapping the Grounds: recording fishermen's use of the seas. A report published for the Finding Sanctuary project. Finding Sanctuary website. Available: http://findingsanctuary.marinemapping.com/06_all%20project%20reports/Fishe rmap%20report%20November%202008.pdf. Accessed 2012 Aug 11.
36. Wheeler M, Chambers FMJ, Sims-Castley R, Cowling RM, Schoeman D (2008) From beans to breams: how participatory workshops can contribute to marine conservation planning. African Journal of Marine Science 30: 475–487.
37. Lee J, South AB, Jennings S (2010) Developing reliable, repeatable, and accessible methods to provide high-resolution estimates of fishing-effort distributions from vessel monitoring system (VMS) data. ICES Journal of Marine Science 67: 1260–1271. Available: http://icesjms.oxfordjournals.org/cgi/doi/10.1093/icesjms/fsq010.
38. Gerritsen H, Lordan C (2010) Integrating vessel monitoring systems (VMS) data with daily catch data from logbooks to explore the spatial distribution of catch and effort at high resolution. ICES Journal of Marine Science 68: 245–252. Available: http://icesjms.oxfordjournals.org/cgi/doi/10.1093/icesjms/fsq137.
39. Hinz H, Murray LG, Lambert GI, Hiddink JG, Kaiser MJ (2013) Confidentiality over fishing effort data threatens science and management progress. Fish and Fisheries 14: 110–117. doi:10.1111/j.1467-2979.2012.00475.x.
40. DETI (2011) Regional Locational Guidance (RLG ) for Offshore Renewable Energy Developments in NI Waters. Department of Enterprise, Trade and Investment. Belfast. Belfast, Northern Ireland. Department of Enterprise, Trade and Investment website. Available:http://www.detini.gov.uk/regional_locational_guidance__rlg__for_offshore_renewable_energy_developments_in_ni_waters. Accessed 2012 Sep 15.
41. DETI (2009) Off shore wind and marine renewable energy in Northern Ireland strategic environmental assessment (SEA) - Non Technical Summary. Joint Nature Conservation Committee (JNCC) website. Available: http://jncc.defra.gov.uk/PDF/0955_Explanatory_Summary.pdf Accessed 2012 Nov 10.
42. The Crown Estate (2012) Successful bidders in Northern Ireland offshore energy leasing rounds. Crown Estate website. Avaliable: http://www.thecrownestate.co.uk/news-media/news/2012/northern-ireland-offshore-energy-successful-bidders/. Accessed 2012 Oct 5.
43. Cappell R, Nimmo F, Rooney L (2012) The value of Irish Sea Conservation Zones to the Northern Irish Fishing Industry. Report to: Seafish Northern Ireland Advisory Committee (SNIAC). POSEIDON Aquatic Resource Management Ltd. Seafish website. Available: http://www.seafish.org/media/Publications/Poseidon_NI_MCZ_valuation_final_report_August_2012__2_.pdf Accessed 2012 Oct 5.
44. ICES (2009) Report of the Benchmark Workshop on Nephrops (WKNEPH ). Aberdeen, UK.
45. AFBI (n.d.) Inshore Fisheries - Scallops. Available: http://www.afbini.gov.uk/index/services/services-specialist-advice/coastal-science/inshore-fisheries/services-inshore-fisheries-bivalve-molluscs.htm. Agri-food and Biosciences Instiute website. Accessed 2012 Nov 10.
46. ICES (2012) ICES IBTSWG Report 2012 Report of the International Bottom Trawl Survey Working Group (IBTSWG ): 27–30. March 2012, Lorient, France. ICES CM 2012/SSGESST: 03. 323 p.
47. Yates KL, Payo Payo A, Schoeman DS (2013) International, regional and national commitments meet local implementation: A case study of marine conservation in Northern Ireland. Marine Policy 38: 140–150. Available: http://linkinghub.elsevier.com/retrieve/pii/S0308597X12001285.
48. Yates K (2012) Mapping the spatial access priorities of the Northern Ireland Fishing Fleet. The Diverse Seas Project, University of Ulster, Northern Ireland. University of Ulster, Environmental Science Research Institute website. Available: http://www.science.ulster.ac.uk/esri/Katherine-Yates.html. Accessed 2013 Jun 12.
49. Kelley K, Clark B, Brown V, Sitzia J (2003) Good practice in the conduct and reporting of survey research. International journal for quality in health care: journal of the International Society for Quality in Health Care/ISQua 15: 261–266. Available: http://www.ncbi.nlm.nih.gov/pubmed/12803354.
50. Pacific States Marine Fisheries Commission (2002) West Coast Catcher Boat Survey Summary 1997–1998. Seattle. Pacific States Marine Fisheries Commission website. Avaliable : http://www.psmfc.org/efin/abstracts-data.html. Accessed 2013 Jan 20.
51. Close CH, Hall GB (2006) A GIS-based protocol for the collection and use of local knowledge in fisheries management planning. Journal of environmental management 78: 341–352. Available: http://www.ncbi.nlm.nih.gov/pubmed/16115723.
52. Thornton TF, Scheer AM (2012) Collaborative Engagement of Local and Traditional Knowledge and Science in Marine Environments: A Review. Ecology and Society 17. Available: http://www.ecologyandsociety.org/vol17/iss3/art8/.
53. Stringer LC, Reed MS, Dougill AJ, Seely MK, Rokitzki M (2007) Implementing the UNCCD: Participatory challenges. Natural Resources Forum 31: 198–211. Available: http://doi.wiley.com/10.1111/j.1477-8947.2007.00154.x.
54. Yaffee SL (1996) Ecosystem management in practice: the importance of human institutions. Ecological Applications 6: 724–727.
55. Duane T (1997) Community Participation in Ecosystem Management. Ecology Law Quarterly 24: 771–797.
56. Smith AH, Berkes F (1991) Solutions to the "Tragedy of the Commons": Sea-urchin Management in St Lucia, West Indies. Environmental Conservation 18: 131. Available: http://www.journals.cambridge.org/abstract_S0376892900021706.
57. Roberts CM, Andelman S, Branch G, Bustamante RH, Castilla JC, et al. (2003) Ecological criteria for evaluating candidate sites for marine reserves. Ecological ApplicationsVol13, suppl1 13: S199–S214.
58. Klein CJ, Chan A, Kircher L, Cundiff AJ, Gardner N, et al. (2008) Striking a balance between biodiversity conservation and socioeconomic viability in the design of marine protected areas. Conservation biology: the journal of the Society for Conservation Biology 22: 691–700. Available: http://www.ncbi.nlm.nih.gov/pubmed/18325043.
59. Richardson EA, Kaiser MJ, Edwards-Jones G, Possingham HP (2006) Sensitivity of Marine-Reserve Design to the Spatial Resolution of Socioeconomic Data. Conservation Biology 20: 1191–1202.
60. Moore J, Balmford A, Allnutt T, Burgess N (2004) Integrating costs into conservation planning across Africa. Biological Conservation 117: 343–350. Available: http://linkinghub.elsevier.com/retrieve/pii/S0006320704000023.
61. Cudney-Bueno R, Basurto X (2009) Lack of cross-scale linkages reduces robustness of community-based fisheries management. PloS one 4: e6253. Available: http://www.pubmedcentral.nih.gov/articlerender.fcgi?artid = 2705799&tool = pmcentrez&rendertype = abstract.
62. Johannes RE, Freeman MMR, Hamilton RJ (2000) Ignore fishers ' knowledge and miss the boat. Fish and Fisheries 1: 257–271.
63. Fischer A, Young JC (2007) Understanding mental constructs of biodiversity: Implications for biodiversity management and conservation. Biological Conservation 136: 271–282. Available: http://linkinghub.elsevier.com/retrieve/pii/S0006320706005313.
64. Silvano RAM, Begossi A (2009) What can be learned from fishers? An integrated survey of fishers' local ecological knowledge and bluefish (Pomatomus saltatrix) biology on the Brazilian coast. Hydrobiologia 637: 3–18. Available: http://www.springerlink.com/index/10.1007/s10750-009-9979-2.
65. Silvano RAM, Valbo-Jørgensen J (2008) Beyond fishermen's tales: contributions of fishers' local ecological knowledge to fish ecology and fisheries management. Environment, Development and Sustainability 10: 657–675. Available: http://www.springerlink.com/index/10.1007/s10668-008-9149-0.
66. Burton P, Goodlad R, Croft J, Abbott J, Hastings A, et al. (2004) What works in community involvement in area-based initiatives? A systematic review of the literature. Available: http://webarchive.nationalarchives.gov.uk/20110218135832/http://rds.homeoffice.gov.uk/rds/pdfs04/rdsolr5304.pdf.
67. National Research Council (2008) Public participation in environmental assessment and decision making. National Acedemy Press. Washington, D.C. The National Academies Press Website. Avaliable: http://www.nap.edu/catalog.php?record_id = 12434. Accessed 2012 Sep 12.
68. Turner M, Weninger Q (2005) Meetings with Costly Participation: An Empirical Analysis. Review of Economic Studies 553: 247–268.
69. Klein CJ, Steinback C, Watts M, Scholz AJ, Possingham HP (2010) Spatial marine zoning for fisheries and conservation. Frontiers in Ecology and the Environment 8: 349–353. Available: http://www.esajournals.org/doi/abs/10.1890/090047.
70. Scholz A, Birney K, Bonzon K, Chan C, Hamrin S, et al. (2006) Commercial fishing grounds and their relative importance off the Central Coast of California. Report to the California Marine Life Protection Act Initiative. Ecotrust. California Department of Fish and Wildlife website. Available: http://www.dfg.ca.gov/marine/pdfs/agenda_workshop041807_report.pdf. Accessed 2012 Feb 15.

# 15

# When Is Spillover from Marine Reserves Likely to Benefit Fisheries?

**Colin D. Buxton[1]*, Klaas Hartmann[1], Robert Kearney[2], Caleb Gardner[1]**

**1** Fisheries Aquaculture and Coasts Centre, Institute for Marine and Antarctic Studies, University of Tasmania, Hobart, Tasmania, Australia, **2** Institute for Applied Ecology, University of Canberra, Canberra, Australian Capital Territory, Australia

## Abstract

The net movement of individuals from marine reserves (also known as no-take marine protected areas) to the remaining fishing grounds is known as spillover and is frequently used to promote reserves to fishers on the grounds that it will benefit fisheries. Here we consider how mismanaged a fishery must be before spillover from a reserve is able to provide a net benefit for a fishery. For our model fishery, density of the species being harvested becomes higher in the reserve than in the fished area but the reduction in the density and yield of the fished area was such that the net effect of the closure was negative, except when the fishery was mismanaged. The extent to which effort had to exceed traditional management targets before reserves led to a spillover benefit varied with rates of growth and movement of the model species. In general, for well-managed fisheries, the loss of yield from the use of reserves was less for species with greater movement and slower growth. The spillover benefit became more pronounced with increasing mis-management of the stocks remaining available to the fishery. This model-based result is consistent with the literature of field-based research where a spillover benefit from reserves has only been detected when the fishery is highly depleted, often where traditional fisheries management controls are absent. We conclude that reserves in jurisdictions with well-managed fisheries are unlikely to provide a net spillover benefit.

**Editor:** Kerrie Swadling, University of Tasmania, Australia

**Funding:** This work was supported by the Fisheries Research and Development Corporation (FRDC 2010/226). http://frdc.com.au/Pages/home.aspx. The funders had no role in study design, data collection and analysis, decision to publish, or preparation of the manuscript.

**Competing Interests:** The authors have declared that no competing interests exist.

* Email: colin.buxton@utas.edu.au

## Introduction

Marine reserves (MR), also known as no-take marine protected areas (MPA), are widely acknowledged as a conservation tool and their utility in a variety of situations is well established [1]. In particular over-exploited fish populations are shown to recover in the absence of fishing and generally become more abundant and attain a larger mean size in the reserve [2]. MPAs are also frequently promoted for the management of fisheries [3–7], even though compelling evidence in support of a net fisheries benefit is lacking [8]. Fisheries are proposed to benefit from reserves through increased production of eggs and larvae from the reserve (recruitment effect) and the net movement of adults into adjacent fishing grounds (spillover effect) [9].

In this study we focus on the spillover effect and, to avoid confusion over the use of terms, we define *spillover* as the net movement of fish across the boundary of a reserve into the fished ground, which would be expected to occur on the basis of fundamental physical principles of random movement. This is in contrast to *net spillover benefit* which involves spillover of sufficient magnitude to compensate for lost productivity due to the closure of fishing grounds, resulting in an overall benefit to the fishery through higher catch or economic yield.

Our review of the extensive literature reporting fisheries benefits reveals that there are surprisingly few empirical studies that attempt to quantify either the recruitment effect or a net spillover benefit. For example, Goni et al. [10] claims to be the first study to demonstrate a net spillover benefit in a fishery. Harrison et al. [11] make a similar claim with respect to the recruitment benefit of reserves in terms of larval export. Whilst spillover has been shown in several other studies, most do not accommodate the reduction in catch that results from reducing the area of the fishery, and consequently do not demonstrate a net spillover benefit.

Fishers are generally opposed to the introduction of reserves because they reduce the size of their fishing grounds, which is inferred to result in a loss of yield. Spillover is a common counter argument from reserve proponents, including Government agencies in the US, Europe and Australia, claiming that it will compensate for the lost fishing grounds to the extent that a net improvement in fisheries yield occurs [12–14].

The impact of the introduction of reserves on yield has been addressed in a number of theoretical studies (e.g., [15–17]), several of which progressively conclude that under broad assumptions well-managed fisheries should not benefit from the introduction of reserves [18–21]. Hart [22] quantifies this result to some degree by using an age-structured model, concluding that a benefit from spillover should not be anticipated unless open area fishing mortality considerably exceeds that which produces MSY.

The assumptions underlying these studies primarily concern the homogeneity of fish stocks and are reasonable for a large range of species. The obvious exception occurs in fish stocks with strong variability in spatial structure, for example where source-sink relationships exist or where reserves may result in the closure of disproportionately productive areas [23]. Such spatial heterogeneity is the basis of traditional spatial management of fisheries, and is a well-established and understood technique. Targeted spatial closures can be expected to benefit fisheries for selected species if the closed area is of disproportionate significance to the productivity of the species in question. Not surprisingly some models have shown that, at least under certain conditions, higher sustainable yields can be achieved with a marine reserve than without, e.g, [17], [23], [24]. But despite the common demonstration that special circumstances are required to achieve a spillover benefit from reserves, the implication of these findings have received limited attention and appear to have contributed little to the international public debate over fisheries benefits and to current management policy.

In this paper we use a widely applied fisheries population dynamics model which minimizes assumptions in order for the outputs to be applicable to a broad range of fisheries in non-structured environments ('normal' or 'average' fisheries). We modify this model to incorporate a MR and consider the management circumstances under which a non-specific reserve is likely to provide a benefit to the fishery. Our work highlights the effect that the degree of mismanagement under conventional fisheries management practices has on the ability of a reserve to provide a net fisheries benefit. It also investigates how this relationship changes with the rate that fish move between the reserve and the main population.

## Methods

### Population Dynamics

The population dynamics were modeled using a deterministic difference equation of the form:

$$N_{t+1}=f(N_t)N_t-C(N_t), \tag{1}$$

where $N_t$ is the stock size at time $t$, $f(N_t)$ is the biological model that defines population growth and $C(N_t)$ is the catch. Common examples for the biological component of this model include the Ricker model:

$$f(N)=e^{r(1-N/K)}, \tag{2}$$

and logistic model:

$$f(N)=1+r(1-N/K). \tag{3}$$

In both models $r$ is the maximal growth rate and $K$ the carrying capacity (maximum population size).

Throughout this analysis we assume that the population is homogenous - a small proportion, $\delta$, of the population will behave identically in isolation to a larger proportion of the population. Mathematically, this implies that the carrying capacity can be reduced to $\delta K$. Alternatively we can consider the biological model to be a function of population density, in this case our model becomes:

$$N_{t+1}=f(N_t/\delta)N_t-C(N_t/\delta). \tag{4}$$

The divisor in the catch term indicates that catches are proportional to the population density (or constant).

Consider splitting a population into two areas: (i) a reserve occupying a proportion, $\alpha$, of the original habitat size and (ii) the remaining fishing grounds of size $1-\alpha$. Denoting the two population sizes by $R_t$ and $M_t$ respectively, the model becomes:

$$\begin{aligned} R_{t+1}&=f(R_t/\alpha)R_t-S_t\\ M_{t+1}&=f(M_t/(1-\alpha))M_t-C(M_t/(1-\alpha),M_t)+S_t \end{aligned} \tag{5}$$

where $S_t$ denotes the spillover from the reserve into the fished population.

### Spillover

We assume that a proportion, $\mu$, of the population in the reserve moves into the fishing ground at each time step. As the population in the reserve is $R_t$, then $\mu R_t$ will migrate out of the reserve. Similarly a proportion, $v$, of the population in the main fishing ground will migrate into the reserve. This results in the net movement from the reserve into the main fishing ground (the spillover) being:

$$S_t=\mu R_t-vM_t. \tag{6}$$

The values $\mu$ and $v$ will depend on both the size and geometry of the reserve, however given the homogeneity of the population we also require that the net spillover is zero ($S_t=0$) when the population density in the reserve and the fishing ground is equal (i.e. $R_t/\alpha=M_t/(1-\alpha)$). With this requirement and (6) we have:

$$\begin{aligned} S_t&=\mu R_t-vM_t\\ 0&=\mu\frac{\alpha}{1-\alpha}M_t-vM_t\\ v&=\frac{\alpha}{1-\alpha}\mu. \end{aligned} \tag{7}$$

As a direct result of the assumption of spatial homogeneity, a single parameter, $\mu$, is sufficient to define the strength of the movement both in and out of the reserve. The net spillover from the reserve therefore becomes:

$$S_t=\mu\left(R_t-\frac{\alpha}{1-\alpha}M_t\right). \tag{8}$$

Note that we assume that $\mu$ (and $v$) are independent of the population density in and outside of the reserve. While there may be evidence to suggest that some individuals do follow a density gradient [25], [26] this does not substantially alter our findings, as it is akin to an increase in $\mu$.

### Fishing

We have specified the catch as a function of the population density and population size, $C_t=C(M_t/(1-\alpha),M_t)$. One common catch model is constant catch, as found, for example, in a subsistence fishery where a certain catch must be obtained each year to feed the population:

$$C(M_t/(1-\alpha),M_t)=P. \tag{9}$$

Well managed fisheries either have natural restrictions that prevent over-exploitation of the fish stock (e.g., limited demand of a niche product) or management controls to prevent over-exploitation. Management controls can be divided into two broad categories – input and output controls. Input controls limit the effort applied in the fishery. Denoting this by $E$ we have:

$$C(M_t/(1-\alpha),M_t)=\frac{qEM_t}{1-\alpha}, \tag{10}$$

where q is a constant of proportionality. With this formulation, catch is directly proportional to the effort and population density (hence division of $M$ by $1-\alpha$ to obtain a density). Other functional forms may be more appropriate for certain fisheries and fishing methods (e.g. purse seining of schooling fish). We considered all effort applied to the fishery to shift instantaneously from the reserve to the open area.

Output controls limit the catch that can be taken from a fishery and were not explored, as the existence of an effective output control (that does not cause a fishery collapse at equilibrium) implies effective fisheries management [27]. In reality there are many examples of ineffective output controls in fisheries that have not collapsed. These fisheries persist as the output controls are adjusted through time or, when the stock is in low abundance, effort controls (whether through management or limited numbers of participating fishers) restrict the fishery. Modeling such systems requires many assumptions; hence we have focused on input controlled fisheries in this analysis.

## Net effect of the reserve on catch

We consider an effort-controlled fishery with a fish stock governed by the Logistic model. Stock size is measured in biomass, consequently growth encompasses both individual growth and recruitment. We assume that the population is homogenous and that introduction of the reserve will concentrate the effort in the remaining fishing grounds. The latter would be expected in a poorly managed fishery.

We assume that the population was at equilibrium prior to the introduction of a reserve and compare this with the post-reserve equilibrium. During the transient time between these two states spillover will be less. Since we are considering the equilibrium states we have $N_{t+1}=N_t$ which we simply denote by $N$, similarly for $M$, $R$ and $S$.

Firstly, consider a fishery with a level of effort corresponding to near extinction, $E=E_E$. Introduction of a reserve will increase surplus production unless the population is beyond recovery.

At the other extreme, consider a pre-reserve fishery that is producing maximum sustainable yield (MSY) from the total area: $E=E_{MSY}$. By definition at this point, surplus sustainable production cannot increase. Therefore introduction of a reserve must decrease overall catch.

At $E_{MSY}$ the spillover effect is less than the lost productivity and at $E_E$ it exceeds the lost productivity. At some level of effort in between, the reserve must switch from having a net negative effect on the fishery to a net positive effect due to spillover. The level of effort at which this occurs is dependent on the model and its parameters. We now establish the point at which this occurs for a logistic model (equation (3)).

If spillover equals lost productivity in the fishing area, the pre-reserve and post-reserve catches must equal $qEN=qEM/(1-\alpha)$; hence $M=N(1-\alpha)$. Simply put, the population density in the fishing grounds must remain unchanged. Substitution in equation (5) yields:

$$N(1-\alpha)=f(N)N(1-\alpha)-qEN+S \tag{11}$$

subtracting equation (1) (at equilibrium) and solving for $S$ gives:

$$S=\alpha qEN \tag{12}$$

Consequently, the spillover must equal the surplus production of the original fishing grounds that has now been encompassed in the reserve.

For a given level of effort, the pre-reserve fishery given by equation (4) will possess a solution, the nature of which depends on the population dynamics model. For example the non-zero solution for the logistic model is:

$$N=(r-qE)K/r \tag{13}$$

Using the full two area logistic model with effort controlled fishing (equations (5), (8) and (10)) and substituting equations (11) and (12) permits us to eliminate several of the unknowns. In this case we choose to eliminate $N$, $M$, $R$ and $S$ since conceptually we consider these to be determined by the remaining parameters. After algebraic manipulation (not shown here) we obtain the level of effort at which the introduction of the reserve does not change the overall catch:

$$qE=\left(r-2\mu+\sqrt{r^2+4\mu^2}\right)/2. \tag{14}$$

Note that $qE=0$ is also a solution (if no fishing is taking place, introduction of a reserve will not reduce the catch). A negative solution also exists but is of no further interest as the population would be extinct and negative densities are merely a mathematical curiosity. The same approach can be used for other population dynamics models, however for some models (e.g. the Ricker model) straight-forward analytic solutions do not exist. Qualitatively we would expect similar results for other population dynamics models and found this to be the case for numerical solutions to the Ricker model (results not shown here).

The optimal effort for this fishery without a reserve is $r=2q$. We divide equation (14) by this and subtract 1 to obtain the minimum excess effort (as a proportion) required for a reserve to be beneficial:

$$\hat{E}=\sqrt{1+(2\mu/r)^2}-2\mu/r. \tag{15}$$

This depends only on the ratio of the movement rate out of the reserve to the growth rate of the stock ($\mu/r$), and not on the proportion of the area dedicated to the reserve ($\alpha$). However, it should be noted that the movement rate out of the reserve, $\mu$, is likely to depend on the reserve size. This link has not been explicitly explored here, however, for a given choice of $\mu$, there is likely to be only a limited range of values of $\alpha$ that is possible.

Equations (14) and (15) are derived in more detail in Appendix S1.

## Results

Figure 1 shows an example where a 10% reserve is introduced with 5% movement out of the reserve ($\mu$) and a maximum growth rate ($r$) of 10%. This figure explores the effect of a reserve for different levels of initial effort applied to the fishery. The maximum sustainable yield (MSY) is obtained with an effort of 0.05 ($E_{MSY}$).

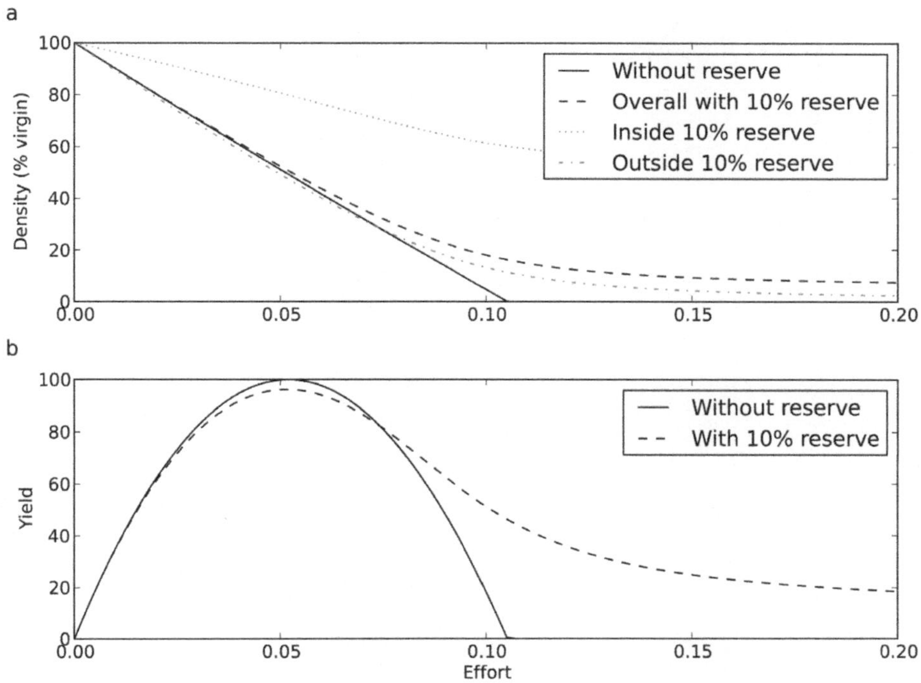


**Figure 1. Changes in population and fishery dynamics resulting from the introduction of a reserve.** (a) The equilibrium biomass density as a function of fishing effort. The density is shown for the whole stock without a reserve and with a 10% reserve. For the reserve scenario the density inside and outside of the reserve is also shown. (b) Yield as a function of fishing effort both with and without a reserve.

Introduction of the reserve decreases the yield at $E_{MSY}$ and by definition there is no alternative effort that produces the same maximal yet sustainable yield. The point of intersection in the bottom panel corresponds to a level of effort, $E_I$, where the yield is the same with or without a reserve. At levels of effort above $E_I$, the introduction of a reserve increases yield. In this scenario, $E_I$ is 150% of $E_{MSY}$, so a fishery would have to have 50% excess effort for the reserve to be beneficial in terms of the yield of the target species. At even higher levels of effort (>150% $E_{MSY}$) the MPA mitigates the impact of overfishing and permits sustainable (but substantially reduced) yield.

The level of excess effort at which a reserve has a neutral impact on fisheries yield depends only on the ratio of movement out of the reserve ($\mu$) to the maximum growth rate ($r$) (equation (15)). This relationship is shown in Figure 2a, when the movement rate is high relative to the growth rate, a reserve is beneficial at low levels of excess effort. The extreme situation where $\mu/r$ approaches infinity corresponds for example to a miniscule reserve, which clearly will have negligible impact on a fishery. At the other extreme, $\mu/r=0$, there is no movement out of the reserve, consequently it will always have a negative impact.

Alternatively we consider the excess effort required for a reserve to be beneficial as a function of the reserve density at equilibrium (Figure 2b). If the reserve is at 50% virgin biomass density it has neutral effect on the fishery. This is because 50% virgin biomass corresponds to MSY in this model and all surplus production is moved to the main population through spillover. At reserve densities above this, a fishery must have more excess effort to benefit from a reserve. In particular if reserves have a high percentage of virgin biomass (a common conservation goal for reserves) they will only benefit fisheries that have greater mismanagement. For example, at 80% virgin biomass a reserve will only benefit fisheries with more than 60% excess effort.

## Discussion

### Model outcomes

The model presented here examines the circumstances under which spillover from a reserve is sufficient to increase fishery yield (thus providing a net spillover benefit). As expected, density of exploited species was higher in the reserve than the fished area, which may be mistaken in itself as evidence that the reserve will create a net beneficial increase through larvae production [28]. However, it is important to consider the net effect, which in our model case was a decline in average density and a loss of yield except where effort exceeded $E_{MSY}$. While models are by necessity a simplification of ecological complexity, we show that the extent to which effort must exceed $E_{MSY}$ for any yield benefit to occur from the reserve depends on the ratio of the rate of movement out of the reserve and the growth rate of the species concerned. Highly mobile/slow growing species received relatively less benefit from reserves where effort was above management targets compared to species with low movement/fast growth.

Our model is a relatively simple one chosen to illustrate a fundamental principle that is applicable across a broad range of fisheries. Different formulations for the biological model, $f(N_t)$, can be specified and similar results were obtained for the Ricker model (not shown here). Three major assumptions were made to maintain model simplicity: spatial homogeneity, density dependence and steady state dynamics.

*Spatial homogeneity* is an inappropriate assumption for some species. For example, where there are clear source-sink relationships protecting the source in a reserve is likely to provide an overall benefit [29]. The location of source areas can be consistent across different species and trophic levels, and in rare cases where these locations are known, it becomes possible to locate reserves that provide benefit to numerous, and theoretically all, species [30].

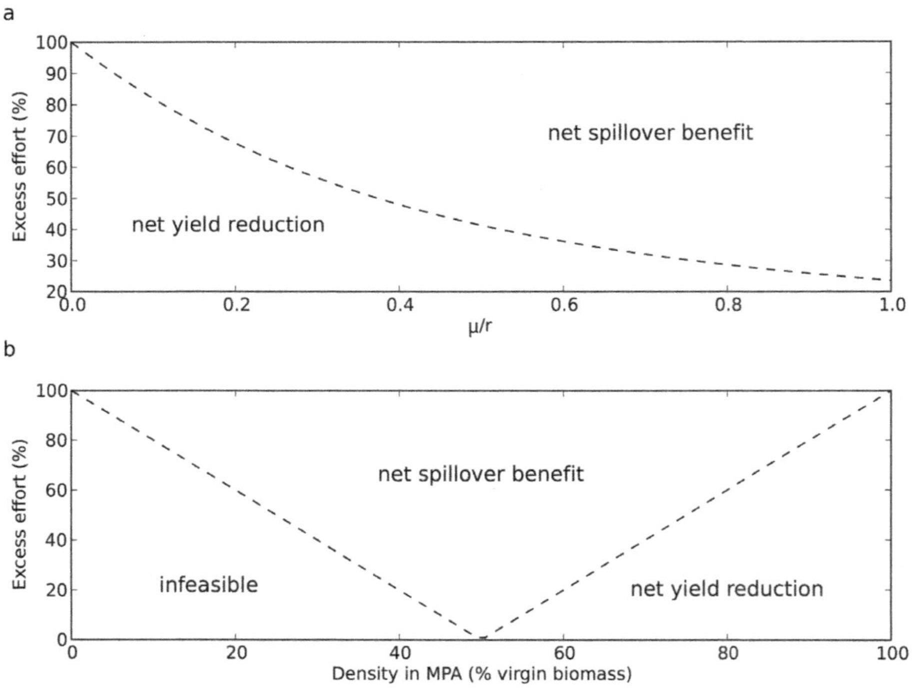


**Figure 2. Characterisation the management and biological circumstances in which a reserve is beneficial.** (a) The excess effort required for a reserve to improve fishery yield. For our simple model this was found to depend only on the ratio of the movement rate out of the reserve (and thus on reserve size) to the growth rate of the stock ($\mu/r$). (b) The excess effort required for optimality as a function of the reserve density (at equilibrium). For example a reserve with 80% virgin biomass at equilibrium will provide a net economic benefit for a fishery that has more than 60% excess effort relative to optimal management. Combinations of excess effort and reserve density that fall in the bottom left region are infeasible; in these situations a reserve would have to decrease in population density after being formed (not possible in our model). Inside the "V" the reserve provides a net increase in fishery yield. In the right region the reserve decreases yield.

*Density dependence* in our model is a function of the total biomass in the local area (i.e. the fished population or the reserve population). This does not adequately capture the dynamics of species where density dependence varies substantially with age (e.g. density dependence occurring primarily during larval stages) and where different age classes have different movement rates across the reserve boundary. In such situations it could be possible for the reserve to provide a greater benefit by providing a recruitment increase to the fished region.

*Steady state dynamics* are widely used to explore fundamental fisheries principles. In the context of reserves, some models have shown that biological stochasticity may lead to theoretical net spillover benefits in fisheries where the biomass can be determined accurately on an annual basis and corresponding perfect catch limits set each year [31], [32]. Given the unrealistic nature of this assumption for most management situations there would be some value in further research that explored reserve benefits in a stochastic setting with realistic management. After the introduction of a reserve, it will take some time for the reserve population to build to the final density. Consequently it is expected that the reduction in yield will initially be much greater than predicted by our steady state model. With the concentration of effort the fished population would initially decrease before increasing some time later due to spillover from the reserve.

Our model did not consider that the introduction of a reserve may result in an effort reduction due, for example, to decreased accessibility or increased fishing costs. This would be beneficial for stock status and overall production in over-exploited fisheries, however, it would result in a reduction of production in well-managed fisheries.

Under our model there were no combinations of growth rate or movement where a net spillover benefit from reserves could occur unless effort exceeded $E_{MSY}$. Where effort is less than $E_{MSY}$, a loss of yield always occurs when reserves are implemented. The level of excess effort beyond $E_{MSY}$ at which a reserve provides net spillover benefits was shown to depend only on the ratio of movement out of the reserve to the rate of growth of the population ($\mu/r$). We also showed that reserve configurations that achieve higher densities of stock are only beneficial for mismanaged fisheries (Figure 2b). For example, a reserve that ultimately increases biomass density to 75% of unfished levels would benefit a fishery if the initial effort exceeds $E_{MSY}$ by more than 50%. These results show that reserves will generally negatively impact yield for well managed fisheries. However reserves could minimize their impact on a well managed fishery by reducing the density increase of the fishery's target species in the reserve. For example, a reserve could be of a sufficient size to protect species with small home ranges whilst being small enough that individuals of the target species frequently move beyond reserve boundaries (a high movement rate, $\mu$). This could also be achieved by having high reserve boundary length to total area ratios. The feasibility of this outcome will depend on the movement characteristics of the species involved.

Our finding that reserves cannot improve the yield of a well-managed fishery is consistent with several other theoretical studies [18–20]. The work here extends these findings by exploring the extent to which a fishery must be mismanaged before introduction of a reserve provides a benefit to the fishery in terms of yield.

Many fisheries have management objectives that constrain catch below the target of MSY assumed here, for example to manage risk from stochastic processes such as recruitment, or where there is an objective to target a maximum economic yield (MEY) that is variant to MSY. In these fisheries, effort and catch are lower than would occur with the MSY target [33], which reduces the negative impact of reserves on total yield, but also shifts the fishery further away from the level of depletion required for a net spillover benefit to occur.

## Empirical context

The results from this study are consistent with other studies that have modeled the impact and/or benefits of reserves on fisheries in terms of improvements in yield. Following the publication of the early models on the potential net spillover benefits from reserves [16], [18], [34] there have been surprisingly few empirical studies that have attempted to demonstrate the effect. Most of the reserve literature has concentrated on the changes within reserves, showing an increase in size and abundance of resident fish and crustaceans, particularly of reef associated species (for a review see [35]). Despite the lack of empirical evidence the argument persists that reserves will confer a net spillover benefit to fisheries [5]. This view is actively promoted by government agencies [12–14]. However, the literature confirms that the evidence for such a benefit is far from conclusive. Several studies report a lack of evidence for spillover due to the low movement at the scale of the reserve [36–38], while others showed that spillover occurred but not that lost yield was compensated to produce a net benefit (e.g., [39–41]).

While density dependent export from reserves is considered to be a rational expectation [42], no studies have been able to conclusively demonstrate a net spillover benefit, and leakage from reserves is probably more related to random movement within species (e.g., [38], [41], [43], [44]). Several studies fail to provide conclusive evidence for net spillover benefits, yet argue that reserves are needed to provide fishery benefits (e.g., [39], [40]). Spillover has been inferred from observations of a density gradient between the reserve and adjacent fished area (e.g., [45], [46]) even though evidence was acknowledged to be equivocal (e.g., [42], [47], [48]), and where confounding factors such as a change in fishing practices (e.g., [49]) or changed fisheries management strategies over the study period were ignored (e.g., [4], [7]). Few of these studies consider whether the purported spillover to the fishery (as inferred from catch rates) has actually resulted in a net spillover benefit for the fishery. Even if CPUE goes up in a fished area it may be insufficient to result in a net production gain for the whole of the fishery.

Several studies have been able to demonstrate that spillover has contributed to an improvement in biomass and thus catch rate adjacent to the reserve [50–53]. These examples, all in the Mediterranean, were conducted in areas where the total fishery had been severely depleted. In this respect they are similar to several studies in other areas that, on multiple lines of evidence, infer a net spillover benefit to fisheries. Examples come from Africa [48] and Asia [7], [42], [47] where the fisheries in question were over-exploited and where there was limited application and/or enforcement of standard fisheries management controls. The result was that the proclamation of a reserve resulted in a recovery of the population in the reserve and a subsequent improvement in catches close to the reserve boundary. This is consistent with our conclusion that reserves can provide a net spillover benefit for severely depleted stocks. It does not, however, provide evidence that the declaration of the reserve was the most efficient means of achieving that benefit.

There are many possible variations on the biological assumptions made in our model. Aspects such as stock heterogeneity and variant density dependence assumptions will influence the impacts of a reserve as well as the level of mismanagement, where a reserve switches from being beneficial to being detrimental for a fishery.

The model results presented here are for a general case, which is appropriate for consideration of reserves where a large number of species with variable life histories and spatial distributions are affected by change in management. Closed areas for traditional fishery management purposes are applied on a species by species basis and may have very different management outcomes to reserves because they can be designed and located to affect an individual stock. There are numerous cases where species with spatial heterogeneity, such as spawning aggregations or larval source-sink dynamics, benefit from fishery closures that target important source areas [54]. A total fishing closure would achieve the same result for those species, but can be expected to have less beneficial results for other exploited species.

## Conclusions

We conclude that in fisheries where there is effective management, marine reserves are unlikely to produce a net spillover benefit for the total fishery, whereas they may be beneficial where the fishery has been mismanaged and stocks severely depleted. These results expand the implications of previous work by providing estimation and evaluation of the degree of mismanagement of fisheries that is necessary for non-specific closures to provide net benefits to fisheries.

The conclusions from the modeling presented here are supported by review of empirical studies, where spillover benefits have only been conclusively demonstrated in highly depleted areas. Together with the combined weight of earlier modeling work, they suggest that a net benefit from spillover should not be expected in areas already benefiting from quality traditional fisheries management.

These generalised findings in relation to reserves should not be confused with the use of targeted spatial closures for single fisheries, where it is possible to increase yield through closures by taking account of the spatial heterogeneity of life history traits.

While reserves may be proclaimed for a range of conservation objectives (including addressing impacts such as the effect of fishing on benthic environments, interactions with threatened species and catch of non-target species), we contend that it is misleading for governments to promote reserves on the basis of net spillover benefit in the context of well-managed fisheries. Reserves are only likely to be an effective strategy for fisheries management where effort is not or cannot be effectively controlled across the wider stock.

## Acknowledgments

This work was supported by a grant from the Fisheries Research and Development Corporation (FRDC 2010/226). The authors would also like to thank Dr Zoe Doubleday for her assistance with the project and Prof Reg Watson for his valuable comments on an earlier draft of the manuscript, and two anonymous referees whose comments substantially improved the paper and made it accessible to a broader audience.

## Author Contributions

Conceived and designed the experiments: CB KH RK CG. Analyzed the data: CB KH. Contributed reagents/materials/analysis tools: KH. Wrote the paper: CB KH RK CG.

## References


1. Edgar GJ, Barrett NS, Stuart-Smith RD (2009) Exploited reefs protected from fishing transform over decades into conservation features otherwise absent from seascapes. Ecological Applications 19: 1967–1974.
2. Lester SE, Halpern BS, Grorud-Colvert K, Lubchenco j, Ruttenberg BI, et al. (2009) Biological effects within no-take marine reserves: a global synthesis. Marine Ecology Progress Series 384: 33–46.
3. Roberts CM (1997) Ecological advice for the global fisheries crisis. Trends in Ecology and Evolution 2(1): 35–38.
4. Roberts CM, Bohnsack JA, Gell FR, Hawkins JP, Goodridge R (2001) Effects of marine reserves on adjacent fisheries. Science 294: 1920–1923.
5. Gell FR, Roberts CM (2003) Benefits beyond boundaries: the fishery effects of marine reserves. Trends in Ecology & Evolution 18(9): 448–455.
6. Halpern BS, Lester SE, Kellner JB (2009) Spillover from marine reserves and the replenishment of fished stocks. Environmental Conservation 36(4): 268–276.
7. Russ GR, Alcala AC (2011) Enhanced biodiversity beyond marine reserve boundaries: The cup spillith over. Ecological Applications 21: 241–250. [doi:10.1890/09-1197.1].
8. Kerwath SE, Winker H, Gotz A, Attwood CG (2013) Marine protected area improves yield without disadvantaging fishers. Nature Communications 4: 2347 DOI:10.1038/ncomms3347.
9. Russ GR (2002) Yet another review of marine reserves as reef fishery management tools. In: Coral Reef Fishes (ed. Sale P) Academic Press, San Diego, CA. p 421–443.
10. Goni R, Hilborn R, Diaz D (2010) Net contribution of spillover from a marine reserve to fishery catches. Marine Ecology Progress Series. 400: 233–243. [doi:10.3354/meps08419].
11. Harrison HB, Wiliamson DH, Evans RD, Almany GR, Thorrold SR, et al. (2012) Larval export from marine reserves and the recruitment benefit for fish and fisheries. Current Biology 22: 1023–1028. [doi:10.1016/j.cub.2012.04.008].
12. DEH (2003) The benefits of marine protected areas. Information Paper - The Commonwealth Department of Environment and Heritage, Australian Government. Available: http://www.environment.gov.au/coasts/mpa/publications/wpc-benefits.html.
13. Revenga S, Badalamenti F (2008) Management of marine protected areas for fisheries in the Mediterranean. Options Mediterraneennes Series B, 62: 107–111.
14. NOAA (2011) Benefits of a National System of Marine Protected Areas. Available: http://www.mpa.gov/resources/publications/factsheets/Accessed 2012 May 31.
15. Sladek Nowlis J, Roberts CM (1999) Fishery benefits and optimal design of marine reserves. Fishery Bulletin 97: 604–616.
16. Sladek Nowlis J (2000) Short- and long-term effects of three fishery management tools on depleted fisheries. Bulletin of Marine Science 66(3): 651–662.
17. Steele JH, Beet AR (2003) Marine protected areas in 'nonlinear' ecosystems. Proceedings of the Royal Society London B (Supp) 270: S230–S233. [doi:1098/rsbl.2003.0074].
18. Polacheck T (1990) Year around closed areas as a management tool. Natural Resource Modeling 4: 327–354.
19. Hilborn R, Micheli F, De Leo GA (2006) Integrating marine protected areas with catch regulation. Canadian Journal of Fisheries and Aquatic Sciences 63: 642–649.
20. Le Quesne WJF, Codling EA (2009) Managing mobile species with MPAs: the effects of mobility, larval dispersal, and fishing mortality on closure size. ICES Journal of Marine Science 66: 122–131.
21. Barnes B, Sidhu H (2013) The impact of marine closed areas on fishing yield under a variety of management strategies and stock depletion levels. Ecological Modelling 269: 113–125.
22. Hart DR (2006) When do marine reserves increase fisheries yield? Canadian Journal of Fisheries and Aquatic Sciences 63: 1445–1449. [doi:10.1139/F06-071].
23. Apostolaki P, Milner-Gulland EJ, McAllister MK, Kirkwood GP (2002) Modelling the effects of establishing a marine reserve for mobile fish species. Canadian Journal of Fisheries and Aquatic Sciences 59: 405–415.
24. Ralston S, O'Farrell MR (2008) Spatial variation in fishing intensity and its effect on yield. Canadian Journal of Fisheries and Aquatic Sciences 65: 588–599.
25. Zeller D, Stoute SL, Russ GR (2003) Movements of reef fishes across marine reserve boundaries: effects of manipulating a density gradient. Marine Ecology-Progress Series 254: 269–280.
26. Abesamis RA, Russ GR, Alcala AC (2006) Gradients of abundance of fish across no-take marine reserve boundaries: evidence from Philippine coral reefs. Aquatic Conservation-Marine and Freshwater Ecosystems 16(4): 349–371.
27. Costello C, Gaines SD, Lynham J (2008) Can catch shares prevent fisheries collapse? Science (Washington), 321(5896): 1678–1681. doi:http://dx.doi.org/10.1126/science.1159478.
28. Bohnsack JA (1998) Application of marine reserves to reef fisheries management. Australian Journal of Ecology 23: 298–304.
29. Stockhausen WT, Lipcius RN, Hickey BM (2000) Joint effects of larval dispersal, population regulation, marine reserve design, and exploitation on production and recruitment in the caribbean spiny lobster. Bulletin of Marine Science, 66(3): 957–990.
30. White JW, Samhouri JF (2011) Oceanographic coupling across three trophic levels shapes source-sink dynamics in marine metacommunities. Oikos 120(8): 1151–1164.
31. Yamazaki S, Grafton QR, Kompas T, Jennings S (2012) Biomass management targets and the conservation and economic benefits of marine reserves. Fish and Fisheries. [doi:10.1111/faf.12008].
32. Grafton RQ, Kompas T, Van Ha P (2006) The economic payoffs from marine reserves: resource rents in a stochastic environment. The Economic Record 82(259): 469–480.
33. Grafton RQ, Kompas T, Hilborn RW (2007) Economics of overexploitation revisited. Science 318(5856): 1601.
34. DeMartini EE (1993) Modelling the potential of fishery reserves for managing Pacific coral reef fishes. Fishery Bulletin 91: 414−427.
35. Halpern BS, Warner RR (2002) Marine reserves have rapid and lasting effects. Ecology Letters 5: 361–366.
36. Davidson RJ, Villouta E, Cole RG, Barrier RGF (2002) Effects of marine reserve protection on spiny lobster (*Jasus edwardsii*) abundance and size at Tonga Island Marine Reserve, New Zealand. Aquatic Conservation - Marine and Freshwater Ecosystems 12: 213–227. [DOI:10.1002/aqc.505].
37. Tewfik A, Bene C (2003) Effects of natural barriers on the spillover of a marine mollusc: implications for fisheries reserves. Aquatic Conservation Marine and Freshwater Ecosystems 13(6): 473–488.
38. Tupper MH (2007) Spillover of commercially valuable reef fishes from marine protected areas in Guam, Micronesia. Fishery Bulletin 105(4): 527–537.
39. Rowe S (2001) Movement and harvesting mortality of American lobsters (*Homarus americanus*) tagged inside and outside no-take reserves in Bonavista Bay, Newfoundland. Canadian Journal of Fisheries and Aquatic Sciences 58(7): 1336–1346.
40. Pillans S, Pillans RD, Johnstone RW, Fraft PG Haywood MDE, et al. (2005) Effects of marine reserve protection on the mud crab *Scylla serrata* in a sex-biased fishery in subtropical Australia. Marine Ecology Progress Series 295: 201–213.
41. Follesa MC, Cuccu D, Cannas R, Sabatini A, Deiana AM, et al. (2009) Movement patterns of the spiny lobster *Palinurus elephas* (Fabricius, 1787) from a central western Mediterranean protected area. Scientia Marina 73(3): 499–506.
42. Abesamis RA, Russ GR (2005) Density-dependent spillover from a marine reserve: Long-term evidence. Ecological Applications 15(5): 1798–1812.
43. Cole RG, Villouta E, Davidson RJ (2000) Direct evidence of limited dispersal of the reef fish *Parapercis colias* (Pinguipedidae) within a marine reserve and adjacent fished areas. Aquatic Conservation: Marine and Freshwater Ecosystems 10(6): 421–436.
44. Kelly S, Scott D, MacDiarmid AB (2002) The value of a spillover fishery for spiny lobsters around a marine reserve in Northern New Zealand. Coastal Management 30(2): 153–166.
45. Ashworth JS, Ormond RFG (2005) Effects of fishing pressure and trophic group on abundance and spillover across boundaries of a no-take zone. Biological Conservation 121(3): 333–344.
46. Abesamis RA, Alcala AC, Russ GR (2006) How much does the fishery at Apo Island benefit from spillover of adult fish from the adjacent marine reserve? Fishery Bulletin 104(3): 360–375.
47. Russ GR, Alcala AC (1996) Do marine reserves export adult fish biomass? Evidence from Apo Island, central Philippines. Marine Ecology Progress Series 132: 1–9.
48. McClanahan TR, Mangi S (2000) Spillover of exploitable fishes from a marine park and its effect on the adjacent fishery. Ecological Applications 10(6): 1792–1805.
49. McClanahan TR, Kaunda-Arara B (1996) Fishery recovery in a coral-reef marine park and its effect on the adjacent fisher. Conservation Biology 10(4): 1187–1199.
50. Stobart B, Warwick R, Gonzalez C, Mallol S, Diaz D, et al. (2009) Long-term and spillover effects of a marine protected area on an exploited fish community. Marine Ecology Progress Series 384: 47–60.
51. Forcada A, Valle C, Bonhomme P, Criquet G, Cadiou G, et al. (2009) Effects of habitat on spillover from marine protected areas to artisanal fisheries. Marine Ecology Progress Series 379: 197–211.
52. Goni R, Quetglas A, Renones O (2006) Spillover of spiny lobsters *Palinurus elephas* from a marine reserve to an adjoining fishery. Marine Ecology Progress Series 308: 207–219.
53. Vandeperre F, Higgins RM, Sanchez-Meca J, Maynou F, Goni R, et al. (2011) Effects of no-take area size and age of marine protected areas on fisheries yields: a meta-analytical approach Fish and Fisheries 12(4): 412–426.
54. Wakefield CB (2010) Annual, lunar and diel reproductive periodicity of a spawning aggregation of snapper *Pagrus auratus* (Sparidae) in a marine embayment on the lower west coast of Australian Journal of Fish Biology 77(6): 1359–1378.

# 16

# Conserving Biodiversity in a Human-Dominated World: Degradation of Marine Sessile Communities within a Protected Area with Conflicting Human Uses

**Valeriano Parravicini[1,2]*, Fiorenza Micheli[3], Monica Montefalcone[4], Carla Morri[4], Elisa Villa[4], Michela Castellano[4], Paolo Povero[4], Carlo Nike Bianchi[4]**

**1** CESAB-FRB, Immeuble Henri Poincaré - Domaine du Petit Arbois, Aix-en-Provence, France, **2** IRD – UR 227 Coreus – Laboratoire Arago, Banyuls/mer, France, **3** Hopkins Marine Station, Stanford University, Pacific Grove, California, United States of America, **4** DiSTAV, Dipartimento di Scienze della Terra, dell'Ambiente e della Vita, Genoa University, Genoa, Italy

## Abstract

Conservation research aims at understanding whether present protection schemes are adequate for the maintenance of ecosystems structure and function across time. We evaluated long-term variation in rocky reef communities by comparing sites surveyed in 1993 and again in 2008. This research took place in Tigullio Gulf, an emblematic case study where various conservation measures, including a marine protected area, have been implemented to manage multiple human uses. Contrary to our prediction that protection should have favored ecosystem stability, we found that communities subjected to conservation measures (especially within the marine protected area) exhibited the greatest variation toward architectural complexity loss. Between 1993 and 2008, chronic anthropogenic pressures (especially organic load) that had already altered unprotected sites in 1993 expanded their influence into protected areas. This expansion of human pressure likely explains our observed changes in the benthic communities. Our results suggest that adaptive ecosystem-based management (EBM), that is management taking into account human interactions, informed by continuous monitoring, is needed in order to attempt reversing the current trend towards less architecturally complex communities. Protected areas are not sufficient to stop ecosystem alteration by pressures coming from outside. Monitoring, and consequent management actions, should therefore extend to cover the relevant scales of those pressures.

**Editor:** Arga Chandrashekar Anil, CSIR- National institute of oceanography, India

**Funding:** The present work is done under the aegis of the project 'The impacts of biological invasions and climate change on the biodiversity of the Mediterranean Sea' (Italy–Israel co-operation) funded by the Italian Ministry of the Environment. FM was partly supported by a Pew Fellowship in Marine Conservation. The funders had no role in study design, data collection and analysis, decision to publish, or preparation of the manuscript.

**Competing Interests:** The authors have declared that no competing interests exist.

* E-mail: valeriano.parravicini@ird.fr

## Introduction

Most marine ecosystems are challenged worldwide by a vast set of human pressures ranging from pollution to habitat modification, overfishing and climate change [1]. The recognition of the profound influence humans exert on marine ecosystems has motivated worldwide conservation efforts mainly aimed at the establishment of Marine Protected Areas (MPAs). Yet, MPAs alone are not sufficient [2,3]. They rarely cover an adequate extent of ecosystem types being typically quite small (ranging approx. $10^{-3}$ to $10^{3}$ $km^{2}$) and are thereby ineffective in halting the effects of pressures acting at scales larger than those encompassed by the protection schemes.

Anthropogenic pressures acting over large scales are common [1], especially along intensely populated coastal areas, where marine ecosystems are challenged by the cumulative effect of multiple pressures and conflicting human uses. The need for complementary measures to the protection by MPAs has called for the development of ecosystem-based management (EBM) to be implemented at regional scales. EBM is an integrated approach that considers the entire ecosystem, including humans [4]. EBM differs from other approaches that focus on a single species or sector because it includes consideration of the interactions among ecosystem components and the cumulative impacts of multiple activities [5]. Some examples of EBM implementation are the zoning of the Great Barrier Reef Marine Park [6] and the establishment of marine sanctuaries in the USA [7]. Similar instruments have been developed in the European Union, where legislation mandates the establishment of a network of *Sites of Community Importance* (SCIs) to preserve habitats and species [8] and *Integrated Coastal Zone Management* (ICZM), whose goal is the sustainable development of coastal areas [9]. Despite the application of EBM approaches to complement MPAs, few studies have assessed whether such plans effectively sustain the conservation of marine coastal ecosystems. Ecosystem-based approaches have recently been advocated also to understand ecosystem shifts due to climate change [10].

Thanks to historical data from 1993 onwards ([11] and unpublished information), we were able to evaluate the results of the management plan applied in an emblematic study case (the Tigullio Gulf) by examining the decadal-scale change of rocky reef communities. The Tigullio Gulf is a densely populated area where conservation measures included the establishment of one MPA in

1998 (designed mainly to regulate fishing) and of six SCIs in 2000 (designed to regulate multiple human pressures on protected habitat and species). The area has been, and still is, subjected to various human uses and associated possible impacts: organic load from sewage outfalls, input of sediments and nutrients from a river, aquaculture, periodic beach replenishments, and small marinas.

Coastal development and related pressures have been shown to affect Mediterranean rocky reef communities by causing the replacement of canopy-forming algae with algal turfs, with consequent habitat homogenization and loss of architectural complexity [12]. In addition to removing or preventing these disturbances directly, through a regulation of development within SCIs, MPAs are expected to maintain or recover algal canopies through top-down control as protection from fishing allows predatory fishes to exert control upon sea urchins, which otherwise may proliferate and overgraze benthic algae [13,14]. Yet, at the same time, protection in MPAs may increase the abundance of herbivorous fish, which may also imply increased grazing pressure on canopy-forming algae [15].

To date, few studies have addressed how concomitant human pressures [16] impact reef communities within the Mediterranean Sea and whether EMB effectively manages non-fishing human uses that are not reduced by MPAs.

Here, we compared surveys of reef communities conducted prior to protection in 1993 and in 2008, 10 years after MPA and 8 years after SCIs establishment. We assessed relevant pressures operating in the study area and we related their intensities and trends to the observed change in reef communities in order to clarify whether the existing protection effort was effective. Results represent the first assessment of the efficacy of current management plans, and provide guidelines for EBM, with particular attention to the design of monitoring plans.

## Methods

### Study area and pressure regime

The Tigullio Gulf is located in the Ligurian Sea, NW Mediterranean Sea (Figure 1). The coastal zone is partially included within the Portofino MPA and six SCIs [3]. According to the standards of the European Water Framework Directive, water quality in the study area is classified as 'good' [17]. However, the area is densely populated by residents and tourists, resulting in comparatively high nutrient inputs into coastal waters.

Since the late 1950s, many studies have been carried out in this area [18]. Historical information on planktonic production [19,20] evidenced that chlorophyll-*a* concentration was significantly lower in the period 2000–2008 than in 1985–1994 (Figure S1a; t-test, df = 8, $P<0.017$). On the contrary, water transparency did not show any significant variation of the annual mean for the same periods (Figure S1b).

Potential anthropogenic pressures on coastal marine ecosystems in the last two decades include the presence of four outfalls discharging above the summer thermocline (i.e. about 25 m depth), the Entella River mouth (with a number of waste water pipes along its way), a fish farm (established in 2000), marinas and tourist harbors (built during the 1970s), and beaches frequently replenished (Table S1). Human population remained virtually stable: residents passed from 52 inhabitants km$^{-2}$ in 1993 to 50 inhabitants km$^{-2}$ in 2008, tourists from 1.34 million in 1998 to 1.36 million in 2007 (ISTAT, the Italian Statistical Institute). Due to change in local industrial activity, the potential organic load computed in term of equivalent individuals [21] showed a slight increase from 1991 to 2001 (Figure S1c). In the same decades, global warming has been showing a large influence on the hydrology and ecology of the Mediterranean basin [22]; the Ligurian Sea, in particular, underwent a regime shift in the late 1990s, two major heat waves in 1999 and 2003 having caused mass mortalities of gorgonians, sponges and other sessile organisms [23–25].

### Sampling

Reef communities were surveyed in summer 1993 [11], before the establishment of the MPA and SCIs, in 10 sites showing similar environmental characteristics (Figure 1). Four sites (G, H, I and J) have subsequently been included within the Portofino MPA in 1998, the remaining six sites within SCIs since 2000. When necessary, all permits needed for performing fieldwork were obtained by the Genoa University from the Portofino MPA Authority. At each site, the percent cover of individual epibenthic taxa was visually estimated within four 1 m$^2$ quadrats randomly placed on vertical rocky walls at 5 m depth. Although the sampling effort employed may be considered rather small, Mediterranean rocky reef communities are 'miniaturized' compared to those of other temperate regions, and the surface of epibenthic communities surveyed at each site is one order of magnitude larger than the recommended minimum area of 2,170 cm$^2$ to be sampled in the Mediterranean [26]. Thus, our sampling units encompassed small-scale heterogeneity, while the four replicates were placed within a habitat (shallow vertical and sub-vertical walls) that exhibits fairly homogeneous assemblages over scales of 10 s m.

Percent cover was quantified by dividing the sampling quadrat into 25 sub-squares through a nylon line and giving each taxon a score from 0 to 4 within each square and then adding up scores for all squares within the quadrat [27]. Species identification was conducted visually at the lower possible taxonomic level; voucher specimens were collected only when in doubt and never involved protected or endangered species. In the same quadrats used for the assessment of epibenthic communities, the total density of sea-urchins was estimated. In 2008, sites were precisely relocated by two of the authors (CNB and CM) that sampled communities in 1993 [11]. The sites were then surveyed by relocating 4 sampling quadrats at 5 m depth and following the same procedure. Vertical rocky walls were about 10 m height and the tidal height in the area is about 30 cm.

### Community descriptors

Architectural aspects were considered in addition to taxonomic composition. For architectural complexity [28], the three-dimensionality of the assemblages was assessed. Based on the ratio between height (h) and radius (r), species were grouped into five categories: 'high' ($h \gg r$, e.g. *Cystoseira zosteroides*), 'medium-high' ($h>r$, e.g. *Dictyopteris polypodioides*), 'medium' ($h \sim r$, e.g. *Reteporella grimaldii*), 'medium-low' ($h<r$, e.g. *Petrosia ficiformis*), and 'low' ($h \ll r$, e.g. *Lithophyllum incrustans*).

### Data analysis

In order to characterize the degree of human pressure upon individual sites, the distances from the MPA (protection) and the main potential sources of impact (i.e. river mouth, outfalls, fish farm, marinas, and replenished beaches) were calculated for each site. As raw explanatory variables were collinear, the matrix of distances from potential impact sources was treated with principal component analysis (PCA) [29]: the 1$^{st}$ and the 2$^{nd}$ axes of the PCA (PCA1 and PCA2) were found to be a good synthesis of the pressure gradient from protection by the MPA and closeness to potential sources of organic load (i.e. outfalls, river mouth, and fish farm), whereas the 3$^{rd}$ axis was correlated with the distance from the marinas and the replenished beaches (Table S2).

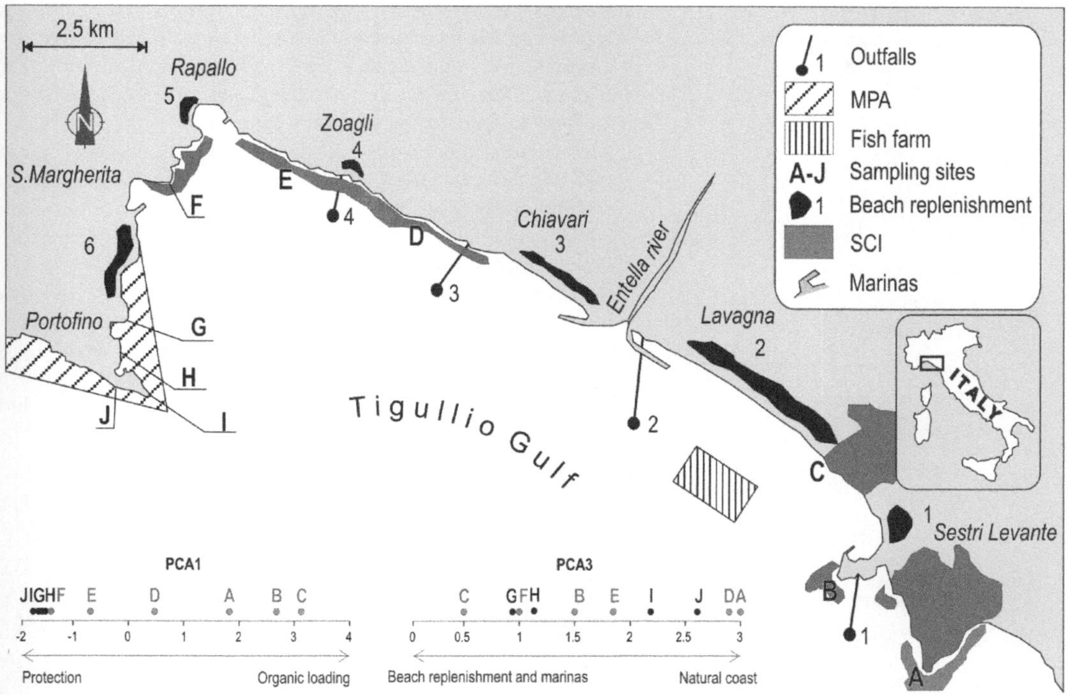


**Figure 1. Position of sampling sites and of the main human-induced pressures in the study area.** On the bottom, the position of individual sites on PCA axes is shown. The first axis represents closeness to sources of organic loading, the third closeness to replenished beaches and marinas.

Two distinct datasets were obtained from communities surveys, the first using species as variables, the second using categories of three-dimensional structure. In order to test whether a significant change occurred in the time considered, permutational multivariate analysis of variance (PERMANOVA) [30] was used on each dataset considering sampling year and sampling site as random, crossed factors. PCA1 and PCA2 were included in the analysis as covariates and their interaction with sampling year were considered in order to test whether different pressures lead to different variation through time. The two sampling years (i.e. 1993 vs. 2008) were considered as random. Protection in the MPA was not included as a categorical factor to avoid redundancy since its potential effect was already considered in the PCA1 as distance from the MPA.

Permutational analysis of multivariate dispersion (PERMDISP) was used to test potential variation in β diversity [31]. Multivariate patterns were visualized through non-metric multidimensional scaling ordination (nMDS). Similarity percentage analysis (SIMPER) [32] was used to identify the taxa and community descriptors mainly responsible for significant differences.

Sea-urchins had low abundance in both sampling occasions, and potential differences through time were assessed using descriptive statistics.

## Results

A total of 88 conspicuous sessile species was found (Table 1). The most speciose groups were algae (35 species, of which 17 rhodophytes, 7 ochrophytes, and 11 chlorophytes) and sponges (22 species); other invertebrate groups were less represented (9 cnidarians, 4 mollusks, 5 serpuloideans, 2 barnacles, 6 bryozoans, and 5 ascidians). Sessile communities had changed significantly between 1993 and 2008 (Table 2). Total species richness passed from 68 species (17±0.78 SE per quadrat) in 1993, to 81 (21±0.63 SE per quadrat) in 2008. Dominant species in 1993 included tall algae such as *Dictyopteris polypodioides* and *Sphaerococcus coronopifolius*, which had nearly disappeared in 2008; amongst the invertebrates, sponges decreased, while hydroids increased.

However, differences between the two sampling years were not consistent among sampling sites (significant Year×Sites) and were affected by the spatial distribution of human pressures, both for taxonomic composition and three-dimensional structure. In particular, the pressures considered by the PCA1 (i.e. protection by the MPA and distance from potential source of organic loading) had a significant role in driving the community change through time (significant Year×PCA1) whereas beach replenishments and tourist harbors played a negligible role (Year×PCA3 not significant). Correspondingly, the MDS plots highlighted marked differences in the direction of change between years (i.e. 1993 vs 2008) among individual sites using both species (Figure 2a) and categories of species three-dimensionality (Figure 2b) as community descriptors. MDS further suggested that sites far from potential sources of organic load or included within the MPA showed a higher degree of change between the two years than the remaining sites.

Although varying significantly among sites, the overall direction of change implied a decreased multivariate dispersion in the pattern of species occurrence and abundance among sites and hence a reduction in β diversity (PERMDISP: $P = 0.04$), indicative of homogenization of the communities through time.

Even within the MPA, the between-year change greatly varied among individual sites (SIMPER results, Table S3). *Dictyopteris polypodioides*, a potential canopy forming algal species, disappeared from site J (the furthest from potential impact sources) while *Dictyota dichotoma*, an erect generalist algal species, increased. The perennial algae *Halimeda tuna*, *Peyssonnelia squamaria* and *Flabellia petiolata* decreased at sites I, H and G, respectively, while the

**Table 1.** Total list of the sessile taxa found, in systematic order by major taxa.

| | |
|---|---|
| **Ochrophyta** | *Crambe crambe* |
| *Cystoseira zosteroides* | *Dysidea avara* |
| *Dictyopteris polypodioides* | *Haliclona cratera* |
| *Dictyota dichotoma* | *Haliclona fulva* |
| *Dictyota fasciola* | *Hemimycale columella* |
| *Dictyota implexa* | *Hymeniacidon perlevis* |
| *Padina pavonica* | *Ircinia oros* |
| *Stypocaulon scoparium* | *Ircinia variabilis* |
| **Rhodophyta** | *Leucosolenia variabilis* |
| *Acrothamnion preissii* | *Petrosia ficiformis* |
| algal turf | *Phorbas tenacior* |
| *Amphiroa rigida* | *Spongia lamella* |
| *Amphiroa cryptarthrodia* | **Cnidaria** |
| *Asparagopsis armata* | *Aiptasia mutabilis* |
| *Bonnemaisonia asparagoides* | *Balanophyllia europaea* |
| *Ellisolandia elongata* | *Caryophyllia inornata* |
| *Jania rubens* | *Cladocora caespitosa* |
| *Laurencia obtusa* | *Clavularia crassa* |
| *Lithophyllum incrustans* | *Eudendrium racemosum* |
| *Lithophyllum stictaeforme* | Hydrozoa indet. |
| *Mesophyllum lichenoides* | *Parazoanthus axinellae* |
| *Peyssonnelia rubra* | *Pennaria disticha* |
| *Peyssonnelia squamaria* | **Mollusca** |
| *Sphaerococcus coronopifolius* | *Arca noae* |
| *Tricleocarpa fragilis* | *Chama gryphoides* |
| *Wrangelia penicillata* | *Thylacodes arenarius* |
| **Chlorophyta** | *Vermetus triquetrus* |
| *Acetabularia acetabulum* | **Serpuloidea** |
| *Caulerpa racemosa* | *Protula tubularia* |
| *Cladophora prolifera* | *Salmacina dysteri* |
| *Cladophora* sp. | *Serpula vermicularis* |
| *Codium bursa* | Serpulidae indet. |
| *Codium coralloides* | Spirorbidae indet. |
| *Codium fragile* | **Cirripedia** |
| *Flabellia petiolata* | *Balanus trigonus* |
| *Halimeda tuna* | *Perforatus perforatus* |
| *Pseudochlorodesmis furcellata* | **Bryozoa** |
| *Valonia utricularis* | *Margaretta cereoides* |
| **Porifera** | *Reptadeonella violacea* |
| *Acanthella acuta* | *Reteporella grimaldii* |
| *Agelas oroides* | *Schizoporella dunkeri* |
| *Axinella damicornis* | *Schizoporella errata* |
| *Axinella polypoides* | *Scrupocellaria reptans* |
| *Axinella verrucosa* | **Ascidiacea** |
| *Chondrilla nucula* | *Clavelina lepadiformis* |
| *Chondrosia reniformis* | *Didemnum candidum* |
| *Clathrina clathrus* | *Diplosoma listerianum* |
| *Clathrina contorta* | *Halocynthia papillosa* |
| *Cliona viridis* | *Microcosmus polymorphus* |

Within major taxa, species or species groups are ordered alphabetically. Species are named according to the World Register of Marine Species (www.marinespecies.org/).

seasonal algae *Stypocaulon scoparium* and *Dictyota dichotoma* increased. Algal turf greatly increased at all sampling sites. The invasive algae *Caulerpa racemosa* and *Acrothamnion preissii* were recorded in 2008, but not in 1993. In particular, *Caulerpa racemosa* appeared only at sites close to potential sources of organic load.

Considering the change in the percentage cover of three-dimensionality categories, all sites within the MPA experienced a reduction of the cover by the categories of 'high' or 'medium-high' three-dimensionality and a parallel increase of the cover of the category 'medium' three-dimensionality (see Figure 3).

Sea urchins were represented in both years by *Paracentrotus lividus* and, especially, *Arbacia lixula*. There were no obvious changes in their densities between 1993 and 2008. Sea-urchins were absent from most quadrats in both years, with few exceptions. The highest sea urchin density in 1993 was recorded at site J (4 ind. $m^{-2}\pm2.5$ SE) and in 2008 at site E (3 ind. $m^{-2}\pm1.1$ SE).

## Discussion

In this study we found that an MPA and a network of six SCIs, two major tools used in the Mediterranean Sea for EBM implementation, were ineffective in halting degradation on rocky reef sessile communities. Species composition and architectural complexity changed significantly between 1993 and 2008, before and after the implementation of these conservation measures. Although we compared only two points in time, too few to assess long-term change and to elucidate cause-effect relationships, the strong association of the observed changes in the communities with the documented pressure regime allows generating hypotheses about possible drivers of change. In particular, our analyses indicate that organic loading from sewage outfalls may be a key driver of the observed change in these communities.

The taxonomic composition of rocky reef communities within the MPA became more similar, in 2008, to that of the communities close to sewage outfalls than in the past, thus implying a reduction of β diversity over the whole area. Three-dimensional structure decreased owing to the reduction of erect forms (e.g. tall macroalgae) and a parallel increase of flatter forms (e.g. turf-forming or encrusting organisms). In 1993, assemblages within the MPA and far from sources of impact were dominated by erect algae, most of which were canopy-formers. Earlier descriptive information from the 1950s through the 1980s [18,33–38] reported the diffuse presence of dense canopies of the brown algae *Dictyopteris polypodioides* and *Sargassum vulgare* (Figure 4), already rare in 1993 and virtually absent in 2008. The loss of habitat complexity is a sign of degradation that has been highlighted as one of the major threats to marine biodiversity worldwide [8,12]. Notwithstanding sessile species diversity showed slightly higher in 2008 than in 1993, reduced habitat complexity is likely to have negatively affected numerous canopy-associated epiphytic and motile organisms, whose diversity and abundance is known to be facilitated by habitat cascades via habitat provision and ecosystem engineering [39,40]. Moreover, we cannot discard the possibility that the occurrence of algal canopies in 1993 hid species that showed obvious only to the visual census of 2008, when canopies were reduced.

In temperate reefs, sea water warming and associated extreme climate events are known to be able to cause extirpation of canopy-forming macroalgae [41–43]. However, lack of erect macroalgae from unprotected sites was already noticed in 1993, i.e. before the heat waves of 1999 and 2003. Similarly, the arrival in the Ligurian Sea of the invasive species *Acrothamnion preissii* and *Caulerpa racemosa* predates the heat waves [44,45]; establishment of invasive species is said to be favored by already stressed ecosystems

**Table 2.** Results of PERMANOVA analyses on individual taxa and three-dimensional categories.

| **Taxonomic composition** | | | | | |
|---|---|---|---|---|---|
| **Source of variation** | **df** | **SS** | **MS** | **pseudo-F** | ***P* (perm)** |
| Year | 1 | 20266 | 20266 | 16.59 | 0.001 |
| Site | 9 | 53914 | 6739 | 5.51 | 0.001 |
| PCA1 | 1 | 21050 | 21050 | 17.23 | 0.001 |
| PCA3 | 1 | 2116 | 2116 | 1.73 | 0.075 |
| Year×Site | 9 | 27053 | 3381 | 2.76 | **0.001** |
| Year×PCA1 | 1 | 5642 | 5641 | 4.61 | **0.001** |
| Year×PCA3 | 1 | 580 | 580 | 0.47 | 0.902 |
| Residuals | 60 | 70853 | 1221 | | |
| Total | 79 | 201440 | | | |
| **Three-dimensional structure** | | | | | |
| **Source of variation** | **df** | **SS** | **MS** | **pseudo-F** | ***P* (perm)** |
| Year | 1 | 1118 | 1118 | 4.66 | 0.008 |
| Site | 9 | 20121 | 2515 | 10.48 | 0.001 |
| PCA1 | 1 | 11193 | 11193 | 46.63 | 0.001 |
| PCA3 | 1 | 526 | 526 | 2.19 | 0.090 |
| Year×Site | 9 | 3788 | 473 | 1.97 | **0.017** |
| Year×PCA1 | 1 | 2313 | 2313 | 9.64 | **0.001** |
| Year×PCA3 | 1 | 214 | 214 | 0.89 | 0.424 |
| Residuals | 60 | 13920 | 240 | | |
| Total | 79 | 53197 | | | |

Bold numbers indicate significant results.

[46,47]. A shift from erect-dominated to turf-dominated algal communities can be produced by nutrient enrichment from sewage outfalls [48]. Our sampling sites were relatively far from sewage outlet; therefore, no indicator species of organic pollution was detected. However, algal turf and many of the algal species first appeared or found more abundant in 2008 are known to be comparatively tolerant to lowered water quality with respect to those dominating in 1993 [49].

In the Tigullio Gulf, the undiminished input of organic load has possibly expanded its effects further away from its sources in recent years. The assemblages close to these chronic sources already showed a bi-dimensional structure in 1993 and have likely changed less as compared to the furthest sites, still architecturally complex in 1993. The sequence of disturbance is known to affect the final structure of communities whose response depends also on their previous history [50]. The spatio-temporal variation of stressors intensities is likely to be the main driver of the delay in degradation observed in the MPA. However, the establishment of the MPA in 1998 and the implementation of SCIs were ineffective in reversing the trend toward the deterioration of communities. Finding the alien alga *Caulerpa racemosa* in 2008 constituted a further signal of alteration. *C. racemosa* has been recognized as a strong invader [51] and its fast spread is favored by the presence of algal turf [52]. Turf development in the protected sites anticipated the settlement of *C. racemosa*, which has been subsequently recorded also in the Portofino MPA [53].

Our results suggest sequential community change under the combined influence of human pressure and invasive species. Reef communities are undergoing a shift from canopy-dominated to turf-dominated, and then to a *C. racemosa*-dominated state. This trajectory towards ecosystem deterioration has not been hampered by protection measures. Paradoxically, protection itself might even have contributed to the observed community shift. Fishing restriction through MPA enforcement may have a dual effect on reef communities: on one hand, it recovers the abundance of large carnivorous fish, which have the desirable effect of controlling sea-urchin overgrazing [54]; on the other hand, it also favors large herbivorous fishes such as *Sarpa salpa*. Sea urchin densities did not change between 1993 and 2008 and, although published quantitative data about their densities prior to 1993 are not available, field observations in the 1980s by two of us (CNB and CM) suggest no obvious change with time. On the contrary, the abundance of *S. salpa* was shown to have increased after the creation of the Portofino MPA [15]. Increased grazing by *Sarpa salpa* is capable of driving algal assemblages towards a turf-dominated state [55].

Anthropogenic stressors and climate change may act in synergy on temperate reef communities, lowering their resilience to perturbations [56]. While climate and other global impacts require international actions, regional management practices may help reducing local impacts. Managing temperate reefs requires efforts to reduce nutrient and organic inputs from land sources and pollution by boat traffic. Best coastal management practices may promote favorable conditions to confer canopy-forming species with resistance and resilience against pressures that cannot be managed locally, such as sea water warming, storms, and introduced species.

MPAs, which are designed mainly to regulate fishing activities, are largely ineffective in managing land-based pressures (e.g. outfalls, river discharge) and the spread of invasive species. In this

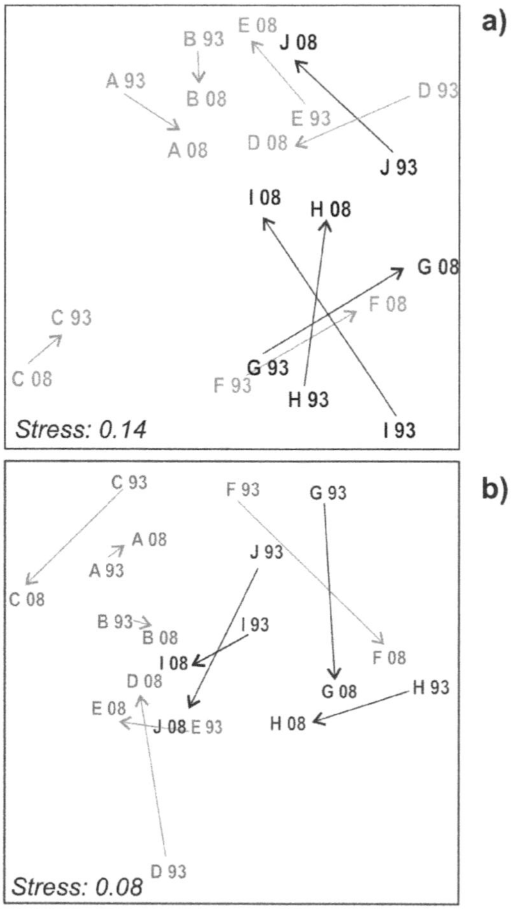


**Figure 2. MDS plots calculated on species (a) and three-dimensional categories (b).** Black characters and arrows refer to sampling sites inside the MPA, grey ones to sampling sites outside the MPA.

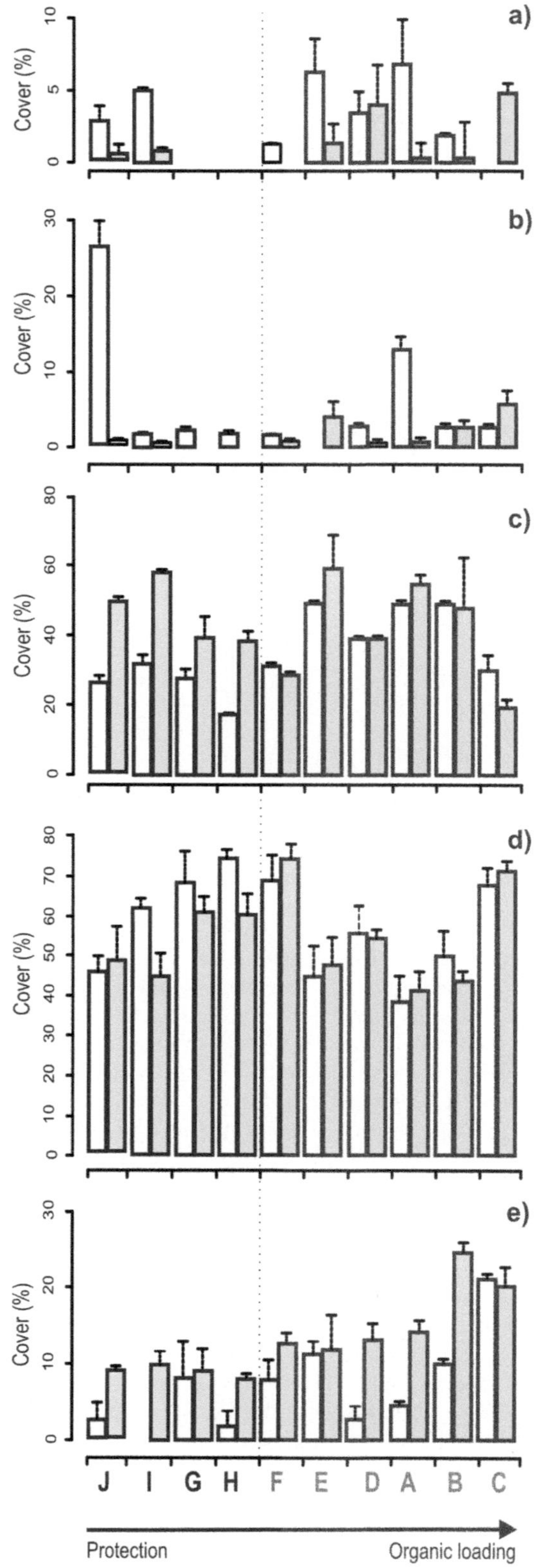


**Figure 3. Cover of three dimensional categories for each sites in 1993 (white) and 2008 (grey). a)**: high; **b)** medium-high; **c)** medium; **d)** medium-low; **e)** low. Note that scales on Y axis are different. The vertical dotted line separates the sampling sites within the MPA (black characters) from those outside (grey characters).

context, SCIs or other complementary measures need to play a significant role in addressing diverse and escalating threats. Yet, SCIs implementation regulated a limited set of pressures (i.e. anchoring and the building of new coastal structures) that may not be the major drivers of change, especially in cases as the Tigullio Gulf where multiple and chronic pressures act together. At present, SCIs represent a 'passive' application of the legislative constraints in the coastal zones. Legislation, of course, plays a crucial role for conservation. However, legislative constraints alone are not enough to achieve effective EBM [3]: managers and scientists should apply and use the legislative instruments in a scientifically sound way. For SCIs to become effective EBM tools, the selection of the crucial pressures to be managed should not be driven by existing laws, but should rather be dictated by the actual pressure regime and local ecological characteristic of different coastal zones (Figure 5).

Small-scale protection measures and SCIs will not be effective if key large scale pressures, such as those coming from adjacent unprotected zones or linked to global change, are not identified. Comprehensive management plans should address multiple pressures through a suite of policy and management tools: carefully designed monitoring is critical for the early detection and eradication of invasive species, for the upgrade of water treatment plants, and for the regulation of coastal development in proximity to sensitive reef and seagrass habitats.

**Figure 4. Change in reef community at Portofino, in the vicinity of sampling site J: a)** ***Sargassum vulgare*** **and** ***Dictyopteris polypodioides*** **canopy in 1981; b)** ***Sarpa salpa*** **grazing in a turf-dominated environment in 2009.**

Monitoring programs that are too narrow in scope, e.g., involving only ecosystems within MPAs, may fail in the identification of relevant pressures that often act at larger scales [57] and in the quantification of the cumulative impacts of multiple pressures [1,7]. Only when pressures are exclusively local, monitoring based on hierarchical designs within the protected area and according to enforcement levels may be effective in providing managers with valuable solutions [58]. When multiple pressures originate outside the boundaries of the protected area and have wider impact, monitoring programs should be designed across gradients at the appropriate scales to address relevant pressures and resulting cumulative impacts (Figure 5). This will help identifying the ultimate causes of deterioration and setting management priorities to reverse ecosystem trajectory.

## Supporting Information

**Table S1** Characteristics of the main pressures in the study area. Information on beach replenishment was available from 2003.

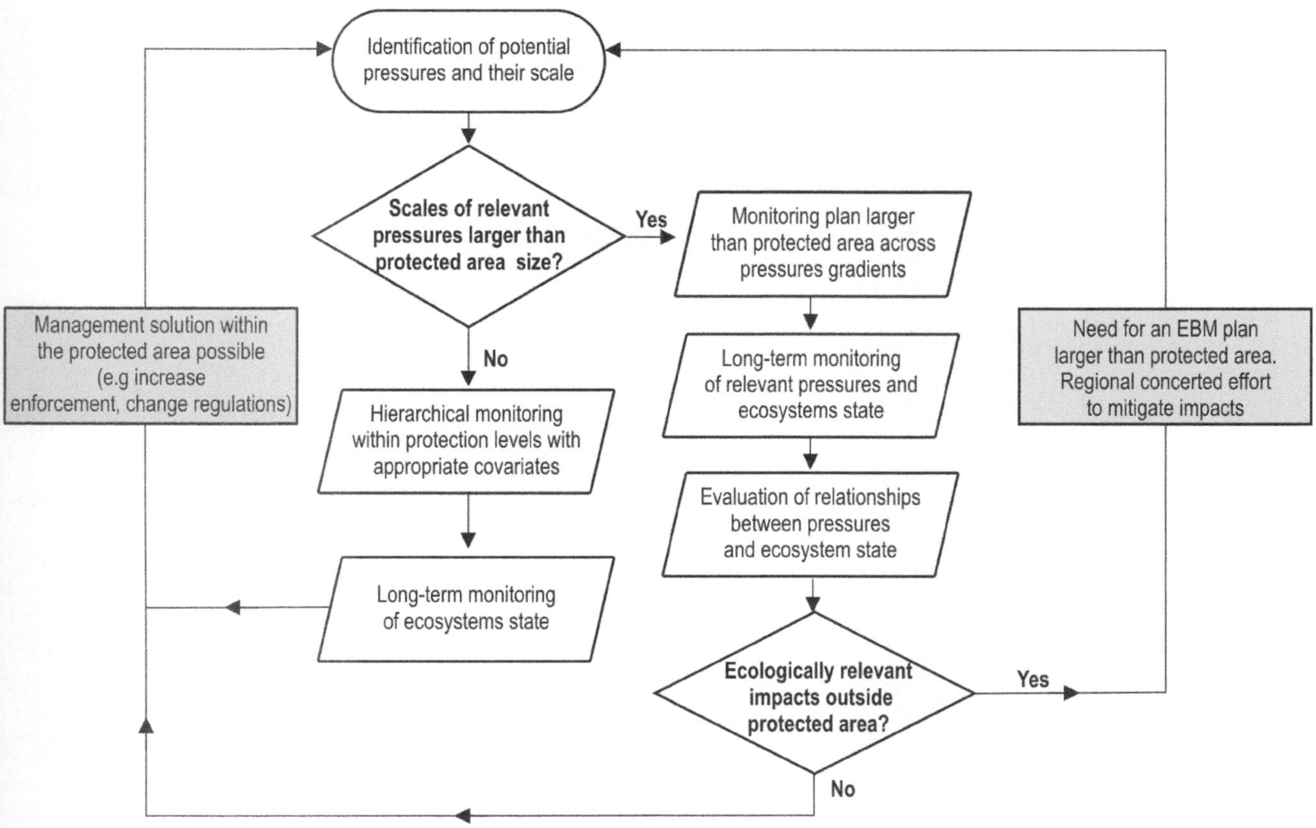


**Figure 5. Decision flow-chart for implementing EBM and choosing effective monitoring design in coastal zones to be protected.** White boxes: steps required for ecosystem based evaluation of relevant pressures; grey boxes: policy consequences of ecosystem-based analysis.

**Table S2** Significant correlations of the individual pressures with the first three axes of the Principal Component Analysis (PCA).

**Table S3** Results of SIMilarity PERcentage (SIMPER) analysis identifying taxa major contributing to differences between Years and among Sites. Grey letters indicate sampling sites outside the MPA while black letters indicate sampling sites within the MPA boundaries.

**Figure S1** **a**) Annual and seasonal mean (+SE) of chlorophyll-a concentration in the period 1985–1994 (grey bars) and in the period 2002–2008 (black bars). **b**) Annual and seasonal mean (+SE) of water transparency (measured with Secchi disk) in the period 1985–1994 (grey bars) and in the period 2002–2008 (black bars). **c**) Potential organic load (expressed in term of equivalent inhabitants) calculated for the 7 administrations facing the coastal zone of the study area following the method described by Lopez y Royo et al. (2010); C: Camogli; L: Lavagna; P: Portofino; R: Rapallo; Sa: Santa Margherita; Se: Sestri Levante; Z: Zoagli; grey bars: data relative to 1991; black bars: data relative to 2001.

## Acknowledgments

Data of 1993 were collected in the frame of an investigation coordinated by the ENEA Marine Environmental Research Centre of La Spezia: thanks are due to S. Cocito, F. Degl'Innocenti, M. Morgigni, A. Peirano and S. Sgorbini, who participated in the field activities. Data used in this study are available from the authors upon request.

## Author Contributions

Conceived and designed the experiments: VP FM CNB MM CM EV MC PP. Performed the experiments: VP MM EV MC CNB CM PP. Analyzed the data: VP FM MM EV CNB. Contributed reagents/materials/analysis tools: CNB FM CM PP MC. Wrote the paper: VP FM MM CM EV MC PP CNB.

## References

1. Halpern BS, Walbridge S, Selkoe KA, Kappel CV, Micheli F (2008) A global map of human impact on marine ecosystems. Science 319: 948–952.
2. Allison GW, Lubchenco J, Carr MH (1998) Marine reserves are necessary but not sufficient for marine conservation. Ecol Appl 8: S79–S92.
3. Montefalcone M, Albertelli G, Morri C, Parravicini V, Bianchi CN (2009) Legal protection is not enough: *Posidonia oceanica* meadows in marine protected areas are not healthier than those in unprotected areas of the northwest Mediterranean Sea. Mar Pollut Bull 58: 515–519.
4. McLeod KL, Lubchenco J, Palumbi SR, Rosenberg AA (200). Scientific consensus statement on marine ecosystem-based management. Communication Partnership for Science and the Sea, www.compassonline.org/pdf_files/EBM_Consensus_Statement_v12.pdf.
5. Leslie HM, McLeod KL (2007) Confronting the challenges of implementing marine ecosystem-based management. Front Ecol Environ 5: 540–548.
6. Olsson P, Folke C, Hughes TP (2008) Navigating the transition to ecosystem-based management of the Great Barrier Reef, Australia. Proc Natl Acad Sci USA 105: 9489–9494.
7. Lester SE, McLeod KL, Tallis H, Ruckelshaus M, Halpern BS, et al. (2010) Science in support of ecosystem-based management for the US West Coast and beyond. Biol Conserv 143: 576–587.
8. Airoldi L, Beck MW (2007) Loss, status and trends for coastal marine habitats of Europe. Ann Rev Mar Sci 45: 345–405.
9. Sano M, Gonzalez-Riancho P, Areizaga J, Medina R (2010) The strategy for coastal sustainability: a Spanish initiative for ICZM. Coast Manag 38: 76–96.
10. Russell BD, Harley CDG, Wernberg T, Mieszkowska N, Widdicombe S, et al. (2012) Predicting ecosystem shifts requires new approaches that integrate the effects of climate change across entire systems. Biol Lett 8: 164–166.
11. Bianchi CN, Cocito S, Morgigni M, Morri C, Peirano A, et al. (1994) Benthos del Golfo Tigullio: fondi duri e prati di *Cymodocea*. ENEA CRAM Santa Teresa, La Spezia.
12. Airoldi L, Balata D, Beck MW (2008) The Gray Zone: relationships between habitat loss and marine diversity and their applications in conservation. J Exp Mar Biol Ecol 366: 8–15.
13. Micheli F, Benedetti-Cecchi L, Gambaccini S, Bertocci I, Borsini C, et al. (2005) Cascading human impacts, marine protected areas, and the structure of Mediterranean reef assemblages. Ecol Monog 75: 81–102.
14. Parravicini V, Thrush SF, Chiantore M, Morri C, Croci C, et al. (2010) The legacy of past disturbance: chronic angling impairs long-term recovery of marine epibenthic communities from acute date-mussel harvesting. Biol Conserv 143: 2435–2440.
15. Guidetti P, Milazzo M, Bussotti S, Molinari A, Murenu M, et al. (2008) Italian marine reserve effectiveness: does enforcement matter? Biol Conserv 141: 699–709.
16. Crain CM, Kroeker K, Halpern BS (2008) Interactive and cumulative effects of multiple human stressors in marine systems. Ecol Lett 11: 1304–1315.
17. Asnaghi V, Chiantore M, Bertolotto RM, Parravicini V, Cattaneo-Vietti R, et al. (2009) Implementation of the European Water Framework Directive: Natural variability associated with the CARLIT method on the rocky shores of the Ligurian Sea (Italy). Mar Ecol 30: 505–513.
18. Morri C, Bianchi CN, Damiani V, Peirano A, Romeo G, et al. (1988) The marine environment between Punta della Chiappa and Sestri Levante (Ligurian Sea): ecotipological outline and proposal of a biocenotic map. Boll Mus Ist Biol Univ Genova 52: 213–231.
19. Danovaro R, Fabiano M (1997) Seasonal changes in quality and quantity of food available for benthic suspension-feeders in the Golfo Marconi (North-western Mediterranean). Estuar Coast Shelf Sci 44: 723–736.
20. Misic C, Covazzi Harriague A (2008) Organic matter recycling in a shallow coastal zone (NW Mediterranean): The influence of local and global climatic forcing and organic matter lability on hydrolytic enzyme activity. Cont Shelf Res 28: 2725–2735.
21. Lopez y Royo C, Silvestri C, Pergent G, Casazza G (2009) Assessing human-induced pressures on coastal areas with publicly available data. J Env Manag 90: 1494–1501.
22. Bianchi CN, Morri C (2003) Global sea warming and "tropicalization" of the Mediterranean Sea: biogeographic and ecological aspects. Biogeographia 24: 319–327.
23. Cerrano C, Bavestrello G, Bianchi CN, Cattaneo-Vietti R, Bava S, et al. (2000) A catastrophic mass-mortality episode of gorgonians and other organisms in the Ligurian Sea (North-western Mediterranean), summer 1999. Ecol Lett 3: 284–293.
24. Garrabou J, Coma R, Bensoussan N, Bally M, Chevaldonné P, et al. (2009) Mass mortality in Northwestern Mediterranean rocky benthic communities: effects of the 2003 heat wave. Glob Change Biol 15: 1090–1103.
25. Parravicini V, Guidetti P, Morri C, Montefalcone M, Donato M, et al. (2010) Consequences of sea water temperature anomalies on a Mediterranean submarine cave ecosystem. Estuar Coast Shelf Sci 86: 276–282.
26. Martí R, Uriz MJ, Ballesteros E, Turon X (2005) Seasonal variation in the structure of three Mediterranean algal communities in various light conditions. Estuar Coast Shelf Sci 64: 613–622.
27. Bianchi CN, Pronzato R, Cattaneo-Vietti R, Benedetti-Cecchi L, et al. (2004) Mediterranean marine benthos: a manual of methods for its sampling and study. Hard bottoms. Biol Mar Medit 11: 185–215.
28. Alvarez-Filip L, Gill JA, Dulvy NK, Perry AL, Watkinson AR, et al. (2011) Drivers of region-wide declines in architectural complexity on Caribbean reefs. Coral Reefs 30: 1051–1060.
29. Mardia KV, Kent JT, Bibby JM (1980) Multivariate Analysis. Academic Press, London.
30. Anderson MJ, Ter Braak CJF (2003) Permutation tests for multi-factorial analysis of variance. J Stat Comp Sim 73: 85–113.
31. Anderson MJ, Ellingsen KE, McArdle BH (2006) Multivariate dispersion as a measure of beta diversity. Ecol Lett 9: 683–693.
32. Clarke KR, Gorley RN (2006) PRIMER v6: User Manual/Tutorial. PRIMER-E Ltd, Plymouth.
33. Tortonese E (1958) Bionomia marina della regione costiera fra Punta della Chiappa e Portofino (Riviera ligure di levante). Archo Oceanogr Limnol 11: 167–210.
34. Tortonese E (1961) Nuovo contributo alla conoscenza del bentos della scogliera ligure. Archo Oceanogr Limnol 12: 163–183.
35. Tortonese E (1962) Recenti ricerche sul bentos in ambienti litorali del mare Ligure. Pubbl Staz Zool Napoli 32: 99–116.
36. Melegari GE (1973) Portofino sub: guida alla conoscenza dei fondali del promontorio di Portofino. ERGA, Genova.
37. Pisano E, Bianchi CN, Matricardi G, Relini G (1982) Accumulo della biomassa su substrati artificiali immersi lungo la falesia di Portofino (Mar Ligure). Atti del Convegno delle Unità Operative afferenti ai sottoprogetti Risorse Biologiche ed Inquinamento Marino. CNR, Roma: 93–105.
38. Pisano E, Bianchi CN, Relini G. (1980) Insediamento su substrati artificiali immersi lungo la falesia di Portofino (Mar Ligure): metodologie e dati preliminari. Mem Biol Mar Oceanogr 10: 269–274.
39. Thomsen MS, Wernberg T, Altieri A, Tuya F, Gulbransen D, et al. (2010) Habitat cascades: the conceptual context and global relevance of facilitation cascades via habitat formation and modification. Integr Comp Biol 50: 158–175.

40. Wernberg T, Thomsen MS, Tuya F, Kendrick GA (2011a) Biogenic habitat structure of seaweeds change along a latitudinal gradient in ocean temperature. J Exp Mar Biol Ecol 400: 264–271.
41. Smale DA, Wernberg T (2013) Extreme climatic event drives range contraction of a habitat-forming species. Proc R Soc B 280: 20122829.
42. Wernberg T, Smale DA, Tuya F, Thomsen MS, Langlois TJ, et al. (2013) An extreme climatic event alters marine ecosystem structure in a global biodiversity hotspot. Nature Clim Change 3: 78–82.
43. Duarte L, Viejo RM, Martínez B, de Castro M, Gómez-Gesteira M, et al. (2013) Recent and historical range shifts of two canopy-forming seaweeds in North Spain and the link with trends in sea surface temperature. Acta Oecol 51: 1–10.
44. Bianchi CN, Morri C (1994) Southern species in the Ligurian Sea (northern Mediterranean): new records and a review. Boll Mus Ist Biol Univ Genova 58–59: 181–197.
45. Bussotti S, Conti M, Guidetti P, Martini F, Matricardi G (1996) First record of *Caulerpa racemosa* (Forsskål) J. Agardh along the coast of Genoa (North-Western Mediterranean). Doriana 6: 1–5.
46. MacDougall AS, Turkington R (2005) Are invasive species the drivers or passengers of change in degraded ecosystems? Ecology 86: 42–55.
47. Didham RK, Tylianakis JM, Gemmell NJ, Rand TA, Ewers RM (2007) Interactive effects of habitat modification and species invasion on native species decline. Trends Ecol Evol 22: 489–496.
48. Claudet J, Fraschetti S (2010) Human-driven impacts on marine habitats: a regional meta-analysis in the Mediterranean Sea. Biol Conserv 143: 2195–2206.
49. Orfanidis S, Panayotidis P, Stamatis N (2001) Ecological evaluation of transitional and coastal waters: a marine benthic macrophytes based model. Med Mar Sci 2: 45–65.
50. Fukami T (2001) Sequence effects of disturbance on community structure. Oikos 92: 215–224.
51. Montefalcone M, Albertelli G, Morri C, Bianchi CN (2010) Pattern of wide-scale substitution within *Posidonia oceanica* meadows of NW Mediterranean Sea: invaders are stronger than natives. Aquat Conserv 20: 507–515.
52. Bulleri F, Benedetti-Cecchi L (2008) Facilitation of the introduced green alga *Caulerpa racemosa* by resident algal turfs: experimental evaluation of underlying mechanisms. Mar Ecol Prog Ser 364: 77–86.
53. Cattaneo-Vietti R, Albertelli G, Aliani S, Bava S, Bavestrello G, et al. (2010) The Ligurian Sea: present status, problems and perspectives. Chem Ecol 26: 319–340.
54. Sala E, Boudouresque CF, Harmelin-Vivien M (1998) Fishing, trophic cascades, and the structure of algal assemblages: evaluation of an old but untested paradigm. Oikos 82: 425–439.
55. Ruitton S, Francour P, Boudouresque CF (2000) Relationships between algae, benthic herbivorous invertebrates and fishes in rocky sublittoral communities of a temperate sea (Mediterranean). Estuar Coast Shelf Sci 50: 217–230.
56. Wernberg T, Russell BD, Moore PJ, Linge SD, Smale DA, et al. (2011) Impacts of climate change in a global hotspot for temperate marine biodiversity and ocean warming. J Exp Mar Biol Ecol 400: 7–16.
57. Hewitt JE, Thrush SF, Dayton PK, Bonsdorff E (2007) The effect of spatial and temporal heterogeneity on the design and analysis of empirical studies of scale-dependent systems. Am Nat 169: 398–408.
58. Greco S, Notarbatolo di Sciara G, Tunesi L (2004) 'Sistema Afrodite': an integrated programme for the inventorying and monitoring of the core zones of the Italian marine protected areas. Aquat Conserv 14: S119–S122.

# 17

# Recovery Trends in Marine Mammal Populations

**Anna M. Magera[1]*, Joanna E. Mills Flemming[2], Kristin Kaschner[3,4], Line B. Christensen[5], Heike K. Lotze[1]**

**1** Department of Biology, Dalhousie University, Halifax, Nova Scotia, Canada, **2** Department of Mathematics and Statistics, Dalhousie University, Halifax, Nova Scotia, Canada, **3** Department of Biometry and Environmental System Analysis, University of Freiburg, Freiburg, Germany, **4** CESAB (Centre de Synthèse et d'Analyse sur la Biodiversité), Immeuble Henri Poincaré, Domaine du Petit Arbois, Aix-en-Provence, France, **5** Fisheries Centre, University of British Columbia, Vancouver, British Columbia, Canada

## Abstract

Marine mammals have greatly benefitted from a shift from resource exploitation towards conservation. Often lauded as symbols of conservation success, some marine mammal populations have shown remarkable recoveries after severe depletions. Others have remained at low abundance levels, continued to decline, or become extinct or extirpated. Here we provide a quantitative assessment of (1) publicly available population-level abundance data for marine mammals worldwide, (2) abundance trends and recovery status, and (3) historic population decline and recent recovery. We compiled 182 population abundance time series for 47 species and identified major data gaps. In order to compare across the largest possible set of time series with varying data quality, quantity and frequency, we considered an increase in population abundance as evidence of recovery. Using robust log-linear regression over three generations, we were able to classify abundance trends for 92 spatially non-overlapping populations as Significantly Increasing (42%), Significantly Decreasing (10%), Non-Significant Change (28%) and Unknown (20%). Our results were comparable to IUCN classifications for equivalent species. Among different groupings, pinnipeds and other marine mammals (sirenians, polar bears and otters) showed the highest proportion of recovering populations, likely benefiting from relatively fast life histories and nearshore habitats that provided visibility and protective management measures. Recovery was less frequent among cetaceans, but more common in coastal than offshore populations. For marine mammals with available historical abundance estimates (n = 47), larger historical population declines were associated with low or variable recent recoveries so far. Overall, our results show that many formerly depleted marine mammal populations are recovering. However, data-deficient populations and those with decreasing and non-significant trends require attention. In particular, increased study of populations with major data gaps, including offshore small cetaceans, cryptic species, and marine mammals in low latitudes and developing nations, is needed to better understand the status of marine mammal populations worldwide.

**Editor:** Konstantinos I. Stergiou, Aristotle University of Thessaloniki, Greece

**Funding:** Financial support was provided by the Census of Marine Life's Future of Marine Animal Populations (FMAP) program (www.fmap.ca) with a grant to HKL, the Natural Sciences and Engineering Research Council (NSERC, http://www.nserc-crsng.gc.ca) with grants to HKL and JMF, and Dalhousie University (www.dal.ca). Financial support for data collection was provided by Pew Charitable Trusts (www.pewtrusts.org) through the Sea Around Us Project (www.seaaroundus.org) for populating the abundance databases provided by KK and LBC. The funders had no role in study design, data collection and analysis, decision to publish, or preparation of the manuscript.

**Competing Interests:** AMM currently works as a Fisheries Management Biologist with the Nunavut Wildlife Management Board. All other authors have declared that no competing interests exist. This does not alter the authors' adherence to all the PLOS ONE policies on sharing data and materials.

* E-mail: ammagera@gmail.com

## Introduction

In the marine realm, mammals appear to be one of the taxa that have benefitted the most from a shift from resource exploitation toward wildlife conservation [1–4]. Marine mammals, a loose grouping of approximately 127 species, include cetaceans (whales, porpoises and dolphins), pinnipeds (true seals, fur seals and sea lions), as well as marine otters and sea otters, sirenians (manatees and dugongs) and polar bears [5]. Throughout history, humans have exploited and often depleted marine mammal populations [3,6–13]. Yet in the 20$^{th}$ century, substantial population declines led to relatively early and widespread reduction or cessation of commercial exploitation and implementation of conservation measures [1,2,4,14]. While several marine mammals have been held up as key conservation success stories (e.g. eastern North Pacific gray whales, *Eschrichtius robustus* [15,16] and sea otter, *Enhydra lutris*, populations [17]), not all marine mammal populations have recovered from earlier exploitation-driven declines. However, with data indicating positive abundance trend estimates for various populations, we now have the opportunity to not only evaluate threat status, decline, and extinction risk in marine mammals, but also increase and recovery.

Threats to marine mammals are numerous and have changed over time. Historically, marine mammals have been prized sources of meat, oil, fur, baleen and ivory [6,7,9,11]. They have also been captured for display in aquariums, culled when declared nuisances, used for bait, and indirectly exploited as bycatch [18–24]. Numerous marine mammal species were reduced to very low abundances by or during the 1900 s [2,7,25], and some almost to extinction, such as Northern elephant seals (*Mirounga angustirostris*) [26] and Guadalupe fur seals (*Arctocephalus townsendi*) [27]. Certain populations were regionally extirpated, including the sea otter throughout most of its range [17], the Atlantic gray whale, and the walrus (*Odobenus rosmarus*) in parts of the Northwest Atlantic [1]. A handful of species have become globally extinct, including the Steller's sea cow (*Hydrodamalis gigas*), sea mink (*Neovison macrodon*),

Caribbean monk seal (*Monachus tropicalis*), and Japanese sea lion (*Zalophus japonicus*) [5,26]. Among the survivors, however, some substantial population recoveries have occurred, for example for gray whales in the eastern North Pacific, multiple populations of humpback whales (*Megaptera novaeangliae*), southern right whales (*Eubalaena australis*) [28], sea otters [29], northern elephant seals [26], grey seals (*Halichoerus grypus*) in the UK [30] and Northwest Atlantic [31], and numerous fur seal species [27]. Yet direct and indirect exploitation, as well as disease, competition for prey, habitat degradation from coastal development and dams, ship traffic, offshore oil and gas exploration, pollution (chemical, physical and acoustic), and climate change continue to impact marine mammal populations today [23,24,32–34].

Although it is a commonly used term, "recovery" can have many definitions in different management and conservation contexts [25,35]. What definition of recovery is chosen depends on the goals of the study and can alter the conclusions. Generally, recovery is taken to be "a return to a normal state of health... or strength" [36]. In analyses of wild animal populations, however, we often do not know the "normal state" of a population [25]. Pre-commercial exploitation abundance or carrying capacity (K) can be used as a reference point, with the increase of a population towards such a reference point indicating recovery [6,25,37]. However, neither pre-exploitation nor K estimates exist for many species that lack records or data on past catches, traded products (e.g. oil, fur), scientific surveys, genetics, life history, or population structure. Furthermore, there is often debate as to whether K estimates should refer to pre-exploitation or current ecosystem conditions [6,38]. Therefore, a basic and practical approach in many data-limited (in terms of time span, number, frequency, and precision of data points) cases has been to view any significant abundance increase as evidence of a recovering population and at least partial recovery [25,39]. This is the definition we used in this study as it allowed us to compare across a maximal number of marine mammal population abundance time series with variable data quality.

Management bodies often judge recovery with respect to a proportion of K or pre-exploitation size. The U.S. Marine Mammal Protection Act specifies management for an "optimal sustainable population" level, which is defined by the U.S. National Marine Fisheries Services (NMFS) as "a population level between carrying capacity and the population size at maximum net productivity" [37]. An optimal sustainable population level for marine mammals is thought to be between 50–85% of K [37], but generally 60% is used [15], and the International Whaling Commission assumes 60% is the level at which whale populations are most productive [28]. In the absence of either K or pre-exploitation abundance, management bodies may use maximum observed population levels, based on survey data for example, as a reference point. Such is the case with Atlantic seal populations in Canada [40]. However, in cases where reports of substantial declines predate quantitative records, maximum observed population levels would not represent pre-exploitation levels, and could result in the underestimation of declines and the overestimation of recoveries. Additional criteria, such as targets for demographic features (e.g. ratios of juveniles to adults, or males to females), social dynamics, or ecological functions [41] may be relevant to evaluating recovery depending on the situation. Recovery may also be measured over different time periods, such as an entire time series of data, a set time period (e.g. 50 years, or 1950–2000), or with respect to the species' life history (e.g. three generations is commonly used by the International Union for Conservation of Nature (IUCN) [42]).

Despite the strong conservation focus on marine mammal management in many parts of the world, abundance data are limited and subject to high uncertainty. Reliable, effort-corrected catch data are absent for many populations, and marine mammals are notoriously difficult to survey accurately for abundance. Current population designations and/or distributions may not match historical records, leading to speculation that some populations have changed their distributions over time [43,44], and complicating assessments of long-term trends. Longer time series with historical population estimates are lacking for many populations that are or were not commercially valuable [2]. Many marine mammal species have elusive behaviors (e.g. extensive migrations or deep diving for prolonged time periods) which, combined with their widespread occurrences in remote areas, represent high logistical and economic challenges for monitoring their population statuses [15,45–47]. With the exception of land-breeding pinnipeds, accurate estimation of abundance trends over relatively short time periods is difficult in many cases [46], and abundance time series often encompass irregular survey intervals. Existing variability in survey coverage and methodology greatly hampers the assessment of population trends [47]. However, data availability has improved greatly with the population monitoring and modeling that have accompanied increased management and conservation efforts internationally and domestically since the 1970 s.

Although marine mammal abundance data are collected and assessed by several organizations for management and conservation purposes (e.g. IUCN, U.S. marine mammal stock assessments), a quantitative global synthesis of marine mammal recoveries on a population level has not been previously attempted. Populations within the same species may show different abundance trends, and population-level abundance monitoring has become increasingly valued. Thus, the objectives of our study were to compile publicly available population-level abundance time series for marine mammals around the world, assess population trends, and classify recovery. Where possible, we also aimed to quantify the relationship between historical population decline and subsequent recovery. Our overall goal was to enhance our understanding of recovery in formerly exploited marine mammals and marine species overall.

## Materials and Methods

### Data Compilation

We expanded a database of marine mammal abundance estimates collected by Kaschner (2004) [48] and Christensen (2006) [11]. We collected population-level abundance data from around the world from publicly available published journal articles, online government documents, stock assessment reports, and recovery plans. Major sources included: Fisheries and Oceans Canada (DFO), Committee on the Status of Endangered Wildlife in Canada (COSEWIC), U.S. National Oceanic and Atmospheric Administration (NOAA) and NMFS technical and administrative reports, U.S. Marine Mammal Stock Assessment Reports, St. Andrew's Sea Mammal Research Unit reports, Australian and New Zealand government documents, International Whaling Commission documents, and numerous published primary sources (Table S1). We collected data up to 2008 or the next most recent data point. We also collected information on generation times, methods of data collection and abundance estimation, and data reliability. For abundance trend analysis, we limited our attention to time series with at least three estimates. Although a population is generally described as a group of interbreeding organisms of the same species in a defined area [49], for the purpose of this study we chose population abundance data according to consistently defined areas described in the source literature. Thus, we did not exclusively adhere to population or stock definitions from monitoring or regulatory agencies, but the population designations often matched.

Population abundance data came from dedicated and opportunistic aerial, land-based and ship-based surveys, extrapolated total population or pup-count data, photo-identification and mark-recapture models, genetic diversity analysis, combined totals from the literature, models that relied heavily on bycatch, catch or catch-per-unit-effort data, various other model-derived estimates from these aforementioned types of data (including age- or stage-based, simple regression, Bayesian or state-space models), and in some cases unknown or unstated methods (Table S1). When year ranges were provided for abundance estimates, we used the mid-point of the range.

Error is present in virtually all marine mammal abundance data, but it may originate from different sources. For example, catch, bycatch, or product data could be subject to intentional or unintentional misreporting. In turn, abundance estimation from inaccurate data may provide faulty estimates of K or the intrinsic rate of increase ($r_{max}$), both of which may vary over time [6]. Historical abundance estimates from DNA have been questioned because of uncertainty over changes in mutation rates, appropriate designation of particular populations, and migration, all of which can affect population estimates [38]. Survey data are affected by biases such as avoidance or attraction behavior of the species surveyed, and estimates may also be skewed because of detectability issues, particularly for more cryptic species. Analysis methods have been developed, however, to correct survey data for unobserved animals that may be at sea or are otherwise not visible [45,50,51].

We collected any reported error information (e.g. coefficient of variation, confidence intervals, standard error, or standard deviation). Where available, we gathered pre-exploitation abundance estimates and K estimates. In cases where multiple historical estimates were found (e.g. catch vs. genetic data) we recorded them all. Not all data sources reported error, so we used a system to incorporate quantitative error information (where available) with qualitative information about the data source for each data point and its reliability into abundance trend estimation (Abundance Confidence ID (ACID), Text S1).

## Data Analysis – Trend Overview

We used a general definition of recovering populations being those that displayed an increase in abundance. Our main objectives were to gather abundance data and statistical evidence so as to broadly classify populations as having either increasing or decreasing trends, rather than attempting to estimate and describe complex population trajectories. In order to do so we estimated a trend for each population over the three most recent generations. This corresponded with the criteria used by the IUCN for assessing population decline [42]. In cases where we did not have data for the entire three-generation time period, we used the available time period with a required minimum of ten years instead (i.e. the minimum time period used by the IUCN for assessing population declines [52]) (Table S2).

## Trend Analysis

Given available data and our aim, we used robust regression to estimate population trends. Robust regression is a powerful tool for fitting linear models that simultaneously identifies and down weights outlying data points [53,54]. Specifically we used the lmRob command in the robust library in R (which makes available an S-estimator [55] with a high breakdown point). We performed a full exploratory analysis using simple and robust linear regression, as well as log-linear robust regression (i.e. with a log transformation of the abundance values) and with and without weighting to include abundance data error information. We fit simple and robust regressions to both regular abundance data (i.e. number of individuals) and standardized abundance data (obtained by subtracting the mean from each data point and dividing by the standard deviation). The standardization facilitated comparison of regression results amongst different populations. We decided to use robust linear and robust log-linear weighted regressions as they best reflected the data. The complete R code is available in Text S2.

Results of fitting the linear and log-linear robust weighted regressions were then used to classify each population as Significantly Increasing, Significantly Decreasing, Non-Significant Change, or Unknown. To start, any populations with insufficient data (i.e. <3 data points over three generations or minimum 10 years) were deemed Unknown. Those remaining were classified according to both the direction and significance of their abundance trend. Specifically, Significantly Increasing and Significantly Decreasing populations had positive or negative slope estimates, respectively, and 95% confidence intervals that did not include zero. Populations with 95% confidence intervals that did include zero were classified as showing Non-Significant Change. Our classifications of population trends allowed for an easily interpretable summary of patterns in the marine mammal population abundance data.

## Trend Comparisons

We examined both the population trend estimates (and the ensuing classifications) for marine mammals overall, as well as for notable taxonomic divisions (cetaceans, pinnipeds, other marine mammals), main habitat type according to Jefferson et al. (2008) [5] (for cetaceans only: coastal, offshore, or both), and other relevant sub-groupings (e.g. mysticetes or baleen whales, odontocetes or toothed whales, a sub-group of just dolphins and porpoises, otariids or eared seals, phocids or true seals). The groupings were not necessarily mutually exclusive. For example, a pantropical spotted dolphin (*Stenella attenuata*) population would be included in the overall marine mammal category, as well as the cetacean, odontocete, and dolphin and porpoise sub-groupings. The groupings for each population are included in Table S2. Many of the populations for which time series data exist overlapped spatially or were nested in each other's range, thus likely representing a similar set of individual animals. To avoid such double counting, we grouped populations into the largest and smallest non-overlapping or non-nested areas and analyzed both sets separately. These groupings by large and small areas are also listed in Table S2.

## Comparison with IUCN Data

We compared our population trend classifications (Significantly Increasing, Significantly Decreasing, Non-Significant Change, Unknown) to those available through the IUCN Red List (Increasing, Decreasing, Stable, Unknown) (www.iucnredlist.org, Table S3). Since our data were mostly collected at a population level, and the IUCN mostly works with species level, we first summarized our data at a species level to facilitate comparison. For each species, we selected from our database those populations that comprised the majority of the overall species abundance. In cases of multiple non-nested and approximately equal-sized populations, we took the most frequently occurring trend classification of the populations of the species. If this did not exist (e.g. as for three populations with different trend classifications), we listed the population classification for that particular species as Unknown.

### Historical Declines and Recent Increases

With the available data, we also had the opportunity to examine another facet of marine mammal recovery. For some populations (n = 47), we had at least one historical estimate, as well as a minimum population estimate and a current population estimate that was above the minimum estimate. Thus, these populations demonstrated both a historical decline (decrease) and evidence of recent recovery (increase) (Table S4). We were interested in whether the magnitude of recent recovery was related to the magnitude of historical decline. Therefore, the historical abundance estimate(s) was compared to the population minimum and the most recent abundance estimates to estimate the magnitude of decline and the magnitude of recovery with respect to the historical population size. For the populations for which there existed multiple historical abundance estimates, we used the mean of the declines and recoveries relative to these historical estimates. In addition to the magnitude of recovery, we were also interested in comparing the magnitude of historical population declines to the rates of recovery (i.e. the slopes of our above regression analysis). Unfortunately, only a small subset of the populations with historical abundance estimates had enough data over three generations to derive a rate of recovery, which provided too small a sample size from which to derive meaningful patterns.

## Results

### State of Available Marine Mammal Population Data

Overall, we were able to compile 182 population abundance time series for 47 species of marine mammals (i.e. 37% of the 127 currently recognized marine mammal species) for which there were at least three abundance estimates over three generations or at least ten years. Originally, we had compiled 198 population abundance time series; however, some time series corresponded to the same population, representing both pup count and regular count (i.e. entire population) data. We used the regular data unless they were much sparser than the pup count data. With these duplicate population indices removed, we had 182 population abundance time series in total.

A breakdown of species by taxonomic groups and different sub-groupings represented in this study compared to all marine mammals is depicted in Figure 1. Groups with high representation (>50% of known species represented) were the baleen whales (mysticetes) and pinnipeds. Groups with low representation in our data (<50% of known species represented) included the sirenians and cetaceans overall, especially toothed whales (odontocetes) within the cetaceans grouping, and dolphins and porpoises within the toothed whales grouping. No species of beaked whales (n = 21 species globally) or river dolphins (n = 4 species globally) were included. It was difficult to obtain time series that met our criteria for many smaller cetacean species (notably porpoises, beaked whales and river dolphins), Antarctic true seals, and sirenians. However, we were able to obtain three-generation abundance data (rather than the 10 years of data) for 95 of the 182 populations.

### Marine Mammal Population Trends

Population trends were estimated for 182 non-duplicated populations from 47 species (see Figure 2 for examples, Figure 3 for population trend classifications, Figure S1 for plots for all populations, and Table S5 for all regression results). The models that best fit the available data were found to be those based on scaled abundance data and estimated via robust log-linear regression (weighted by ACID). We compared the R-squared values for both the robust linear and log-linear models, and found that robust log-linear models provided a better fit to the data for 78 of the populations, while robust linear models were better for 50 of the populations. There was 1 population for which both models produced a similar result, and 33 populations with Unknown trends. However, the resulting trend classifications were largely the same, with the two models producing different results for only 20 populations. In these cases, the robust log-linear model provided a better fit for 13 populations, and for the other 7 populations it found no evidence of trend (i.e. Non-Significant Change) as opposed to a Significantly Increasing or Significantly Decreasing trend provided by the better fitting robust linear model. Since the robust log-linear model best described the data in the majority of cases, we reported these results in the paper. However, plots and results for both the robust linear and log-linear regressions are included in the supplementary materials (Figure S1 and Table S5).

In terms of marine mammal population trends for the largest non-overlapping areas (n = 92 populations), 42% were Significantly Increasing, 10% were Significantly Decreasing, 28% showed Non-Significant Change, and 20% were deemed Unknown (see 'All' in Figure 3). Since some populations were nested within each other and consequently not independent, we analyzed trends for populations chosen by the smallest (n = 135) and the largest (n = 92) non-overlapping areas. Results did not differ substantially so we report results by the largest non-overlapping areas. The number of other marine mammal populations (polar bears, otters and sirenians) in the analysis was low (n = 7), with sea otters comprising the majority (n = 4) of the sample of other marine mammals.

Some populations did have flat slopes (i.e. very close to zero) and small confidence intervals that might suggest stable population abundance, such as the New Zealand sea lion (*Phocarctos hookeri*) (Sandy Bay) and the Southern elephant seal (*Mirounga leonina*) (Kerguelen Isles) (Figure S1). However, they did not have significant trends and rather than include an additional trend classification category, they were grouped with other populations that had larger confidence intervals and/or more variable data as displaying Non-Significant Change.

When comparing different categories of marine mammals (Figure 3), our results indicate that proportionally more sirenian, polar bear and sea otter populations (i.e. "Other", 71%) and pinnipeds (50%) were Significantly Increasing, than marine mammals overall (42%) or cetaceans (31%). For the pinnipeds, eared seals (58%) showed more Significantly Increasing populations compared to true seals (44%). Among the cetaceans, all taxonomic groups showed relatively low numbers of Significantly Increasing populations (39% baleen whales, 23% all toothed whales, 28% just dolphins and porpoises). Primarily coastal cetaceans, however, had proportionally more Significantly Increasing populations (58%) than primarily offshore cetaceans (13%). Overall, pinnipeds (and especially eared seals), other marine mammals, and coastal cetaceans showed the highest proportions of Significantly Increasing populations. Toothed whales, and dolphins and porpoises, had the highest proportions of Significantly Decreasing populations. All categories of cetaceans, except for coastal cetaceans, had approximately one third or more Non-Significant trends. Offshore cetaceans and baleen whales also had over a third of populations with Unknown trends over three generations.

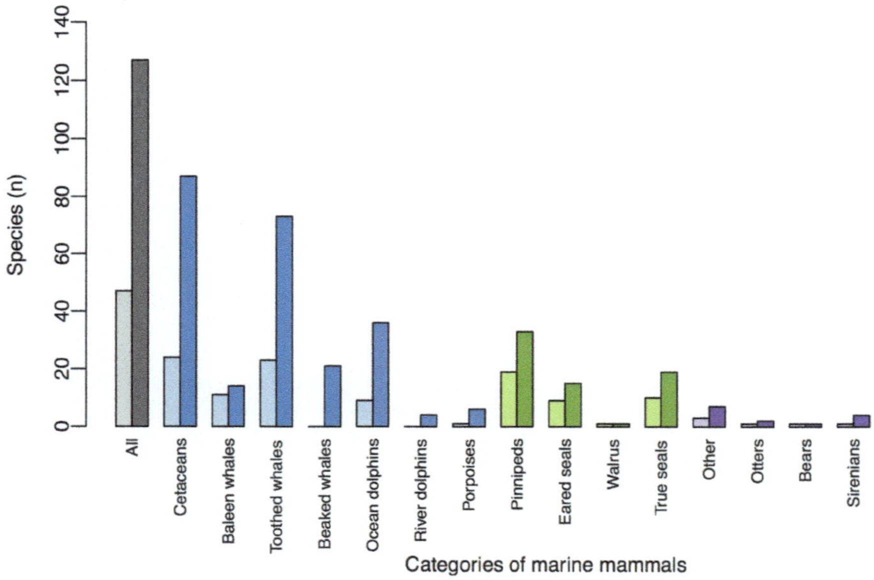


**Figure 1. Species represented in population-level data in this study compared to marine mammal species overall.** This study = light bars, marine mammals species overall = dark bars. Colors represent main taxonomic groups and their relevant sub-groupings: all marine mammals (grey), cetaceans (blue), pinnipeds (green), and other marine mammals (marine and sea otters, polar bears and sirenians) (purple).

## Comparison with IUCN Data

The IUCN lists trend classifications for 127 marine mammal species, while we had data for populations from 47 species. We had enough population-level data to allow for comparison of 27 of the 47 species to the IUCN classifications (Figure 4, Table S3). Most trend classifications were similar, with approximately 37–41% of species showing Increasing trends, 11–19% Decreasing, 7% Non-Significant or 11% Stable (these were two categories that did not match between the IUCN classifications and the ones in our study), and 37% Unknown (Figure 4). However, compared to all 127 mammals in the IUCN database, our data showed a higher proportion of Increasing populations and lower proportions of Decreasing and Unknown populations.

## Historical Declines and Recent Increases

Among all non-duplicated populations in our database (n = 182), we had 47 non-nested populations that included a historical, minimum and recent population estimate (Figure 5, Table S4). Historical abundance declines and recent recoveries, as a proportion of the historical abundance estimate(s), ranged from virtually zero to over 100%, with some population recoveries exceeding the best available historical population

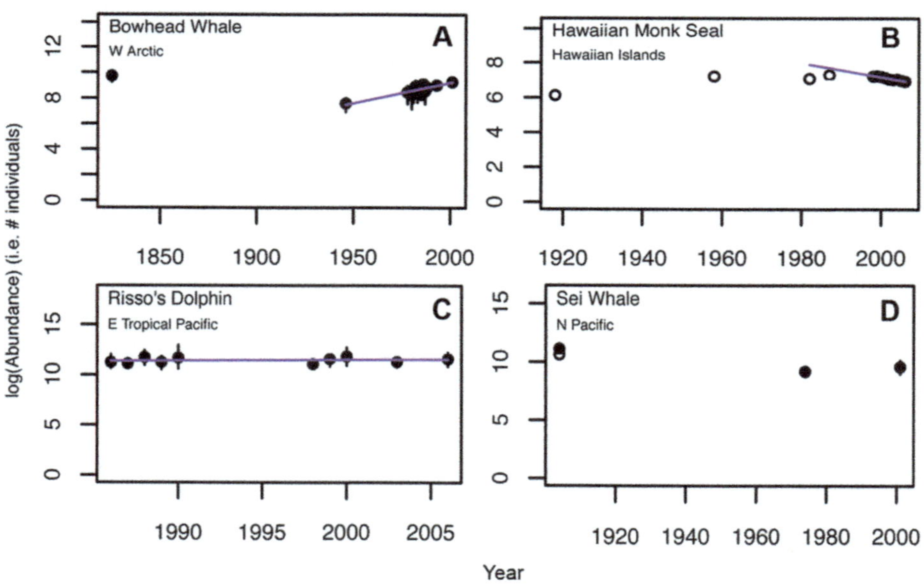


**Figure 2. Examples of Significant Increase (A), Significant Decrease (B), Non-Significant Change (C) and Unknown (D) population abundance trends.** Robust weighted log-linear regression line is depicted over three generations or at least ten years. Solid points = abundance estimates with reported error (95% confidence intervals). Open points = abundance estimates without reported error.

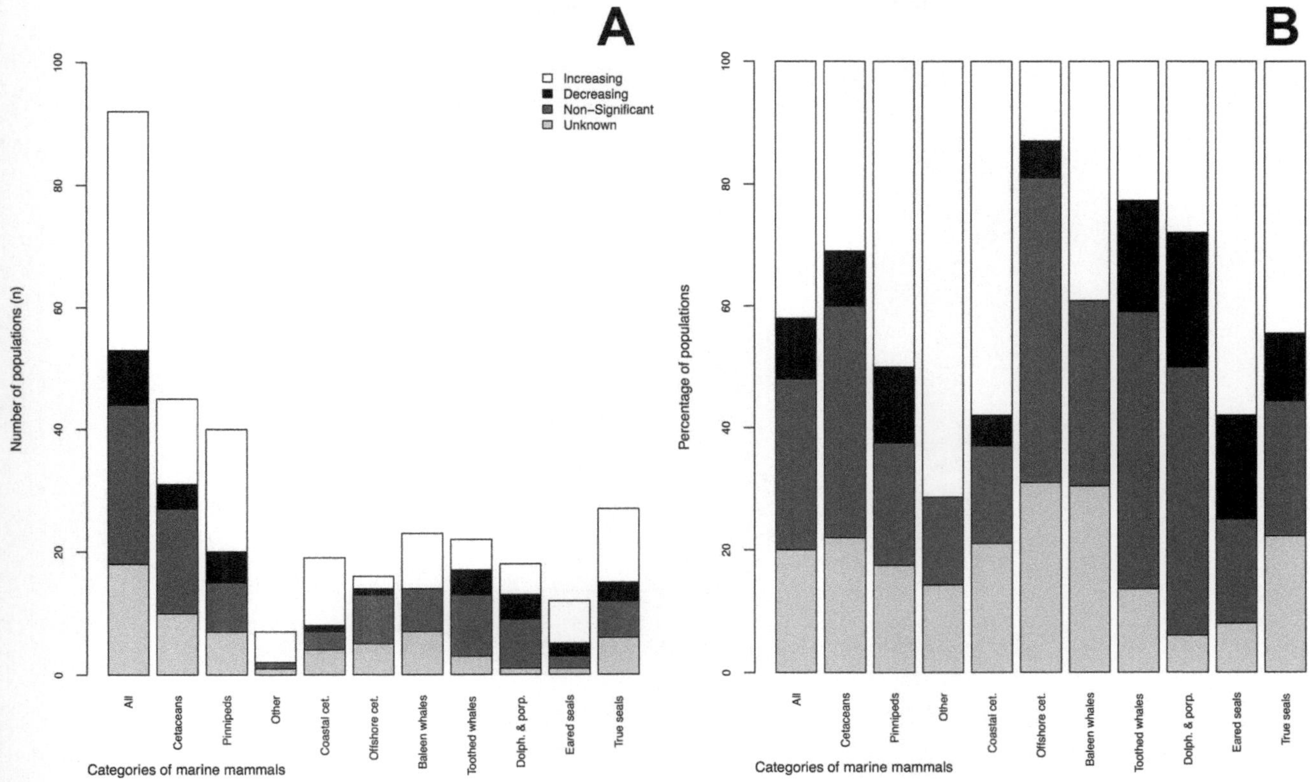


**Figure 3. Trend classification for marine mammal populations by different categories as numbers (A) and proportions (B).** Summary of results from robust weighted log-linear regressions for 92 (non-nested, including the largest possible areas) marine mammal populations. "Other" includes sirenians, polar bears and sea otters.

estimates we could find. One should note that populations with smaller historical declines (e.g. to 90% of historic levels) could only increase over a smaller range (e.g. between 90 and >100%), while those with large historical declines (e.g. to 10% of historic levels) could increase over a much larger range (e.g. between 10 and >100%). Overall, however, there were many populations with very large historical declines >90% that have so far shown only small recent increases (clustered in the lower right corner of Figure 5A on or slightly above the diagonal line). Yet there was also considerable variation in recoveries among populations with declines >80%, and five populations with very high declines to <10% of historical levels also showed very high recoveries to >90% of historical levels. These included South African fur seals (South Africa and Namibia), harbour seals (Washington coast, Oregon), and humpback whales (North Atlantic and North Pacific).

On average, all marine mammal populations declined by 71% and have so far recovered to 61% of their historical abundance (Figure 5B). Recovery responses were generally more variable within each group than declines, and pinnipeds (true seals and eared seals) showed the most variation. Cetaceans that spend the majority of their time in coastal areas instead of offshore waters and other marine mammals (in this case, n = 1 sea otter population) showed the greatest mean declines (93% and 96% respectively), and coastal cetaceans the lowest recoveries (38%). The two groupings that had the lowest mean declines, just dolphins and porpoises (41%) and all toothed whales (49%), showed the highest mean recoveries (89% and 78% respectively).

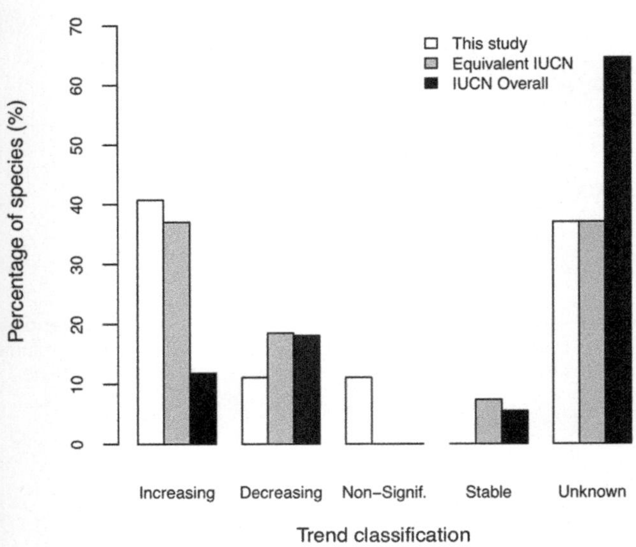


**Figure 4. Comparison of marine mammal population trend classification results from this study with IUCN classifications.** Percentage of marine mammals at the species level in abundance trend categories from this study (n = 27), the equivalent IUCN species (n = 27), and all marine mammal species assessed by IUCN (n = 127). The IUCN does not have a "Non-Significant" category as in this study, but it does have a "Stable" category not used in our results.

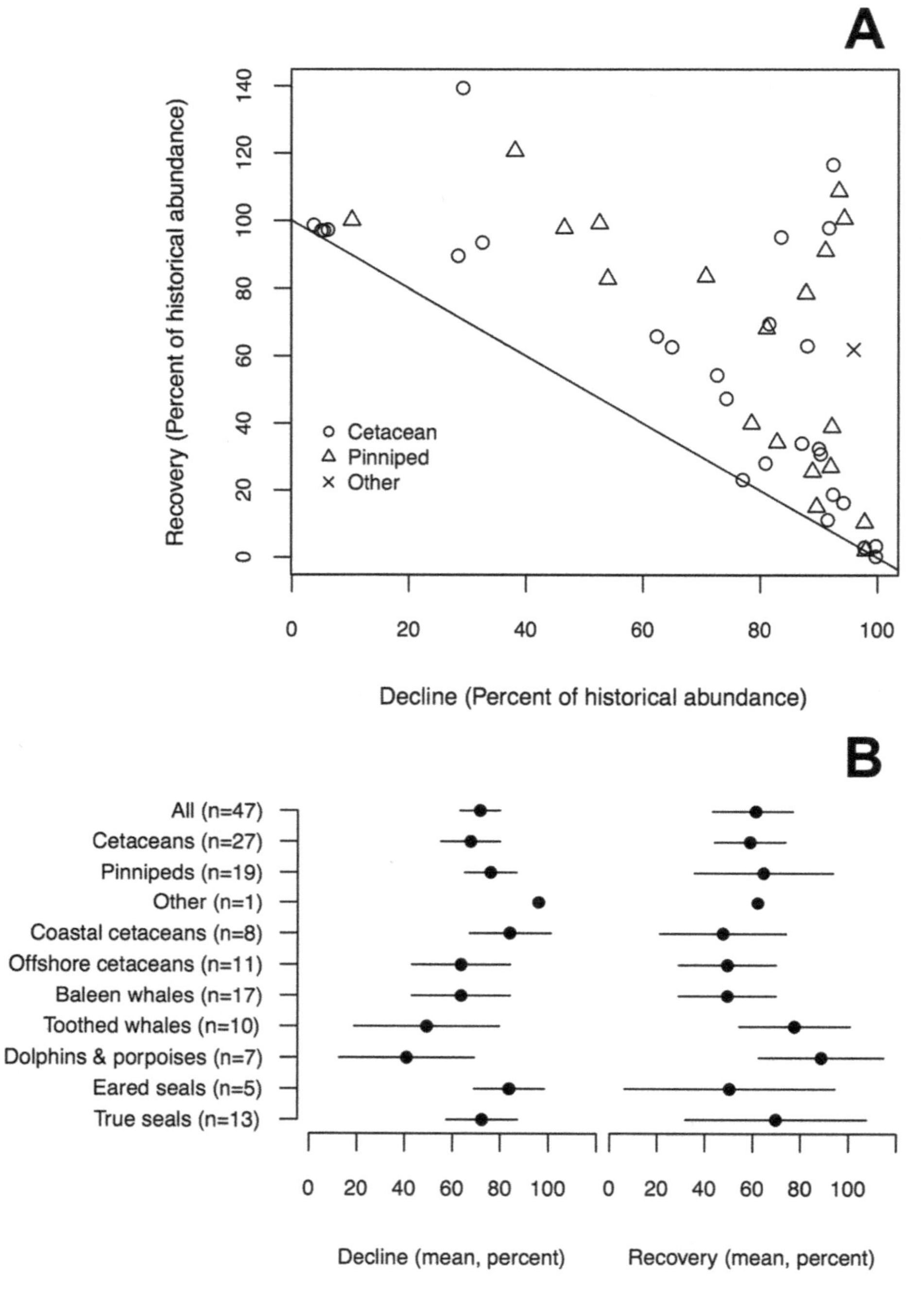


**Figure 5. Historical decline and recent recovery relative to historical level for 47 non-nested marine mammal populations.** Proportional decline and recovery in abundance for individual populations (A), and average decline and recovery across different categories of marine mammals (B). Note that in (A) that recovery must be equal to or greater than decline for a population (i.e. in the area above the diagonal line).

## Discussion

The aim of our study was to perform a quantitative assessment of recovery in marine mammal populations worldwide using the best publicly available abundance data. Overall, our results suggest that more marine mammal populations are recovering than not, especially among pinnipeds, other marine mammals (e.g. sirenians, polar bears and otters) and coastal cetaceans, but a large proportion (43%) of populations have non-significant or unknown trends mainly due to high variability or data scarcity. Our study also suggests that, in general, more historically depleted populations have so far shown either low or variable recovery. These results highlight the many recovery successes while also drawing attention to those populations needing more attention in terms of monitoring, management and conservation.

### Available Marine Mammal Population Data

Notwithstanding that population level abundance data are limited in many ways for marine mammals, we successfully compiled a substantial number of time series that represented 37% of marine mammal species worldwide. These time series provided insight into data availability and quality at the population level, as well as valuable information pertaining to recoveries. Any additional data could easily be included in our database for further investigations.

Populations with better abundance time series were found to share common characteristics. They typically had either present or past commercial or cultural value, as is the case with many large whales and pinnipeds, or they were iconic or charismatic species such as killer whales (*Orcinus orca*) or common bottlenose dolphins (*Tursiops truncatus*). Moreover, species with good data were generally easier to monitor because of some combination of factors that made them accessible and visible. These factors included aspects of their behavior, such as the regular use of haul-out and breeding areas by numerous pinniped species [46], consistent coastal migration routes as for the gray whale on the west coast of North America [56], or long times spent at the water surface, as with North Atlantic right whales (*Eubalaena glacialis*) [57]. Habitat or body size characteristics may have also contributed to better time series information: abundance data collection may be easier for animals in accessible coastal areas, with smaller, well-known ranges, with large body size (e.g. great whales), or with individually identifying markings (e.g. killer whales) [3,26,46].

Definite gaps in our knowledge of marine mammal abundance trends still exist, namely for beaked whales, river dolphins, sirenians and Antarctic true seal populations. Beaked whales are typically pelagic deep-divers that spend very little time at the surface where there can be detected. Moreover, they have large ranges and low densities [46]. Relative lack of commercial value and recognition in the public sphere may also have contributed to the lack of monitoring and data for these species, although short (typically <10 years), sparse, recent time series [46,47] or estimates [58] do exist for some populations. Time series data will likely improve due to interest in the susceptibility of beaked whales to acoustic disturbance, especially from seismic and naval sonar testing [46,59]. Many river dolphins are similarly cryptic, live at low densities, or do not gather in social groups. A lack of conservation and monitoring plans with standardized population and habitat assessment techniques has limited data availability for them [23]. Although intensely studied, West Indian or Florida manatees (*Trichechus manatus*) are difficult to observe and reliable abundance estimates are challenging to obtain [60]. The lack of management has also contributed to a lack of time series data for Amazonian (*T. inunguis*) and West African manatees (*T. senegalensis*) [61,62]. Despite intensive survey efforts in many areas, low densities and large ranges have inhibited reliable abundance estimates for dugongs (*Dugong dugon*) [63]. Many Antarctic true seal populations have relatively recent population estimates and are thought to have healthy populations, but they lack longer time series or historical estimates due to the inaccessibility of their remote breeding habitats until recently [24]. Abundance monitoring of Antarctic species may become more important for assessing and managing the effects of climate change and expanding Antarctic fisheries [64,65,66].

The time spans for which data were available varied between different groupings of marine mammals. Three generations of data were available for 52% of populations, while shorter time series were typically available for dolphins, porpoises, small whales, polar bears (*Ursus maritimus*), and sirenians. These data gaps are not surprising considering that many of these populations were not heavily commercially exploited, and thus abundance records likely only began recently with management and monitoring. Populations of great whales, Pacific dolphin species, northern true seals, belugas (*Delphinapterus leucas*) and narwhals (*Monodon monoceros*) made up the majority of the populations with historical abundance estimates. Historical population estimates for commercially valuable great whales typically came from either (1) catch data or product trade records, or (2) back-casting or genetic analysis techniques [6,67,68]. Numerous dolphin species found in Pacific U.S. waters were impacted by the tuna fishery that was established in the 1960 s [69,70], and they are a management concern under the U.S. Marine Mammal Protection Act. Certain northern marine mammal species (northern true seals, walrus, polar bear, beluga, narwhal, and bowhead whales) are still exploited for subsistence use by aboriginal groups, and in certain cases commercially or for sport hunts. They have been subject to study and management, although survey data collection can be logistically challenging (e.g. [9,40,71–74]). Commercial pinniped hunts exist in Canada, Greenland, Namibia, Norway and Russia for several true and fur seal populations [40,75]. With the lack of historical catch or abundance data for many populations, the development of genetic techniques [67,68] or habitat availability analyses [76] for estimating historic population size may provide more insight into pre-exploitation estimates for some populations in coming years.

We also found geographical biases in the data, which was typically from North America, Europe (especially northern Europe), and to a lesser extent Australia, New Zealand, Japan, South America and southern Africa. This bias is generally mimicked in global marine fish abundance datasets (e.g.[77–79]) and the distribution of worldwide line-transect cetacean survey effort [80]. A recent study by Kaschner et al. found that the Eastern Tropical Pacific has been the most surveyed region in the world via aerial or ship-based line-transects for cetaceans since the 1980 s [80]. The geographical biases in the data also reflect the availability of financial and logistical resources for monitoring and assessment in richer nations. With the increasing interest in the value of global biodiversity, monitoring may expand to other more data-poor areas. Population abundance monitoring is important for both conservation (e.g. recovery plans) and sustainable extractive management to assess trends and reference points for conservation goals and management targets [25,35,77].

## Marine Mammal Population Trends

Our analysis provides a general overview of recovery trends across marine mammal populations for which data was available. Amalgamating data from numerous different sources posed challenges, but we chose analytical techniques and recovery definitions accordingly. Using a robust regression over three generations allowed us to estimate in a statistically sound manner the dominant recent abundance trends for the largest number of marine mammal populations, as scaled to life history and comparable to IUCN methods [52]. In a few cases, historical estimates were included in the 3-generation regression timespan, but the robust regression generally downweighted high leverage points (i.e. abundances that are far from the mean). The analysis of historical decline and recent recovery, however, allowed us to incorporate populations with longer time spans and historical abundance estimates, but insufficient data over three generations, into the study. Our method for identifying population trends and those populations showing signs of increase or recovery – i.e. those showing a statistically significant increase in abundance, robustly estimated using scaled data – also worked well for the challenges we faced, including non-uniform data time-spans and different life histories. Only in a few cases with short and/or highly variable time series did this method not capture what by sight appeared to be the actual trend of the data. In other analyses, depending on the goals, more specific definitions of recovery may be appropriate [25]. In addition, the sample size of other marine mammal populations in this analysis was low (n = 7) and further investigation of this category is recommended as more population data becomes available, particularly for species other than sea otters.

Overall, 42% of populations were Significantly Increasing, 10% were Significantly Decreasing, 28% showed Non-Significant Change, and 20% were deemed Unknown. The somewhat large proportion of Non-Significant and Unknown trends points to the difficulty of studying abundance trends in these animals and estimating their population size with accuracy, stressing the need for better monitoring efforts for certain groups. Improvements in increasingly used techniques such as acoustic monitoring, tagging, photo-identification and mark-recapture, as well as computerized database and analysis technologies, modeling, and data sharing among organizations, including incorporating local community knowledge [81], may improve data quality and quantity.

Despite the data gaps, those populations with sufficient data showed some interesting patterns. Half of the pinniped populations were Significantly Increasing. Their fast life history characteristics (e.g. shorter generation times) may have helped promote recovery. Management and conservation efforts, in terms of limiting direct exploitation, bycatch and trade, as well as either the isolation or protection of important haul-out or breeding habitats, likely also contributed to recoveries in some populations [24].

Coastal cetaceans were also recovering relatively well, possibly because of their early exploitation and subsequent relatively early management, conservation, and attention in the public sphere. Large, coastal cetaceans were often the first species commercially hunted in an area because of their economic value and relatively easy access (e.g. bowhead [82], North Atlantic right [82], gray [56], and humpback whales [82]). International concern over steep declines in numbers by the 1900 s for most populations led to some of the first multilateral conservation agreements, protection from international trade, domestic exploitation bans or regulations, habitat protection, and recovery planning [3,25,26,83]. Numerous marine mammals have also become endearing symbols of the environmental movement [84].

Less visible or charismatic species, such as more predominantly offshore or smaller cetaceans, may have suffered in terms of population recovery for a few reasons. They were often exploited after the depletion of more easily accessible coastal species. Later onset of management or lack of directed management may have also been a contributing factor.

Some toothed cetaceans, such as sperm, pilot (*Globicephala* species) and killer whales, also have highly developed social structures that may be important to survival. As a result they may be more heavily impacted by the effects of selected removal and small population size [85,86]. Critical factors which may determine population recovery have been investigated, and appear to support the assertion that primarily coastal marine mammals had a higher probability of recovering than offshore populations, as did earlier maturing populations and pinnipeds (Anna M. Magera, MSc thesis, Dalhousie University, 2011).

## Comparison with IUCN Data

In order to verify our population-level trends, we compared our results with IUCN assessments for those species for which we had the majority of the species' abundance data at a population level. This included 27 species, approximately 21% of all marine mammal species worldwide. For these, we did have good agreement with the equivalent IUCN trend determination for the same species, which strengthens confidence in our results. However, we clearly had more Increasing, and fewer Decreasing and Unknown species (populations) compared to the global IUCN assessment of all 127 marine mammals (Figure 4). This is primarily attributable to the absence of data that met our criteria for many rare or difficult to monitor species, which resulted in them being excluded from the study (see Methods), combined with the large percentage (58%) of species classified as data-deficient by the IUCN.

## Historical Declines and Recent Increases

Many populations with available time series did not have pre-exploitation or K estimates of historical population size. However, we were able to obtain estimates of historical population size and thus compare the magnitude of historical declines and recent recoveries for 47 of the non-nested populations. As shown for marine fish populations [39,87], smaller historical population declines were generally associated with more successful recent recoveries. In turn, populations with very large historical declines (>90%) in fish [39,87] and many marine mammals (this study) have shown low magnitudes of recovery so far, with some notable exceptions. The large proportion of populations with >60% to >90% declines in this study highlights the substantial historical declines in many marine mammal populations [2].

We also found differences in declines and recoveries between different types of marine mammals, with the smaller, typically less commercially valuable toothed whales, and especially porpoises and dolphins showing the smallest historical declines and greatest recent recoveries. In turn, easily accessible coastal cetaceans and other marine mammals showed the largest historical declines. Sethi et al. (2010) note that in commercial fisheries development is typically driven by which taxa will produce the most profit at the minimal cost [88]. Similar explanations certainly exist for the history of whaling and sealing [7,13]. Recovery among marine mammal populations showed quite high variability. However, on average eared seals, baleen whales, as well as offshore and coastal cetaceans seemed to show the smallest recoveries, likely because of heavy historical exploitation (eared seals, baleen and coastal cetaceans) or lack of management (offshore cetaceans) [7,13,24].

Other studies have quantified historical declines and, in part, recoveries in marine mammals. Christensen (2006) estimated historical baselines for exploited cetaceans and pinnipeds and found a cumulative decline of 22% (range: 0–62%) in numbers between 1800 and 2001. The largest declines were in the great whales (64%, range: 40–79%) during periods of increased catches [11], which is comparable to the declines we estimated for the baleen whales. Christensen's overall marine mammal declines, however, presented a smaller decrease in terms of overall numbers than calculated in our study, likely because Christensen [11] relied mainly on the use of catch data-driven models and limited her assessment of cumulative marine mammal declines to 1800–2001. By this time some populations (e.g. North Atlantic right whales) had already been substantially depleted. A second recent study of historical baselines for large marine animal populations (not limited to a specific period) estimated a historical decline of approximately 96% for pinnipeds, otters and sirenians, and a recovery to approximately 25% of historical abundance [2]. Whales declined by approximately 82% and recovered to approximately 32% of historical levels. The pinniped, sea otter and sirenian category was not directly comparable to any of our categories, but for pinnipeds, our study presented smaller declines and larger recoveries. This may have been because the study included populations that did not exhibit any recovery, including extirpated populations, while our examination of n = 47 populations did not. The whale category only included great whale populations (mainly baleen) and was comparable to our baleen whale decline and recovery results [2].

Although there was relatively high variability in population recoveries after large historical declines in our study, many populations showed minimal population recovery following very high declines despite many decades of protection (e.g. right and bowhead whales since the 1930 s), thus possibly indicating Allee effects [89]. Only five populations out of 18 that underwent very large declines to <10% of the historical population abundance also showed very high (>90%) recoveries. This reflected a similar finding in certain fast-growing clupeid populations, which recovered to levels that were not observed in any other types of fish after declines of similar magnitude [39,89]. Possible explanations include a faster life history and relatively early age at maturity (4–5.5 years) that allowed for more rapid population increase as well as a longer time since exploitation was halted and better protection [26,90–96]. Other possible explanations might be that historical population estimates underestimated true pre-commercial exploitation population size, or ecosystem conditions changed in these areas to favor substantial population growth and larger abundances.

## Conclusion

Despite sparse data for many species and regions, lack of historical abundance estimates and often large error associated with available data, we were able to assess the recovery status for 182 marine mammal populations worldwide. Overall, 42% of populations for which we have good abundance time series are recovering from former depletions, especially pinnipeds, coastal cetaceans and other marine mammals (i.e. polar bears, otters and sirenians). Offshore cetaceans, all toothed whales, and just dolphins and porpoises showed relatively few recovering populations. On a species level, our results were comparable to assessments performed by IUCN, suggesting that our robust weighted regression over three generations is a useful and appropriate method for estimating general population trends. However, compared to all marine mammal species assessed by IUCN, our data over-represented recovering populations due to a lack of data for data-poor populations, most notably the sirenians, river dolphins and beaked whales. We also found that populations with smaller historical population declines were more likely to show stronger population recoveries in more recent times, while those that have been extensively depleted showed more variability in their recovery success, but tended to have smaller recoveries. This synthesis and compiled database are useful tools for other researchers interested in marine mammal population trends as well as decision makers to guide management and conservation efforts towards non-recovering populations. Moreover, our results stress the need for enhanced and increasingly innovative monitoring of offshore and cryptic marine mammals and those in low-latitude and developing nations that have not been extensively studied. Available and reliable abundance data are critical for better management and conservation, and will help to produce more complete and accurate trend estimates and recovery assessments in the future.

## Supporting Information

**Figure S1 Marine mammal population abundances over time and trends over three generations for robust log-linear (A) and robust linear (B) regressions.** Species and population areas are described in the upper left hand corner of each plot (n = 198 populations with duplicates, n = 182 populations without duplicate regular and pup count data). Population robust regression trend classification (long-linear (A) or linear (B)) is indicated in the upper right hand corner: I = Significantly Increasing, D = Significantly Decreasing, NS = Non-Significant Change, NA = Unknown. Solid lines = robust regression weighted by Abundance Confidence ID (ACID). Solid points = abundance data with quantitative error information (95% confidence interval bars). Empty points = abundance data with no stated quantitative error information. Black points = regular data that was collected from the entire population. Grey points = indicate pup count data.

**Table S1 Abundance data source information for study populations.** For each population, denoted by a numeric area code (Population Area ID) and area description, abundance data sources are listed along with the data collection and/or additional analysis methods used to obtain the population abundance estimates. Abundance Confidence ID (ACID) provides an uncertainty rank (1 to 6, 1 = lowest, 6 = highest). CPUE = catch per unit effort.

**Table S2 Overview of population time series: area descriptions, generation times, time spans, and data types.** All time references are in years. Time series refer to population abundance time series. List of abbreviations used in table: Species Type = general species taxon – cetacean, pinniped or other (sirenian, otter or polar bear). Regular Data Type = non-pup count data. Pup Count Data Type = only pup count data. N, S, E, W = North, South, East, West. Include Small/Large Area = was the population included if the smallest or largest non-nested populations were considered for the species? (1 = yes, 0 = no). Dominant Habitat = dominant habitat type. Sub-type = -sub-type of marine mammal within species type. Dolphin or Porpoise? = whether the population is a type of a dolphin or porpoise.

**Table S3 Abundance trends for 127 marine mammal species listed by the IUCN (2008).** Species Type = general species taxon – cetacean, pinniped or other (sirenian, otter or polar bear). Data? = whether or not population level data for the species is included in this study. IUCN Trend = species trend listed on the IUCN Red List of Threatened Species (2008). IUCN typically describes abundance trends on a species level, while this study examines population level trends. Note, IUCN does not have a "Non-Significant" category as in this study, but does have a "Stable" category not used in our results.

**Table S4 Historical population declines and recent recoveries with respect to historical level.** For populations (n = 47) with historical, minimum and recent abundance estimates. Decline and recovery percentages are mean values for populations where more than one historical estimate is present. In some cases, recent abundance values are higher than historical population estimates, leading to recovery values of greater than 100%.

**Table S5 Results from scaled robust log-linear and robust linear regressions for marine mammal populations (n = 198).** The regressions were weighted by Abundance Confidence ID (ACID) over three generation times. Duplicate pup count/regular populations included in blue text. These will not be used in further analyses, resulting in 182 remaining populations (out of 198). Species Type = general species taxon (cetacean, pinniped or other (sirenian, otter or polar bear)). N, S, E, W = North, South, East, West. Include Small/Large Area = was the population included if the smallest or largest non-nested populations were considered for the species. Regression & Data Type = *log* lmRob or lmRob and Regular (i.e. non-pup count data) or Pup Count (i.e. only pup count data). Coefficient or slope = the scaled rate of population growth (i.e. change in abundance over time). SE = standard error. CI95 = 95% confidence interval. tval = t-value. pval = p-value. Increasing_Sig = significantly increasing abundance trend. Increasing_NS = non-significant increasing abundance trend. Decreasing_Sig = significantly decreasing abundance trend. Decreasing_NS = non-significant decreasing abundance trend. NA indicates a lack of sufficient data over the three-generation time period necessary to perform a robust regression analysis. Dominant Habitat = dominant habitat type. Sub-type = sub-type of marine mammal within species type. Dolphin or Porpoise? = whether the population is a type of a dolphin or porpoise.

## Acknowledgments

We would like to thank W. Blanchard and S. Anderson for statistical and graphing advice, and J. Hutchings, H. Whitehead, T. Gerrodette, T. Branch and an anonymous reviewer for valuable comments and discussions. We would also like to thank all the researchers, collaborators and research assistants who contributed to the abundance database, in particular J. Beblow and K. Willis Ladel.

## Author Contributions

Conceived and designed the experiments: AMM JMF HKL. Analyzed the data: AMM JMF HKL. Contributed reagents/materials/analysis tools: AMM KK LBC. Wrote the paper: AMM JMF KK LBC HKL.

## References

1. Lotze HK, Milewski I (2004) Two centuries of multiple human impacts and successive changes in a North Atlantic food web. Ecological Applications 14: 1428–1447.
2. Lotze HK, Worm B (2009) Historical baselines for large marine animals. Trends in Ecology & Evolution 24: 254–262.
3. Aron W (1988) The commons revisited - thoughts on marine mammal management. Coastal Management 16: 99–110.
4. Adams WM (2004) Against extinction - the story of conservation. London: Earthscan. 311 p.
5. Jefferson TA, Webber MA, Pitman RL (2008) Marine mammals of the world: A comprehensive guide to their identification. LondonUK: Elsevier. 573 p.
6. Baker CS, Clapham PJ (2004) Modelling the past and future of whales and whaling. Trends in Ecology & Evolution 19: 365–371.
7. Lotze HK, Lenihan HS, Bourque BJ, Bradbury RH, Cooke RG, et al. (2006) Depletion, degradation, and recovery potential of estuaries and coastal seas. Science 312: 1806–1809.
8. Colten RH (2002) Prehistoric marine mammal hunting in context: Two western North American examples. International Journal of Osteoarchaeology 12: 12–22.
9. Reeves RR (2002) The origins and character of 'aboriginal subsistence' whaling: a global review. Mammal Review 32: 71–106.
10. Smith I (2005) Retreat and resilience: fur seals and human settlement in New Zealand. In: Monks G, editor. The Exploitation and Cultural Importance of Sea Mammals: Oxbow Books. 6–18.
11. Christensen LB (2006) Marine mammal populations: Reconstructing historical abundance at the global scale. Vancouver: University of British Columbia Fisheries Centre. 167 p.
12. Pauly D, Watson R, Alder J (2005) Global trends in world fisheries: impacts on marine ecosystems and food security. Philosophical Transactions of the Royal Society B: Biological Sciences 360: 5–12.
13. Reeves RR, Smith TD (2006) A Taxonomy of world whaling operations and eras. In: Estes JA, DeMaster DP, Doak DF, Williams TM, BrownellJr RL, editors. Whales, whaling, and ocean ecosystems. Los Angeles and London: University of California Press. 82–101.
14. Lotze HK (2007) Rise and fall of fishing and marine resource use in the Wadden Sea, southern North Sea. Fisheries Research 87: 208–218.
15. Gerber LR, DeMaster DP, Roberts SP (2000) Measuring success in conservation. American Scientist 88: 316–324.
16. Angliss RP, Allen BM (2009) Alaska marine mammal stock assessments, 2008. U.S. Department of Commerce, National Oceanic and Atmospheric Administration, National Marine Fisheries Service, Alaska Fisheries Science Center. 269 p.
17. Sea Otter Recovery Team (2007) Recovery strategy for the sea otter (*Enhydra lutris*) in Canada. Vancouver, B. C.: Fisheries and Oceans Canada. vii +56 p.
18. Read AJ (2008) The looming crisis: interactions between marine mammals and fisheries. Journal of Mammalogy 89: 541–548.
19. Bigg MA (1988) Status of the Steller Sea Lion, *Eumetopias jubatus*, in Canada. Canadian Field Naturalist 102: 315–336.
20. Reijnders PJH (1994) Historical population size of the harbour seal, *Phoca vitulina*, in the Delta area, SW Netherlands. Hydrobiologia: 557–560.
21. Kasuya T (2007) Japanese whaling and other cetacean fisheries. Environmental Science & Pollution Research 14: 39–48.
22. Ward EJ, Holmes EE, Balcomb KC (2009) Quantifying the effects of prey abundance on killer whale reproduction. Journal of Applied Ecology 46: 632–640.
23. Reeves RR, Smith BD, Crespo EA, Notarbartolo di Sciara G (2003) Dolphins, whales and porpoises: 2002–2010 conservation action plan for the world's cetaceans Gland, Switzerland and Cambridge, UK: IUCN. 147 p.
24. Reijnders P, Brasseur S, vanderToorn J, vanderWolf P, Boyd I, et al. (1993) Seals, fur seals, sea lions and walrus. GlandSwitzerland: IUCN. 98 p.
25. Lotze HK, Coll M, Magera AM, Ward-Paige C, Airoldi L (2011) Recovery of marine animal populations and ecosystems. Trends in Ecology & Evolution 26: 595–605.
26. Gerber LR (1998) Seeking a rational approach to setting conservation priorities for marine mammals. Integrative Biology 1: 90–98.
27. Gerber LR, Hilborn R (2001) Catastrophic events and recovery from low densities in populations of otariids: implications for risk of extinction. Mammal Review 31: 131–150.
28. Best PB (1993) Increase rates in severely depleted stocks of baleen whales. ICES Journal of Marine Science 50: 169–186.
29. Nichol LM, Watson JC, Ellis GM, Ford JKB (2005) An assessment of abundance and growth of the sea otter population (*Enhydra lutris*) in British Columbia. NanaimoBC: Fisheries and Oceans Canada. 22 p.
30. Thomas L, Harwood J (2008) Estimating the size of the UK grey seal population between 1984 and 2007. St. AndrewsUK: NERC Sea Mammal Research Unit and Centre for Research into Ecological and Environmental Modelling, University of St Andrews. 18 p.
31. Thomas L, Hammill MO, Bowen WD (2007) Estimated size of the Northwest Atlantic grey seal population 1977–2007. In: Fisheries and Oceans Canada, editor: Canadian Science Advisory Secretariat. 2007/082. 31 p.
32. Schipper J, Chanson JS, Chiozza F, Cox NA, Hoffmann M, et al. (2008) The status of the world's land and marine mammals: Diversity, threat, and knowledge. Science 322: 225–230.
33. Kappel CV (2005) Losing pieces of the puzzle: threats to marine, estuarine, and diadromous species. Frontiers in Ecology and the Environment 3: 275–282.
34. Davidson AD, Boyer AG, Kim H, Pompa-Mansilla S, Hamilton MJ, et al. (2012) Drivers and hotspots of extinction risk in marine mammals. Proceedings of the National Academy of Sciences of the United States of America 109: 3395–3400.
35. Hutchings JA, Butchart SHM, Collen B, Schwartz MK, Waples RS (2012) Red flags: correlates of impaired species recovery. Trends in Ecology & Evolution 27: 542–546.
36. Oxford Dictionaries (2010) Recovery. Oxford Dictionaries: Oxford University Press.
37. Gerrodette T, Demaster DP (1990) Quantitative determination of optimum sustainable population level. Marine Mammal Science 6: 1–16.
38. Freeman MMR (2008) Challenges in assessing cetacean population recovery and conservation status. Endangered Species Research 6: 173–184.
39. Hutchings JA (2000) Collapse and recovery of marine fishes. Nature 406: 882–885.
40. Fisheries and Oceans Canada (2011) 2011–2015 Integrated Fisheries Management Plan for Atlantic Seals. 34 p.
41. Sanderson EW (2006) How many animals do we want to save? The many ways of setting population target levels for conservation. Bioscience 56: 911–922.
42. IUCN (2001) IUCN Red List categories and criteria: Version 3.1. GlandSwitzer-Switzerland and Cambridge, UK: IUCN Species Survival Commission. 30 p.
43. Kaschner K, Watson R, Trites AW, Pauly D (2006) Mapping world-wide distributions of marine mammal species using a relative environmental suitability (RES) model. Marine Ecology Progress Series 316: 285–310.
44. Whitehead H (2002) Estimates of the current global population size and historical trajectory for sperm whales. Marine Ecology Progress Series 242: 295–304.

45. Amstrup SC, Laake JL, Manly BFJ, MacDonald LL, Robertson DG (1999) Preface. In: Garner GW, Amstrup SC, Laake JL, Manly BFJ, Macdonald LL et al., editors. Marine mammals survey and assessment methods. Rotterdam: AA Balkema. pp. XI–XII.
46. Taylor BL, Martinez M, Gerrodette T, Barlow J, Hrovat YN (2007) Lessons from monitoring trends in abundance of marine mammals. Marine Mammal Science 23: 157–175.
47. Jewell R, Thomas L, Harris CM, Kaschner K, Wiff R, et al. (2012) Global analysis of cetacean line-transect surveys: detecting trends in cetacean density. Marine Ecology Progress Series 453: 227–240.
48. Kaschner K (2004) Modelling and Mapping Resource Overlap between Marine Mammals and Fisheries on a Global Scale. Vancouver: University of British Columbia. 241 p.
49. Gotelli NJ (1995) A Primer in Ecology. SunderlandMassachusetts: Sinauer Associates, Inc. 206 p.
50. Berta A, Sumich JL, Kovacs KM (2006) Population structure and dynamics. Marine Mammals - Evolutionary Biology. Second Ed. London: Academic Press. 416–455.
51. Forney KA, Wade PR (2006) Worldwide distribution and abundance of killer whales. In: Estes JA, Demaster DP, Doak DF, Williams TM, BrownellJr RL, editors. whales, whaling, and ocean ecosystems. Los Angeles and London: University of California Press. 145–162.
52. IUCN Standards and Petitions Subcommittee (2010) Guidelines for using the IUCN Red List categories and criteria. Version 8.0. IUCN. 85 p.
53. Wright DB, London K (2009) Modern regression techniques using R. London: Sage. viii +204 p.
54. Wilkinson L, Task Force on Statistical Inference (1999) Statistical methods in psychology journals - guidelines and explanations. American Psychologist 54: 594–604.
55. Wang J, Zamar R, Marazzi A, Yohai V, Salibian-Barrera M, et al. (2009) Package 'robust'. CRAN.
56. Swartz SL, Taylor BL, Rugh DJ (2006) Gray whale *Eschrichtius robustus* population and stock identity. Mammal Review 1: 66–84.
57. NMFS (2005) Recovery plan for the North Atlantic right whale (*Eubalaena glacialis*). Silver SpringMD: National Marine Fisheries Service. 137 p.
58. Whitehead H, Hooker SK (2012) Uncertain status of the northern bottlenose whale Hyperoodon ampullatus: population fragmentation, legacy of whaling and current threats. Endangered Species Research 19: 47–61.
59. Jepson PD, Arbelo M, Deaville R, Patterson IAP, Castro P, et al. (2003) Gas-bubble lesions in stranded cetaceans - Was sonar responsible for a spate of whale deaths after an Atlantic military exercise? Nature 425: 575–576.
60. Lefebvre LW, Reynolds III JE, Ragen TJ, Langtimm CA, Valade JA (2009) Manatee population ecology and management workshop proceedings. In: Lefebvre LW, Reynolds III JE, Ragen TJ, Langtimm CA, Valade JA, editors; April 2002; Gainsville, Florida. U.S. Geological Survey.
61. Reeves RR, Leatherwood S, Jefferson TA, Curry BE, Henningsen T (1996) Amazonian manatees, *Trichechus inunguis*, in Peru: Distribution, exploitation, and conservation status. Interciencia 21: 246–254.
62. Silva MA, Araujo A (2001) Distribution and current status of the West African manatee (*Trichechus senegalensis*) in Guinea-Bissau. Marine Mammal Science 17: 418–424.
63. Marsh H, Penrose H, Eros C, Hughes J (2002) Dugong: status report and action plans for countries and territories. UNEP. 172 p.
64. Southwell C, Paxton CGM, Borchers D, Boveng P, De La Mare W (2008) Taking account of dependent species in management of the Southern Ocean krill fishery: estimating crabeater seal abundance off east Antarctica. Journal of Applied Ecology 45: 622–631.
65. Lake S, Burton H, Barker R, Hindell M (2008) Annual reproductive rates of Weddell seals in eastern Antarctica from 1973 to 2000. Marine Ecology Progress Series 366: 259–270.
66. Forcada J, Trathan P, Boveng PL, Boyd IL, Burns JM, et al. (2012) Responses of Antarctic pack-ice seals to environmental change and increasing krill fishing. Biological Conservation 149: 40–50.
67. Roman J, Palumbi SR (2003) Whales before whaling in the North Atlantic. Science 301: 508–510.
68. Alter SE, Rynes E, Palumbi SR (2007) DNA evidence for historic population size and past ecosystem impacts of gray whales. Proceedings of the National Academy of Sciences of the United States of America 104: 15162–15167.
69. Gerrodette T, Watters GM, Perryman W, Ballance L (2008) Estimates of 2006 Dolphin abundance in the Eastern Tropical Pacific, with revised estimates from 1986–2003. U.S. Department of Commerce, National Oceanic and Atmospheric Administration, National Marine Fisheries Service. 43 p.
70. Wade PR, Watters GM, Gerrodette T, Reilly SB (2007) Depletion of spotted and spinner dolphins in the eastern tropical Pacific: modeling hypotheses for their lack of recovery. Marine Ecology Progress Series 343: 1–14.
71. COSEWIC (2004) COSEWIC Assessment and update status report on the beluga whale *Delphinapterus leucas* in Canada. Ottawa: Committee on the Status of Endangered Wildlife in Canada. ix +71 p.
72. COSEWIC (2005) COSEWIC Assessment and update status report on the narwhal *Monodon monoceros* in Canada. Ottawa: Committee on the Status of Endangered Wildlife in Canada. vii +50 p.
73. COSEWIC (2008) COSEWIC Assessment and update status report on the polar bear *Ursus maritimus* in Canada. Ottawa: Committee on the Status of Endangered Wildlife in Canada. vii +75 p.
74. COSEWIC (2006) COSEWIC Assessment and update status report on the Atlantic walrus *Odobenus rosmarus rosmarus* in Canada. Ottawa: Committee on the Status of Endangered Wildlife in Canada. viii +65 p.
75. Kirkman SP, Lavigne DM (2010) Assessing the hunting practices of Namibia's commercial seal hunt South African Journal of Science 106: 1–3.
76. Pyenson ND, Lindberg (2011) What happened to gray whales during the Pleistocene? The ecological impact of sea-level change on benthic feeding areas in the North Pacific Ocean. PLoS One 6: e21295.
77. Worm B, Hilborn R, Baum JK, Branch TA, Collie JS, et al. (2009) Rebuilding global fisheries. Science 325: 578–585.
78. Hutchings JA, Minto C, Ricard D, Baum JK, Jensen OP (2010) Trends in the abundance of marine fishes. Canadian Journal of Fisheries and Aquatic Sciences 67: 1205–1210.
79. Reynolds JD, Dulvy NK, Goodwin NB, Hutchings JA (2005) Biology of extinction risk in marine fishes. Proceedings of the Royal Society B: Biological Sciences 272: 2337–2344.
80. Kaschner K, Jewell R, Williams R, Harris CM (2012) Global coverage of cetacean line-transect surveys: status quo, gaps and future challenges. PLoS One 7: e44075.
81. Dale A, Armitage D (2011) Marine mammal co-management in Canada's Arctic: Knowledge co-production for learning and adaptive capacity. Marine Policy 35: 440–449.
82. Clapham P, Young SB, Brownell RL (1999) Baleen whales: conservation issues and the status of the most endangered populations. Mammal Reviews 29: 35–60.
83. Bhargava M (2005) Of otters and orcas: marine mammals and legal regimes in the North Pacific. Ecology Law Quarterly 32: 939–988.
84. Lavigne DM, Scheffer VB, Kellert SR (1999) The Evolution of North American attitudes toward marine mammals. In: J. R. Twiss Jr., Reeves RR, editors. Conservation and management of marine mammals. Washington and London: Smithsonian Institution Press. 10–47.
85. Whitehead H, Rendell L, Osborne RW, Wursig B (2004) Culture and conservation of non-humans with reference to whales and dolphins: review and new directions. Biological Conservation 120: 427–437.
86. Wade PR, Reeves RR (unpublished) Social and behavioural factors in cetacean responses to over-exploitation: Are odontocetes less 'resilient' than mysticetes?
87. Hutchings JA (2001) Influence of population decline, fishing, and spawner variability on the recovery of marine fishes. Journal of Fish Biology 59: 306–322.
88. Sethi SA, Branch TA, Watson R (2010) Global fishery development patterns are driven by profit but not trophic level. Proceedings of the National Academy of Sciences of the United States of America 107: 12163–12167.
89. Hutchings JA, Reynolds JD (2004) Marine fish population collapses: Consequences for recovery and extinction risk. Bioscience 54: 297–309.
90. Boyd IL (2002) Antarctic marine mammals. In: Perrin WF, Wursig B, Thewissen JGM, editors. Encyclopedia of Marine Mammals. San Diego: Academic Press.
91. Hoezel AR, editor (2002) Marine mammal biology: an evolutionary approach. Oxford: Blackwell Publishing.
92. Whitehead H, Mann J (2000) Female reproductive strategies of cetaceans: Life histories and calf care. In: Mann J, Connor RC, Tyack PL, Whitehead H, editors. Cetacean Societies. Chicago and London: University of Chicago Press. 219–246.
93. Brown RF, Wright BE, Riemer SD, Laake J (2005) Trends in abundance and current status of harbor seals in Oregon: 1977–2003. Marine Mammal Science 21: 657–670.
94. Perry SL, DeMaster DP, Silber GK (1999) Special issue: The great whales: History and status of six species listed as endangered under the U.S. Endangered Species Act of 1973. Marine Fisheries Review 61: 1–74.
95. David J, van Sittert L (2008) A reconstruction of the Cape (South African) fur seal harvest 1653–1899 and a comparison with the 20th-century harvest. South African Journal of Science 104: 107–110.
96. Jeffries S, Huber H, Calambokidis J, Laake J (2003) Trends and status of harbor seals in Washington state: 1978–1999. Journal of Wildlife Management 67: 207–218.

# 18

# Patterns of Dolphin Bycatch in a North-Western Australian Trawl Fishery

**Simon J. Allen[1]*, Julian A. Tyne[1], Halina T. Kobryn[1], Lars Bejder[1], Kenneth H. Pollock[1,2], Neil R. Loneragan[1]**

**1** Murdoch University Cetacean Research Unit, Centre for Fish, Fisheries and Aquatic Ecosystems Research, School of Veterinary and Life Sciences, Murdoch University, Murdoch, Western Australia, Australia, **2** Department of Biology, North Carolina State University, Raleigh, North Carolina, United States of America

## Abstract

The bycatch of small cetaceans in commercial fisheries is a global wildlife management problem. We used data from skippers' logbooks and independent observers to assess common bottlenose dolphin (*Tursiops truncatus*) bycatch patterns between 2003 and 2009 in the Pilbara Trawl Fishery, Western Australia. Both datasets indicated that dolphins were caught in all fishery areas, across all depths and throughout the year. Over the entire datasets, observer reported bycatch rates (n = 52 dolphins in 4,124 trawls, or 12.6 dolphins/1,000 trawls) were ca. double those reported by skippers (n = 180 dolphins in 27,904 trawls, or 6.5 dolphins/1,000 trawls). Generalised Linear Models based on observer data, which better explained the variation in dolphin bycatch, indicated that the most significant predictors of dolphin catch were: (1) vessel - one trawl vessel caught significantly more dolphins than three others assessed; (2) time of day – the lowest dolphin bycatch rates were between 00:00 and 05:59; and (3) whether nets included bycatch reduction devices (BRDs) - the rate was reduced by ca. 45%, from 18.8 to 10.3 dolphins/1,000 trawls, after their introduction. These results indicated that differences among vessels (or skippers' trawling techniques) and dolphin behavior (a diurnal pattern) influenced the rates of dolphin capture; and that spatial or seasonal adjustments to trawling effort would be unlikely to significantly reduce dolphin bycatch. Recent skipper's logbook data show that dolphin bycatch rates have not declined since those reported in 2006, when BRDs were introduced across the fishery. Modified BRDs, with top-opening escape hatches from which dolphins might escape to the surface, may be a more effective means of further reducing dolphin bycatch. The vulnerability of this dolphin population to trawling-related mortality cannot be assessed in the absence of an ongoing observer program and without information on trawler-associated dolphin community size, broader dolphin population size and connectivity with adjacent populations.

**Editor:** Konstantinos I. Stergiou, Aristotle University of Thessaloniki, Greece

**Funding:** The Australian Government's Fisheries Research and Development Corporation (http://frdc.com.au/Pages/home.aspx), the Western Australian Government's Development of Better Industry Fund, and the Nickol Bay Professional Fishers Association Inc. funded this research through the FRDC's Tactical Research Fund (Project No. 2008/048). The funders had no role in study design, data collection and analysis, decision to publish, or preparation of the manuscript.

**Competing Interests:** The Australian Government's Fisheries Research and Development Corporation, the Department of Fisheries Western Australia and the Nickol Bay Professional Fishers Association provided funding for this study. There are no patents, products in development or marketed products to declare. This does not alter our adherence to all the PLOS ONE policies on sharing data and materials.

* E-mail: S.Allen@murdoch.edu.au

## Introduction

Demersal trawl fishing for crustaceans, cephalopods and fish impacts benthic habitats and results in large quantities of incidental catch, or bycatch, of non-targeted species [1–3]. Trawling, gill netting and purse seining are the three largest causes of fisheries-related small cetacean mortalities worldwide [4–7]. Entanglements in fishing gear and large-scale habitat modification have resulted in the extinction of the Yangtze river dolphin (*Lipotes vexillifer*), representing the first loss of a cetacean species directly attributable to human influences [8]. Several other populations and, indeed, species of small cetaceans, such as the Maui's dolphin (*Cephalorhynchus hectori maui*) of New Zealand's North Island and the vaquita (*Phocoena sinus*) of the Sea of Cortez, are at risk of extinction from the cumulative impacts of fishing related mortality and disturbance from gill netting and trawl fisheries [9–12].

As a result of suspected and/or measured declines in dolphin populations due to fisheries bycatch, public concerns and pressure from non-government organizations, major changes in fisheries policy and practice have been implemented in several regions. After the introduction of the *U.S. Marine Mammal Protection Act* (*MMPA*) in 1972, for example, high observer coverage and a variety of bycatch mitigation measures were implemented to quantify and reduce the bycatch of two dolphin species (spotted, *Stenella attenuata*, and spinner dolphins, *S. longirostris*) in the purse seine fishery for tuna in the eastern tropical Pacific [13][14]. Although massive reductions in dolphin capture rates were achieved, the impacted populations have not recovered [15][16].

New Zealand's Department of Conservation administers their *MMPA* of 1978. Numerous protected areas have been established, with time- and area-based restrictions placed on fishing activities that present high entanglement risks to marine mammals. One such protected area was established off the west coast of the North Island to reduce entanglements of the critically endangered Maui's dolphin. However, concerns remain over the efficacy of these measures, as gill netting and trawling are still allowed in certain areas [17]. Surveys of the distribution of the endangered Hector's dolphins (*C. hectori*) off the South Island suggest that restrictions on commercial gill netting protect 60% or less of the dolphin population for three months of the year [18]. New Zealand's endemic dolphin populations are predicted to continue declining under current management, driven primarily by ongoing bycatch in gill net and trawl fisheries [12].

All marine mammals in Australian waters are protected under the *Environmental Protection and Biodiversity Conservation Act* 1999. Fishers are required by legislation to report fatal and non-fatal entanglements of marine mammals to State and Commonwealth fisheries management agencies. The greatest threats to small cetaceans in Australian waters are also associated with gill netting, purse seining operations and trawl fisheries [19]. Here, the species most often affected by fishing mortality include: bottlenose (*Tursiops* spp.), common (*Delphinus delphis*), Australian snubfin (*Orcaella heinsohni*), humpback (*Sousa chinensis*) and spinner dolphins [20][21]. Thousands of dolphins have died in commercial fishing operations over the past three decades [21], the impacts of which are impossible to quantify without baseline data on the abundance and distribution of dolphin populations across the vast majority of Australian waters [22]. The tropical waters of north-western Australia are no exception, where numerous dolphin populations are exposed to commercial fishing, as well as large-scale habitat modification through the proliferation of the oil, gas and mining industries. No population estimates exist for any species in this region [23].

Trawl fisheries operate in many regions of Australian waters [24]. The North West Shelf region of Western Australia (WA) has been trawled since the early 1970s, with the Taiwanese pair-trawl fishery catching in excess of 100,000 tons of fish, cephalopods and other invertebrates in the mid-1970s [25]. Catches declined to less than 10,000 tons per annum by the mid-1980s, when Chinese and Korean trawlers also fished the area and an experimental management regime was introduced [26]. Shortly after this new regimen commenced, the foreign fleet diminished and a domestic fishery developed [25]. Since the early 1990s, catches in the Pilbara Trawl Fishery (PTF) have fluctuated between 2,000 and 3,500 tons per annum, dropping to <1,400 in the last five years, associated with a decline in trawl effort [27].

The bycatch of a number of protected species (dolphins, sea snakes, turtles and sawfish) in the PTF was first documented in 2002 [28]. A variety of bycatch mitigation techniques were pursued, focussed primarily on reducing dolphin bycatch [27][29]. The efficacy of acoustic deterrents, or 'pingers', for reducing dolphin interactions with fisheries has continued to be evaluated in this and other regions, with inconsistent results depending on the dolphin species involved, type of fishery and the type and number of pingers deployed [30]. Pingers proved ineffective in deterring bottlenose dolphins from interacting with trawl gear in the PTF [31]. Field trials of bycatch reduction devices (BRDs) resulted in a reduction in the number of dolphins landed on deck and they were made compulsory across the fishery in March 2006 [29].

The dolphins subject to bycatch in the PTF are common bottlenose dolphins (*T. truncatus*, 'bottlenose dolphins' hereafter) [32], a globally widespread species, occurring in tropical and temperate latitudes [33]. Bottlenose dolphins are thought to be a widely distributed in Australian pelagic waters [21][34], mixing with and/or being replaced by Indo-Pacific bottlenose dolphins (*T. aduncus*) in shallow, coastal areas, including those of north-western Australia [23][32]. In the *Action Plan for Australian Cetaceans*, bottlenose dolphins are listed as 'no category assigned because of insufficient information' [20] and very little is known of the populations off north-western Australia or, indeed, any pelagic population around Australia. Due to this broad lack of even baseline data, assessments of the status of individual bottlenose dolphin stocks, or populations, are not yet possible.

Previous studies in the PTF by the Department of Fisheries WA have been based primarily on trialling the efficacy of pingers [31] and various BRDs [29] in reducing dolphin bycatch. While some aspects of the geographical and temporal nature of incidental dolphin captures were evaluated, these assessments were based on data collected during ca. six- and 18-month trials between 2004 and 2006 [29][31]. Here, we used six years of data from skippers' logbooks and independent observer records collected from August 2003 until September 2009 to build upon this earlier research and investigate the spatial and temporal patterns of dolphin bycatch across the PTF. We aimed, firstly, to assess the spatial, daily and seasonal data on fishing effort and dolphin bycatch, and secondly, to evaluate the effectiveness of different net designs (those with and without BRDs) in reducing dolphin bycatch.

## Materials and Methods

### Characteristics of the Pilbara Trawl Fishery

The PTF, within the broader Pilbara Demersal Scalefish Fishery (which also includes trap and line fisheries), is bound by longitudes of 116° to the west and 120° to the east, and by an approximation of the 50 m depth contour to landward and the 100 m depth contour to seaward (Fig. 1). Since being gazetted in 1998, four Management Areas have been open to trawling, comprising a total fishing area of ca. 23,000 $km^2$ (6,900 $nm^2$). The equivalent of 4.3 full-time vessels operated year-round in the PTF between 2003 and 2009, with slightly reduced effort from December to March when cyclones are more frequent. The trawlers generally stay at sea for five to 12 days at a time, fishing throughout the day and night. Individual trawls ranged in duration from 30 min to five h, with a median trawl time of ca. 2.7 h. The total trawling effort ranged from 4,500 to 6,000 trawls (ca. 11,000 to 16,000 h) per annum (Table 1) between 2003 and 2009, though it has averaged ca. 9,000 h per annum from 2010 to 2012 [27].

Data for this fishery are reported in three different 12-month intervals: The annual State of the Fisheries reports by the Department of Fisheries summarise data from January to December (calendar years); skippers' logbook data are summarised from July until the following June (Australian financial years) by industry for the Australian Taxation Office; and observer records for the fishery were summarised from October to September. In this study, we have presented data from the Department of Fisheries by calendar years for ease of comparison, but based our analyses of logbook and observer data on the industry format of the financial year (see Table 2).

Trawl vessels in the PTF tow a single net at a speed of ca. three to three and a half knots (5.6–6.5 km/h), with twin otter boards maintaining the net spread (see also [35]). Most nets in the PTF consist of four main sections: the wings, which form the opening or mouth of the net; the belly and neck, which are immediately behind the mouth of the net and where the net tapers; the extension, a tubular section; and the codend, where the catch is collected (Fig. 2). The diameter and mesh size decrease in each panel with distance from the opening of the net. The length of the head rope must not exceed 36.6 m, while the total length of the net, including cables, sweeps and bridles, is limited to 274.3 m. The footrope is weighted and contains bobbins (<35 cm in diameter) that are spaced about 30 cm apart and roll along the sea floor.

Nets used in the PTF consist of diamond mesh. The first section of the net belly measures 4.8 m in length when the net is stretched.

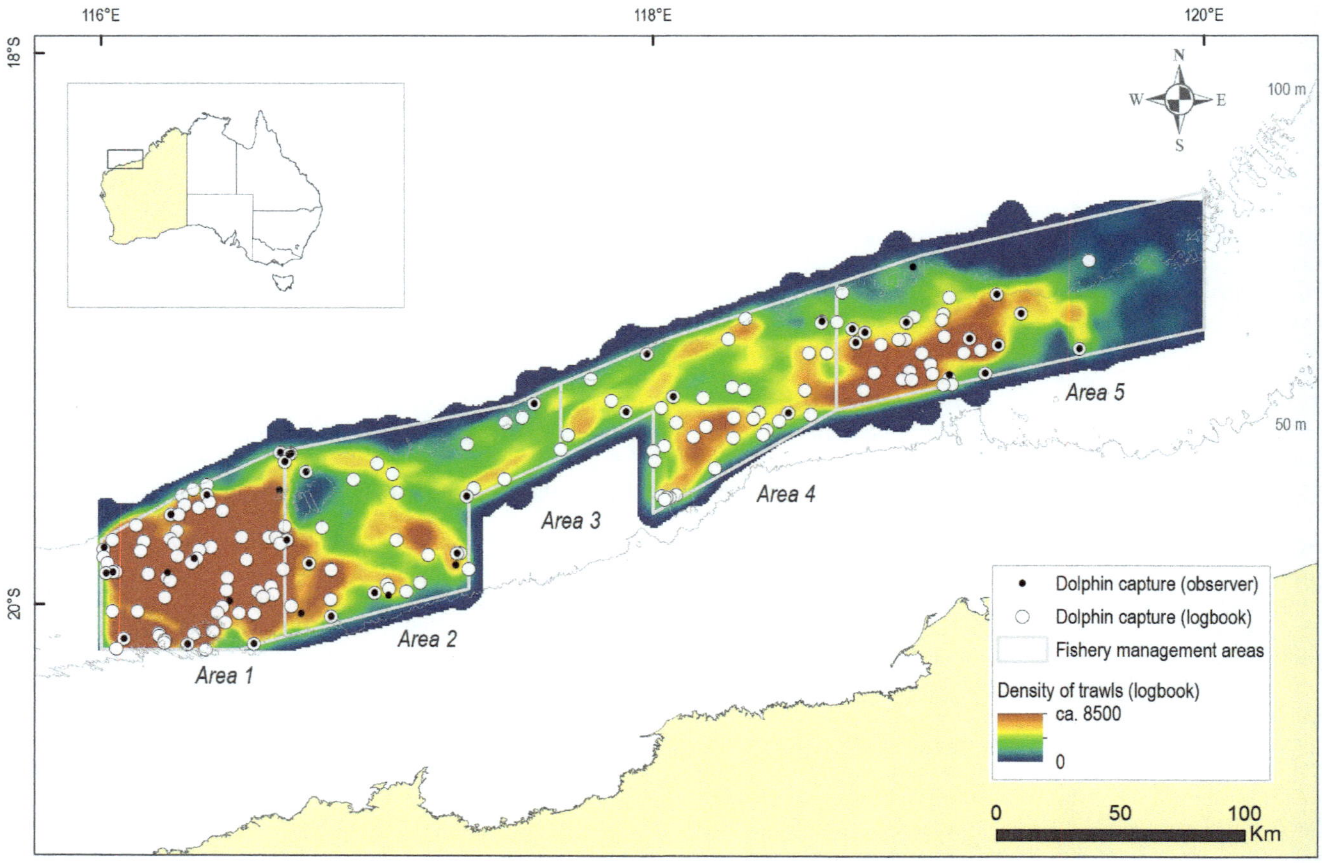


**Figure 1. The Pilbara Trawl Fishery off the north-western coast of Australia.** Figure includes: a) the 50 and 100 m depth contours; b) the boundaries of the fishery management areas (Areas 1, 2, 4 and 5; Area 3 in the central southern region of the fishery is closed to trawling); c) the spatial density of fishing effort based on logbook trawls; and d) the location of dolphin bycatch events reported in skippers' logbooks and by independent observers (August 2003 to September 2009).

In October 2008, the belly and neck sections of the nets were shortened to allow for a shorter escape route for dolphins that enter the net and interact with the exclusion grid (Fig. 2). Based on stretch mesh measurements, the nets are approximately 44 m long from the footrope to the start of the codend and, when trawling/ fishing, they are likely to be about 60–70% of this length.

**Table 1.** Total numbers of trawl days, hours and individual trawls in the PTF.

| Year | Trawl days | Trawl hours | Trawl numbers |
|---|---|---|---|
| 2003 | 1,014 | 14,663 | 1,107* |
| 2004 | 953 | 15,372 | 5,591 |
| 2005 | 886 | 14,721 | 5,500 |
| 2006 | 914 | 15,792 | 5,882 |
| 2007 | 841 | 14,197 | 5,204 |
| 2008 | 831 | 11,966 | 4,533 |
| 2009 | 713 | 10,605 | 2,845* |

*Not all trawl data for calendar year provided. These figures represent the subset subject to analyses, not the annual totals.
Numbers refer to those conducted by calendar year (January to December) in the Pilbara Trawl Fishery from 2003 until 2009 (source = Department of Fisheries [27]).

Bycatch exclusion grids and escape hatches were trialled in 2004 and 2005, then fitted into all nets used in the PTF in March 2006 [29][31]. The BRDs in use at the time of this study consisted of a semi-flexible metal grid and a bottom-opening escape hatch (through which large animals could leave the net), with a loose skirt of netting to prevent the loss of target species covering the hatch (see also [35]). The exclusion grid was held upright by a number of floats. The grid lay at an angle with the float-equipped top section anterior to the lower section, so that bycatch and benthos were deflected down toward the bottom-opening escape hatch. In June 2008, the BRDs were moved forward in the net, from just before the codend, to the start of the net extension. This was done to prevent dolphins from backing down into the extension and to provide a shorter escape route between the BRDs and the opening of the net. All grids featured vertical bars made of stainless tube and central sections of braided stainless wire.

The trawl data were categorised into three broad net types: 1) before the introduction of the BRDs (August 2003 until February 2006; excluding the BRD trials) – "No BRD"; 2) BRD trials from the previous period, after the compulsory introduction of the BRDs and before they were moved forward (primarily March 2006 to May 2008) – "BRD"; and 3) after the BRDs were moved forward in the net (June 2008 until September 2009) – "BRD forward". The total number of trawls, number of dolphin bycatch events and dolphin catch rate per 1,000 trawls were calculated for each category (Table 2).

**Table 2.** Dolphin bycatch rates reported by skippers and observers.

| Period | Skippers' logbook | | | Independent observer | | |
|---|---|---|---|---|---|---|
| | # dolphins | # trawls | #/1,000 trawls | # dolphins | # trawls | #/1,000 trawls |
| *a) No BRD* | | | | | | |
| Aug03–Jun04 | 19 | 3,373 | 5.6 | 1 | 46 | 21.7 |
| Jul04–Jun05 | 48 | 4,793 | 10.0 | 9 | 481 | 18.7 |
| Jul05–Feb06 | 32 | 3,002 | 10.7 | 10 | 537 | 18.6 |
| Total No BRD | 99 | 11,168 | 8.9 | 20 | 1,064 | 18.8 |
| *b) BRD* | | | | | | |
| Jan05–Feb06 | 5 | 854 | 5.9 | 3 | 298 | 10.1 |
| Mar06–Jun06 | 9 | 1,569 | 5.7 | 7 | 657 | 7.6 |
| Jul06–Jun07 | 31 | 5,345 | 5.8 | 10 | 1,055 | 9.5 |
| Jul07–May08 | 16 | 3,871 | 4.1 | 5 | 429 | 11.7 |
| Total BRD | 61 | 11,639 | 5.2 | 25 | 2,439 | 10.3 |
| *c) BRD forward* | | | | | | |
| Jun08–Jun09 | 18 | 4,365 | 4.1 | 7 | 621 | 11.3 |
| Jul09–Sep09 | 2 | 732 | 2.7 | | | |
| Total BRD forward | 20 | 5,097 | 3.9 | 7 | 621 | 11.3 |
| *d) BRD+BRD forward* | | | | | | |
| Jan05–Sep09 | 81 | 16,736 | 4.8 | 32 | 3,060 | 10.5 |
| **TOTAL** | **180** | **27,904** | **6.5** | **52** | **4,124** | **12.6** |

Number of dolphins caught, number of trawls observed, and dolphin bycatch rate/1,000 trawls in Australian financial years of July to June (as per presented by industry in logbooks) and divided by net type (i.e. No BRD, after the introduction of BRDs, and then BRDs being moved forward in the extension).

## Data analyses

The skippers' logbook and independent observer data from October 2003 to August 2009 for the PTF were provided by the Department of Fisheries and stored in a Microsoft Access database. The skippers' logbook data for this period of trawl fishing activity comprised information on targeted catch and bycatch from 30,684 trawls and the observer data set contained similar details from 4,939 trawls. Structured query language (SQL) queries were written to filter the dolphin bycatch data and location data were examined in ArcGIS. Summary figures and binary logistic generalised linear models were run in SPSS 16.01.

Excel files containing latitudes and longitudes of trawls were used to create point files in ArcGIS. The start and end latitudes and longitudes were combined using the Merge function and used to calculate the straight-line distance (line segment) of each trawl (in nautical miles, nm). Data were screened for trawls that were either outside the trawling management areas and/or line segments longer than 21 nm (39 km, equivalent to a trawl duration of about six to eight h). After removing these data, along with duplicate records and those with missing values, ca. 90% of the logbook data (or n~11,200 trawls for the No BRD net type, ~11,700 for the BRD net type and n~5,100 trawls for the BRD forward category) and 85% of observer data remained for further analyses. The density of lines (trawls) was calculated using the Line Density function in ArcGIS.

As with net type, each of the trawl variables used in the analyses were placed into the following categories: Time of day (morning {06:00–11:59}/afternoon {12:00–17:59}/night {18:00–23:59}/early morning {0:00–05:59}); Area (1/2/4/5, fishery management area 3 is closed to trawling); Season (wet {December–April}/dry {May–November}); Vessel (1/2/3/4); Trawl duration (0.1–1.0 h/1.1–2.0 h/2.1–3.0 h/3.1–4.0+ h); Trawl distance (0.1–5.0 nm/5.1–10.0 nm/10.1–15.0 nm/15.1–20.0 nm); Trawl depth (51–60 m/61–70 m/71–80 m/81–90 m/91–100+ m); and dolphin bycatch (present/absent).

## Binary logistic generalised linear models

The categorical variables net type, time of day, management area, season, vessel and trawl duration were fitted as individual predictors for the presence of dolphin bycatch in separate binary logistic generalised linear models (GLMs) [36] for the logbook data and the observer data. These models were used to determine which variables were significant in predicting the presence of dolphin bycatch. The unit of measure for the presence of dolphin bycatch was the individual trawl and each is assumed to be independent. The significant predictors, and the interactions between them, were then used in combination in multi-predictor binary logistic GLMs. The multi-predictor GLMs were used to determine which combination of predictors accounted for the highest probability of the presence of dolphin bycatch. Akaike's Information Criterion (AIC) [37], which selects the most parsimonious model that best fits the data by taking into account the variation explained and the number of terms in the model, was used to select the best model. The lower the AIC value, the better the fit of the model [38].

Given the relatively low number of trawls in the observer data set for the BRD and BRD forward net categories, the data for these categories were pooled into a BRD category for No BRD versus BRD comparisons. For ease of comparison with earlier bycatch mitigation research by the Department of Fisheries [29][31], the results of dolphin capture rates under the various conditions assessed are presented in dolphin captures per 1,000 trawls.

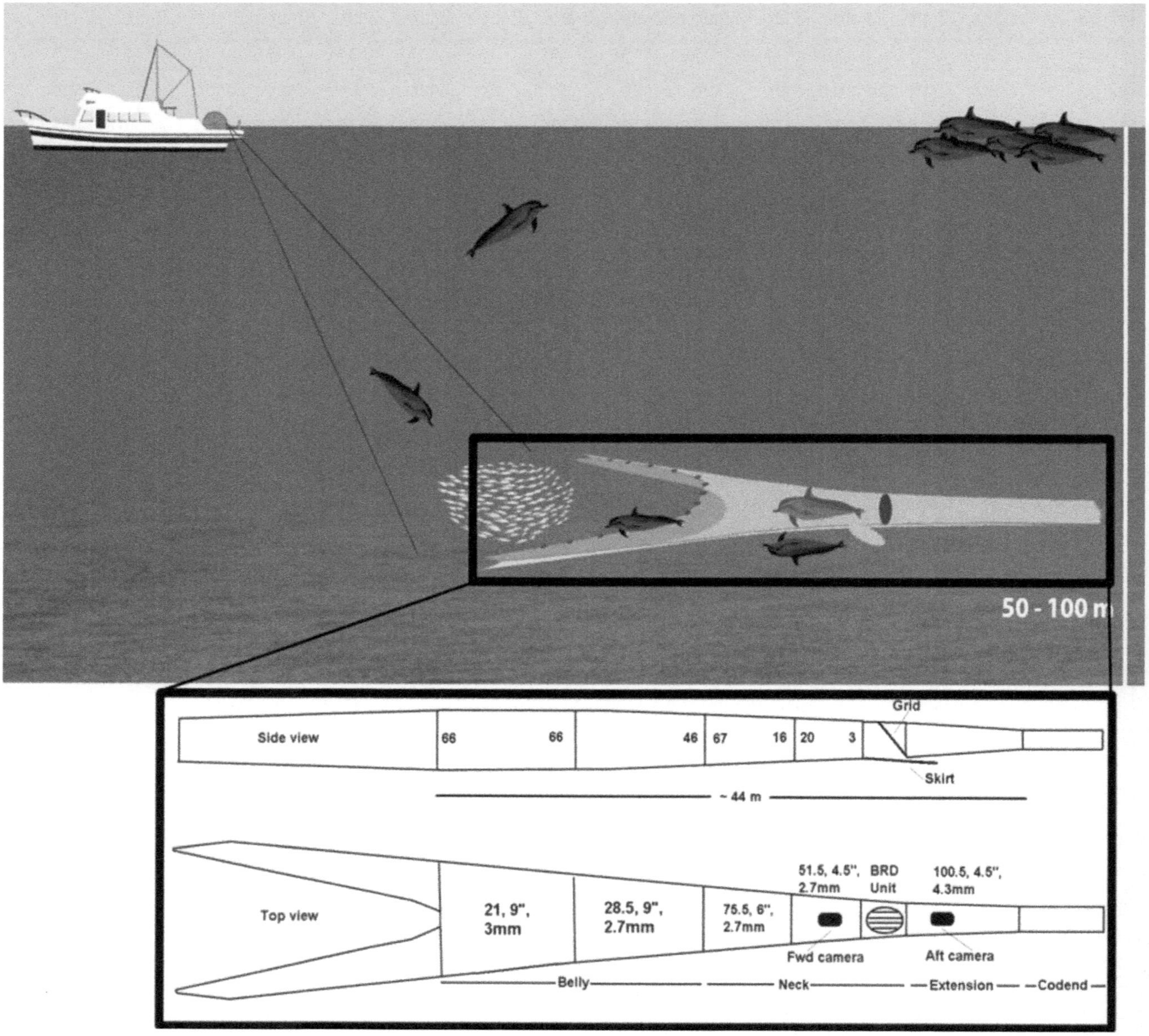


**Figure 2. Schematic of trawler and trawl net on or near the seabed.** Also illustrated are the typical positions of dolphins in and around the net, as well as following the trawler on the surface. The detailed net diagram represents the typical net specifications used in the Pilbara Trawl Fishery in 2008 and 2009, showing the side and top views, location of the Bycatch Reduction Device and the skirt covering the escape hatch. The lengths of the different panels are given as number of meshes, mesh length (in inches) and diameter of twine (in mm). In the side view, the height of each panel is given as the number of meshes. Diagram not to scale. Modified from Stephenson et al. [29] and Jaiteh et al. [35] following plans by H. McKenna for the 'Magnet Box Diamond Net' with short neck.

## Independent observer coverage levels

After the introduction of BRDs across the PTF in March 2006, scientific advice from within the Department of Fisheries suggested that minimum observer coverage from 2006–2007 onward should be at least 22% of total fishing effort and be representative of the operations of the fishery [29]. We calculated observer coverage rates in Australian financial years from the datasets provided. We also assessed what sample size of trawls should be monitored by observers to obtain an estimate of dolphin bycatch rates with a relative proportional standard error (Coefficient of Variation) of 20%. We used standard estimation equations for a population total based on simple random sampling from a finite population [39].

# Results

## Overall dolphin bycatch rates

The total number of trawls for a full calendar year in the PTF data provided was highest in 2006 (5,882 trawls; 15,792 h of trawling) and lowest in 2008 (4,533 trawls; 11,966 h of trawling). A total of 171 dolphin capture events, involving 180 dolphins, were recorded in the skippers' logbook data of 27,904 trawls from 2003 to 2009, at an overall rate of 6.5 dolphins/1,000 trawls (Table 2). Observers reported 48 dolphin capture events, involving 52 dolphins, in the observed subset of 4,124 trawls at an overall rate of 12.6 dolphins/1,000 trawls (Table 2). Note, however, that the dolphin bycatch rates varied among the broad categories of net type and were lower after Bycatch Reduction Devices (BRDs) were installed (Table 2, see also below).

In general, one dolphin was caught in the net, except on nine occasions in which skippers reported two dolphins caught in one trawl and four occasions in which observers reported two dolphins caught in one trawl. These data were too sparse to model actual counts, so the presence of a dolphin bycatch event ('at least one dolphin caught') was the measure used for the generalised linear models. Underwater video observations of dolphins in 44 operating trawls (ca. 1% of 2008–2009 effort) found that one or two dolphins typically swim inside the actively fishing trawl nets at a time, although up to nine dolphins have been recorded in the net at any one time [35].

## Spatial dolphin bycatch and fishing effort

The spatial distribution of lines, representing trawls from August 2003 until September 2009, indicated that fishing effort was most intense in Management Area 1 and least intense in the most remote (in terms of distance from home ports) northern and eastern regions of Management Area 5 (Fig. 1). The catch of dolphins appeared largely to reflect the intensity of fishing effort, with most dolphins captured in Area 1 (Fig. 1). From the logbook data, dolphin capture rates were greatest in Management Area 4, but for observer data, they were highest in Area 2. These differences in dolphin catch rates among areas were, however, not significant (Table 3).

## Predictors of dolphin bycatch

Binary logistic generalised linear models fitting single predictors found that vessel, net type (BRD and BRD forward pooled), and time of day were each significant in predicting the occurrence of dolphin bycatch in the PTF for both the skippers' logbook and observer data (Table 3). Trawl duration was also a significant predictor in the logbook data only (higher bycatch rates in longer trawls). In contrast, management area and season (wet versus dry) were not significant in predicting the occurrence of dolphin bycatch (Table 3). For the independent observer data, only vessel, time of day and net type (BRD and BRD forward pooled) were significant in predicting the occurrence of dolphin bycatch (Table 3). Rates of dolphin bycatch were significantly higher in one vessel than the other three and capture rates were significantly lower in the early morning than at other times of the day (Fig. 3a and b). While the magnitudes of dolphin capture rates were consistently higher for the observer than the logbook data, they followed a similar pattern of change for each factor (Fig. 3a and b, Table 3). The predicted dolphin capture rates did not differ significantly between different trawl durations, trawl distances, seasons or between depths (Table 3).

The mean rates of dolphin bycatch differed markedly between skippers' logbooks and observer reports. Observer reported dolphin bycatch rates were 2.1, 2.0 and 2.9 times higher than those from logbooks in the No BRD, BRD and BRD forward periods of trawling activity, respectively (Table 2; Fig. 4). The number of observed trawls for the BRD forward category (n = 621) was much lower than in the prior two periods (n = 1,064 and 2,439, respectively). Observer reported catch rates were 2.2 times higher than the logbook reported rates in the BRD and BRD forward pooled category.

After the introduction of BRDs, the rate of dolphin bycatch from both the skipper and observer records declined by ca. 45% (Fig. 4). After the BRDs were moved forward in the nets, the logbook data showed a further slight decline in dolphin capture rates and the observer data indicated a slight increase in dolphin catch rates (Table 2; Fig. 4).

From the skippers' logbook data, time of day, net type and vessel were significant in predicting dolphin bycatch, as was the net type×vessel interaction, although the effect was not nearly as strong as the main factors (Tables 4a and 5a). This interaction was due to the fact that dolphin capture events were lower in the BRD than No BRD net type category for three vessels, but remained the same for one vessel (Fig. 5). For the independent observer data, vessel and time of day were the strongest predictors of dolphin bycatch, while the net type effect was close to significance ($P = 0.06$, Tables 4b and 5b).

## Independent observer coverage levels

From the 2003–2009 data provided for this study, the overall observer coverage was ca. 16%. Observer coverage levels attained after the 22% minimum was recommended by internal Department of Fisheries scientific advice were 17% in 2006–2007, 13% in 2007–2008, 13% in 2008–2009 and ca. 8% from July to September 2009.

Given a "population" of 3,000 trawls (six-eight months of trawling in the PTF), with a dolphin bycatch rate of 20 dolphins/1,000 trawls (roughly equivalent to the rate reported by observers prior to the introduction of BRDs), about 30% observer coverage

**Table 3.** Presence of dolphin bycatch by individual factors.

| Factor | | Skippers' logbook | | | Independent observer | | |
|---|---|---|---|---|---|---|---|
| | df | AIC | Likelihood ratio ($\chi^2$) | *P* | AIC | Likelihood ratio ($\chi^2$) | *P* |
| Time of day | 3 | 29.07 | 44.03 | <0.001 | 23.53 | 8.39 | 0.039 |
| Net type (separate) | 2 | 22.93 | 18.18 | <0.001 | 19.25 | 5.18 | 0.075 |
| Net type (pooled) | 1 | 16.45 | 17.89 | <0.001 | 13.87 | 5.06 | 0.025 |
| Vessel | 3 | 29.97 | 8.20 | 0.042 | 23.74 | 11.76 | 0.008 |
| Trawl duration* | 3 | 24.41 | 12.22 | 0.007 | 24.07 | 3.48 | 0.323 |
| Trawl area | 3 | 30.00 | 3.95 | 0.267 | 24.81 | 2.87 | 0.413 |
| Season | 1 | 16.46 | 0.01 | 0.904 | 13.87 | 0.34 | 0.853 |

*Trawl duration was estimated using a reduced dataset (AIC not directly comparable).

Results from binary logistic generalised linear models to predict the presence of dolphin bycatch in trawl nets by individual factors, based on skippers' logbook data and independent observer data (skippers' logbook number of trawls, n = 27,914; independent observer, n = 4,178, except for the predictor trawl duration, where logbook n = 27,489 and observer n = 4,153). df = degrees of freedom. Net type (separate) = analysis of three net types (No BRD, BRD, BRD forward). Net type (pooled) = analysis of two net types (data for BRD and BRD forward pooled).

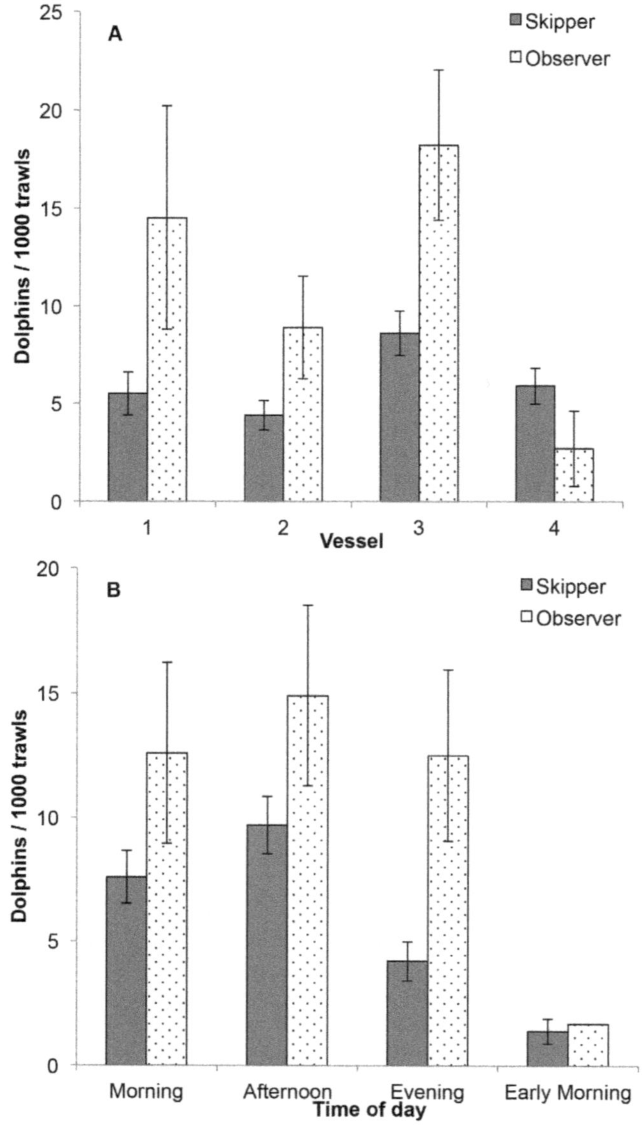


**Figure 3. Dolphin bycatch rates by a) vessel and b) time of day.** Mean (±1 SE) dolphin bycatch rates by a) vessel (1–4) and b) time of the day (Morning = 06:00–11:59; Afternoon = 12:00–17:59, Evening = 18:00–23:59; Early Morning = 00:00– 05:59) in the Pilbara Trawl Fishery. For skippers' logbook, n = 27,914; for independent observer, n = 4,178.

would be required in order to estimate bycatch rates with a relative proportional standard error of 20%. For dolphin bycatch rates of 10 dolphins/1,000 trawls (approximating the rate reported by observers after the introduction of BRDs), 46% of the 3,000 trawls would need to be monitored and, for five dolphins/1,000 trawls (the rate reported by skippers after the introduction of BRDs), 62% of the 3,000 trawls would need to be observed.

## Discussion

In this study, we evaluated detailed, long-term catch and effort data from the Pilbara Trawl Fishery to assess patterns of dolphin bycatch and determine whether the introduction of spatial and/or temporal fishery management measures might contribute to mitigating dolphin bycatch. We also aimed to quantify differences in dolphin bycatch rates by net type. In general, the trends in dolphin bycatch rates from skippers' logbooks and independent observer datasets followed similar patterns of variation with vessel, time of day and the net type (with or without Bycatch Reduction Device) in use. The reported rates of dolphin capture from the logbooks were consistently lower (< ½) than those reported by observers. This is consistent with the reporting of bycatch in a number of other fisheries; particularly where the capture of marine mammals is illegal [5][40]. Assuming the number of dolphin capture events reported in the 2010–2012 logbooks [27][41] was ca. ½ the number observers would have reported, had they

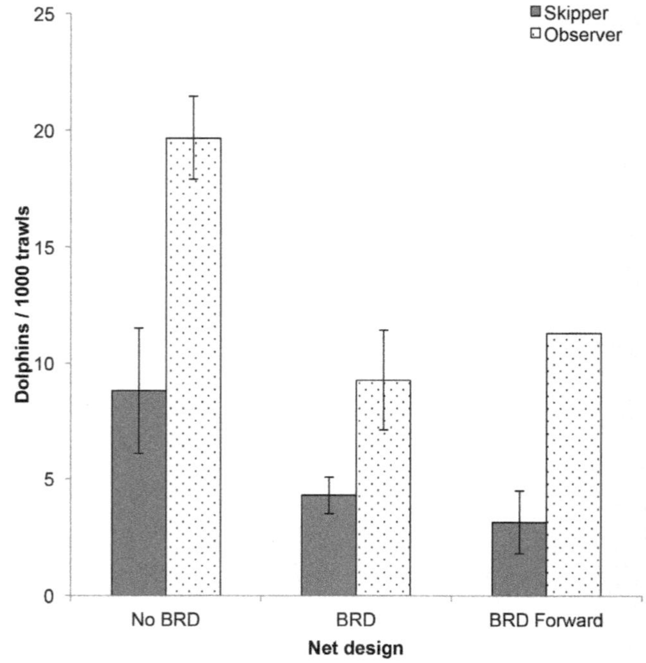


**Figure 4. Dolphin bycatch rates by differing net design in the Pilbara Trawl Fishery.** Mean (±1 SE) dolphin bycatch rates by differing net designs. BRD = BRD fitted just forward of the codend and at the aft end of the tubular net extension; BRD forward = BRD moved to the forward end of the extension.

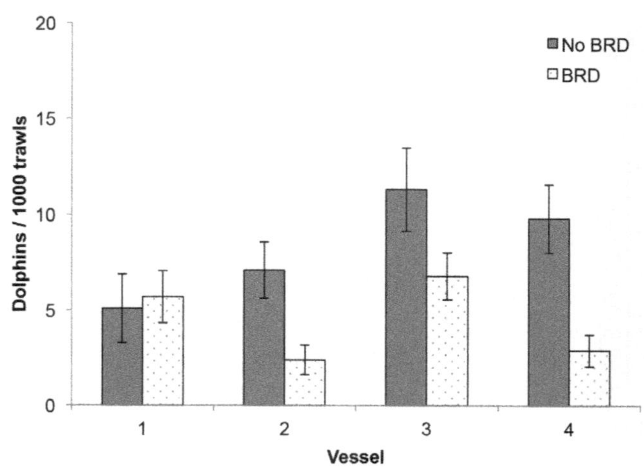


**Figure 5. Dolphin bycatch rates net type x vessel interaction.** Mean (±1 SE) dolphin bycatch rates illustrating the net type (No BRD vs BRD) x vessel (1-4) interaction based on skippers' logbook data.

**Table 4.** Summary of full models to predict dolphin bycatch by fitting multiple factors for a) skippers' logbook data and b) independent observer data.

| Full models | AIC | df | Model Log-likelihood | Likelihood ratio ($\chi^2$) | P |
|---|---|---|---|---|---|
| *a) Skippers' logbook data* | | | | | |
| $\beta_0+\beta_1V1+\beta_2V2+\beta_3V3+\beta_4TOD1+\beta_5TOD2+\beta_6TOD3+\beta_7NT1+\beta_8(V1*NT1)+\beta_9(V2*NT1)+\beta_{10}(V3*NT1)$ | 130.16 | 10 | −54.08 | 82.63 | <0.001 |
| *b) Independent observer data* | | | | | |
| $\beta_0+\beta_1V1+\beta_2V2+\beta_3V3+\beta_4TOD1+\beta_5TOD2+\beta_6TOD3+\beta_7NT1$ | 87.81 | 7 | −35.89 | 24.27 | 0.001 |

Results from binary logistic generalised linear models for predicting the occurrence of dolphin bycatch in the Pilbara Trawl fishery using time of day, net type, vessel and the net type×vessel interaction as predictors from a) skippers' logbook data (number of trawls, n = 27,914) and b) observer data (n = 4,178).

continued after 2009, a minimum of ca. 500 bottlenose dolphins were caught in the PTF in the decade 2003–2012.

The analysis of dolphin bycatch patterns on spatial, daily and seasonal scales, and comparisons among different net designs, contradicted some of our expectations. For example, the greatest source of variation was not the net type (No BRD vs BRD/BRD forward). Most of the variation in dolphin bycatch was explained by the predictor variables of vessel and time of day. In the full model, net type was significant for the skippers' data and close to significance for the observer data. While the logbook data was a much larger dataset (ca. seven times more trawls than the observer dataset), the observer data better explained variation in dolphin bycatch, perhaps because skippers missed some bycatch events (i.e. when dolphins fell out of the bottom-opening escape hatch on winch-up and prior to being landed on deck). Accordingly, most of the discussion focuses on the results from the observer data.

## Temporal patterns of dolphin bycatch and fishing effort

Observer reported dolphin catch rates in the six-hour period of early morning (00:00–05:59), when the least fishing occurs, were up to 85% lower than in the other three 6 h periods of the day. Logbook records also indicated a similar pattern, although the difference between periods was not as marked as those from the observer data. This result is reasonably consistent with the smaller subset of data summarised by Stephenson and Wells [31]. Bycatch records collected between January 2004 and June 2005 indicated that 92% of dolphins were caught between 7am and 8pm [31]. It is difficult to determine why dolphins might be less likely to be caught in the late evening and early morning. Bottlenose dolphins were seen foraging around the trawlers in the PTF throughout the day and night (SA, pers. obs.), though it was not possible to determine if this occurred to the same extent at night as it did during the day. Bottlenose dolphins (*Tursiops* spp.) also foraged around trawlers throughout the day and night in Moreton Bay, Queensland [42], and Spencer Gulf, South Australia [43]. In a study of two fisheries off the north-eastern United States, Waring et al. [44] noted that the bycatch of common dolphins (*Delphinus* sp.) and pilot whales (*Globicephala* sp.) tended to follow a diel pattern, with common dolphins being caught at night and pilot whales caught during the day.

**Table 5.** Model comparison with the full model (Table 4) for predicting dolphin bycatch by fitting multiple predictors for a) skippers' logbook data and b) independent observer data.

| Predictors | df | Likelihood ratio ($\chi^2$) | *P* |
|---|---|---|---|
| *a) Skippers' logbook data* | | | |
| Intercept | 1 | 33,323.66 | <0.001 |
| Time of day (*TOD*) | 3 | 46.99 | <0.001 |
| Net type (*NT*) | 1 | 15.15 | <0.001 |
| Vessel (*V*) | 3 | 12.58 | 0.006 |
| Net type * Vessel | 3 | 8.57 | 0.035 |
| *b) Independent observer data* | | | |
| Intercept | 1 | 3,375.33 | <0.001 |
| Vessel (*V*) | 3 | 11.05 | 0.011 |
| Time of day (*TOD*) | 3 | 8.89 | 0.031 |
| Net type (*NT*) | 1 | 3.49 | 0.062 |

Model comparison compares the full model (from Table 4) with the reduced model to indicate the significance of the additional predictors from a) skippers' logbook data (number of trawls, n = 27,914) and b) observer data (n = 4,178).

Bottlenose dolphins interacting with trawlers in the PTF may have adopted a diurnal pattern of behavior in response to this foraging association, just as the behavioral budgets and social structure of bottlenose dolphins have adapted to other circumstances in which anthropogenic activities mediate a particular schedule or regimen [45]. Numerous studies have demonstrated the adaptability of bottlenose dolphin behavior to human activity, for example: dolphins spent less time in Milford Sound, New Zealand, during periods of intense tour boat activity [46]; free-ranging dolphins adopt a daily activity pattern to take advantage of provisioning by tourists in Shark Bay, Western Australia [47]; and artisanal fishing in Laguna, Brazil, and trawl fishing in Moreton Bay, Queensland, have documented influences on the activity budgets and association patterns of bottlenose dolphins [42][48][49]. This rapid learning ability and behavioral flexibility means that the chances of dolphins interacting with boats and fishing gear are greatly increased, particularly when food is an incentive [50].

Underwater video data collected in daylight hours suggests that dolphins spend considerable time foraging in and around trawl nets in the PTF [35]. If this foraging effort is sufficient to procure their daily food requirements, the dolphins may be less inclined to do so late at night and in the early morning, opting instead to forage less and remain vigilant against predation. Known predators of dolphins, such as sandbar (*Carcharhinus plumbeus*), oceanic white tip (*C. longimanus*) and tiger (*Galeocerdo cuvier*) sharks, also follow trawlers in the PTF (SA, pers. obs.). Dolphins have been shown to modify their habitat use in response to the presence of tiger sharks [51]. Regardless of the reasons behind the lower bycatch rate in the early mornings, a restriction of fishing activity in the daytime and concomitant increase at night would be unlikely to reduce dolphin bycatch rates in the medium- to long-term, as the dolphins may be able to adapt their behavior to this change in fishing activity.

Both skippers' logbook and independent observer data suggested that season had little influence on the likelihood of dolphin bycatch in the PTF. Although relatively little is known of the ecology and movements of bottlenose dolphins interacting with the PTF (but see [35]) or, indeed, those in any of Australia's extensive pelagic waters [21], this lack of an effect is not surprising. Prevailing winds and rainfall levels do change on a seasonal basis in this region, but there are no marked changes in the physical or biological conditions (such as water temperature or prey abundance/density) in the pelagic environment that might be expected to result in seasonal fluctuations in the numbers of dolphins in the area. Furthermore, underwater video footage, photographic identification and genetic evidence indicates that at least some individual dolphins show fidelity to foraging around trawlers for periods of weeks to years [32][35].

## Vessel and net type effects on the probability of dolphin bycatch

A strong vessel effect was evident in both the logbook and observer data, and in both single- and multi-predictor generalised linear models (GLMs). One vessel had higher bycatch rates than the other three assessed. The difference in dolphin catch rates among vessels is difficult to interpret, especially given the similarities in boat configurations and nets in this small fishery. It may be attributable to the different fishing practises employed by different skippers in the fleet. For example, some skippers tend to conduct their fishing operations in a very consistent manner over time, while others tend to modify how they are fishing on a frequent basis. Furthermore, some skippers tend to take more risks than others in terms of trawling over or near benthic structures, such as rocky reefs or pipelines associated with the offshore oil and gas industry. It is likely that more consistent trawling, and therefore fewer instances of rapid winch-up or net collapse, results in fewer dolphin captures.

Net type (No BRD vs BRD vs BRD forward) was also significant in predicting dolphin bycatch in the single and multi-predictor GLMs using logbook data. In the observer data, net type (No BRD vs BRD/BRD forward pooled) was significant in the individual model, and close to significance in the multi-predictor model. The relatively small sample size of observer coverage for the BRD forward design, along with the relatively low incidence of dolphin capture, reduced the power to detect any change/effect among the three net designs in the observer data. Differences in skipper behavior and detail of reporting dolphin bycatch are likely to account for the interaction detected between vessel and net type detected in the skippers' logbook dataset.

Both the skipper and observer reported rates of dolphin bycatch dropped by ca. 45% after the introduction of BRDs, consistent with the earlier assessment of the smaller subset of data [29]. The trends after the forward movement of the BRDs, however, were inconsistent between skipper (a slight further decrease in dolphin catch rates) and observer data (a slight increase) and no significant changes were detected in either dataset. Exclusion grids and escape hatches of various forms have been trialled to reduce bycatch of marine mammals, turtles and other megafauna in numerous trawl fisheries around Australia and the world. While detailed measures of their efficacy, including long-term follow-up, are scarce, those that have met with some success include the following: Northridge et al. [52] have experimented with exclusion grids and top-opening escape hatches in an English pelagic bass pair-trawl fishery, reporting reductions in common dolphin bycatch without the loss of target species; Zeeberg et al. [53] report on the use of escape hatches to reduce the bycatch of a number of species of small cetaceans and other megafauna in the Dutch trawl fleet fishing off Mauritania; top-opening escape hatches and exclusion grids have reduced the bycatch of turtles, large sharks and rays in Australia's Northern Prawn Fishery [54]; the bycatch and mortality rates of fur seals (*Arctocephalus* spp.) were reduced with the use of large, bottom-opening escape hatches in a pelagic, mid-water trawl fishery off Tasmania [55].

Since the reported reduction in dolphin bycatch rates in the PTF after the introduction of BRDs in early 2006 (this study, [29]), annual fishing effort has declined by ca. 35% (2006 = 15,792 h; 2012 = 10,269 h) [27]. According to the latest skippers' logbook data, however, dolphin bycatch rates have increased above those reported immediately after the BRDs were made mandatory (2006 = 2.2 dolphins/1000 h trawling; 2012 = 2.8 dolphins/1000 h trawling) [27]. This minimum estimate, combined with underwater video footage showing a proportion of incidentally caught megafauna falling out of the bottom-opening escape hatch before being landed on deck [32], indicates that bycatch rates reported by both skippers and observers are invariably underestimates and that BRDs are unlikely to be as effective as first presumed. Similarly, in South Australia, an unknown proportion of endangered Australian sea lions (*Neophoca cinerea*) sustain life-threatening injuries or die in gill nets and drop out before being detected, even by vigilant onboard observers [56][57]. Furthermore, bottom-opening escape hatches are not well suited to dolphins and other air-breathing animals in the PTF, which tend to swim upward and push on the upper ceiling of the net (in an attempt to get to the surface) when trying to escape [32].

## Spatial patterns of dolphin bycatch and fishing effort

Logbook data suggested that dolphin capture rates were highest in Management Area 4 of the PTF, while observer data indicated the highest rate was in Area 2. These differences were not significant in predicting dolphin capture in either single- or multi-predictor GLMs based on observer data. Nor were there any marked differences in capture rates by depth in the logbook or observer data. These results, based on six years of data, corroborate the earlier study by Stephenson and Wells [31] from an 18-month subset of data and are to be expected, due to the broad extent of interactions between dolphin and trawlers operating in the PTF in both space and time [32][35]. Dolphin bycatch events are spread across the four Management Areas open to trawling and across all depths (50–100+ m) in the fishery. Fernández-Contreras et al. [58] suggested that limiting trawling to deeper waters would reduce common dolphin bycatch in the pelagic trawl fishery off north-western Spain. The operators in the PTF have periodically undertaken spatial restrictions in fishing effort in response to declining stocks of target species in the past [27], but the results of our study suggest that such a spatial restriction within the PTF would be highly unlikely to reduce dolphin bycatch.

## Skippers' logbook data and independent observer coverage

The extensive databases of logbook and observer records formed the basis of this assessment of spatial and temporal patterns of dolphin bycatch in the PTF, but some problems were evident in the quality of the data. Due to errors such as blank fields and erroneous location data in the logbook dataset, only ca. 90% of the trawl records could be analysed. More blank fields and errors were seen in the observer data, particularly from 2004 to 2006, and only ca. 85% of these records were of sufficient quality for analyses.

Sound reporting practices and validation checks would improve the quality of these data sources and their value for interpreting patterns of bycatch.

The dolphin bycatch rates from the independent observer coverage data were over double those reported in the skippers' logbooks. The under reporting of bycatch by skippers is not unusual [5][40], but highlights the importance of having enough observer coverage to provide robust estimates of dolphin bycatch and other incidental catches in non-selective fisheries. The Department of Fisheries specified that minimum observer coverage of 22% of total fishing effort was required from 2006–2007 onward [29], but the 2003–2009 data indicated that overall coverage was ca. 16% and declined over time from 17% in 2006–2007 to 8% in late 2009, when it ceased. This low and declining coverage, combined with the relatively infrequent incidence of dolphin capture, means that estimates of dolphin bycatch rates are imprecise and that the comparisons of dolphin bycatch rates between the different net designs have low power [32].

We calculated the amount of observer coverage required at between 30% and 62% of total fishing effort. Due to the high financial cost of independent observer programs, an electronic observer system involving deck-mounted video cameras was trialled in the PTF as an alternative to human observers [59]. The evaluation of this system concluded that it should not be used to replace independent observers, because of: the technical difficulties associated with maintaining the system in such a remote fishery; the system's lack of capability in differentiating between species of both targeted catch and incidental bycatch; the system's inability to detect dolphins that are caught and then fall from the BRD's escape hatch prior to being landed on deck; and the fact that the system is only of moderately lower cost than human observers when compared over a five-year period with observer coverage rates of ca. 60% or more [59]. The cost advantage of an electronic observer system should also be considered against the human observer's better reliability (far lower chance of data loss), ability to collect more detailed and accurate data on both target and non-target catch and capability to perform other tasks (such as otolith collection) for research and management purposes [59]. The independent observer program has not recommenced since its cessation in September 2009 and the Department of Fisheries adopted an electronic observer system for further trials in the PTF in 2012. Thus, more recent comparisons of the dolphin bycatch rates reported by skippers and independent observers cannot be made.

## Acoustic pingers as an alternative strategy for mitigating dolphin bycatch

Acoustic alarms or deterrents, "pingers", were designed to alert marine mammals to the presence of fishing gear and/or deter them from approaching fishing gear and aquaculture operations. They are often deployed in static fisheries, such as gill nets and long lines. The Department of Fisheries conducted trials of trawl nets equipped with active and inactive pingers and monitored by underwater video cameras in the PTF, yielding no differences in the number of dolphins swimming into the nets [31]. Pinger trials were subsequently abandoned in favour of the compulsory introduction of BRDs across the fishery [29]. The Department of Fisheries have subsequently (2012) commenced another trial of larger, louder pingers in the PTF [27], but results are yet to be reported. Pingers have been shown to reduce the bycatch of some cetaceans, including harbor porpoises (*P. phocoena*), Franciscana dolphins (*Pontoporia blainvillei*) and common dolphins (reviewed in [30]). However, they do not elicit consistent responses in all small cetacean species, nor do they have the same affects across all types of fisheries. For example: gill nets equipped with active pingers induced only subtle behavioral changes, rather than an avoidance response, in bottlenose dolphins [60]; a more recent study found fewer bottlenose dolphins approaching within 100 m of pinger-equipped gill nets, suggesting that pingers reduce the frequency of, but do not eliminate, interactions [61]; and Berg Soto et al. [62] found that pingers elicited only subtle behavioral responses in Australian snubfin and Indo-Pacific humpback dolphins, suggesting they may not be effective in reducing bycatch of these species in gill nets or anti-shark meshing for bather protection and that alternative mitigation measures should be explored.

Pingers deployed in a pelagic pair trawl fishery did not reduce common dolphin bycatch [63], although more recent trials of louder pingers showed some promise [64]. The sample sizes used in the recent trials were, however, too small to provide statistically robust evidence of their efficacy [64]. Entanglements of bottlenose dolphins in various pinger-equipped fishing nets suggest that they are not an effective means of bycatch mitigation for this species [30]. The bottlenose dolphins interacting with the PTF exhibit a number of attributes that suggest that pingers are unlikely to deter them from interacting with the trawl nets or to reduce dolphin bycatch: for example, bottlenose dolphins are known to be behaviorally flexible [45]; they are not only aware of the presence of trawl fishing gear, but highly motivated by foraging and socializing opportunities to interact with the gear [35]; and, they appear to show fidelity to the region and foraging around trawlers [32][35].

## Conclusions and recommendations

1. A minimum of ca. 500 bottlenose dolphins were incidentally caught in the Pilbara Trawl Fishery in the last decade;
2. Spatial and/or temporal fisheries management adjustments to fishing effort are unlikely to be effective in significantly reducing dolphin bycatch, as the extent of interactions between the dolphins and the PTF are great, the motivations for the dolphins to interact with the PTF and undertake risky behavior are considerable and bottlenose dolphins are behaviorally adaptable;
3. Pingers are unlikely to be effective in reducing the level of interactions between bottlenose dolphins and the PTF, due to the active and already-noisy nature of trawl fisheries (i.e. dolphins are aware of the presence of the fishing gear irrespective of the pingers), as well as the reasons listed above regarding dolphin behavior;
4. There has been no further reduction in dolphin bycatch since the Bycatch Reduction Devices were introduced, with an unknown quantity of bycatch falling out of bottom-opening escape hatches and, thus, not being reported;
5. Modified BRDs, with top-opening escape hatches, may be a more effective means of reducing dolphin bycatch;
6. Extensive independent observer coverage, as well as in-net video collection, are essential in order to quantify bycatch and estimate any reductions in bycatch with greater precision and statistical power following modifications to BRDs;
7. Data on dolphin population size and connectivity with adjacent populations are essential in order to calculate acceptable levels of human-caused dolphin mortality and determine the effectiveness of modified BRDs.

## Acknowledgments

We thank the Steering Committee of the project (Dr. Lindsay Joll, Shane O'Donoghue and Graeme Stewart) for helpful discussions and advice during the research. Assoc. Prof. Elisabeth Slooten, Dr. Konstantinos Stergiou and Dr. Brett Molony provided comments that greatly improved the manuscript. Our thanks also go to Vanessa Jaiteh and various members of the Pilbara Trawl Fishery, especially the skippers and crews, for in-kind and logistical support. The Department of Fisheries WA provided the updated skippers' logbook and observer data upon which this research was based. Dr. John Henstridge of Data Analysis Australia provided statistical advice. We also thank Geoff Diver and Gavin Kewan for independent observer data collection.

## Author Contributions

Conceived and designed the experiments: SJA LB NRL. Analyzed the data: SJA JAT HTK KHP NRL. Contributed reagents/materials/analysis tools: JAT HTK. Wrote the paper: SJA JAT HTK LB KHP NRL.

## References

1. Kennelly SJ (1995) The issue of bycatch in Australia's demersal trawl fisheries. Reviews in Fish Biology and Fisheries 5: 213–234.
2. Pauly D, Christensen V, Guénette S, Pitcher TJ, Sumaila UR, et al. (2002) Towards sustainability in world fisheries. Nature 418: 689–695.
3. Kelleher K (2005) Discards in the world's marine fisheries: An update. FAO Fisheries Technical Paper 470. Rome: FAO. 131 p. Available: http://www.fao.org/docrep/008/y5936e/y5936e00.htm.
4. Northridge SP (1991) An updated world review of interactions between marine mammals and fisheries. FAO Fisheries Technical Paper 251, Suppl. 1. Rome: FAO. pp. 1–58. ISBN 92-5-103054-5.
5. Read AJ, Drinker P, Northridge S (2006) Bycatch of marine mammals in U.S. and global fisheries. Conservation Biology 20: 163–169.
6. Read AJ (2008) The looming crisis: interactions between marine mammals and fisheries. Journal of Mammalogy 89: 541–548.
7. Reeves RR, McClellan K, Werner TB (2013) Marine mammal bycatch in gillnet and other entangling net fisheries, 1990 to 2011. Endangered Species Research 20: 71–97.
8. Turvey ST, Pitman RL, Taylor BL, Barlow J, Akamatsu T, et al. (2007) First human-caused extinction of a cetacean species? Biology Letters 3: 537–540.
9. Perrin WF, Donovan GP, Barlow J (1994) Mortality of cetaceans in passive fishing nets and traps. Reports of the International Whaling Commission 15 (special issue): 1–53.
10. Dawson S, Pichler F, Slooten E, Russell K, Baker CS (2001) The North Island Hector's dolphin is vulnerable to extinction. Marine Mammal Science 17: 366–371.
11. Jaramillo-Legorreta A, Rojas-Bracho L, Brownell Jr RL, Read AJ, Reeves RR, et al. (2007) Saving the vaquita: immediate action, not more data. Conservation Biology 21: 1653–1655.
12. Slooten E (2013) Effectiveness of area-based management in reducing bycatch of the New Zealand dolphin. Endangered Species Research 20: 121–130.
13. Hall MA (1998) An ecological view of the tuna-dolphin problem: impacts and trade-offs. Reviews in Fish Biology and Fisheries 8: 1–34.
14. Hall MA, Alverson DL, Metuzals KI (2000) By-catch: problems and solutions. Marine Pollution Bulletin 41: 204–219.
15. Lewison RL, Crowder LB, Read AJ, Freeman SA (2004) Understanding impacts of fisheries bycatch on marine megafauna. Trends in Ecology and Evolution 19: 598–604.
16. Cramer KL, Perryman WL, Gerodette T (2008) Declines in reproductive output in two dolphin populations depleted by the yellowfin purse-seine fishery. Marine Ecology Progress Series 369: 273–285.
17. Slooten E, Dawson S, Rayment W, Childerhouse S (2006) A new abundance estimate for Maui's dolphin: What does it mean for managing this critically endangered species? Biological Conservation 128: 576–581.
18. Rayment W, Clement D, Dawson S, Slooten E, Secchi E (2011) Distribution of Hector's dolphin (*Cephalorhynchus hectori*) off the West Coast, South Island, New Zealand, with implications for the management of bycatch. Marine Mammal Science 27: 398–420.
19. Shaughnessy P, Kirkwood R, Cawthorn M, Kemper C, Pemberton D (2003) Pinnipeds, cetaceans and fisheries in Australia: a review of operational interactions. In: Gales N, Hindell M, Kirkwood R, editors. Marine Mammals: Fisheries, Tourism and Management Issues. Collingwood: CSIRO Publishing. pp. 136–152.
20. Bannister JL, Kemper CM, Warneke RM (1996) The Action Plan for Australian Cetaceans. Canberra, Australian Capital Territory 2601, Australia: Australian Nature Conservation Agency. 242 p. ISBN 0-642-21388-7.
21. Ross GJB (2006) Review of the Conservation Status of Australia's Smaller Whales and Dolphins. Canberra: Australian Government. 124 p. Available: http://www.environment.gov.au/coasts/publications/pubs/conservation-smaller-whales-dolphins.pdf.
22. Hamer DJ, Ward TM, McGarvey R (2008) Measurement, management and mitigation of operational interactions between the South Australian Sardine Fishery and short-beaked common dolphins (*Delphinus delphis*). Biological Conservation 141: 2865–2878.
23. Allen SJ, Cagnazzi DDB, Hodgson AJ, Loneragan NR, Bejder L (2012) Tropical inshore dolphins of north-western Australia: Unknown populations in a rapidly changing region. Pacific Conservation Biology 18: 56–63.
24. Larcombe J, Charalambou C, Herreria E, Casey AM, Hobsbawn P (2006) Marine Matters National - Australian Marine Fishing and Coastal Communities. Canberra: Bureau of Rural Sciences. 60 p. ISBN 0-642-47559-8.
25. Althaus F, Woolley K, He X, Stephenson P, Little R (2006) The spatial distribution of commercial fishery production on Australia's North West Shelf. Technical Report No. 10 of the North West Shelf Joint Environmental Management Study. 130 p. Available: http://www.cmar.csiro.au/nwsjems/reports/NWSJEMS_TR10.pdf.
26. Sainsbury KJ (1987) Assessment and management of the demersal fishery of the continental shelf of Northwestern Australia. In: Polovina JJ, Relston S, editors. Biology and Fisheries Management. Boulder and London: Westview Press.
27. Department of Fisheries (2013) Status Reports of the Fisheries and Aquatic Resources of Western Australia 2012/13: The State of the Fisheries. Fletcher WJ, Santoro K, editors. 377 p. Available: http://www.fish.wa.gov.au/Documents/sofar/status_reports_of_the_fisheries_and_aquatic_resources_2012-13.pdf.
28. Stephenson PC, Chidlow J (2003) Bycatch in the Pilbara Trawl Fishery. Final report to the Natural Heritage Trust. Perth: Natural Heritage Trust. 74 p.
29. Stephenson PC, Wells S, King JA (2006) Evaluation of exclusion grids to reduce the catch of dolphins, turtles, sharks and rays in the Pilbara trawl fishery. Fisheries Research Report No. 171. Perth: Department of Fisheries. 24 p. ISBN: 1-921258-17-9.
30. Dawson S, Northridge S, Waples D, Read AJ (2013) To ping or not to ping: the use of active acoustic devices in mitigating interactions between small cetaceans and gillnet fisheries. Endangered Species Research 19: 201–221.
31. Stephenson PC, Wells S (2006) Evaluation of the effectiveness of reducing dolphin catches with pingers and exclusion grids in the Pilbara trawl fishery. Final report to Fisheries Research and Development Corporation, Project No. 2004/068. Fisheries Research Report No. 173. Perth: Department of Fisheries. 44 p.
32. Allen S, Loneragan N (2010) Reducing dolphin bycatch in the Pilbara Finfish Trawl Fishery. Final Report to the Fisheries Research and Development Corporation. Perth: Murdoch University. 59 p. ISBN: 978-0-86905-926-5.
33. Rice DW (1998) Marine Mammals of the World: Systematics and distribution. Special publication No. 4. The Society for Marine Mammalogy. 231 p.
34. Hale P (2008) Offshore Bottlenose Dolphin. In: Van Dyck S, Strahan R editors. The Mammals of Australia, 3rd edition. Australia: New Holland Publishers. pp. 839–840.
35. Jaiteh VF, Allen SJ, Meeuwig JJ, Loneragan NR (2013) Subsurface behaviour of bottlenose dolphins (*Tursiops truncatus*) interacting with fish trawl nets in north-western Australia. Marine Mammal Science 29: E266–E281.
36. Quinn GP, Keough MJ (2002) Experimental Design and Data Analysis for Biologists. Cambridge: Cambridge University Press. 537 p.
37. Akaike H (1974) A new look at the statistical model identification. IEEE Transactions on Automatic Control 19: 716–723.
38. Burnham KP, Anderson DR (2002) Model selection and multimodel inference: a practical information - theoretic approach. New York: Vol. Springer science and business media.
39. Thompson SK (2002) Sampling, 2nd Edition. Wiley Series in Probability and Mathematical Statistics. New York: Wiley.
40. Moore JE, Cox TM, Lewison RL, Read AJ, Bjorkland R, et al. (2010) An interview-based approach to assess marine mammal and sea turtle captures in artisanal fisheries. Biological Conservation 143: 795–805.
41. Department of Fisheries (2011) State of the Fisheries and Aquatic Resources Report 2010/11. Fletcher WJ, Santoro K, editors. Perth: Department of Fisheries. 359 p. Available at http://www.fish.wa.gov.au/About-Us/Publications/Pages/Annual-Report.aspx.
42. Chilvers BL, Corkeron PJ (2001) Trawling and bottlenose dolphins' social structure. Proceedings of the Royal Society, London B 268: 1901–1905.
43. Svane I (2005) Occurrence of dolphins and seabirds and their consumption of by-catch during prawn trawling in Spencer Gulf, South Australia. Fisheries Research 76: 317–327.
44. Waring GT, Gerrior P, Payne PM, Parry BL, Nicolas JR (1990) Incidental take of marine mammals in foreign fishery activities off the northeast United States, 1977–88. Fisheries Bulletin 88: 347–360.
45. Shane SH, Wells RS, Würsig B (1986) Ecology, Behaviour and Social Organization of the Bottlenose Dolphin: A Review. Marine Mammal Science 2: 34–63.
46. Lusseau D (2005) Residency pattern of bottlenose dolphins (*Tursiops* spp.) in Milford Sound, New Zealand, is related to boat traffic. Marine Ecology Progress Series 295: 265–272.

47. Smith H, Samuels A, Bradley S (2008) Reducing risky interactions between tourists and free-ranging dolphins (*Tursiops* sp.) in an artificial feeding program at Monkey Mia, Western Australia. Tourism Management 29: 994–1001.
48. Ansmann IC, Parra GJ, Chilvers BL, Lanyon JM (2012) Dolphins restructure social system after reduction of commercial fisheries. Animal Behaviour 84: 575–581.
49. Daura-Jorge FG, Cantor M, Ingram SN, Lusseau D, Simões-Lopes PC (2012) The structure of a bottlenose dolphin society is coupled to a unique foraging cooperation with artisanal fishermen. Biology Letters 8: 702–705.
50. Donaldson R, Finn H, Calver M (2010) Illegal feeding increases risk of boat-strike and entanglement in bottlenose dolphins. Pacific Conservation Biology 16: 157–161.
51. Heithaus MR, Dill LM (2002) Food availability and tiger shark predation risk influence bottlenose dolphin habitat use. Ecology 83: 480–491.
52. Northridge S, Mackay A, Cross T (2005) Dolphin bycatch: observations and mitigation work in the UK bass pair trawl fishery 2004–2005 season. Occasional Report to DEFRA. St. Andrews: University of St. Andrews, U.K.
53. Zeeberg JJ, Corten A, De Graaf E (2006) Bycatch and release of pelagic megafauna in industrial trawler fisheries off Northwest Africa. Fisheries Research 78: 186–195.
54. Brewer D, Heales D, Milton D, Dell Q, Fry G, et al. (2006) The impact of turtle excluder devices and bycatch reduction devices on diverse tropical marine communities in Australia's northern prawn trawl fishery. Fisheries Research 81: 176–188.
55. Lyle JM, Willcox ST (2008) Dolphin and seal interactions with mid-water trawling in the Commonwealth Small Pelagic Fishery, including an assessment of bycatch mitigation strategies. Final Report Project R05/0996. Canberra: Australian Fisheries Management Authority. 49 p. ISBN 978-1-86295-441-0.
56. Hamer DJ, Ward TM, Shaughnessy PD, Clark SR (2011) Assessing the effectiveness of the Great Australian Bight Marine Park in protecting the endangered Australian sealion *Neophoca cinerea* from bycatch mortality in shark gillnets. Endangered Species Research 14: 203–216.
57. Hamer DJ, Goldsworthy SD, Costa DP, Fowler SL, Page B, et al. (2013) The endangered Australian sea lion extensively overlaps with and regularly becomes bycatch in demersal shark gill-nets in South Australia. Biological Conservation 157: 386–400.
58. Fernández-Contreras MM, Cardona L, Lockyer CH, Aguilar A (2010) Incidental bycatch of short-beaked common dolphins (*Delphinus delphis*) by pairtrawlers off northwestern Spain. ICES Journal of Marine Science 67: 1732–1738.
59. Diver G (2012) Development and cost-benefit analysis of an electronic observer system to monitor a remote small vessel commercial fishery. Final report to FRDC 2009/048. Perth: Diversity. 31 p. ISBN 978-0-646-58029-6.
60. Cox TM, Read AJ, Swanner D, Urian K, Waples D (2003) Behavioural responses of bottlenose dolphins, *Tursiops truncatus*, to gillnets and acoustic alarms. Biological Conservation 115: 203–212.
61. Waples DM, Thorne LH, Hodge LEW, Burke EK, Urian KW, et al. (2013) A field test of acoustic deterrent devices used to reduce interactions between bottlenose dolphins and a coastal gillnet fishery. Biological Conservation 157: 165–171.
62. Berg Soto A, Cagnazzi D, Everingham Y, Parra GJ, Noad M, et al. (2013) Acoustic alarms elicit only subtle responses in the behaviour of tropical coastal dolphins in Queensland, Australia. Endangered Species Research 20: 271–282.
63. Northridge S, Sanderson D, Mackay A, Hammond P (2003) Analysis and mitigation of cetacean bycatch in UK fisheries. Final report to the Department for Environment, Food and Rural Affairs. Project No. MF0726. London: DEFRA.
64. Northridge S, Kingston A, Mackay A, Lonergan M (2011) Bycatch of vulnerable species: understanding the process and mitigating the impacts. Department for Environment, Food and Rural Affairs. Project No. MF1003. London: DEFRA. 99 p. Available: http://sciencesearch.defra.gov.uk/Document.aspx?Document=MF1003-FINALRevisedAugust2011.pdf.

# Recovery of a Temperate Reef Assemblage in a Marine Protected Area following the Exclusion of Towed Demersal Fishing

**Emma V. Sheehan[1]*, Timothy F. Stevens[2], Sarah C. Gall[1], Sophie L. Cousens[1], Martin J. Attrill[1]**

1 Marine Institute, Plymouth University, Drake Circus, Plymouth, United Kingdom, 2 Griffith School of Environment and Australian Rivers Institute - Coasts and Estuaries, Griffith University, Gold Coast, Queensland, Australia

**Abstract**

Marine Protected Areas MPA have been widely used over the last 2 decades to address human impacts on marine habitats within an ecosystem management context. Few studies have quantified recovery of temperate rocky reef communities following the cessation of scallop dredging or demersal trawling. This is critical information for the future management of these habitats to contribute towards conservation and fisheries targets. The Lyme Bay MPA, in south west UK, has excluded towed demersal fishing gear from 206 $km^2$ of sensitive reef habitat using a Statutory Instrument since July 2008. To assess benthic recovery in this MPA we used a flying video array to survey macro epi-benthos annually from 2008 to 2011. 4 treatments (the New Closure, previously voluntarily Closed Controls and Near or Far Open to fishing Controls) were sampled to test a recovery hypothesis that was defined as 'the New Closure becoming more similar to the Closed Controls and less similar to the Open Controls'. Following the cessation of towed demersal fishing, within three years positive responses were observed for species richness, total abundance, assemblage composition and seven of 13 indicator taxa. Definitive evidence of recovery was noted for species richness and three of the indicator taxa (*Pentapora fascialis, Phallusia mammillata* and *Pecten maximus*). While it is hoped that MPAs, which exclude anthropogenic disturbance, will allow functional restoration of goods and services provided by benthic communities, it is an unknown for temperate reef systems. Establishing the likely timescales for restoration is key to future marine management. We demonstrate the early stages of successful recruitment and link these to the potential wider ecosystem benefits including those to commercial fisheries.

**Editor:** Christopher J. Fulton, The Australian National University, Australia

**Funding:** The Department for Environment, Food and Rural Affairs (United Kingdom) funded this research. The funders had no role in study design, data collection and analysis, decision to publish, or preparation of the manuscript.

**Competing Interests:** The authors have declared that no competing interests exist.

* E-mail: emma.sheehan@plymouth.ac.uk

## Introduction

Management of marine environments has historically been targeted towards maintaining commercial fish stocks, with conservation objectives coming second to economic imperatives [1]. Over the past two decades, studies have increasingly attempted to understand the wider effects of fishing and other human activities on the marine environment, resulting in a shift from fisheries-centred management to an ecosystem management approach [1,2]. This type of management should not only benefit marine biodiversity, but should also feedback and benefit commercial fisheries by increasing the abundance of target species [3].

Marine Protected Areas (MPAs) can provide an effective ecosystem management approach to reducing the damaging effects of fishing on benthic assemblages and habitat [1,4–9]. They can meet both fisheries management and conservation goals [1,7] by protecting important and/or fragile habitat and preventing overfishing. Over time, well planned and managed MPAs can eventually enhance fisheries and facilitate the recovery of previously fished areas, known as spillover [10–12].

The performance of these MPAs must be assessed not only for management effectiveness, but also to ensure that governments comply with their management responsibilities. For example, EU countries are committed to establishing ecologically coherent networks of MPAs to enhance ecosystem health (Habitats Directive (92/43/EEC) and Birds Directive 79/409/EEC), and monitoring is therefore crucial to any assessment of their success. Establishing, enforcing and monitoring MPAs is costly and it is therefore also important to report their effectiveness to governments and to the public to encourage support for their use as marine conservation tools.

Lyme Bay, on the south west coast of the UK (Fig. 1), is an area of high-biodiversity reefs formed of mudstone, limestone, chalk and granite outcrops, pebbles, cobbles and boulders, listed under Annex I of the Habitats Directive. These reefs are home to species including the iconic *Eunicella verrucosa* (Pallas, 1766) pink sea fan (listed under Schedule 5 of the UK Wildlife and Countryside Act 1981), the habitat-forming *Pentapora fascialis* (Pallas, 1766) ross coral and the commercially fished *Pecten maximus* (Linnaeus, 1758) [13] scallop. These constituent elements have allowed the site to be designated as an Annex 1 habitat 'reefs'. Concerns have been raised over many years about the effects of towed demersal fishing gear, particularly scallop dredging that break up or overturn sections of fragile reef habitat and remove sessile fauna [3,14,15]. Many temperate reef sessile species are long lived and slow growing, and fishing disturbance is consequently long lasting and has been shown to have a substantial negative influence on benthic communities through changes in assemblage composition, trophic structure and habitat complexity [15–20].

Concerns raised over the impacts of towed demersal gear on Lyme Bay reef habitats were initially addressed through the creation of four small voluntary closures (totalling 22 $km^2$), which were implemented in 2001 and 2006. Variable adherence to the voluntary agreements spurred continued support for one large MPA with greater levels of protection. In 2008, Lyme Bay became the UK's largest MPA under a Statutory Instrument (SI) protecting marine biodiversity through the exclusion of towed demersal fishing gear (scallop dredging and trawling) from a 206 $km^2$ (60 $nm^2$) area of seabed. Static gear fisheries, including potting and netting, were permitted to continue, along with diving for scallops and recreational activities, such as sea angling and SCUBA diving. The Lyme Bay closure is widely regarded as an important test site for UK and European marine conservation policy [21].

Effects of MPAs have been well reported for tropical systems [22,23]. This is less well documented for temperate systems [19,24–26], and this lack of information on the response of temperate reef fauna to protection meant that the recovery of Lyme Bay reef biodiversity was far from certain [27,28]. Knowledge of recovery trends is, however, essential if MPAs are to be managed effectively to achieve conservation goals and be used as a tool to enhance fish stocks.

The aim of this study was to determine whether the biota of reef habitats within Lyme Bay showed evidence of recovery once the effects of scallop dredging and demersal trawling were removed ("passive recovery", using the terminology of Elliott et al [29]).

## Materials and Methods

### Field surveys

Lyme Bay (Fig. 1) has a diverse range of benthic habitats, from rocky and cobble reefs to mixed pebbly sand and gravel sediments and muddy soft substrata. This study focused on those reefs defined by Annex I of the Habitats Directive as 'habitats where animal and plant communities develop on rock or stable boulders and cobbles' [30]. Annual surveys took place over the summer months from 2008–2011. The 2008 baseline survey took place six weeks after the implementation of the Statutory Instrument SI; however the anticipated changes in the benthic assemblage were expected to occur over annual or decadal time spans [31] so this was considered an adequate baseline. No specific permissions were required for these locations or activities as while some protected organisms were enumerated, no physical samples were taken, only video images. Field studies, therefore, did not involve sampling

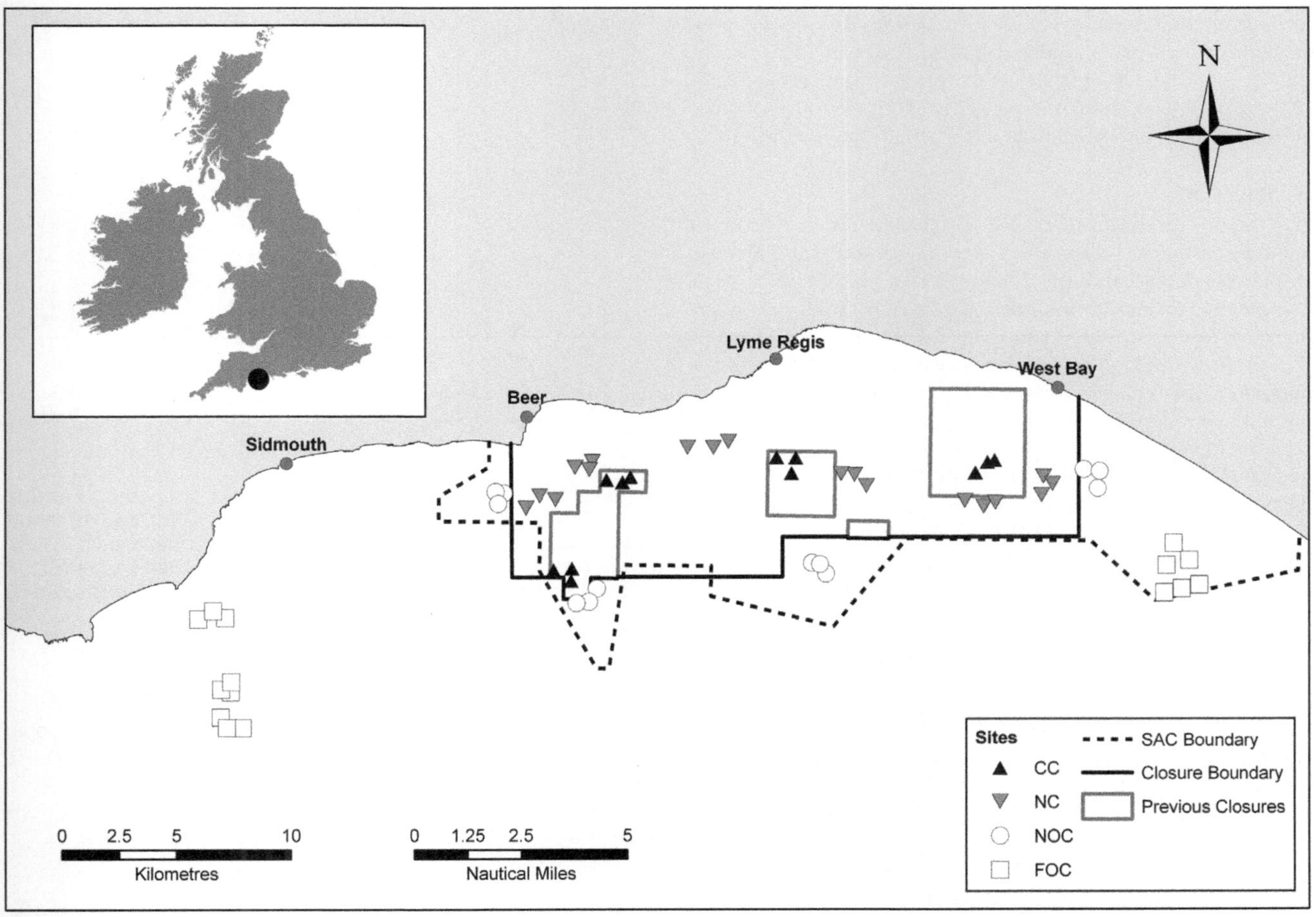


**Figure 1. Locations of sites in Lyme Bay.** Some symbols overlap at this scale.

endangered or protected species across Lyme Bay (50° 34′ N, 3° 24′ W to 50° 37′ N, 2° 35′ W).

## Survey Design

The design of the study considered changes in abundance of epibenthic taxa annually from 2008 to 2011 within four treatment types. The "experimental" treatment was the new MPA, called 'New Closure' (NC), and this was compared to sites that continued to be fished; 'Near Open Controls' (NOC) within 5 km of the closure boundary, and 'Far Open Controls' (FOC) more than 5 km from the closure boundary. Sites were organised in Areas that were nested in Treatment. While there were no areas of Lyme Bay that could be considered "pristine", the previously voluntarily protected areas had been nominally closed to dredging since either 2001 or 2006, so represented 'Closed Controls' (CC) for the purpose of this study. It is important to note that we do not assume that CC sites were completely unimpacted before the start of the study, but they represent areas of reef with the lowest past fishing activity (see site selection, below).

To assess recovery we tested the hypothesis that, subsequent to the closure of the Lyme Bay reefs to towed demersal fishing in 2008, the reef biota (measured as assemblage composition, species richness, total abundance, and abundance of pre-selected indicator taxa [32] in the NCs would increase relative to the open control sites (NOC, FOC) and would become more similar to the closed control sites (CC). Indicator taxa were selected based on life history, tolerance to disturbance and recoverability to represent the range of benthic fauna found in Lyme Bay. In addition to this narrow definition, and bearing in mind that the CC sites might benefit from the buffering effects of the statutory closure now surrounding them, we considered that increases in reef biota in both the NC and CC relative to the open control sites (NOC, FOC) would also constitute success of the MPA; we have, however kept these two scenarios separate in the results and discussion.

## Site Selection

To select candidate sites we conducted spatial analyses combining historical fishing effort, benthic substrate and biotope distribution, depth, and the boundaries of the SI and areas previously closed under voluntary agreements. Information on patterns of historical fishing effort was derived from vessel patrol sightings from 2005–2008 provided by Devon Sea Fisheries Committee (DSFC) and over-flight sightings from 2001–2007 provided by the Marine and Fisheries Agency (MFA). These data were used to construct a composite density plot of relative towed demersal fishing effort in five classes [21].

Data on benthic substrate and biotope distributions were provided by the Devon Biodiversity Records Centre, so that both reference and treatment sites could be located on similar substrates to avoid any habitat bias. Depth data was obtained from published admiralty charts. The boundaries of the SI and previous voluntary closures were added, since they in part define current patterns of use. These layers were merged to provide a single layer of polygons incorporating all the attributes of the source layers, enabling selection of those that met the necessary criteria. All sites were located on hard or "mixed" substrates (rock, boulders or cobbles). All sites were located between 15 and 25 m depth. Newly closed or open sites were located where scalloping effort was historically moderate to high, whereas closed sites were located where it was low (because they were within the voluntary closures) [21].

Final selection of four areas per treatment was conducted after ground-truthing at the commencement of the first sampling period; for example, local knowledge allowed the selection of sites of suitable habitat not identified in the existing habitat classification. Individual video frames (see below) were discarded if they were not located on rock or mixed boulders and cobble habitat. For this reason, while the target was to survey 3 sites for each Area, the number of sites suitable for analysis ranged from 2 to 5 sites per area. 60 useable video transects were analysed from 2008–2010, while 56 transects were analysed for 2011.

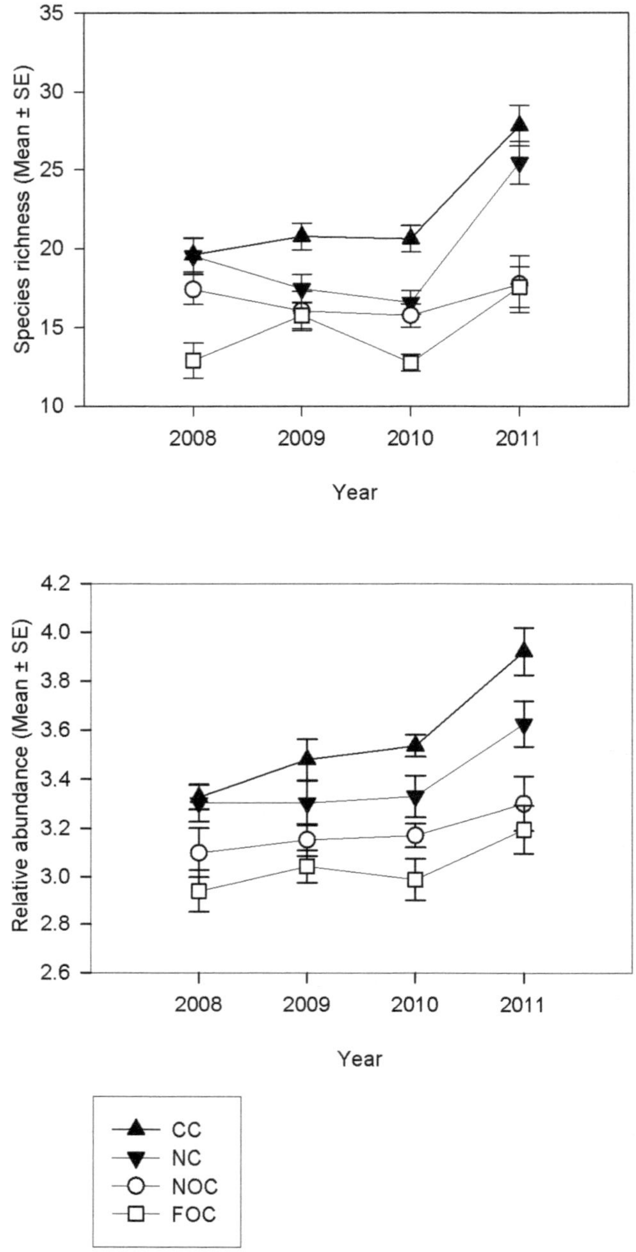


**Figure 2. Univariate diversity measures to assess benthic recovery a) Species richness (mean $m^{-2}$ ± SE) and b) Total abundance of all taxa within frame grabs, (mean $m^{-2}$ ± SE), over time (2008, 2009, 2010, 2011) and treatment type (CC = closed control, NC = new closure, NOC = near open control, FOC = far open control).**

## Video surveys

A towed flying video array was developed to survey a 200 m×0.5 m video transect at each site in a non-destructive and cost-effective way [33]. In summary, the High Definition (HD)

**Table 1.** PERMANOVA of species richness based on Euclidean distance measure.

a)

| Source | *df* | SS | MS | *F* | P |
|---|---|---|---|---|---|
| Year Ye | 3 | 1213.30 | 404.43 | 20.52 | **0.0001** |
| Treatment Tr | 3 | 1697.20 | 565.74 | 10.61 | **0.0002** |
| Area Ar (Tr) | 15 | 713.08 | 47.539 | 2.04 | **0.0358** |
| YexTr | 9 | 410.27 | 45.585 | 2.55 | **0.0171** |
| Site(Ar(Tr)) | 50 | 1061.30 | 21.225 | 1.72 | **0.0117** |
| YexAr(Tr) | 45 | 690.03 | 15.334 | 1.25 | 0.1819 |
| Residual | 110 | 1354.70 | 12.315 | | |
| Total | 235 | 7139.80 | | | |

b)

| | 2008 | | 2009 | | 2010 | | 2011 | |
|---|---|---|---|---|---|---|---|---|
| Groups | t | P | t | P | T | P | t | P |
| CC, NC | 0.73 | 0.5506 | 1.81 | 0.093 | 2.57 | **0.0233** | 0.94 | 0.3938 |
| CC, NOC | 1.71 | 0.1275 | 2.33 | 0.0541 | 3.47 | **0.0091** | 4.66 | **0.0274** |
| CC, FOC | 4.04 | **0.0032** | 2.46 | **0.0382** | 5.08 | **0.001** | 4.14 | **0.0043** |
| NC, NOC | 2.57 | **0.0229** | 0.97 | 0.3862 | 0.86 | 0.4705 | 2.73 | **0.0271** |
| NC, FOC | 5.58 | **0.0001** | 1.08 | 0.3225 | 3.09 | **0.0088** | 2.79 | **0.0216** |
| NOC, FOC | 3.12 | **0.0134** | 0.40 | 0.8904 | 3.08 | **0.0149** | 0.11 | 0.9891 |

a) Main test and b) Pairwise testing for the interaction Year × Treatment. Data were Log (x+1) transformed. Bold type denotes a significant result.

video system included a camera (Surveyor-HD-J12 colour zoom titanium, 720p), LED lights (Bowtech Products limited, LED-1600-13), two green laser pointers (Z-bolt Scuba-1) and a mini CTD profiler (Valeport Ltd). The umbilical was connected topside to a Bowtech System power supply/control unit allowing control of light intensity and camera focus, zoom and aperture. The camera was positioned at an oblique angle to the seabed, with the three lights fixed in front and below the camera to provide improved image definition and colour. The lasers were positioned parallel to each other at a known distance apart, so changes in the field of view with varying height above the substrate [34] could be quantified by measuring the apparent distance between the laser dots. This permits accurate determination of organism densities, without the need for a heavy and potentially damaging benthic sled [21].

## Video data extraction

Analysis of the video transects was conducted in two stages [33]. Firstly, infrequent/conspicuous fauna were counted from each entire video transect. Taxon counts were determined by viewing the video at normal speed, and recording each identifiable organism as it passed through the "gate" formed by the two laser dots. The position of the lasers in the field of view was noted during data extraction, and combined with the length of the tow from GPS positions, allowing the area surveyed to be calculated giving taxon abundance as density (individuals $m^{-2}$). Secondly, frame grabs were extracted from the video at five second intervals (Cybertronix frame extractor) and a digital 0.25 $m^2$ quadrat overlaid. Frame grabs were discarded if they were not in focus, overlapped each other, were not on the appropriate habitat or if the lasers were not within the acceptable margins of the quadrat overlay. Images would therefore only be selected if the camera was at an oblique angle to the seabed, which reduces potential error that may be introduced as a result of changing seabed slope. Analyses of a trial dataset comprising all possible frames from 12 video transects determined that using 30 frames gave equivalent result to extracting data from all frames, but with a substantial saving in processing time [21]. Individual or discrete colonial organisms counted within the 30 frames sub-sampled from each video transect were expressed as densities (individuals $m^{-2}$). The quadrat overlay contained 16 dots. Cover-forming colonial taxa were quantified as percent cover by dividing the number of dots overlying that taxon by the total number of dots for the quadrat.

All organisms present were identified to the highest taxonomic level possible and their abundance recorded. Taxonomically similar species, which could not be distinguished with confidence, were grouped. Such groups included: *Inachus* spp. and *Macropodia* spp. (identified to genus level); Gobies; Hydroids (excepting *Nemertesia antennina* (Linnaeus, 1758), *Gymnangium montagui* (Billard, 1912) and *Nemertesia ramosa* (*Lamouroux, 1816*)) and Branching sponges. The category Turf incorporated hydroids and bryozoans that were $<1$ cm.

## Data analyses

Permutational Multivariate Analysis of Variance (PERMANOVA+, in the PRIMER v6 software package [35,36]) was used to test for changes in the response variables (species richness, total abundance, assemblage composition and the abundance of pre-selected indicator taxa [32]) in the NC relative to the CC, NOC and FOC, over temporal and spatial scales. Analyses of species richness, total abundance and assemblage composition used frame grab data. For analyses of the 13 indicator taxa, five taxa used frame grab data, while the remainder used data from the entire video transect. PERMANOVA is robust to datasets with many zeros, and allows the testing of interactions in complex multifactorial designs with multivariate or univariate data. It has significant advantages over conventional MANOVA in that it makes no

**Table 2.** PERMANOVA of abundance based on Euclidean distance measure.

a)

| Source | *df* | SS | MS | *F* | P |
|---|---|---|---|---|---|
| Year Ye | 3 | 3.79 | 1.26 | 12.97 | **0.0001** |
| Treatment Tr | 3 | 8.44 | 2.81 | 5.13 | **0.01** |
| Area Ar (Tr) | 15 | 7.35 | 0.49 | 4.60 | **0.0001** |
| YexTr | 9 | 0.53 | 0.06 | 0.81 | 0.6059 |
| Site(Ar(Tr)) | 50 | 4.75 | 0.10 | 1.45 | 0.0626 |
| YexAr(Tr) | 45 | 3.34 | 0.07 | 1.13 | 0.2934 |
| Residual | 110 | 7.19 | 0.07 | | |
| Total | 235 | 35.39 | | | |

b)

| | Tr | | | Ye | |
|---|---|---|---|---|---|
| Groups | t | P | Groups | t | P |
| CC, NC | 1.23 | 0.2574 | 2008, 2009 | 1.08 | 0.2928 |
| CC, NOC | 3.75 | **0.0091** | 2008, 2010 | 1.87 | 0.0793 |
| CC, FOC | 3.52 | **0.009** | 2008, 2011 | 4.77 | **0.0001** |
| NC, NOC | 1.61 | 0.1326 | 2009, 2010 | 0.45 | 0.6634 |
| NC, FOC | 2.30 | **0.0473** | 2009, 2011 | 4.35 | **0.0002** |
| NOC, FOC | 1.13 | 0.3055 | 2010, 2011 | 4.40 | **0.0005** |

a) Main test and b) Pairwise testing for the interactions Treatment and Year. Data were Log (x+1) transformed. Bold type denotes a significant result.

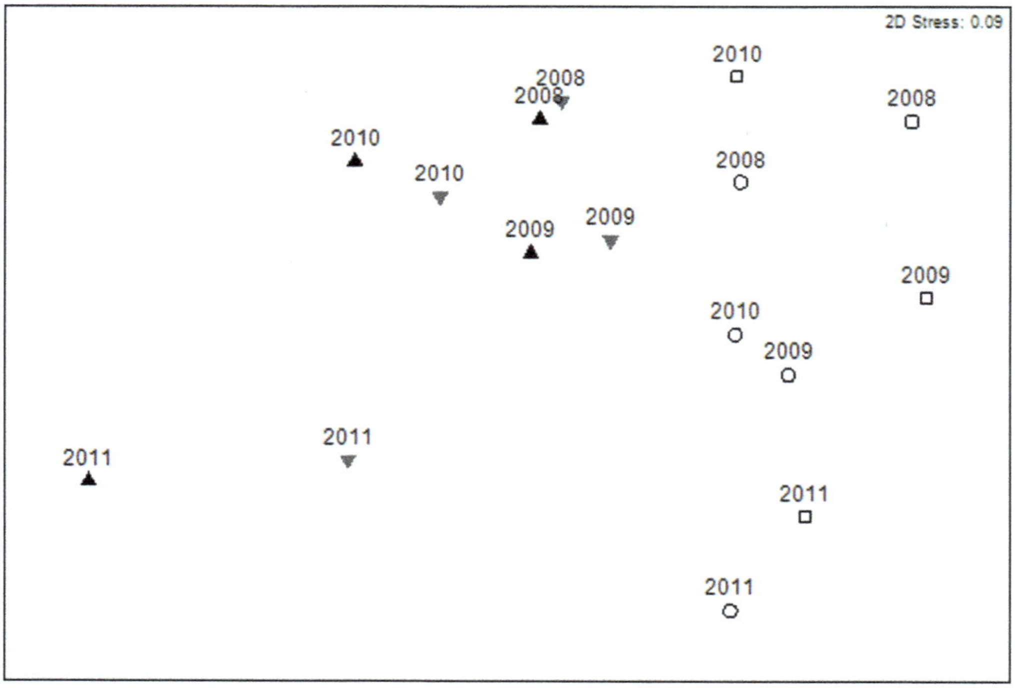


**Figure 3. nMDS plot illustrating similarities in assemblage composition between Treatments (averaged for site within treatment), (closed control = filled black triangles, new closure = filled grey triangles, near open control = open circle, far open control = open square), over time (2008, 2009, 2010 and 2011).** Data were dispersion weighted and square root transformed. Trajectories over time are indicated with lines from 2008 to 2011 for each treatment.

**Table 3.** PERMANOVA of assemblage composition based on Bray Curtis similarity measure.

a)

| Source | *df* | SS | MS | *F* | P |
|---|---|---|---|---|---|
| Year Ye | 3 | 45921 | 15307 | 7.40 | **0.0001** |
| Treatment Tr | 3 | 48855 | 16285 | 3.31 | **0.0006** |
| Area Ar (Tr) | 15 | 66234 | 4415.6 | 3.36 | **0.0001** |
| YexTr | 9 | 24506 | 2722.9 | 1.45 | **0.0055** |
| Site(Ar(Tr)) | 50 | 59146 | 1182.9 | 1.46 | **0.0001** |
| YexAr(Tr) | 45 | 73742 | 1638.7 | 2.03 | **0.0001** |
| Residual | 110 | 88847 | 807.7 | | |
| Total | 235 | 407250 | | | |

b)

| | 2008 | | 2009 | | 2010 | | 2011 | |
|---|---|---|---|---|---|---|---|---|
| Groups | t | P | t | P | t | P | t | P |
| CC, NC | 0.77 | 0.91 | 0.86 | 0.7216 | 1.04 | 0.3788 | 1.33 | 0.0959 |
| CC, NOC | 1.03 | 0.3914 | 1.59 | **0.0384** | 1.85 | **0.01** | 2.46 | **0.0269** |
| CC, FOC | 1.57 | **0.029** | 1.70 | **0.0266** | 1.72 | **0.021** | 2.36 | **0.0077** |
| NC, NOC | 0.99 | 0.4642 | 1.59 | **0.02** | 1.67 | **0.0069** | 1.95 | **0.0299** |
| NC, FOC | 1.70 | **0.0056** | 1.60 | **0.0268** | 1.50 | **0.0348** | 2.02 | **0.0138** |
| NOC, FOC | 1.17 | 0.1774 | 1.43 | 0.0565 | 1.18 | 0.1606 | 1.01 | 0.4466 |

a) Main test and b) Pairwise testing for the interaction Year × Treatment. Data were dispersion weighted and square root transformed. Bold type denotes a significant result.

assumptions about underlying data distributions, and is robust to unbalanced designs [37].

Multivariate data (assemblage) were dispersion weighted and square root transformed. Bray-Curtis similarity indices were calculated from Sites × Taxa abundance data to construct a similarity matrix between sites [38]. Dispersion weighting was employed to down-weight taxa with large and erratic numbers without 'squashing' other taxa [39] and a square root transformation was then applied to allow the rare taxa to contribute to the outcome, and further down-weight high-abundance taxa. Visualisation of the dissimilarity matrices was achieved using non-metric Multi-Dimensional Scaling (nMDS). Univariate data (species richness, total abundance and indicator taxa) were $\log_{10}(x+1)$ transformed and Euclidean distance indices were used to construct similarity matrices between sites [40].

The analytical design had four factors: Year (fixed: 2008, 2009, 2010, 2011), Treatment (fixed: CC, NC, NOC, FOC), Area (random and nested in Treatment), and Site (random and nested in Treatment and Area). Within-transect variation was not of interest given the scale of the study, so the 30 replicate frame grabs were averaged to avoid pseudoreplication. This also increased the precision at which the epibenthic assemblage was quantified.

Each term in the analyses used 9999 permutations of the appropriate units [38]. Multi-level significant interactions were tested using PERMANOVA pairwise tests.

## Results

A total of 136 taxa from 9 phyla were recorded in the surveys: 125 taxa in the frame grab analysis and 46 in the video analysis. While frame grabs were only analysed if they were on 'reef' habitat (which constituted seabed with rock, boulders and cobbles), reef associated fauna, such as soft corals *Alcyonium digitatum* (Linnaeus, 1758) and upright bryozoans *Pentapora fascialis*, were also observed on sediments that appeared to overlay bedrock [41].

### Species richness

Species richness was greatest in the CC in 2011 (27.8 $m^{-2}\pm1.32$) and lowest in the FOC in 2010 (12.77 $m^{-2}\pm0.53$) (Fig. 2a; Table 1). A significant Year × Treatment interaction indicated that species richness differences between treatments varied over time. Clear trends were not apparent for the first two years of the study, but by 2011 the species richness in the NC (25.44 $m^{-2}\pm1.37$) was greater than in both the NOC and FOC (NOC: 17.75 $m^{-2}\pm1.8$; FOC: 17.57 $m^{-2}\pm1.28$) and was not different to the CC (27.83 $m^{-2}\pm1.32$). Significant variation was identified between sites nested within area ($P=0.012$), demonstrating the high degree of small scale spatial variation across the study site (Table 1). Perhaps surprisingly, for both species richness and total abundance (below), NC and CC values were very similar at the outset (2008), and diverged thereafter, although they both diverged further from the open control sites.

### Total abundance

Total abundance calculated from the frame grabs was greatest in the CC in 2011 (3.9 $m^{-2}\pm0.1$) and lowest in the FOC in 2008 (2.94 $m^{-2}\pm0.09$), (Fig. 2b). Abundance differed between treatments and years ($P<0.05$) and was significantly greater in the CC (3.57 $m^{-2}\pm0.07$) than the NOC (3.18 $m^{-2}\pm0.0.08$) or FOC (3.04 $m^{-2}\pm0.08$). Abundance in the NC (3.39 $m^{-2}\pm0.08$) was also greater than the FOC, and was greater in 2011 (3.51 $m^{-2}\pm0.1$) than any other year (2008 = 3.17 $m^{-2}\pm0.08$; 2009 = 3.25 $m^{-2}\pm0.08$; 2010 = 3.26 $m^{-2}\pm0.07$; all $P<0.001$; Fig. 2b; Table 2). While there appears to be increased abundance in the NC and CC relative to the fished treatments (Fig. 2b), there

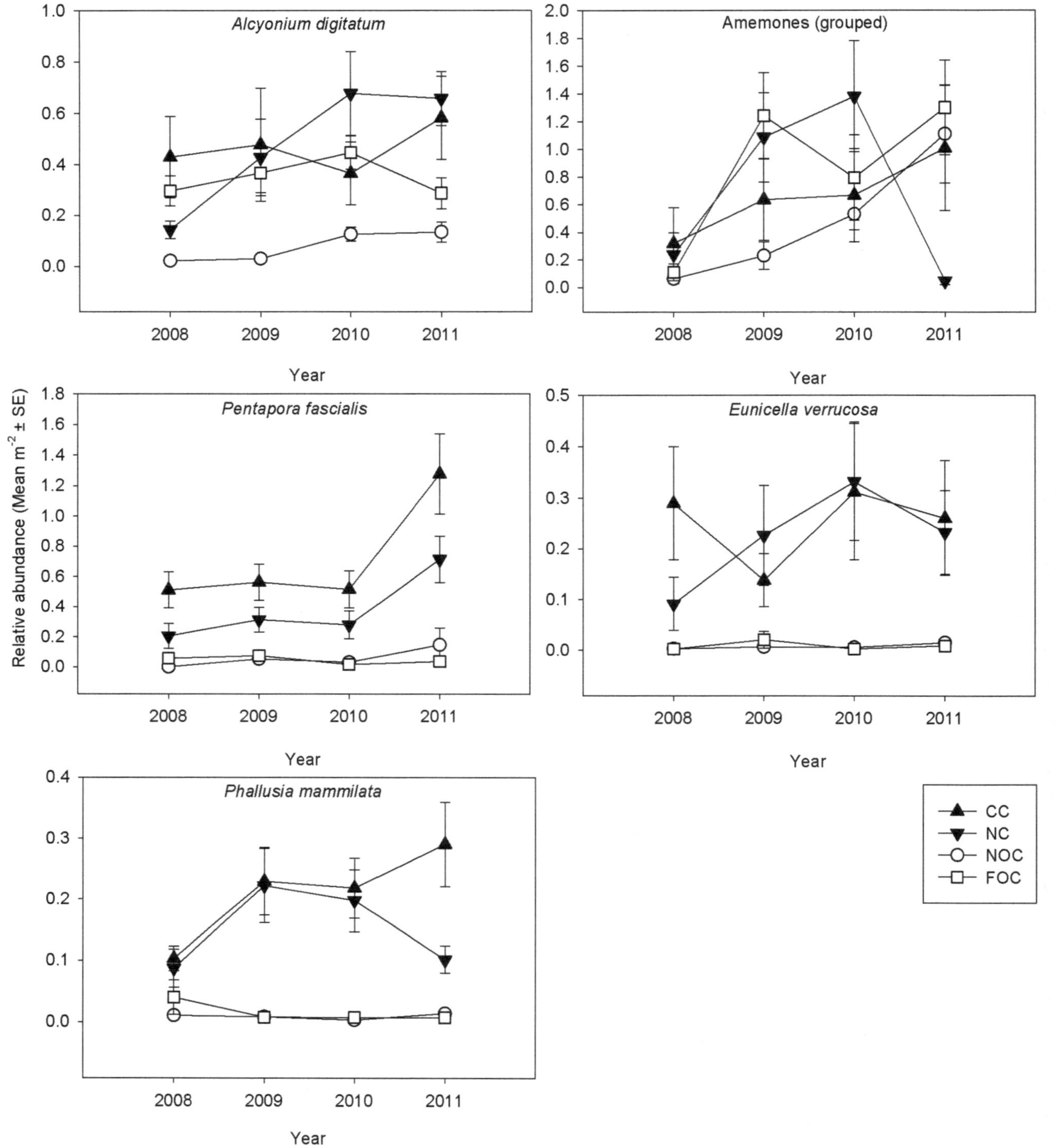


**Figure 4. Relative abundance of sessile indicator species (mean $m^{-2}$ ± SE) per treatment (CC = closed control (black triangle), NC = new closure (grey triangle), NOC = near open control (white circle), FOC = far open control (white square), per year (2008, 2009, 2010, 2011) identified through frame grabs.**

was no Year × Treatment interaction and so differences were not yet a significant indication of recovery as defined.

## Assemblage composition

Assemblage composition was significantly different for every factor tested (Table 3a). Pairwise tests for Year × Treatment interaction showed significant differences for all years between the NC and FOC and the CC and FOC (all $P<0.01$, Table 3b). In 2008, the assemblages in the NC and NOC and CC were not different but became significantly different between protected and fished treatments by 2009. These differences remained consistent into 2010 and 2011. However, the nMDS (Fig. 3) showed that the

**Table 4.** Summary of recovery status with evidence from pairwise statistical tests (in supporting information).

| Response metric | Data Type | Recovery | Positive response |
|---|---|---|---|
| Species richness | Frames | Yes | Yes |
| Total Abundance | Frames | No | Yes |
| Assemblage composition | Frames | No | Yes |
| Sessile indicator taxa | | | |
| Branching sponges | Video | No | No |
| *Phallusia mammillata* | Video | Yes | Yes |
| *Alcyonium digitatum* | Video | No | Yes |
| *Eunicella verrucosa* | Video | No | Yes |
| *Chaetopterus variopedatus* | Frames | No | No |
| Hydroids (grouped) | Frames | No | Yes |
| *Cellepora pumicosa* | Frames | No | No |
| *Pentapora fascialis* | Frames | Yes | Yes |
| Anemones (grouped) | Frames | No | No |
| Mobile indicator taxa | | | |
| *Asterias rubens* | Video | No | No |
| *Necora puber* | Video | No | Yes |
| *Cancer pagurus* | Video | No | No |
| *Pecten maximus* | Video | Yes | Yes |

*Data Type* refers to data quantified from the 30-frame subsample (*Frames*) or counts over the entire video transect (*Video*). *Recovery* is used in the narrow sense where NC increases relative to NOC & FOC, and approaches CC. *Positive response* indicates that NC increases relative to NOC & FOC, but does not necessarily converge with CC, in that CC may also increase, or show wide variability.

while assemblage composition in the NCs continued to diverge from the open controls and shift toward the CCs, the CCs themselves diverged even further from the open controls, with the result that the NCs also became less similar to the CCs over time (Fig. 3).

## Indicator taxa

**Sessile indicator taxa.** Despite marked spatial variation across the bay, there was clear evidence of recovery for two of the nine sessile indicator taxa (*P. fascialis* and *Phallusia mammillata* (Cuvier, 1815), (Fig. 4, Table 4), and evidence of a positive response in a further three taxa (*A. digitatum*, *E. verrucosa* amd Grouped Hydroids; Figs. 4 and 5, Table 4). The spatial variation detected within treatment for the random area and site factors will not be further interpreted as hypotheses were specific to relative change in treatment over time. Overall *A. digitatum* dead man's fingers and Grouped Anemones were significantly more abundant in 2011 than 2008, but there was substantial spatial variation unrelated to treatment (Fig.4, Tables S1 and S2). Signs of recovery for *P. fascialis* ross coral were indicated by a significant Year × Treatment interaction ($P<0.05$) (Fig. 4; Table S3), and over time, abundance increased in both protected treatments. By 2011, there was a greater abundance in the CC than in the NC and similarly more in the NC than the fished treatments (Table S3).

Substantial spatial variation was detected for the abundance of *E. verrucosa* pink sea fan. The trend shows a marked increase in *E. verrucosa* in the non-fished treatments compared to those that continued to be fished between 2010 and 2011 (Fig. 4) but there was no Year × Treatment interaction to determine a formal recovery trend as defined (Table S4).

While the null hypothesis of no recovery cannot be rejected, there was a strong signal ($P=0.53$; Table S5) and trend (Fig. 4) that populations of *P. mammillata* which were distributed evenly across treatments in 2008 in the bay were increasing in the NC and CC relative to fished controls.

The abundance of *C. pumicosa*, a small, relatively tough bryozoan, differed significantly with Treatment and Area nested within Treatment ($P<0.05$; Table S6), which could be attributed to spatial differences rather than those associated with the closure (Fig. 5). The overall trend suggests that *C. pumicosa* is increasing in the protected treatments relative to the controls, but there is substantial variability in this population.

A Year × Treatment interaction of the abundance of *Chaetopterus variopedatus* (Renier, 1804) parchment worm indicated a difference between treatments over time (Fig. 5; Table S7). By 2011 the abundance of the polychaete was significantly greater in the NC than the NOC sites ($P<0.05$). Generally, however, the pairwise tests did not show a clear recovery trend.

There were significantly more Grouped Hydroids in 2011 than in 2008 ($P=0.0006$) but treatment differences did not vary over time (Table S8). The graph shows an increasing abundance of Hydroids in the protected treatments but there was great spatial variation, which makes any recovery trends difficult to detect at present (Fig. 5).

The abundance of Branching Sponges varied over years and treatment (Fig. 5; Table S9) and appeared to show relative positive change in the CC compared to all other treatments, but there was no Year × Treatment interaction.

**Mobile indicator taxa.** Significant evidence of recovery was apparent for one of the four mobile indicator taxa (*P. maximus*; Fig. 6, Table 4), and evidence of a positive response for another two (*A. rubens* and *Necora puber*) (Fig. 6, Table 4). The great scallop *P. maximus*, one of the main commercial target species in Lyme Bay, was observed in similar abundances across the treatments at the time of the baseline survey in 2008. By 2010, however, there were more *P. maximus* in the NC than in both of the open controls NOCs and FOCs (Ye × Tr interaction $P<0.05$, followed by pairwise tests; Fig. 6; Table S10).

There were significantly more velvet swimming crabs *N. puber* in 2011 than in 2008 ($P=0.0175$), and evidence of an increasing trend in closed areas, but there was great spatial variation (Fig. 6; Table S11).

The abundance of the common sea star *A. rubens* was found to differ significantly between Area nested within Treatment ($P=0.0001$), but no treatment or year effects were found (Fig. 6; Table S12).

A significant treatment effect was found for the edible crab *Cancer pagurus* ($P<0.05$) whose abundance was found to differ significantly, with more crabs in the NOC than the FOC sites ($P<0.05$) (Fig. 6; Table S13).

## Discussion

In 2008, when the MPA in Lyme Bay was designated and the first survey was undertaken, boulders and cobbles inside the newly closed areas had limited sessile life growing on them. This was most likely a result of the scraping action of destructive fishing gear that overturns boulders thereby crushing or removing the attached sessile, slow growing organisms. Three years later, there were significant relative changes indicating some recovery of the epi-benthic fauna.

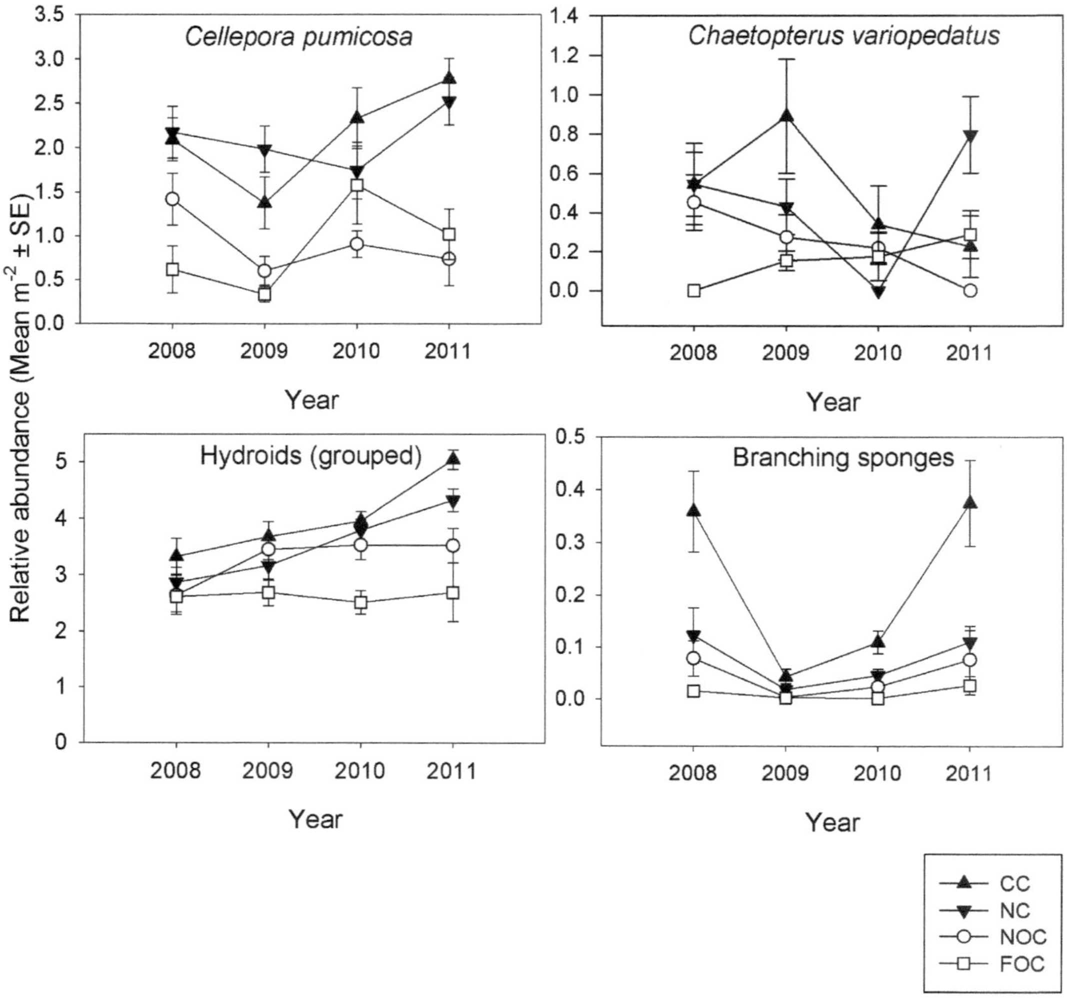


**Figure 5. Relative abundance of sessile indicator species (mean $m^{-2}$ ± SE) per treatment (CC = closed control, NC = new closure, NOC = near open control, FOC = far open control), per year (2008, 2009, 2010, 2011) identified through video transects.**

Our definition for "recovery" required sites within the new closure (NC) to become more similar to those within the closed controls (CC) and less similar those that remained open to towed demersal fishing (NOC, FOC). We also considered a less stringent test, in the context of possible buffering of the CC sites by the NC which now surrounds them, characterized as a "positive response" where both the CC and the NC sites increase (or change, in the case of assemblage composition) relative to controls. Changes in the metrics measured in each treatment were used to determine whether this hypothesis could be accepted. These showed that species richness in the NC became significantly greater within 3 years of protection than that in the NOC and FOC; the abundance of fauna increased over the 3 years, but did not change significantly within the NC compared to the NOC and FOC; and the species assemblage in the NC became less similar to the fished treatments, but also less similar over time to the CCs.

Taken broadly, we conclude that positive changes were occurring within the NC, and that CC sites were also changing, perhaps benefitting from the buffering effect of the NC and the added protection offered by the SI rather than the previous voluntary agreements [42,43]. This meets our definition of positive response but not recovery at this time. It is expected that the assemblage structure in the CC and the NC will eventually converge and remain dissimilar to the NOC and FOC allowing the formal recovery hypothesis to be accepted. Determining how long this will take is very important for marine ecosystem management. Evidence suggests decadal timespans may be required [44].

Within the first three years of the MPA three out of the 13 indicator species (*Pentapora fascialis*, *Phallusia mammilata* and *Pecten maximus*) showed recovery in the new closure. This is particularly of note for *P. fascialis*, a species that was previously known to be impacted by scallop dredging [20], with apparent low recoverability, as it is a functionally important bioconstructor which plays a key role in the formation of biogenic reef [45,46]. Such species are known to improve survivorship of taxa such as juvenile fish through the provision of a structurally complex habitat [47], so its increased abundance is particularly encouraging for the recovery of closed sites. By 2011, *P. fascialis* presence will therefore help to create important fishery nursery areas and feeding grounds [26,47–49].

A further five taxa showed a positive response: Grouped Hydroids, *Alcyonium digitatum*, *Eunicella verrucosa*, *Asterias rubens* (Linnaeus, 1758) and *Necora puber* (Linnaeus, 1767). There was considerable variation across the study area, but with time the

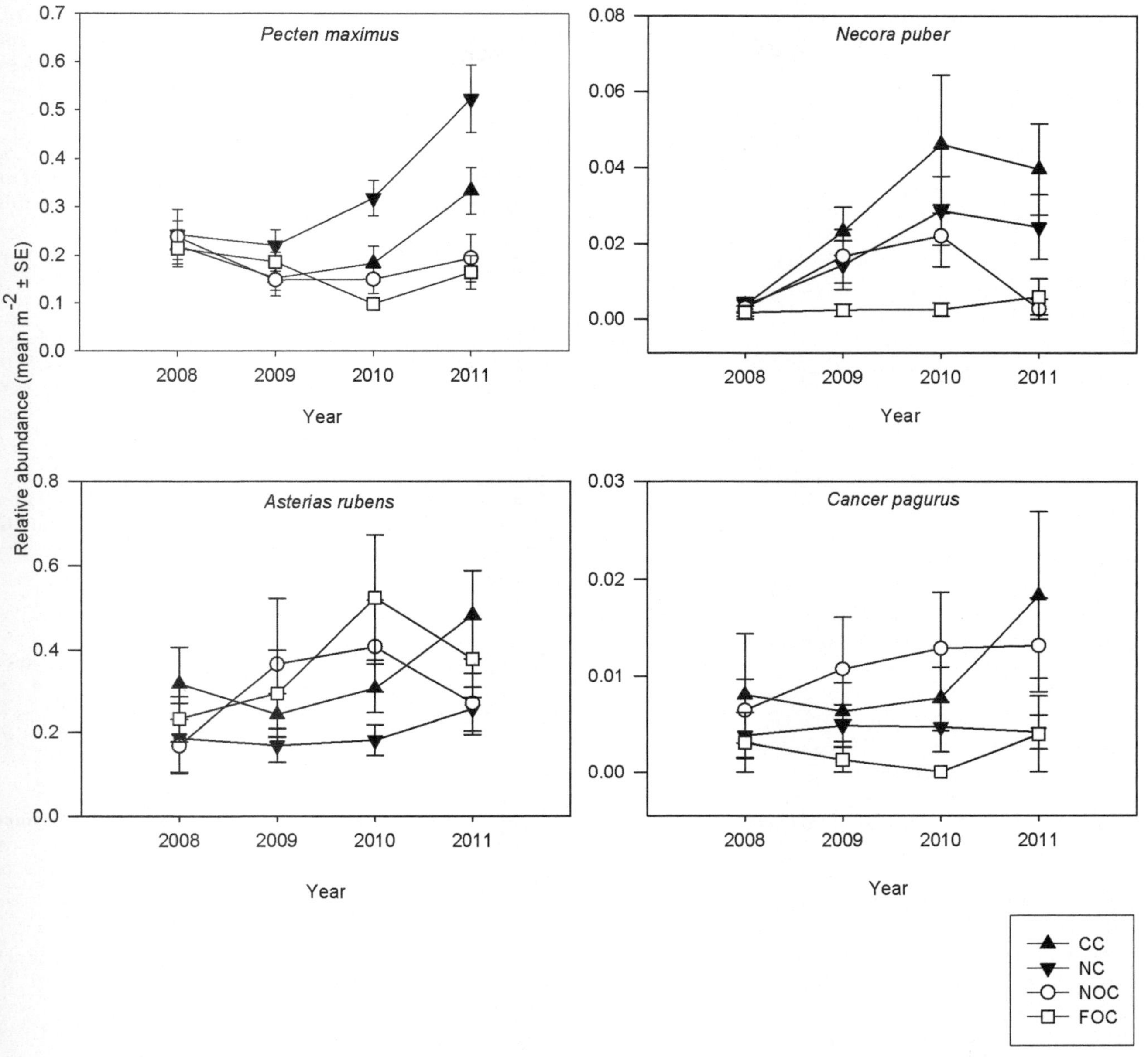


**Figure 6. Relative abundance (mean $m^{-2}$ ± SE) of mobile indicator species per treatment (CC = closed control, NC = new closure, NOC = near open control, FOC = far open control), per year (2008, 2009, 2010, 2011).**

early trends apparent for these or other taxa may consolidate the recovery picture.

The enhanced structural complexity of biogenic reefs, including hydroids, bryozoans and seafans, slows water movement and helps stabilise sediments [47]. Increased structural complexity supports both greater productivity and biodiversity by increasing the surface area and the range of habitat types available for settlement [47]. In turn, as assemblage diversity increases so does resilience to future impacts (including climate change) because of redundancy in trophic structure. More productive assemblages capture and recycle water column nutrients through filter feeding [50], and produce planktonic larvae that supports higher trophic levels. This bentho-pelagic coupling through a range of trophic links provides prey for birds [51], and commercially important fishes such as cod (*Gadus morhua*) [52].

It is important to note that the main target species of the excluded fishery, the commercially valuable great scallop *P. maximus* (DEFRA, 2012) was also found to be in a state of recovery inside the MPA despite a previous study concluding that scallops were not affected by bottom fishing in Lyme Bay [20]. Survey work by Hinz et al [20], which took place a year before the statutory instrument was introduced found no difference of *P. maximus* abundance between fished and non-fished treatments. The present study also found no difference in *P. maximus* abundance between all four treatments in 2008, but by 2011 abundance was significantly greater in the new closure than all other treatments. This suggests that *P. maximus* was impacted across the bay before the statutory instrument was in place but Hinz et al [20] were unable to detect this due to a lack of suitable controls. A similar study [53], assessing the north-east American *Placopecten magellanicus* population, identified a greater abundance of scallops within areas closed to mobile fishing gear. It would be expected that, with time, the protection of the SI will, in the long term, increase the survival of *P. maximus*, leading to a more stable and fecund population as large individuals become more abundant [19,54]. This could result in spillover of individuals from the SI into the fished areas, benefitting the scallop dredge fishery in the bay. Variable results for the abundance of the edible crab *Cancer pagurus* (Linnaeus, 1758) suggest early evidence of spillover as abundance in the NOC increased in 2010 and was greater than in the closed treatments where intensive potting continued. This was also in stark contrast to abundances within the FOC, suggesting that crabs could be moving out to habitats close to the edge of the SI from within the MPA.

In summary, the results after three years of protection are broadly consistent with the international experience. A range of MPA-related studies have reported detectable trends towards recovery within the space of a few years e.g. [54–56], but in many cases more complete recovery occurs at decadal time-scales e.g. [44]. It is, therefore, critical that the closure remains in place while the long term study continues, to determine the time spans of recovery for benthic assemblages.

## Supporting Information

**Table S1 PERMANOVA of *Alcyonium digitatum* abundance based on Bray Curtis similarity measure and b) Pairwise testing for the interaction Ye.** Data were dispersion weighted and square root transformed. Bold type denotes a significant result.

**Table S2 PERMANOVA of anemone abundance based on Bray Curtis similarity measure and b) Pairwise testing for the interaction Ye.** Data were dispersion weighted and square root transformed. Bold type denotes a significant result.

**Table S3 PERMANOVA of *Pentapora fascialis* abundance based on Bray Curtis similarity measure and b) Pairwise testing for the interaction YexTr.** Data were dispersion weighted and square root transformed. Bold type denotes a significant result.

**Table S4 PERMANOVA of *Eunicella verrucosa* abundance based on Bray Curtis similarity measure.** Data were dispersion weighted and square root transformed. Bold type denotes a significant result.

**Table S5 PERMANOVA of *Phallusia mammillata* abundance based on Bray Curtis similarity measure.** Data were dispersion weighted and square root transformed. Bold type denotes a significant result.

**Table S6 PERMANOVA of *Cellepora pumicosa* abundance based on Bray Curtis similarity measure.** Data were dispersion weighted and square root transformed. Bold type denotes a significant result.

**Table S7 PERMANOVA of *Chaetopterus variopedatus* abundance based on Bray Curtis similarity measure and b) Pairwise testing for the interaction YexTr.** Data were dispersion weighted and square root transformed. Bold type denotes a significant result.

**Table S8 PERMANOVA of hydroid abundance based on Bray Curtis similarity measure and b) Pairwise testing for the interactions Tr and Ye.** Data were dispersion weighted and square root transformed. Bold type denotes a significant result.

**Table S9 PERMANOVA of branching sponge abundance based on Bray Curtis similarity measure and b) Pairwise testing for the interaction YexTr.** Data were dispersion weighted and square root transformed. Bold type denotes a significant result.

**Table S10 PERMANOVA of *Pecten maximus* abundance based on Bray Curtis similarity measure and b) Pairwise testing for the interaction YexTr.** Data were dispersion weighted and square root transformed. Bold type denotes a significant result.

**Table S11 PERMANOVA of *Necora puber* abundance based on Bray Curtis similarity measure and b) Pairwise testing for the interaction Ye.** Data were dispersion weighted and square root transformed. Bold type denotes a significant result.

**Table S12 PERMANOVA of *Asterias rubens* abundance based on Bray Curtis similarity measure.** Data were dispersion weighted and square root transformed. Bold type denotes a significant result.

**Table S13 PERMANOVA of *Cancer pagurus* abundance based on Bray Curtis similarity measure and b) Pairwise testing for the interaction Tr.** Data were dispersion weighted and square root transformed. Bold type denotes a significant result.

## Acknowledgments

We are grateful to everyone that helped in the field including Kilian Stehfest, Adam Rees, Sean Lindsay-Leake and our video analysts. We are sincerely thankful to our skipper John Walker who is responsible for the success of the flying array. We are also grateful for the technical support from John Hawker at Cybertronix and Brian Hector at Bowtech. We thank DWT and IFCA for provision of data on habitats and fishing effort.

## Author Contributions

Conceived and designed the experiments: EVS TFS MJA. Performed the experiments: EVS SCG SLC. Analyzed the data: EVS. Wrote the paper: EVS TFS SCG SLC MJA.

## References

1. Roberts CM, Hawkins JP, Gell FR (2005) The role of marine reserves in achieving sustainable fisheries. Philosophical Transactions of the Royal Society B - Biological Sciences 360: 123–132.
2. Botsford LW, Castilla JC, Peterson CH (1997) The management of fisheries and marine ecosystems. Science 277: 509–515.

3. Jennings S, Kaiser MJ (1998) The effects of fishing on marine ecosystems. Advances in Marine Biology 34: 201–352.
4. Agardy TM (1994) Advances in marine conservation: the role of marine protected areas. Trends in Ecology & Evolution 9: 267–270.
5. Babcock RC, Kelly S, Shears NT, Walker JW, Willis TJ (1999) Changes in community structure in temperate marine reserves. Marine Ecology Progress Series 189: 125–134.
6. Auster PJ, Shackell NL (2000) Marine protected areas for the temperate and boreal Northwest Atlantic: the potential for sustainable fisheries and conservation of biodiversity. Northeastern Naturalist 7: 419–434.
7. Murawski SA, Brown R, Lai HL, Rago P, Hendrickson L (2000) Large-scale closed areas as a fishery-management tool in temperate marine systems: The Georges Bank experience. Bulletin of Marine Science 66: 775–798.
8. Gell FR, Roberts CM (2003) Benefits beyond boundaries: the fishery effects of marine reserves. Trends in Ecology & Evolution 18: 448–455.
9. Halpern BS (2003) The impact of marine reserves: Do reserves work and does reserve size matter? Ecological Applications 13: S117–S137.
10. Jennings S, Polunin NVC (1996) Impacts of fishing on tropical reef ecosystems. Ambio 25: 44–49.
11. Allison GW, Lubchenco J, Carr MH (1998) Marine reserves are necessary but not sufficient for marine conservation. Ecological Applications 8: 79–92.
12. Halpern BS, Gaines SD, Warner RR (2003) Export of production and the displacement of effort from marine reserves: effects on fisheries and monitoring programs. Oceans 2003 Proceedings. Vol. 1. IEEE.
13. Hiddink JG, Rijnsdorp AD, Piet G (2008) Can bottom trawling disturbance increase food production for a commercial fish species? Canadian Journal of Fisheries and Aquatic Sciences 65: 1393–1401.
14. Hiscock K, Breckels M (2007) Marine biodiversity hotspots in the UK. A report identifying and protecting areas for marine biodiversity. UK: WWF.
15. Hiddink JG, Jennings S, Kaiser MJ (2006) Indicators of the ecological impact of bottom-trawl disturbance on seabed communities. Ecosystems 9: 1190–1199.
16. de Groot SJ, Lindeboom HJ (1994) Environmental impact of bottom gears on benthic fauna in relation to natural resources management and protection of the North Sea. Texel: Netherlands Institute for Sea Research.
17. Dayton PK, Thrush SF, Agardy MT, Hofman RJ (1995) Environmental effects of marine fishing. Aquatic Conservation-Marine and Freshwater Ecosystems 5: 205–232.
18. Kaiser MJ, Spencer BE (1996) The effects of beam-trawl disturbance on infaunal communities in different habitats. Journal of Animal Ecology 65: 348–358.
19. Blyth RE, Kaiser MJ, Edwards-Jones G, Hart PJB (2004) Implications of a zoned fishery management system for marine benthic communities. Journal of Applied Ecology 41: 951–961.
20. Hinz H, Tarrant D, Ridgeway A, Kaiser M, Hiddink J (2011) Effects of scallop dredging on temperate reef fauna. Marine Ecology Progress Series 432: 91–102.
21. Stevens T, Sheehan E, Gall S, Attrill M, (in press) Monitoring benthic biodiversity restoration in Lyme Bay Marine Protected Area: design, sampling and analysis.
22. McClanahan TR, Graham NAJ (2005) Recovery trajectories of coral reef fish assemblages within Kenyan marine protected areas. Marine Ecology Progress Series 294: 241–248.
23. McClanahan TR, Graham NAJ, Calnan JM, MacNeil MA (2007) Toward pristine biomass: Reef fish recovery in coral reef marine protected areas in Kenya. Ecological Applications 17: 1055–1067.
24. Shears N, Babcock R (2002) Marine reserves demonstrate top-down control of community structure on temperate reefs. Oecologia 132: 131–142.
25. Parsons DM, Shears NT, Babcock RC, Haggitt TR (2004) Fine-scale habitat change in a marine reserve, mapped using radio-acoustically positioned video transects. Marine and Freshwater Research 55: 257–265.
26. Bradshaw C, Veale LO, Hill AS, Brand AR (2001) The effect of scallop dredging on Irish Sea benthos: experiments using a closed area. Hydrobiologia: 129–138.
27. Foden J, Rogers SI, Jones AP (2010) Recovery of UK seabed habitats from benthic fishing and aggregate extraction-towards a cumulative impact assessment. Marine Ecology Progress Series 411: 259–270.
28. Langmead O, Jackson EL, Bayley DTI, Marshall CE, Gall SC (2010) Assessment of the long-term effects of fishery area closures on long-lived and sessile species. Report to Defra from the Marine Life Information Network (MarLIN). Plymouth: Marine Biological Association of the UK.
29. Elliott M, Burdon D, Hemingway KL, Apitz SE (2007) Estuarine, coastal and marine ecosystem restoration: Confusing management and science – A revision of concepts. Estuarine, Coastal and Shelf Science 74: 349–366.
30. Jackson D, Mcleod C (2000) Handbook on the status on the EC Habitats Directive interest features:provisional data on the UK distribution and extent of Annex I habitats and the UK distribution and population size of Annex II species. Joint Nature Conservation Committee, Peterborough, p 180.
31. Glasby TM (1997) Analysing data hom post-impact studies using asymmetrical analyses of variance: A case study of epibiota on marinas. Australian Journal of Ecology 22: 448–459.
32. Jackson EL, Langmead O, Barnes M, Tyler-Walters H, Hiscock K (2008) Identification of indicator species to represent the full range of benthic life history strategies for Lyme Bay and the consideration of the wider application for monitoring of Marine Protected Areas. Report to the Department of Environment, Food and Rural Affairs from the Marine Life Information Network (MarLIN). Plymouth: Marine Biological Association of the UK. Defra contract No. MB101 Milestone 2 Defra contract No. MB101 Milestone 2.
33. Sheehan EV, Stevens TF, Attrill MJ (2010) A quantitative, non-destructive methodology for habitat characterisation and benthic monitoring at offshore renewable energy developments. PLoS ONE 5.
34. Freese L, Auster PJ, Heifetz J, Wing BL (1999) Effects of trawling on seafloor habitat and associated invertebrate taxa in the Gulf of Alaska. Marine Ecology Progress Series 182: 119–126.
35. Anderson MJ (2001) Permutation tests for univariate or multivariate analysis of variance and regression. Canadian Journal of Fisheries and Aquatic Sciences 58: 629–639.
36. Clarke KR, Warwick RM (2001) Change in marine communities: an approach to statistical analysis and interpretation. PRIMER-E: Plymouth.
37. Walters K, Coen LD (2006) A comparison of statistical approaches to analyzing community convergence between natural and constructed oyster reefs. Journal of Experimental Marine Biology and Ecology 330: 81–95.
38. Anderson MJ, Ter Braak CJF (2003) Permutation tests for multi-factorial analysis of variance. Journal of Statistical Computation and Simulation 73: 85–113.
39. Clarke KR, Chapman MG, Somerfield PJ, Needham HR (2006) Dispersion-based weighting of species counts in assemblage analyses. Marine Ecology Progress Series 320: 11–27.
40. Anderson MJ, Millar RB (2004) Spatial variation and effects of habitat on temperate reef fish assemblages in northeastern New Zealand. Journal of Experimental Marine Biology and Ecology 305: 191–221.
41. Sheehan EV, Cousens SL, Nancollas SJ, Stauss C, Royle J, et al. (2013) Drawing lines at the sand: Evidence for functional vs. visual reef boundaries in temperate Marine Protected Areas. Marine Pollution Bulletin 76: 194–202.
42. Claudet J, Osenberg CW, Benedetti-Cecchi L, Domenici P, Garcia-Charton JA, et al. (2008) Marine reserves: size and age do matter. Ecology Letters 11: 481–489.
43. Gaston KJ, Jackson SE, Cantu-Salazar L, Cruz-Pinon G (2008) The Ecological Performance of Protected Areas. Annual Review of Ecology Evolution and Systematics. pp. 93–113.
44. Babcock RC, Shears NT, Alcala AC, Barrett NS, Edgar GJ, et al. (2010) Decadal trends in marine reserves reveal differential rates of change in direct and indirect effects. Proceedings of the National Academy of Sciences 107: 18256–18261.
45. McKinney FK, Jackson JBC (1989) Bryozoan Evolution. USA: The University of Chicago Press.
46. Cocito S, Ferdeghini F (2001) Carbonate standing stock and carbonate production of the bryozoan *Pentapora fascialis* in the north-western Mediterranean. Facies 45: 25–30.
47. Bradshaw C, Collins P, Brand AR (2003) To what extent does upright sessile epifauna affect benthic biodiversity and community composition? Marine Biology 143: 783–791.
48. Auster PJ, Malatesta RJ, Langton RW, Watting L, Valentine PC, et al. (1996) The impacts of mobile fishing gear on seafloor habitats in the gulf of Maine (Northwest Atlantic): Implications for conservation of fish populations. Reviews in Fisheries Science 4: 185–202.
49. Smith F, Witman JD (1999) Species diversity in subtidal landscapes: maintenance by physical processes and larval recruitment. Ecology 80: 51–69.
50. Beaumont NJ (2009) Modelling the transport of nutrients in early animals. Evolutionary Biology 36: 256–266.
51. Grecian WJ, Inger R, Attrill MJ, Bearhop S, Godley BJ, et al. (2010) Potential impacts of wave-powered marine renewable energy installations on marine birds. Ibis 152: 683–697.
52. Gauthier C, Campbell PGC, Couture P (2008) Physiological correlates of growth and condition in the yellow perch (*Perca flavescens*). Comparative Biochemistry and Physiology Part A: Molecular & Integrative Physiology 151: 526–532.
53. Stokesbury KDE, Harris BP, Marino MC, Nogueira JI (2004) Estimation of sea scallop abundance using a video survey in off-shore US waters. Journal of Shellfish Research 23: 33–40.
54. Beukers-Stewart BD, Vause BJ, Mosley MWJ, Rossetti HL, Brand AR (2005) Benefits of closed area protection for a population of scallops. Marine Ecology Progress Series 298: 189–204.
55. Idjadi J, Lee S, Bruno J, Precht W, Allen-Requa L, et al. (2006) Rapid phase-shift reversal on a Jamaican coral reef. Coral Reefs 25: 209–211.
56. Mumby PJ, Harborne AR (2010) Marine Reserves Enhance the Recovery of Corals on Caribbean Reefs. PLoS ONE 5: e8657.

# Permissions

All chapters in this book were first published in PLOS ONE, by The Public Library of Science; hereby published with permission under the Creative Commons Attribution License or equivalent. Every chapter published in this book has been scrutinized by our experts. Their significance has been extensively debated. The topics covered herein carry significant findings which will fuel the growth of the discipline. They may even be implemented as practical applications or may be referred to as a beginning point for another development.

The contributors of this book come from diverse backgrounds, making this book a truly international effort. This book will bring forth new frontiers with its revolutionizing research information and detailed analysis of the nascent developments around the world.

We would like to thank all the contributing authors for lending their expertise to make the book truly unique. They have played a crucial role in the development of this book. Without their invaluable contributions this book wouldn't have been possible. They have made vital efforts to compile up to date information on the varied aspects of this subject to make this book a valuable addition to the collection of many professionals and students.

This book was conceptualized with the vision of imparting up-to-date information and advanced data in this field. To ensure the same, a matchless editorial board was set up. Every individual on the board went through rigorous rounds of assessment to prove their worth. After which they invested a large part of their time researching and compiling the most relevant data for our readers.

The editorial board has been involved in producing this book since its inception. They have spent rigorous hours researching and exploring the diverse topics which have resulted in the successful publishing of this book. They have passed on their knowledge of decades through this book. To expedite this challenging task, the publisher supported the team at every step. A small team of assistant editors was also appointed to further simplify the editing procedure and attain best results for the readers.

Apart from the editorial board, the designing team has also invested a significant amount of their time in understanding the subject and creating the most relevant covers. They scrutinized every image to scout for the most suitable representation of the subject and create an appropriate cover for the book.

The publishing team has been an ardent support to the editorial, designing and production team. Their endless efforts to recruit the best for this project, has resulted in the accomplishment of this book. They are a veteran in the field of academics and their pool of knowledge is as vast as their experience in printing. Their expertise and guidance has proved useful at every step. Their uncompromising quality standards have made this book an exceptional effort. Their encouragement from time to time has been an inspiration for everyone.

The publisher and the editorial board hope that this book will prove to be a valuable piece of knowledge for researchers, students, practitioners and scholars across the globe.

# List of Contributors


**Marilyn Deas**
UR-CoRéUs, Institut de Recherche pour le Développement, Laboratoire d'Excellence, CORAIL, Noumé a, New Caledonia

**Serge André fouët**
UR-CoRéUs, Institut de Recherche pour le Développement, Laboratoire d'Excellence, CORAIL, Noume´ a, New Caledonia

**Marc Léopold**
UR-CoRéUs, Institut de Recherche pour le Développement, Laboratoire d'Excellence, CORAIL, Port-Vila, Vanuatu
Fisheries Department of Vanuatu, Port-Vila, Vanuatu

**Nicolas Guillemot**
UR-CoRéUs, Institut de Recherche pour le Développement, Laboratoire d'Excellence, CORAIL, Noumé a, New Caledonia
Nicolas Guillemot Consultant, Noumea, New Caledonia

**Arthur L. Schultz**
National Marine Science Centre, Southern Cross University, Charlesworth Bay, Coffs Harbour, New South Wales, Australia

**Hamish A. Malcolm**
Marine Ecosystem Research, New South Wales Department of Primary Industries, Coffs Harbour, New South Wales, Australia

**Daniel J. Bucher**
Marine Ecology Research Centre, School of Environment, Science and Engineering, Southern Cross University, Lismore, New South Wales, Australia

**Michelle Linklater**
Coastal and Marine Unit, Science Division, Office of Environment and Heritage, Woolongong, New South Wales, Australia

**Stephen D. A. Smith**
National Marine Science Centre, Southern Cross University, Charlesworth Bay, Coffs Harbour, New South Wales, Australia
Marine Ecology Research Centre, School of Environment, Science and Engineering, Southern Cross University, Lismore, New South Wales, Australia

**Bárbara Horta e Costa**
Eco-Ethology Research Unit, ISPA – Instituto Universitário, Lisboa, Portugal
Centre of Marine Sciences - CCMAR, University of Algarve, Campus de Gambelas, Faro, Portugal

**Marisa I. Batista**
Centro de Oceanografia, Faculdade de Ciê ncias, Universidade de Lisboa, Lisboa, Portugal

**Leonel Gonc¸alves**
Eco-Ethology Research Unit, ISPA – Instituto Universitário, Lisboa, Portugal

**Karim Erzini**
Centre of Marine Sciences - CCMAR, University of Algarve, Campus de Gambelas, Faro, Portugal

**Jennifer E. Caselle**
Marine Science Institute, University of California, Santa Barbara, California, United States of America

**Henrique N. Cabral**
Centro de Oceanografia, Faculdade de Ciências, Universidade de Lisboa, Lisboa, Portugal

**Emanuel J. Gonc¸alves**
Eco-Ethology Research Unit, ISPA – Instituto Universitário, Lisboa, Portugal
Centre of Marine Sciences - CCMAR, University of Algarve, Campus de Gambelas, Faro, Portugal

**Ludvig A. Krag**
DTU Aqua, Technical University of Denmark, Hirtshals, Denmark

**Bent Herrmann**
SINTEF Fisheries and Aquaculture, Fishing Gear Technology, Hirtshals, Denmark

**Svein A. Iversen**
Institute of Marine Research, Bergen, Norway

**Arill Engås**
Institute of Marine Research, Bergen, Norway

**Sigve Nordrum**
Aker BioMarine Antarctic AS, Oslo, Norway

**Bjørn A. Krafft**
Institute of Marine Research, Bergen, Norway

**Kyung-Mi Jung**
Institute of Marine Research (IMR) and Hjort Centre for Marine Ecosystem Dynamics, Bergen, Norway
Department of Biology, University of Bergen, Bergen, Norway

**Arild Folkvord**
Institute of Marine Research (IMR) and Hjort Centre for Marine Ecosystem Dynamics, Bergen, Norway
Department of Biology, University of Bergen, Bergen, Norway

**Olav Sigurd Kjesbu**
Institute of Marine Research (IMR) and Hjort Centre for Marine Ecosystem Dynamics, Bergen, Norway
Centre for Ecological and Evolutionary Synthesis (CEES), Department of Biology, University of Oslo, Oslo, Norway

**Svein Sundby**
Institute of Marine Research (IMR) and Hjort Centre for Marine Ecosystem Dynamics, Bergen, Norway

**Carlos Carreras**
Centre for Ecology and Conservation, University of Exeter, Penryn, United Kingdom

**Brendan J. Godley**
Centre for Ecology and Conservation, University of Exeter, Penryn, United Kingdom

**Yolanda M. Leo´n**
Grupo Jaragua, Santo Domingo, Dominican Republic, Instituto Tecnoló gico de
Santo Domingo, Santo Domingo, Dominican Republic

**Lucy A. Hawkes**
Centre for Ecology and Conservation, University of Exeter, Penryn, United Kingdom

**Ohiana Revuelta**
Cavanilles Institute of Biodiversity and Evolutionary Biology (Scientific Park), University of Valencia, Valencia, Spain

**Juan A. Raga**
Cavanilles Institute of Biodiversity and Evolutionary Biology (Scientific Park), University of Valencia, Valencia, Spain

**Jesú s Tomá s**
Centre for Ecology and Conservation, University of Exeter, Penryn, United Kingdom
Cavanilles Institute of Biodiversity and Evolutionary Biology (Scientific Park), University of Valencia, Valencia, Spain

**Rozaimi Che Hasan**
Deakin University, Centre for Integrative Ecology, School of Life and Environmental Sciences, Faculty of Science and Built Environment, Warrnambool, Victoria, Australia
UTM Razak School of Engineering and Advanced Technology, Universiti Teknologi Malaysia, Kuala Lumpur, Malaysia

**Daniel Ierodiaconou**
Deakin University, Centre for Integrative Ecology, School of Life and Environmental Sciences, Faculty of Science and Built Environment, Warrnambool, Victoria, Australia

**Laurie Laurenson**
Deakin University, Centre for Integrative Ecology, School of Life and Environmental Sciences, Faculty of Science and Built Environment, Warrnambool, Victoria, Australia

**Alexandre Schimel**
Deakin University, Centre for Integrative Ecology, School of Life and Environmental Sciences, Faculty of Science and Built Environment, Warrnambool, Victoria, Australia

**Abigail S. Golden**
Dept. of Ecology, Evolution and Environmental Biology, Columbia University, New York, New York, United States of America

**Waisea Naisilsisili**
Wildlife Conservation Society Fiji, Suva, Fiji

**Isikele Ligairi**
University of the South Pacific, Suva, Fiji

**Joshua A. Drew**
Dept. of Ecology, Evolution and Environmental Biology, Columbia University, New York, New York, United States of America

**Cole C. Monnahan**
Quantitative Ecology and Resource Management, University of Washington, Seattle, Washington, United States of America

**Trevor A. Branch**
School of Aquatic and Fishery Sciences, University of Washington, Seattle, Washington, United States of America

**Kathleen M. Stafford**
Applied Physics Lab, University of Washington, Seattle, Washington, United States of America

**Yulia V. Ivashchenko**
National Marine Mammal Laboratory, Alaska Fisheries Science Center, Seattle, Washington, United States of America

School of Environmental Science and Management, Southern Cross University, Lismore, New South Wales, Australia

**Erin M. Oleson**
Protected Species Division, Pacific Islands Fisheries Science Center, Honolulu, Hawaii, United States of America

**Zhitao Wang**
The Key Laboratory of Aquatic Biodiversity and Conservation of the Chinese Academy of Sciences, Institute of Hydrobiology, Chinese Academy of Sciences, Wuhan, China
University of Chinese Academy of Sciences, Beijing, China

**Tomonari Akamatsu**
National Research Institute of Fisheries Engineering, Fisheries Research Agency, Ibaraki, Japan
CREST, Japan Science and Technology Agency, Tokyo, Japan

**Kexiong Wang**
The Key Laboratory of Aquatic Biodiversity and Conservation of the Chinese Academy of Sciences, Institute of Hydrobiology, Chinese Academy of Sciences, Wuhan, China

**Ding Wang**
The Key Laboratory of Aquatic Biodiversity and Conservation of the Chinese Academy of Sciences, Institute of Hydrobiology, Chinese Academy of Sciences, Wuhan, China
University of Chinese Academy of Sciences, Beijing, China

**Robert Cook**
School of Ocean Sciences, Bangor University, Menai Bridge, Anglesey, United Kingdom
School of Life Sciences, Heriot-Watt University, Edinburgh, United Kingdom

**Jose M. Fariñ as-Franco**
School of Life Sciences, Heriot-Watt University, Edinburgh, United Kingdom

**Fiona R. Gell**
Fisheries Directorate, Department of Environment Food and Agriculture, Isle of Man Government, St John's, Isle of Man, United Kingdom

**Rohan H. F. Holt**
Natural Resources Wales, Bangor, Gwynedd, United Kingdom

**Terry Holt**
Center for Marine and Coastal Studies Ltd. (CMACS), Port Erin, Isle of Man, United Kingdom

**Charles Lindenbaum**
Natural Resources Wales, Bangor, Gwynedd, United Kingdom

**Joanne S. Porter**
School of Life Sciences, Heriot-Watt University, Edinburgh, United Kingdom

**Ray Seed**
School of Ocean Sciences, Bangor University, Menai Bridge, Anglesey, United Kingdom

**Lucie R. Skates**
Natural Resources Wales, Bangor, Gwynedd, United Kingdom

**Thomas B. Stringell**
Natural Resources Wales, Bangor, Gwynedd, United Kingdom

**William G. Sanderson**
Natural Resources Wales, Bangor, Gwynedd, United Kingdom
School of Life Sciences, Heriot-Watt University, Edinburgh, United Kingdom

**Susie Brown**
School of Biological, Earth and Environmental Sciences, Cork Enterprise Centre, University College Cork, Distillery Fields, North Mall, Cork, Ireland

**David Reid**
Marine Institute Headquarters, Rinville, Oranmore, Galway, Ireland

**Emer Rogan**
School of Biological, Earth and Environmental Sciences, Cork Enterprise Centre, University College Cork, Distillery Fields, North Mall, Cork, Ireland

**Peter B. Marko**
Department of Biology, University of Hawai'i at Ma¯noa, Honolulu, Hawai'i, United States of America,

**Holly A. Nance**
Indian River State College, Fort Pierce, Florida, United States of America

**Peter van den Hurk**
Department of Biological Sciences, Clemson University, Clemson, South Carolina, United States of America

**Katherine L. Yates**
Environmental Science Research Institute, University of Ulster, Coleraine, United Kingdom

**David S. Schoeman**
Faculty of Science, Health, Education and Engineering, University of the Sunshine Coast, Maroochydore DC, Queensland, Australia

**Colin D. Buxton**
Fisheries Aquaculture and Coasts Centre, Institute for Marine and Antarctic Studies, University of Tasmania, Hobart, Tasmania, Australia

**Klaas Hartmann**
Fisheries Aquaculture and Coasts Centre, Institute for Marine and Antarctic Studies, University of Tasmania, Hobart, Tasmania, Australia

**Robert Kearney**
Institute for Applied Ecology, University of Canberra, Canberra, Australian Capital Territory, Australia

**Caleb Gardner**
Fisheries Aquaculture and Coasts Centre, Institute for Marine and Antarctic Studies, University of Tasmania, Hobart, Tasmania, Australia

**Valeriano Parravicini**
CESAB-FRB, Immeuble Henri Poincaré - Domaine du Petit Arbois, Aix-en-Provence, France
IRD – UR 227 Coreus – Laboratoire Arago, Banyuls/mer, France

**Fiorenza Micheli**
Hopkins Marine Station, Stanford University, Pacific Grove, California, United States of America

**Monica Montefalcone**
DiSTAV, Dipartimento di Scienze della Terra, dell'Ambiente e della Vita, Genoa
University, Genoa, Italy

**Carla Morri**
DiSTAV, Dipartimento di Scienze della Terra, dell'Ambiente e della Vita, Genoa
University, Genoa, Italy

**Elisa Villa**
DiSTAV, Dipartimento di Scienze della Terra, dell'Ambiente e della Vita, Genoa
University, Genoa, Italy

**Michela Castellano**
DiSTAV, Dipartimento di Scienze della Terra, dell'Ambiente e della Vita, Genoa
University, Genoa, Italy

**Paolo Povero**
DiSTAV, Dipartimento di Scienze della Terra, dell'Ambiente e della Vita, Genoa
University, Genoa, Italy

**Carlo Nike Bianchi**
DiSTAV, Dipartimento di Scienze della Terra, dell'Ambiente e della Vita, Genoa
University, Genoa, Italy

**Anna M. Magera**
Department of Biology, Dalhousie University, Halifax, Nova Scotia, Canada

**Joanna E. Mills Flemming**
Department of Mathematics and Statistics, Dalhousie University, Halifax, Nova Scotia,Canada

**Kristin Kaschner**
Department of Biometry and Environmental System Analysis, University of Freiburg, Freiburg, Germany
CESAB (Centre de Synthèse et d9Analyse sur la Biodiversité ), Immeuble Henri Poincaré , Domaine du Petit Arbois, Aix-en-Provence, France

**Line B. Christensen**
Fisheries Centre, University of British Columbia, Vancouver, British Columbia, Canada

**Heike K. Lotze**
Department of Biology, Dalhousie University, Halifax, Nova Scotia, Canada

**Simon J. Allen**
Murdoch University Cetacean Research Unit, Centre for Fish, Fisheries and Aquatic Ecosystems Research, School of Veterinary and Life Sciences, Murdoch University
Murdoch, Western Australia, Australia,

**Julian A. Tyne**
Murdoch University Cetacean Research Unit, Centre for Fish, Fisheries and Aquatic Ecosystems Research, School of Veterinary and Life Sciences, Murdoch University, Murdoch, Western Australia, Australia

**Halina T. Kobryn**
Murdoch University Cetacean Research Unit, Centre for Fish, Fisheries and Aquatic Ecosystems Research, School of Veterinary and Life Sciences, Murdoch University, Murdoch, Western Australia, Australia

**Lars Bejder**
Murdoch University Cetacean Research Unit, Centre for Fish, Fisheries and Aquatic Ecosystems Research, School of Veterinary and Life Sciences, Murdoch University Murdoch, Western Australia, Australia

**Kenneth H. Pollock**
Murdoch University Cetacean Research Unit, Centre for Fish, Fisheries and Aquatic Ecosystems Research, School of Veterinary and Life Sciences, Murdoch University, Murdoch, Western Australia, Australia

Department of Biology, North Carolina State University, Raleigh, North Carolina, United States of America

**Neil R. Loneragan**
Murdoch University Cetacean Research Unit, Centre for Fish, Fisheries and Aquatic Ecosystems Research, School of Veterinary and Life Sciences, Murdoch University, Murdoch, Western Australia, Australia

**Emma V. Sheehan**
Marine Institute, Plymouth University, Drake Circus, Plymouth, United Kingdom

**Timothy F. Stevens**
Griffith School of Environment and Australian Rivers Institute - Coasts and Estuaries, Griffith University, Gold Coast, Queensland, Australia

**Sarah C. Gall**
Marine Institute, Plymouth University, Drake Circus, Plymouth, United Kingdom

**Sophie L. Cousens**
Marine Institute, Plymouth University, Drake Circus, Plymouth, United Kingdom

**Martin J. Attrill**
Marine Institute, Plymouth University, Drake Circus, Plymouth, United Kingdom

# Index